Micro and Nanoplastics in Soil

Naga Raju Maddela
Kondakindi Venkateswar Reddy • Pabbati Ranjit
Editors

Micro and Nanoplastics in Soil

Threats to Plant-Based Food

Springer

Editors
Naga Raju Maddela
Departamento de Ciencias Biológicas
Universidad Técnica de Manabí
Portoviejo, Ecuador

Kondakindi Venkateswar Reddy
Centre for Biotechnology
Institute of Science and Technology,
Jawaharlal Nehru Technological University
Kukatpally, Telangana, India

Pabbati Ranjit
Centre for Biotechnology
Institute of Science and Technology,
Jawaharlal Nehru Technological University
Kukatpally, Telangana, India

ISBN 978-3-031-21194-2 ISBN 978-3-031-21195-9 (eBook)
https://doi.org/10.1007/978-3-031-21195-9

This Springer imprint is published by the registered company Springer Nature Switzerland AG
The registered company address is: Gewerbestrasse 11, 6330 Cham, Switzerland

Foreword

Annual plastic release into the terrestrial system is much higher than that of aquatic systems. In the recent times, there is a huge focus on the distribution and occurrence of micro and nanoplastics (MPs/NPs) in soil and their effect on ecosystems. This is attributed to unique features of microplastics, such as longer persistence in the environment under low-light and low-oxygen conditions, propensity to accumulate in the biological systems (e.g. phytoaccumulation), endocrine disruptive and carcinogenic nature, and recalcitrance to microbial degradation. Importantly, occurrence of micro and nanoplastics in fruits, vegetables, cereals, and other food stuff is widely reported. Thus, mobilization of micro and nanoplastics in the soil, and their subsequent migration to plant systems and then to the terrestrial food chain has received much attention by researchers, and now it has become a hot topic across the world. Awareness over food insecurity and human-health risk due to the micro and nanoplastics is highly needed in order to mitigate this problem, especially in the developing and underdeveloped countries.

Centering around the above issue, this volume has been well designed to address the latest issues on soil pollution by micro and nanoplastics and their consequences four plant produce and human health. This edited volume has 4 Parts – (i) Soil Pollution by Micro and Nanoplastics, (ii) Trophic Transfer of Micro and Nanoplastics, (iii) Toxicity of Micro and Nanoplastics, and (iv) Bioremediation of Micro and Nanoplastics-Polluted Soil. This volume has been edited by 3 subject experts, and has a total of 17 chapters, contributed by 54 researchers and academicians from 12 different countries across the world. There is in-depth emphasis on diverse topics related to the theme of *micro and nanoplastics in soil and the threats to plant-based food*. Topics include a global scenario on MPs/NPs in soil and their persistence and practical approaches for assessing them; MPs/NPs as carriers of other pollutants and antibiotic resistance genes; phytoaccumulation of MPs/NPs; toxicity of MPs/NPs; and bioremediation approaches for the removal of MPs/NPs in soil. I truly believe that this volume will have a wider readership and will serve researchers, environmental policy makers, industrialists, technicians, and students for a considerable length of time.

Santiago Quiroz Fernández,
Rector, Universidad Técnica de Manabí
Portoviejo, Ecuador
05 June 2022

Preface

The book entitled *Micro and Nano Plastics in Soil: Threats to Plant-Based Food* typically aims at the advances made by the allied fields of microbiology, biotechnology, environmental science, pedology, health science, polymer science, material science, nano technology, and hazardous waste management. The main purpose of this book is to give a brief introduction on the terrestrial MPs and NPs and their effects mainly on terrestrial plants, which indirectly affect the human population through trophic transfer, and how bioremediation can be used to avoid soil contamination due to MPs and NPs.

Earlier, terrestrial MPs and NPs have received little attention, but currently, micro plastics are regarded as emerging pollutants, and as a result, microplastic research has developed at an exponential rate in the previous decade. The effects of MPs and NPs on terrestrial plants and aquatic macrophytes are being reviewed nowadays, with a particular emphasis on adsorption, uptake, and toxicological effects. Plants and aquatic macrophytes are at the base of food webs and constitute an important part of human nutrition. As a result, a better understanding of micro- and nanoplastic adsorption, uptake, and effects, as well as the implications for trophic transmission, food safety, and security, is crucial. Previously, books, book chapters, research papers, and review articles on micro and nano plastics; sources, detection, and identification in soil; as well as the impact of micro and nano plastics on human health were published, but there were a few gaps identified, such as trophic transfer and how to remediate micro and nano plastics in soil samples. In our book, we filled in all of the holes mentioned above. The main purpose of this volume, *Micro and Nano Plastics in Soil: Threats to Plant-Based Food*, as the name suggests, is to give a brief introduction on the terrestrial MPs and NPs and their effects mainly on the terrestrial plants, which indirectly affect the human population through trophic transfer, and how bioremediation can be used to avoid soil contamination due to MPs and NPs.

This book is intended for researchers, scientists, NGOs, authorities, policy makers, and industry professionals in the fields of environmental science, pedology, health science, polymer science, material science, nanotechnology, and hazardous waste management. It will also provide insight into dynamic fields of polymer and health sciences for graduate and postgraduate students.

The book focuses on topics that comprise agricultural, environmental, health and safety pedology, polymer, nano and waste management fields. This book has 4 parts comprising 17 chapters. Part I, Soil Pollution by Micro and Nanoplastics, comprises seven chapters mainly focusing on micro and nano plastics distribution, methodology, assaying, and persistence of micro and nano plastics in the soil. Micro and nano plastics act as carriers for other soil pollutants and antibiotic resistance genes. Part II, Trophic Transfer of Micro and Nanoplastics, comprises one chapter discussing the trophic transfer of MPs and NPs from root uptake. This section mainly focuses on 'phytoaccumulation of micro and nanoplastics: root uptake,' where it describes the phyto availability of micro and nanoplastics as well as the factors (both biotic and abiotic) that promote root uptake of micro and nanoplastics. Part III, Toxicity of Micro and Nanoplastics, comprises Chaps. 9, 10, 11 and 12 which mainly focus on toxicity and impacts of micro and nano plastics in terrestrial, agro ecosystem, plants, animals, and humans. Part IV, Bioremediation of Micro and Nanoplastics-Polluted Soil, comprises Chaps. 13, 14, 15, 16, 17 and gives a brief introduction to phytoremediation, and bacterial and mycoremediation techniques that can be employed in order to bio-remediate the soil polluted with MPs and NPs

Towards the end of this book, micro and nano plastics in agricultural soils and their challenges and future directions have been discussed in detail. The chapters were contributed by 54 academicians, researchers, and scientists from 12 different countries (Argentina, Algeria, Bangladesh, Canada, Chile, Ecuador, Ethiopia, India, Iran, Nigeria, South Africa, the USA) across the world. We strongly believe that this volume could be a single source of information that provides latest information regarding micro and nanoplastics, trophic transfer, and remediate the micro and nanoplastics in soil samples.

Portoviejo, Ecuador — Naga Raju Maddela
Kukatpally, Hyderabad, Telangana, India — Kondakindi Venkateswar Reddy
Kukatpally, Hyderabad, Telangana, India — Pabbati Ranjit

Acknowledgments

We eminently address the support of the contributors to the book for their precious contribution and timely feedback during this project. Without their eagerness and help, this volume would not be prepared in the planned time; therefore, we really prize their coordination and teamwork. We are also grateful to anonymous reviewers for their effective assessment, which helped us in improving the quality of this book by inviting experts to contribute additional chapters. We greatly acknowledge the springer editorial and production teams for their precious support; without their guidelines, this project would not have been complete in such a short duration. It is our privilege to work with them fairly. Finally, yet importantly, we are very much thankful to the colleagues at Universidad Técnica de Manabí (Ecuador) and Jawaharlal Nehru Technological University Hyderabad for their wholehearted support and for the provision of precious recommendations at the time of book proposal and final book preparation.

Universidad Técnica de Manabí
Portoviejo, Ecuador
Naga Raju Maddela

Jawaharlal Nehru Technological University
Kukatpally, Hyderabad, Telangana, India
Kondakindi Venkateswar Reddy

Jawaharlal Nehru Technological University
Hyderabad, Telangana, India
Pabbati Ranjit

Contents

About the Editors

Naga Raju Maddela received his MSc (1996–1998) and PhD (2012) in microbiology from Sri Krishnadevaraya University, Anantapuramu, India. During his doctoral program in the area of environmental microbiology, he investigated the effects of industrial effluents/insecticides on soil microorganisms and their biological activities and has been working as a faculty in microbiology since 1998, teaching undergraduate and postgraduate students. He received "Prometeo Investigator Fellowship" (2013–2015) from Secretaría de Educación Superior, Ciencia, Tecnología e Innovación (SENESCYT), Ecuador, and "Postdoctoral Fellowship" (2016–2018) from Sun Yat-sen University, China. He also received external funding from China Postdoctoral Science Foundation in 2017, internal funding from Universidad Técnica de Manabí in 2020, worked in the area of environmental biotechnology, participated in 20 national/international conferences, and presented research data in China, Cuba, Ecuador, India, and Singapore. Currently, he is working as a full-time Professor at the Facultad de Ciencias de la Salud, Universidad Técnica de Manabí, Portoviejo, Ecuador. To his credit, there are 10 Books, 45 Chapters and 60 research papers.

Kondakindi Venkateswar Reddy obtained his MSc (Microbiology) in 2006–2008 from Srikrishnadevyaraya University, Anantapur, Andhrapradesh, and MTech (Environmental Management) and PhD (2017) in Environmental Science and Technology from Jawaharlal Nehru Technological University Hyderabad, Kukatpally, Hyderabad, Telangana. During his doctoral program in the area of industrial microbiology, he worked on isolation, characterization, and molecular identification of cellulase producing bacterial isolates used in aquaculture wastewater treatment. He has been working as a faculty in microbiology and biotechnology for past 13 years, teaching undergraduate and postgraduate students. His forte is industrial microbiology, biotechnology, and environmental management. He has successfully completed four AICTE-approved faculty development programs in the areas of applied environmental microbiology, experimental biotechnology, biochemistry, and cell biology. He has published 9 book chapters (Springer/ Taylor and Francis) and 30 research papers in reputed journals. He has actively participated in 20 national/international conferences, and presented various research papers in prestigious universities. Currently, he is working as an assistant professor at the Center for Biotechnology, Institute of Science and Technology, Jawaharlal Nehru Technological University Hyderabad, Kukatpally, Hyderabad.

Pabbati Ranjit completed BTech (2003–2007), MTech (2007–2009), and PhD in biotechnology from Jawaharlal Nehru Technological University, Hyderabad, India. His research work focused on the area of medicinal plant biotechnology for evaluation of its antimicrobial activity on various pathogenic bacteria, and antioxidant activity. He also worked to understand the secondary metabolite biosynthesis pathways for mapping of genes involved in antimicrobial activity by transcriptome analysis using next-generation sequencing technique. He has 10 years of teaching experience, teaching undergraduate and postgraduate students. He has mentored undergraduate and postgraduate students for their projects. He has received internal grant form Technical Education Quality Improvement Programme of the Government of India (TEQIP) and has worked in the area of plant biotechnology. Currently, he is working as an assistant professor at the Centre for Biotechnology, Institute of Science and Technology, Jawaharlal Nehru Technological University, Hyderabad, India. He has published 9 book chapters (Springer and Taylor & Francis) and 20 research papers according to research findings.

Contributors

O. P. Abioye Bioresources Development Centre, National Biotechnology Development Agency, Ogbomoso, Nigeria

A. E. Ajiboye Minerals and Metals Initiative Department of Chemical Engineering, University of Cape Town, Cape Town, South Africa

Dandu Akhil Indian Immunologicals Limited, Hyderabad, India

Javier Iván Haro Alvarado Departamento Salud Publica, Facultad de Ciencias de la Salud, Universidad Técnica de Manabí, Portoviejo, Ecuador

Mounia Arab Faculty of Biological Sciences, University of Sciences and Technology Houari Boumediene, Algiers, Algeria

S. A. Aransiola Bioresources Development Centre, National Biotechnology Development Agency, Ogbomoso, Nigeria

B. R. Babaniyi Bioresources Development Centre, National Biotechnology Development Agency, Ogbomoso, Nigeria

Hafida Baoune Faculty of Natural and Life Sciences, Université Kasdi Merbah Ouragla, Ouragla, Algeria

Suresh Babu Bastipati Centre for Biotechnology, IST, JNTUH, Hyderabad, India

D. Bhagawan Centre for Environment, IST, JNTUH, Hyderabad, India

Amira Bouaziz Laboratoire de Biotechnologie des Molécules Bioactives et de la Physiopathologie Cellulaire (LBMBPC), Faculté des sciences de la nature et de la vie, Université Batna 2, Fesdis, Batna, Algeria

N. Chaitanya Centre for Environment, IST, JNTUH, Hyderabad, India

Kalyani Chepuri Centre for Biotechnology, Jawaharlal Nehru Technological University Hyderabad, Institute of Science & Technology, Hyderabad, Telangana, India

N. O. Falade Bioresources Development Centre, National Biotechnology Development Agency, Ogbomoso, Nigeria

Adityah Ganesh Department of Biotechnology, National institute of Technology, Warangal, Telangana, India

Sneha Ganguly Excelra Knowledge Solutions Pvt Ltd, Hyderabad, Telangana, India

Jaime Humberto Flores Garcia Departamento Salud Publica, Facultad de Ciencias de la Salud, Universidad Técnica de Manabí, Portoviejo, Ecuador

Md. Yousuf Gazi Department of Geology, University of Dhaka, Dhaka, Bangladesh

Dayakar Govindu Department of Biotechnology, Kakatiya University, Wrangal, Telangana, India

Istuti Gupta Department of Biotechnology, Meerut Institute of Engineering and Technology, Meerut, Uttar Pradesh, India

Md. Bayazid Hossain Department of Soil, Water & Environment, University of Dhaka, Dhaka, Bangladesh

Aicha Asma Houfani Michael Smith Laboratories and Department of Biochemistry & Molecular Biology, University of British Columbia, Vancouver, BC, Canada

U. J. J. Ijah Department of Microbiology, Federal University of Technology, Minna, Nigeria

A. Ikhumetse Department of Microbiology, Federal University of Technology, Minna, Nigeria

Srinivas Jukuri Centre for Biotechnology, Institute of Science and Technology, JNTUH, Hyderabad, India

Nahid Khoshnamvand Department of Environmental Health Engineering, School of Public Health, Tehran University of Medical Sciences, Tehran, Iran

Palakeerti Srinivas Kumar Centre for Biotechnology, Institute of Science and Technology, JNTUH, Hyderabad, India

Saida Lavudi Centre for Biotechnology, Institute of Science and Technology, JNTUH, Hyderabad, India

Maheswara Reddy Mallu Department of Biotechnology, Koneru Lakshmaiah Education Foundation, Guntur, Andhra Pradesh, India

D. Manikantha Centre for Biotechnology, Jawaharlal Nehru Technological University Hyderabad, Institute of Science & Technology, Hyderabad, Telangana, India

Tappa Mohammad Munawar Department of Biological and Chemical Engineering, Mekelle Institute of Technology, Mekelle University, Mekelle, Ethiopia

O. J. Oyedele Bioresources Development Centre, National Biotechnology Development Agency, Ogbomoso, Nigeria

Rami Reddy Pallerla Indian Immunologicals Limited, Hyderabad, India

Chilumoju Sai Pavan Centre for Biotechnology, Institute of Science and Technology, JNTUH, Hyderabad, India

Mahendar Porika Department of Biotechnology, Kakatiya University, Warangal, Telangana, India

Yeddanapalli Prathima St Anna's High School, Guntur, Andhra Pradesh, India

D. Rathnaprabha Department of Biotechnology, National institute of Technology, Warangal, Telangana, India

A. Vamshi Krishna Reddy Centre for environment, Institute of Science and Technology, Jawaharlal Nehru Technological University, Hyderabad, India

María Antonieta Riera Doctorado en Ingeniería Industrial. Universidad Nacional de Cuyo, Mendoza, Argentina

Malay Sahasrabudhe Excelra Knowledge Solutions Pvt Ltd, Hyderabad, Telangana, India

Golla Shankaraiah Centre for Environment, Institute of Science and Technology, Jawaharlal Nehru Technological University, Hyderabad, India

Shamiha Shafinaz Shreya Department of Oceanography, University of Dhaka, Dhaka, Bangladesh

Nalam Renuka Satya Sree Centre for Biotechnology, Institute of Science and Technology, JNTUH, Hyderabad, India

Srividya Department of Biotechnology, National institute of Technology, Warangal, Telangana, India

Syam Prasad Sura Department of Botany, Government Degree College, Bhupalpally, India

D. V. Surya Prakash Department of Biotechnology, Meerut Institute of Engineering and Technology, Meerut, Uttar Pradesh, India

Mahir Tajwar Department of Geosciences, Auburn University, Auburn, AL, USA

Radhika Tippani Department of Biotechnology, Kakatiya University, Wrangal, Telangana, India

C. D. S. L. N. Tulasi Centre for Biotechnology, Jawaharlal Nehru Technological University Hyderabad, Institute of Science & Technology, Hyderabad, Telangana, India

Chittari Amaravathi Sneha Latha Varma Excelra Knowledge Solutions Pvt Ltd, Hyderabad, Telangana, India

Donthamalla Venkatanarsaiah Indian Immunologicals Limited, Hyderabad, India

M. O. Victor-Ekwebelem Department of Microbiology, Alex Ekwueme Federal University, Abakaliki, Ebonyi State, Nigeria

Medardo Anibal Zambrano-Arcentales Facultad de Ciencias Físicas y Matemáticas, Universidad de Chile, Santiago de Chile, Chile

S. S. Leh-Togi Zobeashia National Biotechnology Development Agency, Abuja, Nigeria

Part I
Soil Pollution by Micro and Nanoplastics

Chapter 1
Soil Pollution by Micro- and Nanoplastics: An Overview

Kondakindi Venkateswar Reddy, Pabbati Ranjit, Javier Ivan Haro Alvarado, Jaime Humberto Flores Garcia, and Naga Raju Maddela

Abstract Annual releases of plastic to the terrestrial environment are 4 to 23 times as high as releases to the marine environment. Microplastics can enter the soil by many routes, for example, compost and sewage sludge as fertilizer, plastic mulching, irrigation and flooding, and atmospheric deposition. The process of top-down irrigation into the soil causes microplastics/nanoplastics (MPs/NPs) to be transported downwards along with soil cavities and eventually possibly into groundwater. Contact of toxic and harmful metal pollutants with MPs/NPs will inevitably occur during the migration process in the environment. Various factors are considered in their transportation such as microplastic properties, pore water forms, and properties of packing materials to influence microplastic transport that can indicate the environmental chance of MPs in soil conditions. Among the important roles in the environmental behavior of MPs/NPs are absorption and migration. Microplastic or nanoplastic particles as a carrier adsorb contaminants and increase or decrease their transportation. The transfer of MPs in the soil environment occurs in the form of vertical and horizontal migration and nonliving transport. MPs are known to adsorb toxic chemicals such as PCBs, PAHs, DDTs, PFASs, PPCPs, and heavy metals. Overall, this chapter provides introductory information about the terrestrial pollution of MPs and the structure of this edited volume.

Keywords Microplastic · Nanoplastic · Soil pollutants · Transportation · Carriers

---

K. Venkateswar Reddy (✉) · P. Ranjit
Centre for Biotechnology, Institute of Science and Technology, JNTUH, Hyderabad, Telangana, India

J. I. H. Alvarado · J. H. F. Garcia
Departamento Salud Publica, Facultad de Ciencias de la Salud, Universidad Técnica de Manabí, Portoviejo, Ecuador

N. R. Maddela
Departamento de Ciencias de Biológicas, Facultad de Ciencias de la Salud, Universidad Técnica de Manabí, Portoviejo, Ecuador

N. R. Maddela et al. (eds.), *Micro and Nanoplastics in Soil*, https://doi.org/10.1007/978-3-031-21195-9_1

1.1 Historical Background

Plastic, a synthetic material, is an aggregate of polymers. They are classified into microplastics which are >25 mm, mesoplatic with 5–25 mm, microplastics with 0.1–5 mm, and nanoplastics having <100 mm (Azeem et al., 2021).

After World War II, Age of Plastics began, and tons of plastic production happened worldwide. Since 1950, nearly 200 million tons of plastic material have been dumped into marine accidentally or intentionally (Brack, 2015). In the early 1960s, the awareness raised regarding the plastic waste contamination of the environment when seabirds perished having their gut piled up with plastic debris (Buks & Kaupenjohann, 2020). In 1968, the term microplastic was mentioned by the US Air Force Materials Laboratory in a publication. Microplastic was then described as the deformed plastic on a scale of microinches per inch. These terms were no longer in use as the scientists opted for a new denotation that generally describes the size of plastic piece. This was since the 1970s due to the discovery of minute plastic pieces in the aquatic ecosystems. The awareness of microplastic across the aquatic ecosystem was first identified by the world in 1972 due to the report provided on plasticles, the term given for small plastic particles that are floating on the Sargasso Sea surface (Crawford & Quinn, 2017). Early research related to MPs and NPs focused only on marine ecosystem but ignored soils (Buks & Kaupenjohann, 2020).

From the last decade, study on microplastics has increased exponentially. Though plants are the base of food web, they have been hugely overlooked in examining and studying ecotoxicology of microplastics (MPs) and nanoplastics (NPs). Recently, the knowledge base of MPs and NPs interaction with plants is rapidly mounting though few crucial gaps persist. The data observed from few decades regarding internalization and external adsorption in plants poses an alarming perspective that MPs and NPs may enter the food web which disrupts a broad range of species including humans (Mateos-Cárdenas et al., 2021).

1.2 Scope and Importance

Need for plastic has increased rapidly due to urbanization (Crawford & Quinn, 2017). Modern-day agricultural methods use plastic polymers such as polypropylene, polyolefin, polyvinyl chloride, polythene, acetate copolymer, and ethylene-vinyl for different applications like plastic reservoirs, boxes, mulching films, packaging materials, silage films, harvesting nets, tunnel green house, and irrigation system. These plastic materials are regularly used in agriculture to provide peculiar microclimate conditions that are required for plant growth (Campanale et al., 2022). Besides the abovementioned plastic polymers, the huge widespread usage of plastic materials like plastic resins, packaging material, is on the rise, and this phenomenon is referred to as "white pollution" which turns out to be a serious environmental issue (Azeem et al., 2021). What happens to all the nondegradable plastic products having a long-term lifetime? European nations lead the world in safe disposal and recycling plastics (Brack, 2015). For a certain period of time, MPs on land and on ocean suffer the exposure of high ultraviolet radiation due to the direct exposure of sunlight and ultimately undergo photooxidative degradation (Crawford & Quinn, 2017). Besides being recycled, the remaining plastic is still being dumped as municipal waste. Including New York, most mega cities are not up to the point in achieving plastic degradation. Even though plastic to some extent is being combusted and recycled then, where the remaining plastic goes is a big query. General household waste is a whole world problem, particularly in third world countries that lack landfills for municipal waste. Having industrial revolution and development, huge plastic is generated. What about the plastic waste developed from industrial society and urbanization? Quantum dots, carbon nanotubes, fullerenes, nano squares, nano boxes, nano crystals, and nanowires are intentionally produced nanoplastics. Nanotechnology uses different nano tools in revolutionizing field such as personal care products, medicines, and packaging. What about all these plastic nanoparticle disappearance (Brack, 2015)? Because of weathering, biodegradation or chemical degradation, and abrasion, plastic polymers further transform to microplastics and nanoplastics (Campanale et al., 2022).

Any piece of plastic having size 5 mm to 1 μm which is considered as microplastic and plastic having size less than 1 μm is termed as nanoplastic. MPs and NPs in terrestrial zone have diverse toxic effects based on the exposure medium and interaction with varied contaminants. The interplay with such pollutants can cause major modifications in the surface properties of plastic, due to which agroecosystem can incorporate these MPs and NPs which results in synergistic, antagonistic, or additive effects. In spite of their potential entry into agroecosystems, the data availability on MP and NP is scarce. The soil presence of plastic pollutes plant and soil organisms. Accumulation of plastic in soil occurs in different ways such as atmospheric deposition, through plastic packing, by wastewater treatment plants, daily use of plastic products, and mulching. In several countries, plastic mulching has been widely implemented in agriculture to enhance vegetable and fruit production leading to instant economic profits. Specifically for mulching, low-density polythene (LDPE) is used. Though mulching gives short-term profits to farmers, soil resources are hugely being exposed to plastic. Once plastic aggregates in soil, it's difficult to remove or recycle due its small size and surface-area-to-volume ratio (Azeem et al., 2021). The degraded plastic persists in the surface of soil and incorporated deeply through soil horizons or drive by water erosion or by wind and transfers into different ecosystems (Campanale et al., 2022). But how these MP and NP are being settled in deep sub soils. Agricultural methods like fertilization, moldboard, deep ploughing, and deep tillage methods disrupt soil layers and stimulate deep dispersion of MP and NP in layers of soil as mentioned in Fig. 1.1. Organisms in soil including mites, collembolans, and earthworms facilitate transportation through ingestion, casting, egestion, and adhesion to their peripheral skin. In addition to these methods, rhizomes pruning can also promote migration of MPs and NPs downwards. Besides that, wet-dry circle procedure also supports their movement downward (Azeem et al., 2021).

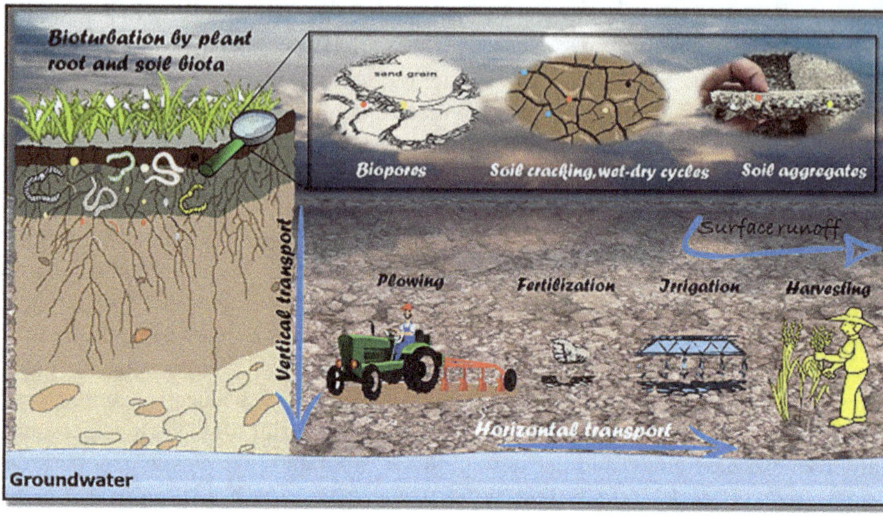

Fig. 1.1 MP and NP transfer in subsoils. (Adapted from Campanale et al., 2022)

The impact of plastics on plants can be internal or external. There is a strong requirement to monitor and study plant as a capable microplastic vector in the environment. Three types of plastic polymers – polypropylene (PP), polyethylene (PE), and polystyrene (PS) – are mostly observed to adsorb on plant surfaces. Numerous mechanisms have been stated to elucidate the spotted adherence of MPs & NPs on plants as mentioned in Fig. 1.2. These mechanisms mainly categorized into entrapment on surface structures and adsorption to surfaces. These mechanisms differs from species to species. Adsorption and internalization of these degraded plastics have a huge impact on the environment. On incorporation they delay germination by adsorption resulting in blockage of pores on surface of spores or in capsules of seeds. Chemical leaching caused from these plastics or by physical presence of MPs and NPs affects germination process. These plastics also have an impact on growth elongation in plants effecting growth between shoot and root, root thickness, and root elongation. These mixed effects of macro- and nanoplastics share similar effects of stressors like "stress-induced morphogenic responses" (SIMR) which cause reactive oxygen species production that results in plant hormone level difference which ultimately terminates growth of few tissues and accelerates other tissues growth. Algae and plants can be utilized as plastic bioindicators that help to recognize plastic hotspots (Mateos-Cárdenas et al., 2021).

Plant roots takes the first contact point with MPs and NPs, and root hairs of plant are majorly observed to be involved. These are absorbed by the plant root system by either symplastic transport through crossing the Casparian strip or apoplastic way which follows cell walls and extracellular spaces (Campanale et al., 2022). This uptake and translocation of these plastics happen in plants majorly by transpiration pull. These particles inside the central cylinder can travel to the plant aerial parts through the xylem. Another pathway for these particles to enter the leaf is possibly

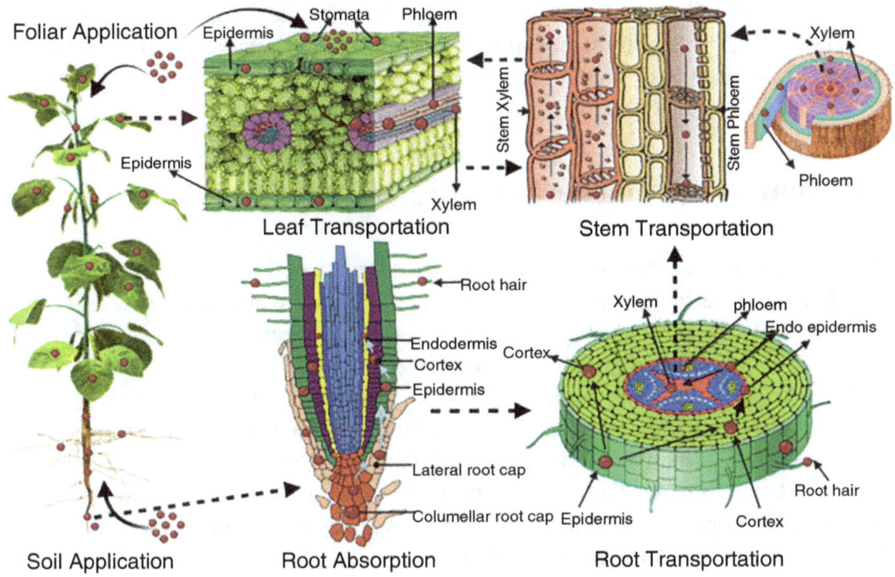

Fig. 1.2 Mechanism indicating plastic uptake by plant. (Adapted from Azeem et al., 2021)

through the stomata by foliar application (Azeem et al., 2021). Plant responses in the presence of NPs include alteration of performance of photosynthesis, pigment reduction, and biochemical changes. NPs induce oxidative stress and rises in reactive oxygen species. Data on accumulation of macro- and nanoplastics in plants, particularly in edible crops demonstrates how we are consuming plastics through food. Compared to vegetables, fruits have been mentioned to have high plastic content due to their greater size, high pulp vascularity, tree age, and complex root system. Few research data demonstrates that when NPs come into contact with heavy metals, they might affect the plant mineral absorption which reduces the plant nutritive value (Campanale et al., 2022). MP and NP abundance in soil also alters microbial population and raises the MP- and NP-favored microbes (Azeem et al., 2021).

So what type of solution has to be expected to prevent the abovementioned issues? To relieve from these problems, biodegradable plastics are the right option. Nowadays mulches are designed in such a way that at the end of harvest season, after being tilled into soil, they are biodegraded and the marked period of time to complete mineralization by microbes is less than 2 years (Douglas Hayes, 2019). Bioplastics are partially or completely biodegradable and are classified into totally biological, partially biological, and synthetic. Polycaprolactone (PCL) and polybutylene adipate-co-terephthalate (PBAT) are the major biodegradable fossil-based synthetic plastics. Besides that, biobased plastics like polylactic acid (PLA) and polyhydroxyalkanoic acids are also produced. All these plastics can replace conventional plastics in packaging and agricultural field. Another promising solution to lower the white pollution in soil is to enhance the in situ degradation by mounting the population of bacteria, fungi, and other organisms in soil. Organic compounds also help in degradation of plastic. Knowing the ecological connection of MP and NP fate and their interactions in soil will provide a better picture of threat to human food chain and health (Pathan et al., 2020).

1.3 Outlines of Volume: Sections, Chapters, and Parts

The main purpose of this book is to give a brief introduction on the terrestrial MPs and NPs and its effects mainly on the terrestrial plants which indirectly affect the human population through trophic transfer and how bioremediation can be used to avoid soil contamination due to MPs and NPs.

This book has four sections in total not only pointing out the problems/concerns of MPs and NPs mainly on plants and humans but also providing the solutions on how to clean the already contaminated soil. Part I comprises one to seven chapters. Chapter 1 has two parts. The first part provides an overview of the soil pollution by micro- and nanoplastics and its impact on ecosystem including human health risk. The second part describes the purpose and sections and the contents of this volume. Chapter 2 focuses on the distribution and occurrence of micro- and nanoplastics in different soil systems (urban, industrial, domestic, and agricultural soils) across the world. Likewise, this chapter will provide the latest insights over the global soil burden by micro- and nanoplastics. Chapter 3 provides latest insights over the soil burden of microplastics and nanoplastics in different regions across the world. A special importance will be given to agricultural ecosystems, because these ecosystems are especially likely to be contaminated with microplastics by multiple sources of plastics used in agricultural practice. Chapter 4 provides that microplastic contamination of the terrestrial ecosystem is a priority research area; however, there is no availability of standard methodologies for the quantification of microplastics and nanoplastics separately. This causes misinterpretations in the analysis of soil burden of MPs and NPs. Therefore, this chapter has been designed exclusively to review the literature in the area of emerging methodologies that are useful for the quantification of MPs and NPs in soil. Chapter 5 has been devoted to providing new insights over the persistence of micro- and nanoparticles in the soil system. This includes interactions between soil particles and micro-/nanoplastics and the impact of micro- and nanoplastics on soil properties. Chapter 6 addresses the following line: interaction of micro- and nanoplastics with different emerging contaminants (e.g., heavy metals, flame retardants, nanoparticles, PFOS, PFOAs, etc.) in the soil system. Thus, this chapter explains how micro- and nanoplastics act as carriers of other pollutants in the soil system and subsequent impact of immobilization of pollutants by micro- and nanoplastics. Chapter 7 is an emerging topic. This chapter provides relevant insights in understanding how micro- and nanoplastics are responsible in making the soil system as a significant reservoir of antibiotic resistance genes. Also, this chapter emphasizes on the spreading of antibiotic resistance genes between different ecosystems and acquisition by pathogens threating human as well as animal health.

Part II comprises one chapter discuss about the trophic transfer of MPs and NPs from root uptake. Chapter 8 describes about the phyto-availability of the micro- and nanoplastics, and factors (both biotic and abiotic) enhance the root uptake of micro- and nanoplastics.

The Part III focuses on all the threats posed to plants and humans in detail. It comprises 9–12 chapters. Chapter 9 provides the details on toxicological effects of different types (qualitatively) of micro- and nanoplastics on microorganisms and flora and fauna in the soil system. Chapter 10 describes toxic effects of MPs and NPs on the environment. Chapter 11 provides information about the effects of micro- and nanoplastics on the stress tolerance in plants and plant growth responses. Chapter 12 aims to provide the deeper insights over the toxic effects of micro- and nanoplastics. In vitro and in vivo experiments using cell cultures and whole animals will be discussed. We are in the opinion that such information aid in the better understanding of micro- and nanoplastics for their cytotoxicity, endocrine disruption propensity, genotoxicity, and carcinogenicity.

The Part IV gives a brief introduction to bioremediation techniques that can be employed in order to bioremediate the soil polluted with MPs and NPs. It comprises 13–17 chapters. Chapter 13 describes different types of plants that exhibit hyperaccumulation of micro- and nanoplastics will be reviewed. Factors that favor the hyperaccumulation will also be covered. This sheds lights on the designing of optimum conditions for the phytoremediation. Chapter 14 which describes potentialities of different bacterial species in the remediation of soils and water contaminated by micro- and nanoplastics will be discussed. Special importance will be given to the diversity in the microbial enzymes that attack these pollutants. Chapter 15 Soil which explains potentialities of different fungal species in the remediation of soils contaminated by micro- and nanoplastics will be discussed. Special importance will be given to the diversity in the microbial enzymes that attack these pollutants. In Chap. 16, a special attention will be paid to recent advances in the remediation of micro- and nanoplastic-polluted sites. Also, some of the recent case studies will be discussed in order to have an idea over the cost analysis for the remediation. Chapter 17 is the concluding chapter of this volume. In this chapter, importance will be given to the main challenges that we are facing in the control or mitigation of micro- and nanoplastics in agricultural soils. Towards the end of the chapter, future guidelines will be suggested.

This book is intended for researchers, scientists, NGOs, authorities, policymakers, and industry professionals in the fields of environmental science, pedology, health science, polymer science, material science, nanotechnology, and hazardous waste management. It will also provide insight into dynamic fields of polymer and health sciences for graduate and postgraduate students.

1.4 Contributors

This volume has been designed with 14 sections having 17 chapters. Overall, the contributors of all 17 chapters are subject experts in their concerned chapters. Professionally contributors are academicians, researchers, and scientists and are geographically belonging to different regions. Overall 54 contributors of 12 countries (Argentina, Algeria, Bangladesh, Canada, Chile, Ecuador, Ethiopia, India, Iran, Nigeria, South Africa, and the USA) have been involved in this volume. We strongly believe that this volume could be a single source of information that provides the latest information regarding micro- and nanoplastics and trophic transfer and remediate the micro- and nanoplastics in soil samples.

References

Azeem, I., Adeel, M., Ahmad, M. A., Shakoor, N., Jiangcuo, G. D., Azeem, K., Ishfaq, M., Shakoor, A., Ayaz, M., & Xu, M. (2021). Uptake and accumulation of Nano/microplastics in plants: A critical review. *Nanomaterials, 11*, 2935. https://doi.org/10.3390/nano11111293

Brack, H. G. (2015). Where have all the plastics gone. *Phenomology of Biocatstrophe, 5,* 1–435.

Buks, F., & Kaupenjohann, M. (2020). Global concentrations of microplastic in soils, a review. *Soil Discussions*. https://doi.org/10.5194/soil-2020-50

Campanale, C., Galafassi, S., Savino, I., Massarelli, C., Ancona, V., Volta, P., & Uricchio, V. F. (2022). Microplastics pollution in the terrestrial environments: Poorly known diffuse sources and implications for plants. *Science of the Total Environment, 805*, 150431. https://doi.org/10.1016/j.scitotenv.2021.150431

Crawford, C. B., & Quinn, B. (2017). *Microplastic pollutants*. Elsevier. https://doi.org/10.1016/B978-0-12-809406-8.00001-3

Hayes, D. (2019). *Micro- and nanoplastics in soil: Should we be concerned?* Performance and Adoptability Biodegradable Mulch, Report No. PA-2019-01.

Mateos-Cárdenas, A., van Pelt, F. N. A. M., O'Halloran, J., & Jansen, M. A. K. (2021). Adsorption, uptake and toxicity of micro- and nanoplastics: Effects on terrestrial plants and aquatic macrophytes. *Environmental Pollution, 284*, 117183. https://doi.org/10.1016/j.envpol.2021.117183

Pathan, S. I., Arfaioli, P., Bardelli, T., Ceccherini, M. T., Nannipieri, P., & Pietramellara, G. (2020). Soil pollution from micro- and nanoplastic debris: A hidden and unknown biohazard. *Sustainability, 12*, 7255. https://doi.org/10.3390/su12187255

Chapter 2
Soil Pollution by Micro- and Nanoplastics: Sources, Fate, and Impact

D. V. Surya Prakash, Istuti Gupta, Maheswara Reddy Mallu, and T. Mohammad Munawar

Abstract Plastic pollution in the soil presents a major intimidation to soil fertility and soil health that is directly associated with food security and human health. The properties, fate, and analysis of microplastics and nanoplastics in soil are known scantily. In actual fact, the majority of 300 million tons of plastic engendered every year is turned out in the ecosystem, and soil serves as a deep-rooted sink for those plastic rubbles. The fate of soil MPs and NPs is convincingly governed by the physical characteristics of the plastic, whereas their chemical constructs wield a marginal effect. The plastic degradation procedure, called aging, not only generates micro- and nano-sized debris, but may stimulate noticeable variations in their physical and chemical properties with pertinent influence on their reactivity. Additionally, these processes can trigger emancipation of noxious monomeric and oligomeric components from the plastics, in addition to poisonous additives that may enter into the food chain, indicating a potential threat to human health and also to the flora and fauna present in the environment. Concerning their persistence in the soil, the number of bacteria, fungi, and insects living in soil and eating plastics is increasing every day. Nevertheless, the key ecological impact of NPs lies in their ability to travel across the membrane of eukaryotic as well as prokaryotic cells. Soil biota, like collembola and earthworms, are the carriers of MPs and NPs via soil. The application of molecular techniques can provide information about the impact of MPs and NPs on the constitution and action of microbial communities residing in the soil as well as in those inhabiting on MP surface and in the gut of the soil plastic-ingesting fauna.

D. V. Surya Prakash · I. Gupta
Department of Biotechnology, Meerut Institute of Engineering and Technology, Meerut, Uttar Pradesh, India

M. R. Mallu
Department of Biotechnology, Koneru Lakshmaiah Education Foundation, Guntur, Andhra Pradesh, India

T. M. Munawar (✉)
Department of Biological and Chemical Engineering, Mekelle Institute of Technology, Mekelle University-1632, Mekelle, Ethiopia

N. R. Maddela et al. (eds.), *Micro and Nanoplastics in Soil*,
https://doi.org/10.1007/978-3-031-21195-9_2

Keywords Soil Pollution · Microplastic · Nanoplastic · Agricultural soil · Urban soil

2.1 Introduction

Irrespective of the fact that the yearly release of plastic in the soil is around 4–23 times greater than that discharged in the sea, analysis of oceanic plastic pollution preceded that of the soil contamination by plastics (Wong et al., 2018). There has been an increasing curiosity and concern regarding plastic pollution, as it is revealed that plastics build up and stay in the setting for a few hundred years under low-oxygen and low-light circumstances. Moreover, their integral monomers and oligomers, for instance, styrene and bisphenol A, are noxious since, as the monomers of PVC, they are carcinogenic and endocrine system disrupting (Demello, 2006). Some plasticizers and plastic additives are toxic materials like brominated flame retardants and phthalates (diesters of 1,2–benzenedicarboxylic acid). Both noxious additives and monomers are emancipated in the gradual process of plastic break-down in soil and may enter the aquatic environs. Additionally, plastic rubble serves as a carrier that absorbs hydrophobic inorganic and organic pollutants in addition to pathogens, which reside in water and soil, consequently escalating the noxiousness of these environments. Nevertheless, their mobility and adsorption characteristics rely on the surface-area-to-volume ratio that is high in MPs and NPs (Echevarria et al., 2016). Amelung and Bläsing studied the techniques for estimating plastics along with their input and fate in the soil, while Horton et al. reevaluated the occurrence, fate, and behavior of MPs in terrestrial environments and freshwater. Given the high ratio of surface area to volume, MPs and NPs are acutely noxious as they may enter in the food chain easily, as they can be consumed by the animals due to their diminutive dimensions (Faivre & Bennet, 2016). Plastic materials are employed as a bulk product of our routine life and economy because of their wide spectrum of favorable characteristics. The deterioration of larger particles of plastic into smaller yet highly persistent ones in the range of nanometer to micrometer intensifies the existing sink issue. Fibers and particles less than five millimeters in size are generally referred to as MPs. Soils have a crucial role in the environment (Freiberg & Zhu, 2004). The pollutants like MP and NP particles which are presented into the soil can collector be freed from the soil through, for example, deep displacement or erosive processes, and therefore be relocated to other environmental sections like the oceans. Damaging outcomes of MPs on the structure of soil and consecutively on soil water balance, soil life, soil microbiology, and soil chemistry as well as on tissue and root characteristics of the plants are scientifically postulated (Garrigue et al., 2004). NP particles can be picked up by the microorganisms or they attach themselves to the tissue of the root or penetrate it and thus alter the cell structure of the roots of plant. As a consequence, NP particles may enter the human food chain during the harvest of plants which have captivated these particles. Plastic particles of size greater than 10 μm are usually seeped in the soil; petite particles of plastic however possess the capacity to travel within the soil (Guo et al., 2014). Bulkier plastics present in soil will decompose to form MPs and NPs over a period of time by erosion, causing plastics to become more susceptible to subsurface transportation. Furthermore, interactions with suspended organic particles and microorganisms may provide MPs and NPs more hydrophilic, thus enabling subsurface transportation. Moreover, soil creatures may relocate plastic particles by the process of bioturbation, and plastic may influence the soil hydraulic characteristics themselves (Handy et al., 2008). Even though many former kinds of research have concentrated on the transfer of immaculate plastic particles, emphasis must be on environmentally pertinent plastics, because of the intricacies of uneven shape, heterogeneous surface properties and polydisperse size, and also the progressive variations of these characteristics instigated by continued environmental amendments (Castelli & Sulis, 2017). The fragmentation and surface crumbling of plastics by the virtue of weathering and human degradative activities create both MPs and NPs, whose size relies on the surface heterogeneity besides layer thickness (Heinze, 2019). The bare groups of chemicals bind the exogenic chemicals by mechanical modification, along with an effect on the rate of plastic degradation. Conclusively, the development of potentially damaging MPs and NPs causes abiotic hydrolysis which may take place in the course of gradual mineralization of biodegradable plastics (Bahadar et al., 2016).

2.1.1 Soil Pollution by Microplastics and Nanoplastics: A Global Scenario

The agglomeration of mismanaged plastic waste (MPW) in the atmosphere is a growing universal concern. Recognizing accurately where clutters are engendered is central for focusing critical vicinities for the administration of improved ment policies (Gautam et al., 2020). In 2005, MPW ranging in between 60 and 99 million metric tons (Mt) was generated globally. The MPW load is expected to remain unduly soaring in Asian and African continents in the coming years also. Commercial manufacture of plastics which initiated around the 1950s has witnessed exceptional progress, to reach the current global annual production of 330 million metric tons (Mt) for the year 2016. At the current growth rate, plastic manufacturing is expected to twofolding coming 20 years. The projected rise in plastic use in the future will result in a concomitant augmentation of post-consumer plastic waste. For example, the global urban population is approximated to generate solid waste >6 Mt on daily basis. Even with the current use of about 10% of the plastic in solid waste pill, this represents more than 200 Mt of plastic waste, which is the absolute plastic manufacture in 2002 globally (Jones et al., 2008). The unpromisingly slow escalation in recycling rates and the increased likelihood of single use products both aggravate this situation. The forthcoming rise in the production of plastic waste at the regional or even national level is spatially heterogeneous (Acharya et al., 2010a). Coastal communities in these localities will create a lopsided burden of environmental plastic waste. Understanding these spatial variations in the plastic incursion into the environment demands the creation of high-resolution maps of global plastic consumption (Liu et al., 2015) which would indicate geographical bias in future trends of plastic waste. Microplastics or tiny fragments are pervasive in soil, lakes, rivers, and also the oceans. Soil can receive microplastics and nanoplastics from a variety of daily activities of humans as well as natural means. The major sources include agricultural procedures like mulching, the use of plastic-containing soil conditioners, and irrigation using water polluted with plastic. Other sources include cluttering on roadsides and tracks, unlawful discarding of wastes, and road spillages (Huerta Lwanga et al., 2016). Natural sources include atmospheric influx and inundation from lake water and river water. (Navarro et al., 2008). The soil contamination naturally caused by undating polluted water bodies was estimated to be 0.82–4.42 plastic fragments per cubic meter. Till now there is no proper estimation methods of soil pollution is caused by NPs and also no techniques for their identification of NPs in the soil owing to their size have been identified. Thus, these days such pollution serves as an obscure and unidentified ecological biohazard (Howdle et al., 2001). Majority of the 300 million tons of plastic manufactured each year is discharged into the atmosphere, while soil behaves like a continuing sink for that waste. The future of microplastics and nanoplastics in the soil is largely governed by the material characteristics of the plastic, while their chemical structures have negligible effect. The process of decomposition of plastics, known as aging, can stimulate significant alterations on their physical and chemical properties that have pertinent results on their activity (Rawat et al., 2011), besides the generation of micro- and nano-sized debris. In addition, these processes may perhaps instigate the emission of poisonous monomeric as well as oligomeric components from the plastics and also harmful additives that might enter in the food chain, posing a potential threat to the health of humans and possibly concerning the flora and fauna. The list of bacteria, fungi, and insects that inhabits and eats plastic in the soil is expanded daily (Hwisa et al., 2013). One of the most important ecological functions that can be attributed to microplastics is linked to their role as a vector of microbes within soil. Nevertheless, the major environmental influence of nanoplastics depends on their competence for crossing prokaryotic as well as the eukaryotic cell membrane. Soil organisms, predominantly ground worms and collembolan (Sigmund et al., 2006), can be carriers of microplastics and nanoplastics across soil. However, the annual release of the plastic in the soil estimates around 4–23 times the amount released in water bodies; the research on oceanic plastic effluence preceded that of soil contamination. Interest and consternation about plastic pollution have increased, as plastic was proclaimed to accrue and persevere in nature for a few hundred years in dim light and low oxygen conditions. Furthermore, their integrant monomers and oligomers, styrene and bisphenol A (Wan et al., 2009), are noxious as like the monomer units of polyvinyl chloride, these are carcinogenic and also disturbing for the endocrine system. Some plastic accompaniments and plasticizers are toxicants, for example, phthalates and bromine flame inhibitors. Both poisonous accompaniments and monomers are emancipated in the course of the gradual disintegration of plastic in soil and could enter the water habitats via leaching. In addition, plastic rubbles act as transporters that pick up hydrophobic pollutants and pathogens, thereby augmenting ecological deadliness. Howbeit, the properties like surface adhesion and kinesis rely on the ratio of their surface area to the volume that is higher for microplastics and nanoplastics (Kapoor et al., 2015). Regardless of the fact that chemical composition is varied in plastics, its nature in soil is largely dependent on its physical characteristics. Amorphous plastic particles have higher reactivity than the crystalline ones, which may be because of the greater pore dimensions and more chemical adsorption that could improve deterioration as well as the genesis of supplementary secondary microplastics and nanoplastics (Kestens et al., 2016). As already mentioned, microplastics and nanoplastics vary in size and hence the ratio of their surface area to volume, with microplastics possessing a greater chemical activity and kinesis than that of microplastics in addition to varying colloidal properties. Colloidal characteristic affects steady or unsteady hetero-accumulation of nanoplastics that also rely on pH value and ionic force of the solution and therefore on organic matter content as well as soil mineral composition (Möller et al., 1994). Ecological-corona or eco-corona or the microenvironment of the plastics' surface area leads to an organic surface layer called corona that alters the properties of plastics and also their interaction with soil constituents and living organisms. Furthermore, the components of the ecological-corona plastic film may be degraded by the organisms. Ecological-corona could be pliable or firm, based on its affinity for getting adsorbed to the intended molecule. The hard ecological-corona possesses a higher affinity for binding, slow exchange time, and extended residence period and might cause substantial structural alterations in existing contaminating particles (Khlebtsov & Dykman, 2010). Soft eco-coronas, on the other hand, are composed of layers of exogenous molecules that are loosely bound and rapidly exchangeable, resulting in a small level of structural changes. Microplastics exist in the environment as spheres or microbeads, fibers, granules, and fragments, whereas due to the methodological issues related to the detection and characterization of nanoplastics, their shapes are relatively unknown in the environment. Microplastic dissemination from the dumping ground to the surrounding soil can be caused by wind, storms, and water disasters (Qi et al., 2016). The operations underlying degradative procedures that produce secondary microplastics in dumping ground rely on the plastic locale. The high adsorption and scattering of UV radiations cause tainting of those particles that are located on the surface level; however, those stationed in more profound landfill layers are debased by leached acidity and chemical activity of the molecules extant in concerning layers. The European Commission has suggested eradicating plastics in the landfills by the year 2025 (Kosmala et al., 2011). The annual estimate of MPs appended to farmlands in North America and Europe are 44 thousand to 300 thousand and 63 thousand 430 thousand tons, respectively, either via the usage of waste procured from processed biosolids or the direct administration of sewage sludge. The main vectors for MPs are represented by the wastewaters from treatment plants, derived from landfills, industry, stormwater, and domestic wastewater. Therefore, these wastewaters ought to be purified before being utilized to water farming land. The productivity of those cycles is determined by the employed technological advancement, whereas plastics' threshold aggregation is set up by law controlling the biosolid utilization and wastewaters on farming grounds (Shaalan et al., 2016). At present time, no policy has been designed in Europe to ward off microplastics and nanoplastics from polluting the environment. The efficiency to adequately oust microplastics from water is reliant upon the size, as elimination efficacy lessens with particle size. After the wastewater treatment, a significant part of the eliminated microplastics get collected in the sewage sludge; henceforth, utilizing it as fertilizer may acquaint soil with microplastics (Watson et al., 2007). Present-day technologies may decrease the concentration of plastic in sewage slop while keeping other nutrients in it intact. An all-inclusive information set of municipality-level trash production records for several nations is presently unavailable. Increased migration into metropolitan areas is an important trend that would likely aggravate evolving hot-spots, thus making use of high-resolution population density along with divisions of GDP to demonstrate the information of waste in a precise geographical network (Zhang et al., 2012). The employment of both of these markers allows to denote plastic trash production proximate to vast metropolitan regions as well as to probably anticipate the probable aggregation around key carriage axes like roadways and rail routes that might not be delineated by municipality-level records.

A study conducted in 2019 calculated the mismanaged plastic waste per year in million metric tons (Mt):

- New Zealand, Australia, etc. – 0.1 Mt
- US & Canada – 0.3 Mt
- Europe – 3.3 Mt
- Latin America and the Caribbean – 7.9 Mt
- Africa – 17 Mt
- Asia – 52 Mt

Top 12 mismanaged plastic waste polluters are China, 27.7%; Indonesia, 10.1%; the Philippines, 5.9%,;Vietnam, 5.8%; Sri Lanka, 5.0%; Thailand, 3.2%; Egypt, 3.0%; Malaysia, 2.9%; Nigeria, 2.7%; Bangladesh, 2.5%; South Africa, 2.0%; India, 1.9%; and the rest of the world, 27.3% (as shown in Fig. 2.1).

2.1.2 *Transport of Micro- and Nanoplastics*

Due to the huge wealth of information available on the movement of microplastics and nanoplastics a porous media, these particles are commonly employed as prototypical colloids to assess simple percolation and shipping mechanisms. Researchers have employed immaculate spherical polystyrene particles, explicitly primary microplastics and nanoplastics, having varied dimensions and surface qualities, as well as glass beads or sand as porous medium, in the majority of experiments. These well-controlled researches laid the groundwork for the development of particle-collector interaction theories, which were then confirmed using microscopic

Fig. 2.1 Mismanaged plastic waste polluter

imaging, verified in soil columns and on-site field experiments (Jordan et al., 2010). More recent experiments have used sophisticated non-spherical polystyrene units to test the influence of the shape of particle on conveyance. Microplastics and nanoplastics are also being utilized as tracers in soils and sediments to assess their transport paths and distances. Under conductive conditions, the attachment takes place in the primary energy minimum, while when the conditions are adverse, the attachment takes place largely in the secondary energy minimum (Lee et al., 2013). The adhesion of element particles on the solid-water interface is enhanced by the surface heterogeneity of the particles and collectors and it may head to the addition in primary energy minimum under adverse conditions of attachment also. The buildup of particles on solid-water interface can increase or decrease the adhesion due to ripening or blocking. Even the physical factors influence the transit of plastic fragments (Koushik & Kompella, 2004). Wedging and straining of pores capture colloidal-sized and bigger particles in tiny pores of the porous media (Table 2.1). For colloid-sized particles, reduced repulsive interfacial interactions ameliorate their pore straining and wedging. Size exclusion accounts for the prompt influx of those particles in the effluent as contrasted to conventional tracers. The air-water interface in unsaturated porous media which provides a supplementary attachment locus for colloidal-sized microplastics and nanoplastics (Li et al., 2016). Particles can adhere to water-air interfaces directly via hydrophobic contacts and can even breach the water-air interface and are then pinned to the interface by capillary forces. In addition, wedging and straining are accentuated, and slim water film straining becomes effective.

Table 2.1 Transport of micro- and nanoplastics by soil fauna

Micro plastic or Nano plastic			Experimental		
Type	Size	Shape	Setup	Major results	References
Polyethylene	Not applicable	<50–<400 (μm)	Mesocosm packed with sandy soil	Microplastics were transported downward in a size-selective manner by earthworms, with smaller particles travelling further than bigger particles	Huerta Lwanga et al. (2017)
Polyethylene	Not applicable	140–1000 (μm)	Column filled with sandy soil	Earthworms carried microplastics vertically and leached them out of the soil	Yu et al. (2019)
Polystyrene	Spherical	0.157 (μm)	Column filled with sandy loam soil	Microplastics were mixed into lower soil depths by earthworms	Heinze (2019)
Polyethylene	Not applicable	<150 (μm)	Mesocosm packed with sandy soil	Microplastics were deposited on the walls of earthworm burrows by earthworms.	Huerta Lwanga et al. (2016)
Polyvinyl chloride	Not applicable	70–240 (μm)	Petri dishes filled with charcoal and plaster of paris	*Damaeus exspinosus*, *Hypoaspis aculeifer*, and *Folsomia candida* horizontally scattered microplastics up to 8–9 cm	Zhu et al. (2018)
Polyethylene	Spherical	600–2700 (μm)	Plant pot filled with sandy soil	Earthworms carried smaller microplastics down to a greater level than larger ones	Rillig et al. (2017)

2.1.3 Sources of Soil Contamination

Soils can obtain microplastics and nanoplastics as a result of several of natural processes and human activities. Farming practices like mulching of plastic, employment of plastic-containing soil enhancers, and irrigation with wastewaters contaminated with plastic represent key human sources (Millstone et al., 2010). Further anthropogenic sources include landfills, illegal waste dumping, littering along streets and trails, and road overspill. Natural supplies are characterized by flooding with river or lake water and atmospheric inputs. The average soil contamination induced by overflowing of river and lake water is estimated to be 0.82–4.42 plastic particles per cubic meter. Lastly, because of their tiny size, there are no approaches for detecting nanoplastic pollution in soil; hence there are no estimates of nanoplastic pollution in soil. Thus, this pollution nowadays represents an unknown and hidden environmental biohazard (Lu et al., 2014).

2.1.3.1 Landfills

Microplastic dispersion from landfills of the surrounding soils can be caused by wind, storms, and droughts. The procedures underlying indiscriminate degradative processes which produce consequent micro plastics in dumping grounds are based on plastics' locale. The high adsorption and scattering of UV radiations cause tainting of those particles that are located on the surface level (Cao, 2002); however, those stationed in more profound landfill layers are debased by leached acidity in concerning layers. Chemicals discharged from the breakdown of plastic might scatter in the environment, and their circulations rely on the size of the pore of plastic and molecular size of additives (Mazurais et al., 2015).

2.1.3.2 Floods, Rise Up of Salt Water in Coastal Soil, and Aeolian Transport

Approximately 80% plastic debris in water bodies are emanated by land sources, primarily via soil erosion and leaching. Coastal regions are susceptible to plastic pollution caused by human activities in addition to the pollution from the sea (Martis et al., 2012). In case of seawater debasing the groundwater, a pertinent cause of microplastics and nanoplastics in it, the farming soil along the coastal regions would be inundated by salt water. As evidenced from the occurrence of microplastics in Swiss floodplain soil stationed away from metropolitan regions, aeolian transit represents a primary use of plastics.

2.1.3.3 Soil Fertilized with Sewage Sludge or Irrigated with Wastewater

Wastewater from treatment plants embodies key carriers for microplastics originating from landfill, industries, stormwater, and household water. Therefore, these wastewaters must be purified before being used to irrigate farming land. Efficacy of those procedures relies upon the used technology, while the maximum amount of plastic content is determined by legislature administering the usage of biosolids as well as wastewaters to agronomic fields (Oprea et al., 2015). No specialized strategy has been intended to ward off the ecological effluence caused by microplastics and nanoplastics, as of now. Since elimination effectiveness lowers as the size of the object enlarges, the efficient reduction of microplastics from watercourses is proportional to the size of the object. The efficiency to adequately oust microplastics from water is reliant upon the size, as exclusion efficacy lessons in accordance with the particle size plus it is less for tinier particles (Mabena et al., 2011). After the wastewater treatment, a significant part of the eliminated microplastics get collected in the sewage sludge; henceforth, utilizing wastewater for irrigation or sludge as a soil fertilizer may lead to the introduction of microplastics in soil and later into water bodies.

2.1.3.4 Soil Under Plastic Mulching

Mulching of plastic is a ubiquitous agronomic procedure that regulates the temperature of soil, improves the efficiency of water usage, and controls pathogens to ameliorate crop quality and yield. Globally, the surface area masked by this practice is anticipated to increase 5.7% each year (Manucci & Franchini, 2017). PVC and low-density polystyrene are the most commonly used polymers, given their expensive rates. Mulching operations create plastic waste and liberate toxic compounds such as phthalates. Both additives and microplastics have the potential to enter the food chain through the contaminated soil and hence pose serious health hazards to humans (Mansha et al., 2017). The use of bioplastics and biodegradable plastics can help mitigate the environmental risks associated with mulching of plastic, although their application is restricted because of the expensive price.

2.1.4 Fate of Microplastics and Nanoplastics in Soil

Plastic detrition in topsoil essentially relies upon plastic's physicochemical characteristics, type of soil, presence of a functioning microbial community, and ecological conditions (Mashaghi et al., 2013). For instance, weathering forces and exposure to UV radiation speed up plastic composition that is greater in clayey soil as compared to sandy soil, which could be an effect of dissimilarities in microbial action of two soils. Biodegradation of plastics occurs in two stages in soil: surface decomposition of polymer is trailed by decomposition of particles deriving from the initial stage (Park et al., 2014). The early plastic biodegradation rate relies upon the accessible surface area of plastics and is represented as a dense growth of mycelia on plastic's surface in addition to the development of bacterial biofilms. A few microorganisms residing in the soil can partially or wholly debase engineered plastics by co-metabolism as the key degradative pathway. Nutrient accessibility is not significantly restricting as pectin plastic decomposition in soil; however, the detrition of microplastics and nanoplastics relies upon hetero- as well as auto-accumulation, henceforth on surface hydrophobicity (Leon et al., 2015), as recounted in water bodies. As previously stated, even decomposition of additional carbon-based substances may take place in the course of plastic degradation, initiating the creation of tiny fragments which may blow out in ecosystem. Ecological-corona can pick up bacteria which colonize the outside surface of plastics. As a matter of fact, the instance of impurities adsorbed on the surface of plastic and metagenomic investigations of surface colonizers may give rise to unique microorganisms that degrade pollutants. Coming investigation making use of amplicon as well metagenome sequencing might offer discernments on the occurrence of decomposers of plastics in soil (Rochman et al., 2014). The fate and impacts of microplastics and nanoplastics in soil rely upon ecological-corona characteristics that, for instance, may influence the interaction of plastics with carbon-based substance in addition to mud minerals, and plastic consumption and noxiousness by soil eaters, like annelids. Moreover, it's imperative to have understanding on:

(a) Adsorption of significant biomolecules, like root exudates, deoxyribonucleic acid, enzymes, etc. on the surface of microplastics and the impacts they show on ecological-corona characteristics.
(b) How various properties of ecological-corona influence environmental conduct of microplastics and nanoplastics, that is, their interactivity with the constituents of soil, therefore on the mobility, endurance, toxicity, and biological availability (Mueller & Nowack, 2008). As it is related to the interfaces of microplastics and nanoplastics with surface-receptive soil units, for example, soil organic matter (SOM) and clays, soil pH of soil may influence the charge of the surface of those plastics, excluding the ones distinguished by hydrophobic surfaces (Siepmann et al., 2004) that do not have any charge. In untouched environmental soils, descending travel of microplastic is sought to be preferred by the incidence of preferential path flows as well as macro-pores, like cracks and bio-pores, and restricted by microporosity through microplastic buildup on soil top deposit. Of course, soil plastic mobility additionally relies on clay minerals and dissolved organic matter since particles of plastic may combine to those components of soil, as debated underneath. Also, capillary percolation or transport of microplastics and nanoplastics might take place via soil, as it happens for the compounds to higher molecular weight (Holzinger et al., 2014); however, this must be experimentally demonstrated. Siegfried et al. and Nizzetto et al. developed model frameworks for the transit of microplastics by draining and erosion of soil to measure the microplastic division in terrestrial as well as aquatic settings. Lack of investigational records, however, implied that the precision of suggested prototypes cannot be substantiated (Nagarajan et al., 2014). Tillage exercises have a positive impact on the surface soil porosity and accumulation, consequently further developing percolation. Scientists have detected a great measure of plastic wastes correlated with 72% accumulates in addition to the occurrence of the fibers of microplastics in microaggregates of modified soil. Plastic waste incorporation in aggregates may advance their buildup. This might influence accumulated incomings as well as an exchange of biological components residing in accumulates along with soil components (Kappos et al., 2004). The end product relies upon the kind of microplastics as, for instance, polypropylene and polyethylene augmented cluster development. Thus, microplastics may influence the structure of soil and hence its function. Nevertheless, tilling might even restrict microplastic and nanoplastic portability in top soil because of the development of plough pan, which may augment the plastic intensity in top soil layers. The analysis of plastic waste present in the river sediments along with the soil overflowed by these ashore might offer understandings on plastic waste movement as well as its destiny in that soil (Yuan et al., 2015). The accumulation of microplastics from water tanks on the deposits is delayed because of small solidity of microplastics, yet the rate of accumulation grows with hetero-accretion of microplastics by particulate inor-

ganic and organic substance because of the heightened solidity of those hetero-particles in comparison with distinct fragments.

2.1.5 Impacts of Microplastics and Nanoplastics on the Properties of Soil

2.1.5.1 Physical-Chemical Properties of Soil

Manifestation of soil microplastics and nanoplastics may modify physical, chemical, and biological characteristics of the soil and alter the approximation of carbon segregation in soil (Nixon et al., 2010). Plastics affect the soil aggregate formation and also the properties of humic acid. Certainly, some researchers perceived that plastic granule appendage augmented overall biological carbon matter of soil because the recent techniques applied for reckoning soil biological carbon additionally govern the imperceptible microplastics portion of soil accumulates (Thote & Gupta, 2005). Consequently, Rillig suggested reconsidering the "true" soil repository of carbon in soil polluted with plastics. Effects of microplastics exhibiting various shapes; densities and chemical composition, on water holding capacity; bulk density; microbial activities; and water-stable aggregates of the soil, were studied by De Souza Machado et al. It leads to the inference that microplastic can presumably initiate working modifications in the soil which are tough to anticipate because of the intricacy of soil structure (Nikalje, 2015).

2.1.5.2 Active Extracellular Molecules of Soil

Decay of extreme molecular-weight natural polymers is caused by the activities of extracellular enzymes and thus plays an important role in soil functioning (Wang et al., 2013). Hydrophobic microplastics and nanoplastics adsorb extracellular enzymes which may extend the enzyme's half-life attributable to the shield counter to proteolysis and decline in thermal denaturation. Frang et al. suggested that after 28 days of incubation, polystyrene nanoplastics reduced the extracellular enzyme activity of soil. The genesis of the enzyme activity, however, is ambiguous. The measurement carried out by a few scientists can be both from extracellular and intracellular enzymes. Moreover, an unconstructive effect of polystyrene nanoplastics was perceived on microbial biomass. Howbeit, this impact would not endure when their clusters are formed.

2.1.5.3 Soil Microbial Community

Due to methodological issues, the transfer of invasive microbial entities using plastic trash, particularly the function of microplastics in the transportation of microbes, is inadequately recognized. According to Sanna et al. (2015) pesticides could migrate via soil structure with microplastics. Concerns about such cases would spur upcoming studies to better realize the task of microplastics as vectors to contaminants as well as other impurities (Sanna et al., 2015). Microplastics can influence several microbial properties; for instance, bacteria associated with microplastics exhibited greater rates of plasmid transfer as compared with free-living bacteria. As the community of bacteria that lives in a biofilm can develop a vast range of resistance against antibiotics, it's plausible to believe them to develop resistance to a wide range of antibiotics. DNA transmission in the biofilm may occur through both conjugation and transformation. Nanoplastics can easily infiltrate lipid membranes in cells, influencing the functionality of cells (Fang et al., 2013). Microorganisms, on the other hand, can prevent NPs from entering into the cells by employing various self-protective mechanisms, like changes in cellular membrane structure, the secretion of contaminants-neutralizing molecules, and obstructions imposed by any kind of biofilm matrix or bacterial cell walls.

2.1.5.4 Soil Fauna

Soil biota, predominantly the collembolan and annelids, may absorb both microplastics and nanoplastics, hence transporting those across soil settings, as earthworms are mostly efficient due to great soil filtering ability (Wang et al., 2015). Given their abundance, which ranges between 10 thousand and 100 thousand individuals per square meters of the soil in the top 10 cm of soil setting, the outcome of collembolan on microplastics and nanoplastics movement in soil is significant. Nevertheless, the impact of more naturally viable biota of soil, for instance protists, which are primary soil bacteria consumers, remains unknown. These organisms can be the important carriers for microplastic delivery in the food chain in soil. Protists may differentiate among various bacteria types other than in between bacterial cells and latex microplastic spheres. The rate of microplastic absorption and incorporation by protists is determined by species, their age, nutritional state, and the concentration of microplastics (Dreaden et al., 2012). Feeders of plastic appear to favor older microplastics due to microorganism's residence in them. The feeders take up evenly shaped microplastics more readily in comparison to the uneven ones. Consumption of microplastics and nanoplastics by the fauna of soil may alter the constitution of microflora in collembolan gut as well as the oligochaete *Enchytraeus crypticus*.

2.1.5.5 Pedogenesis of Soil

An interesting element discussed in the above-stated consequence of microplastics and nanoplastics on soil characteristics, resulting from their protracted period of habitation along with strong reactivity, relates with their potential impacts on pedological developments of soil. Incidence of microplastics and nanoplastics as distinguishing features in classifying top soil along with the soil of subsurface layers may be conceivably postulated (Torchynska et al., 2016). Furthermore, how this waste could alter the pedological processes is still a point of discussion. This prospect is distinguished as well as the herald of exciting advancements. It is critical to contemplate newly found pyroplastics, which are derived from the widely used technique of blazing trash. These types of plastics may become a part of the geological cycle of soil, due to their resistance to degradation (Reiss & Hutten, 2005).

2.1.5.6 Plants

Plastic pollution in the soil may have both indirect and direct impacts on grown flora by the virtue of root absorption or consequences of biological and physicochemical properties of soil; correspondingly about the straight consequences, there has been a rise in the total of indications of microplastic and nanoplastic pollution in flora in the previous 2 years; metabolism of contaminants in plants or storing of resistant impurities might be the primary cause of concern (Hajipour et al., 2012). The physiological and anatomical properties of the plants, the properties of plastics, and environmental conditions impacting surface chemistry and behavior altogether influence the absorption of MPs and NPs by plants. The key issues to consider when discussing microplastics' and nanoplastics 'secondary impacts on farmed plants are their pollutant adsorption and diffusion, influences on the structure of the soil, soil microbial community, immobilization of nutrients, root symbionts, as well as root-associated microbiome (Rillig et al., 2017).

2.1.6 Agricultural Soils

The microplastics and nanoplastics are released by various polluters from a variety of materials. Therefore, the particles differ in their physicochemical properties as well as in their life cycle and consequences on organisms and environmental systems (Rogozea et al., 2016). Agricultural production is also conflicted when it comes to the subject of microplastic and nanoplastic pollution problem. This plastic is released into agricultural soils from littering and tire wear by runoff and aerial dispersal. Compost and sewage sludge contaminated with microplastics and nanoplastics are used in agriculture as fertilizers. As a result, agricultural soils serve as sinks for microplastic particles, which may have harmful impacts on the organisms and soil structure (Zhu et al., 2018). Moreover, soils polluted with microplastics and nanoplastics are in danger from unknown adulterants in plastic fragments. Changing the biophysicochemical properties of soil may influence its ecology and efficacy. During the application of plastic film, secondary microplastic is accidentally discharged into the environment through the fragmentation process. Also, microplastics are released from agricultural soil into other environmental systems. Henceforth, the agriculture industry also contributes to pollution. Microplastic and nanoplastic particles are potentially transported into surface and ground water bodies and drainage by leaching via tiles and soil pores (Ullah et al., 2017). Microplastics are transported into surface water as well as other environmental systems by soil erosion caused by wind or water. Agricultural production necessitates the use of natural resources, which has both constructive and destructive effects. Society, on one hand, associates agricultural output with environmental services. While on other hand, society blames the agriculture industry for adverse environmental effects. Intensive agricultural practice generates a variety of pollutants and negative environmental repercussions (Gore et al., 2016). Microplastic as a pollutant is not the only challenging problem due to the number of polluters and victims. The partially known attributes and the presumed characteristics (but not verified with evidence) are comparable to the properties of finer recognized pollutants. Because of a number of properties of microplastics and nanoplastics, it is challenging to comprehend the fate and effect of this pollutant. For instance, nitrate is a water-soluble material that enters the land and surface water bodies via runoff and leaching. Phosphate is linked to the particles of soil and is so transferred into surface waters through soil erosion. The amount of leaching into the ground water body is comparatively negligible. The transport of microplastics into groundwater is yet to be proved (Saud et al., 2012).

2.1.6.1 Reduction of the Input of Microplastic and Nanoplastic in Agricultural Soils

The reduction or elimination of microplastics and nanoplastics in soils, particularly agricultural land, is critical to minimize contamination of the food chain and other ecosystems, including humans. Plastic content in waters, composts, and wastewater sludge should be reduced through the use of procedures (Sikora et al., 2016). Also significant is the promotion of the use of degradable plastics, for instance, bioplastic or biodegradable plastic for mulching like poly-(butylene adipate-co-terephthalate) for mulching. Disposable and bio-based plastics are becoming increasingly popular.

Bio-Based and Biodegradable Plastics

Bioplastics are partly or wholly perishable materials may be classified into three types, namely, synthetic, partially biological, and completely biological. Polycaprolactone and poly(butyl adipate-co-terephthalate) are the chief biodegradable synthetic plastics whose acyclic molecular component is liable for their compostability (Barrow, 2004). Conversely, bio-based plastics are basically composed of polylactic acid, poly hydroxyl alkanoic acids, PBS-co-adipate, and poly-butylene succinate. There are numerous applications for these plastics, including the replacement of traditional ones in the agricultural and medical industries, as well as in the milk industry. Actinomycetes, bacteria, and fungi may destroy both synthetics and natural plastic, and they do so by causing alterations in the physical and chemical characteristics of the substances. For the most part, biodegradation occurs under aerobic conditions, although it can also occur under anaerobic environments in sediments and landfills, as well as under partial oxygen concentration in soils and compost (Thomas et al., 2015). The chemical composition of bioplastics and the bacterial biomass available in the soils govern the rate at which bioplastics degrade, but not the biodiversity of bacteria. The rate of microbial degradation rises with the surge in surface-area-to-volume ratio, enhancing the consumption of water and oxygen, along with the stimulation of hydrolytic and oxidative processes. Generally, biodegradable products are more costly than non-perishable products, but the prolonged consequences of their non-usage, like environment affluence and greater landfills exploitation, lay the amount into a prospect. Additionally, the utmost appropriate resolution is not based merely on the properties of plastics but even on their amount in the marketplace, the collection presented, and the ground work processing (Tratnyek & Johnson, 2006). Howbeit, it is imperative that existing industries must be restructured in place to enable the production and maintenance of biodegradable plastics as well as their emissions.

Cleanup and Bioremediation Technology Development

A fascinating and potentially effective technique for lowering soil contamination of microplastics and nanoplastics includes the augmentation of on-site abasement by boosting engagement of soil organisms, such as fungus, bacterium, as well as other microorganisms (Huerta Lwanga et al., 2017). *Enterococcus* sp., *Alcaligenes* sp., *Corynebacterium sedlakii*, *Citrobacter sedlakii*, and *Brevundimonas diminuta* were identified and proven to be proficient in the degradation of polystyrene and also microelectronic plastics that contain antimony trioxide and decabromodiphenyl oxide. Muenmee et al. examined the bioremediation of discarded plastic products by carrying out a lysimeter experiment. Pre-aged UV light plastics and in sanitary steadied carbon-based trash taken from uncovered landfills were combined by them. Heterotrophs, autotrophs, and methanotrophs were shown to be the predominant detritivores of plastic pollution in a bacterium community that was identified in plastic waste (Acharya et al., 2010a, b). Screening of plastic decomposing organisms from several discarded soils underlined that the plastics had been completely digested by *Bacillus* species, *Aspergillus* species, *Streptococcus* species, and two *Fusarium* species. Contemplating that the biodecomposition of plastics occurs on the surface, it's rational to postulate that the efficacy of biodegradation might be governed by the ratio of surface area to volume of plastic waste (Della Porta et al., 2013; Demello, 2006). Howbeit, the disintegration rate of polymers can also be influenced by a wide range of physical properties like crystallinity, glass transition point, melting point, and modulus of storage as well as by chemical nature of polymers, activity of the microbial degraders, and environmental conditions. Moreover, degradation of microplastics and nanoplastics happens in hot spots, like the gut of worms, where the quantity and activity of microbial decomposers are greater than in other sites. Bacteria discovered in the gut of mealworms may disintegrate plastics, and the wax moth larvae and bacteria *Ideonella sakaiensis* 201-F6 may degrade polyethylene and polyethylene terephthalate, respectively (Wilson & Suh, 1997). *Lysinibacillus, Bacillus* sp. $T_{2,}$ and other gut bacteria have been found to help termites chew and eat plastics indirectly way. Carbon-based compounds may escalate the speed of biodegradation of plastics. Usually, the end products of plastic decomposition formed through hydroxylation and/or carboxylation are organic compounds having small molecular mass that can be biodegraded easily as an outcome of plastic rubbles elimination in setting. There are, however, certain exceptions to this rule.

2.1.7 Urban Soil

Soil constitutes a combination of a number of liquids, gases, minerals, and carbon-based materials capable of supporting natural life. Soil is a channel for a variety of functions, including carbon sequestration, biogeochemical cycling, and promotion of biodiversity (Yu et al., 2013). Soil serves as a prospective ecological reservoir of microplastics and may instigate a number of land-dwelling problems. Microplastics are capable of making their way amidst waterways via soil. For instance, numerous coastal areas as well as beaches are being exploited as landfills, and also uprising oceanic levels ensue the wearing away. As a result, microplastics in coastal landfills are expected influence waterways. The world's urban landfills, which are used to dispose of garbage, can retain 21–42% of plastic trash produced around the globe (Zhang et al., 2012). Hence, trash disposal at landfills, agricultural technology development, and industrial manufacturing are all linked with the emancipation of primary as well as secondary microplastics that eventually enter the earthly situation via physical drift and energy drift. Given their absorption ability, microplastics does not merely filtrate the soil; rather they take up natural contaminants, and they also function like a catalytic agent to integrate heavy metal availability in the soil. Consequently, the microplastics collected in soil in a greater amount may be consumed by the biological entities residing in soil. Physical plus chemical characteristics of microplastics and nanoplastics make them further more harmful for the environment than bigger plastic wastes. Therefore, the physical and chemical properties of the soil can be changed that may present a negative influence on biological diversity and also different soil procedures as breakdown fashion of carbon-based substance (Draheim et al., 2015). Key cause of microplastic and nanoplastic pollution is wearing of tire as it is very copious as compared to another type of plastic units. Demolition of those plastic fragments causes the creation of fragments along with fibers. Primary microplastics have the potential to alter terrestrial ecosystems by entering the environment. Machado et al. visibly indicated that microplastics and nanoplastics may alter the characteristics of the soil and their effect on the plant performance. According to He et al., microplastics were discovered in several soil samples of landfill with 99.36% of microplastics originating from landfill plastic trash fragmentation (Gross et al., 2016). The process of degradation of plastic relies on several factors like the type of polymer and its age as well as some environmental

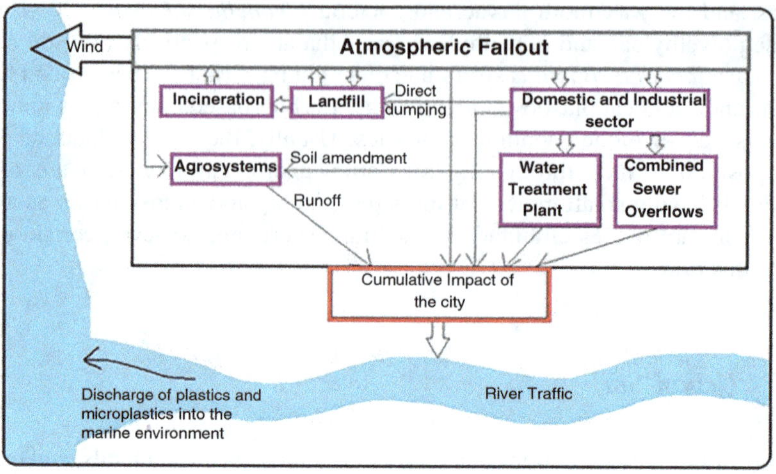

Fig. 2.2 MPs released into the water bodies from urban water system

processes including acidity, alkalinity, weathering processes, and temperature. Plastic sources in catchments subjected to significant anthropogenic influences, particularly urban soil regions, should be given special consideration. Only main plastics and microplastics should be allowed to enter the city, which can be termed as a closed system. During their life span, the produced secondary plastics and MPs may make their way into the three elements of environment: atmosphere, soil, and water. Plastic pollution in the atmosphere can come from particle resuspension, industrial emissions, and other anthropogenic sources such as buildings, urban infrastructure, traffic, etc. Plastics can also enter the aquatic element due to negligent conduct or the urban water system, whether "combined" or "separate." Conclusively, because wastewater treatment sludge, which may comprise plastics, is frequently used on agricultural lands, the soil element is also at a risk of contamination (Falco et al., 2012). Furthermore, plastic particle fallout in the atmosphere may also lead to soil pollution. Almost everything is obscure about the behavior of plastics in these elements and the dynamics between or within them, so as the first step, it was chosen to focus on the channels interacting directly with the collecting water system, namely:

(a) MPs released into the water bodies from urban water system, particularly from the wastewater treatment plants (Fig. 2.2)
(b) MPs arising from the atmosphere

2.1.8 *Other Soils*

2.1.8.1 Domestic Soil

The varieties of MPs and NPs identified with higher affluence in domestic land include polystyrene (PS), polyethylene (PE), polyvinyl chloride (PVC), polypropylene (PP), polyamide (PA), polyethylene terephthalate (PET), polyurethane (PU), and acrylonitrile-butadiene-styrene (ABS). The main shapes of these plastics are fibers, films, spheres. and fragments (Abulateefeh & Alkilany, 2016).

2.1.8.2 Industrial Soil

Synthetic polymers are manufactured using basic raw materials such as coal, natural gas. and oil and are labeled as plastics (Freitas et al., 2005). Both of these varieties of plastic have been tagged as substitutes for synthetic plastic because; as their names suggest, they will biodegrade more promptly. Nevertheless, there is no concrete affirmation that either biodegradable plastics or bioplastics will disintegrate any better in the natural environment than synthetic plastics (Yu et al., 2019).

Important examples of such synthetic polymers include:

- High-density polyethylene (HDPE), polyvinyl chloride (PVC), polyethylene terephthalate (PET), polyurethane (PUR), polystyrene (PS), low-density polyethylene (LDPE), and polypropylene (PP)

2.2 Conclusion

Regardless of the fact that soil pollution caused by microplastics and nanoplastics is a crucial subject today, various information loopholes regarding their effects and fate are present which future researches need to address. A fundamental precondition in such framework is to advance the currently used methods which are exemplified by superior class level for microplastics, yet they turned out to be a tough technical task for nanoplastics. Its striving aim is the syndication of effectiveness, time request, standardization, and minimal cost regardless of extreme variability and complex soil setting. The significance of such a goal is validated by the increasing number of articles published on that topic in the past several years. Forthcoming research undertakings would need to be capable of covering the microplastic and nanoplastic extensive dissemination top soil from the scale of nanometers to micrometers and commencing quite a lot of complicated interactions. Such interfaces encompass entire abiotic as well as biotic constituents of soil and frequently instigate noticeable impacts on the reactivity and properties. To accomplish this object, the research approach would appeal an all-inclusive methodology that is capable of syndicating information from explicit facets in a common structure which review the outcomes at ecosystem level. The abovementioned methodology would additionally consent to gauge the activities as well as actions of microplastics and nanoplastics in a better way and will also make available a well-defined representation of its significance at the level of bio network. Preceding information would signify primary preconditions to deep-seated analysis on microplastics and nanoplastics effluence of soil and also neutralize the perilous outcomes they have on ecosystem of soil. A collective application of the abovementioned methodologies would promote the investigations of primary obstruction associated with microplastics and nanoplastics in soil.

References

Abulateefeh, S. R., & Alkilany, A. M. (2016). Synthesis and characterization of PLGA shell microcapsules containing aqueous cores prepared by internal phase separation. *AAPS PharmSciTech, 17*, 891–897. https://doi.org/10.1208/s12249-015-0413-y

Acharya, G., Shin, C. S., Vedantham, K., Mcdermott, M., Rish, T., Hansen, K., et al. (2010a). A study of drug release from homogeneous PLGA microstructures. *Journal of Controlled Release, 146*, 201–206. https://doi.org/10.1016/j.jconrel.2010.03.024

Acharya, G., Shin, C. S., Mcdermott, M., Mishra, H., Park, H., Kwon, I. C., et al. (2010b). The hydrogel template method for fabrication of homogeneous nano/microparticles. *Journal of Controlled Release, 141*, 314–319. https://doi.org/10.1016/j.jconrel.2009.09.032

Bahadar, H., Maqbool, F., Niaz, K., & Abdollahi, M. (2016). Toxicity of nanoparticles and an overview of current experimental models. *Iranian Biomedical Journal, 20*, 1–11. https://doi.org/10.7508/ibj.2016.01.001

Barrow, W. W. (2004). Microsphere technology for chemotherapy of mycobacterial infections. *Current Pharmaceutical Design, 10*, 3275–3284. https://doi.org/10.2174/1381612043383197

Cao, Y. C. (2002). Nanoparticles with Raman spectroscopic finger-prints for DNA and RNA detection. *Science, 80*, 1536–1540. https://doi.org/10.1126/science.297.5586.1536

Castelli, F., & Sulis, G. (2017). Migration and infectious diseases. *Clinical Microbiology and Infection, 23*, 283–289. https://doi.org/10.1016/j.cmi.2017.03.012

Della Porta, G., Falco, N., Giordano, E., & Reverchon, E. (2013). PLGA microspheres by supercritical emulsion extraction: A study on insulin release in myoblast culture. *Journal of Biomaterials Science. Polymer Edition, 24*, 1831–1847. https://doi.org/10.1080/09205063.2013.807457

Demello, A. J. (2006). Control and detection of chemical reactions in microfluidic systems. *Nature, 442*, 394–402. https://doi.org/10.1038/nature05062

Draheim, C., De Crécy, F., Hansen, S., Collnot, E. M., & Lehr, C. M. (2015). A design of experiment study of nanoprecipitation and nano spray drying as processes to prepare PLGA nano- and microparticles with defined sizes and size distributions. *Pharmaceutical Research, 32*, 2609–2624. https://doi.org/10.1007/s11095-015-1647-9

Dreaden, E. C., Alkilany, A. M., Huang, X., Murphy, C. J., & El-Sayed, M. A. (2012). The golden age: Gold nanoparticles for biomedicine. *Chemical Society Reviews, 41*, 2740–2779. https://doi.org/10.1039/C1CS15237H

Echevarria, M. L., Wolfe, G. V., & Taylor, A. R. (2016). Feast or flee: Bioelectrical regulation of feeding and predator evasion behaviors in the planktonic alveolate *Favella* sp. (Spirotrichia). *Journal of Experimental Biology, 219*, 445–456. https://doi.org/10.1242/jeb.121871

Faivre, D., & Bennet, M. (2016). Materials science: magnetic nanoparticles line up. *Nature, 535*, 235–236. https://doi.org/10.1038/535235a

Falco, N., Reverchon, E., & Della Porta, G. (2012). Continuous supercritical emulsions extraction: Packed tower characterization and application to poly(lactic-co-glycolic acid) plus insulin microspheres production. *Industrial and Engineering Chemistry Research, 51*, 8616–8623. https://doi.org/10.1021/ie300482n

Fang, X. Q., Liu, J. X., & Gupta, V. (2013). Fundamental formulations and recent achievements in piezoelectric nano-structures: A review. *Nanoscale, 5*, 1716. https://doi.org/10.1039/c2nr33531j

Freiberg, S., & Zhu, X. X. (2004). Polymer microspheres for controlled drug release. *International Journal of Pharmaceutics, 282*, 1–18. https://doi.org/10.1016/j.ijpharm.2004.04.013

Freitas, S., Merkle, H. P., & Gander, B. (2005). Microencapsulation by solvent extraction/evaporation: Reviewing the state of the art of microsphere preparation process technology. *Journal of Controlled Release, 102*, 313–332. https://doi.org/10.1016/j.jconrel.2004.10.015

Garrigue, P., Delville, M. H., Labrugere, C., et al. (2004). Top-down approach for the preparation of colloidal carbon nanoparticles. *Chemistry of Materials, 16*, 2984–2986. https://doi.org/10.1021/cm049685i

Gautam, A., Gore, P. M., & Kandasubramanian, B. (2020). Nanocluster materials in photosynthetic machines. *Chemical Engineering Journal, 385*, 123951–123956. https://doi.org/10.1016/j.cej.2019.123951

Gore, P. M., Dhanshetty, M., & Kandasubramanian, B. (2016). Bionic creation of nano-engineered Janus fabric for selective oil/organic solvent absorption. *RSC Advances, 6*, 111250–111260. https://doi.org/10.1039/C6RA24106A

Gross, J., Sayle, S., Karow, A. R., Bakowsky, U., & Garidel, P. (2016). Nanoparticle tracking analysis of particle size and concentration detection in suspensions of polymer and protein samples: Influence of experimental and data evaluation parameters. *European Journal of Pharmaceutics and Biopharmaceutics, 104*, 30–41. https://doi.org/10.1016/j.ejpb.2016.04.013

Guo, D., Xie, G., & Luo, J. (2014). Mechanical properties of nanoparticles: Basics and applications. *Journal of Physics D: Applied Physics, 47*, 13001. https://doi.org/10.1088/0022-3727/47/1/013001

Hajipour, M. J., Fromm, K. M., Ashkarran, A. A., et al. (2012). Antibacterial properties of nanoparticles. *Trends in Biotechnology, 30*, 499–511. https://doi.org/10.1016/j.tibtech.2012.06.004

Handy, R. D., Von Der Kammer, F., Lead, J. R., et al. (2008). The ecotoxicology and chemistry of manufactured nanoparticles. *Ecotoxicology, 17*, 287–314. https://doi.org/10.1007/s10646-008-0199-8

Heinze, W. M. (2019) *Nanoplastic transport in soils by advection and bioturbation* (Doctoral dissertation). Swedish University of Agricultural and Sciences, Uppsala, Sweden.

Holzinger, M., Le Goff, A., & Cosnier, S. (2014). Nanomaterials for biosensing applications: A review. *Frontiers in Chemistry, 2*, 63–68. https://doi.org/10.3389/fchem.2014.00063

Howdle, S. M., Watson, M. S., Whitaker, M. J., Popov, V. K., Davies, M. C., Mandel, F. S., et al. (2001). Supercritical fluid mixing: Preparation of thermally sensitive polymer composites containing bioactive materials. *Chemical Communications, 2001*, 109–110. https://doi.org/10.1039/b008188o

Huerta Lwanga, E., Gertsen, H., Gooren, H., Peters, P., Salánki, T., Van Der Ploeg, M., & Geissen, V. (2016). Microplastics in the terrestrial ecosystem: Implications for Lumbricus terrestris (Oligochaeta, Lumbricidae). *Environmental Science & Technology, 50*, 2685–2691.

Huerta Lwanga, E., Gertsen, H., Gooren, H., Peters, P., Salánki, T., Van der Ploeg, M., & Geissen, V. (2017). Incorporation of microplastics from litter into burrows of Lumbricus terrestris. *Environmental Pollution, 220*, 523–531. https://doi.org/10.1016/j.envpol.2016.09.096

Hwisa, N. T., Katakam, P., Chandu, B. R., & Adiki, S. K. (2013). Solvent evaporation techniques as promising advancement in microencapsulation. *VRI Biological Medicinal Chemistry, 1*, 8–22. https://doi.org/10.14259/bmc.v1i1.29

Jones, N., Ray, B., Ranjit, K. T., & Manna, A. C. (2008). Antibacterial activity of ZnO nanoparticle suspensions on a broad spectrum of microorganisms. *FEMS Microbiology Letters, 279*, 71–76. https://doi.org/10.1111/j.1574-6968.2007.01012.x

Jordan, F., Naylor, A., Kelly, C. A., Howdle, S. M., Lewis, A., & Illum, L. (2010). Sustained release hGH microsphere formulation produced by a novel supercritical fluid technology: *in vivo* studies. *Journal of Controlled Release, 141*, 153–160. https://doi.org/10.1016/j.jconrel.2009.09.013

Kapoor, D. N., Bhatia, A., Kaur, R., Sharma, R., Kaur, G., & Dhawan, S. (2015). PLGA: a unique polymer for drug delivery. *Therapeutic Delivery, 6*, 41–58. https://doi.org/10.4155/tde.14.91

Kappos, A. D., Bruckmann, P., Eikmann, T., Englert, N., Heinrich, U., Höppe, P., et al. (2004). Health effects of particles in ambient air. *International Journal of Hygiene and Environmental Health, 207*, 399–407. https://doi.org/10.1078/1438-4639-00306

Kestens, V., Roebben, G., Herrmann, J., et al. (2016). Challenges in the size analysis of a silica nanoparticle mixture as candidate certified reference material. *Journal of Nanoparticle Research, 18*, 171–174. https://doi.org/10.1007/s11051-016-3474-2

Khlebtsov, N., & Dykman, L. (2010). *Plasmonic Nanoparticles, 4*, 37–85. https://doi.org/10.1201/9781439806296-c2

Kosmala, A., Wright, R., Zhang, Q., & Kirby, P. (2011). Synthesis of silver nano particles and fabrication of aqueous ag inks for inkjet printing. *Materials Chemistry and Physics, 129*, 1075–1080. https://doi.org/10.1016/j.matchemphys.2011.05.064

Koushik, K., & Kompella, U. B. (2004). Preparation of large porous deslorelin-PLGA microparticles with reduced residual solvent and cellular uptake using a supercritical carbon dioxide process. *Pharmaceutical Research, 21*, 524–535. https://doi.org/10.1023/B:PHAM.0000019308.25479.A4

Lee, Y. S., Johnson, P. J., Robbins, P. T., & Bridson, R. H. (2013). Production of nanoparticles-in-microparticles by a double emulsion method: A comprehensive study. *European Journal of Pharmaceutics and Biopharmaceutics, 83*, 168–173. https://doi.org/10.1016/j.ejpb.2012.10.016

Leon, R. A. L., Somasundar, A., Badruddoza, A. Z. M., & Khan, S. A. (2015). Microfluidic fabrication of multi-drug-loaded polymeric microparticles for topical glaucoma therapy. *Particle and Particle Systems Characterization, 32*, 567–572. https://doi.org/10.1002/ppsc.201400229

Li, D., Baydoun, H., Verani, C. N., & Brock, S. L. (2016). Efficient water oxidation using CoMnP nanoparticles. *Journal of the American Chemical Society, 138*, 4006–4009. https://doi.org/10.1021/jacs.6b01543

Liu, J., Liu, Y., Liu, N., Han, Y., Zhang, X., Huang, H., Lifshitz, Y., Lee, S. T., Zhong, J., & Kang, Z. (2015). Metal-free efficient photocatalyst for stable visible water splitting via a two-electron pathway. *Science, 80*, 970–974. https://doi.org/10.1126/science.aaa3145

Lu, Y., Sturek, M., & Park, K. (2014). Microparticles produced by the hydrogel template method for sustained drug delivery. *International Journal of Pharmaceutics, 461*, 258–269. https://doi.org/10.1016/j.ijpharm.2013.11.058

Mabena, L. F., Sinha Ray, S., Mhlanga, S. D., & Coville, N. J. (2011). Nitrogen-doped carbon nanotubes as a metal catalyst support. *Applied Nanoscience, 1*, 67–77. https://doi.org/10.1007/s13204-011-0013-4

Mansha, M., Khan, I., Ullah, N., & Qurashi, A. (2017). Synthesis, characterization and visible-light-driven photo electro chemical hydrogen evolution reaction of carbazole-containing conjugated polymers. *International Journal of Hydrogen Energy, 3*, 53. https://doi.org/10.1016/j.ijhydene.2017.02.053

Manucci, P. M., & Franchini, M. (2017). Health effects of ambient air pollution in developing countries. *International Journal of Environmental Research and Public Health, 14*, 1048. https://doi.org/10.3390/ijerph14091048

Martis, E., Badve, R., & Degwekar, M. (2012). Nanotechnology based devices and applications in medicine: An overview. *Chronicles of Young Scientists, 3*, 68. https://doi.org/10.4103/2229-5186.94320

Mashaghi, S., Jadidi, T., Koenderink, G., & Mashaghi, A. (2013). Lipid nanotechnology. *International Journal of Molecular Sciences, 14*, 4242–4282. https://doi.org/10.3390/ijms14024242

Mazurais, D., Ernande, B., Quazuguel, P., et al. (2015). Evaluation of the impact of polyethylene microbeads ingestion in European sea bass (*Dicentrarchus labrax*) larvae. *Marine Environmental Research, 112*, 78–85. https://doi.org/10.1016/J.MARENVRES.2015.09.009

Millstone, J. E., Kavulak, D. F. J., Woo, C. H., et al. (2010). Synthesis, properties, and electronic applications of size-controlled poly (3-hexylthiophene) nanoparticles. *Langmuir, 26*, 13056–13061. https://doi.org/10.1021/la1022938

Möller, L., Schuetzle, D., & Autrup, H. (1994). Future research needs associated with the assessment of potential human health risks from exposure to toxic ambient air pollutants. *Environmental Health Perspectives, 102*, 193–210. https://doi.org/10.1289/ehp.94102s4193

Mueller, N. C., & Nowack, B. (2008). Exposure modeling of engineered nanoparticles in the environment. *Environmental Science & Technology, 42*, 4447–4453. https://doi.org/10.1021/es7029637

Nagarajan, P. K., Subramani, J., Suyambazhahan, S., & Sathyamurthy, R. (2014). Nanofluids for solar collector applications: A review. *Energy Procedia, 61*, 2416–2434. https://doi.org/10.1016/j.egypro.2014.12.017

Navarro, E., Piccapietra, F., Wagner, B., et al. (2008). Toxicity of silver nanoparticles to Chlamydomonas reinhardtii. *Environmental Science & Technology, 42*, 8959–8964. https://doi.org/10.1021/es801785m

Nikalje, A. P. (2015). Nanotechnology and its applications in medicine. *Medicinal Chemistry, 5*. https://doi.org/10.4172/2161-0444.1000247

Nixon, S. W., Oviatt, C. A., Frithsen, J., & Sullivan, B. (2010). Nutrients and the productivity of estuarine and coastal marine ecosystems. *Journal of the Limnological Society of Southern Africa, 12*, 43–71. https://doi.org/10.1080/03779688.1986.9639398

Oprea, B., Martínez, L., Roman, E., et al. (2015). Dispersion and functionalization of nanoparticles synthesized by gas aggregation source: Opening new routes toward the fabrication of nanoparticles for biomedicine. *Langmuir, 31*, 13813–13820. https://doi.org/10.1021/acs.langmuir.5b03399

Park, K. M., Sung, H., Choi, S. J., Choi, Y. J., & Chang, P. S. (2014). Double-layered microparticles with enzyme-triggered release for the targeted delivery of water-soluble bioactive compounds to small intestine. *Food Chemistry, 161*, 53–59. https://doi.org/10.1016/j.foodchem.2014.03.125

Qi, M., Zhang, K., Li, S., Wu, J., Pham-Huy, C., Diao, X., Xiao, D., & He, H. (2016). Super paramagnetic Fe3O4 nanoparticles: Synthesis by a solvo thermal process and functionalization for a magnetic targeted curcumin delivery system. *New Journal of Chemistry, 4480*, 4480–4491. https://doi.org/10.1039/c5nj02441b

Rawat, M. K., Jain, A., Singh, S., Mehnert, W., et al. (2011). Studies on binary lipid matrix based solid lipid nanoparticles of repaglinide: In vitro and in vivo evaluation. *Journal of Pharmaceutical Sciences, 100*, 2366–2378. https://doi.org/10.1002/jps.22435

Reiss, G., & Hutten, A. (2005). Magnetic nanoparticles: Applications beyond data storage. *Nature Materials, 4*, 725–726. https://doi.org/10.1038/nmat1494

Rillig, M. C., Ziersch, L., & Hempel, S. (2017). Microplastic transport in soil by earthworms. *Scientific Reports, 7*. https://doi.org/10.1038/s41598-017-01594-7

Rochman, C. M., Kurobe, T., Flores, I., & The SJ. (2014). Early warning signs of endocrine disruption in adult fish from the ingestion of polyethylene with and without sorbed chemical pollutants from the marine environment. *Science of the Total Environment, 493*, 656–661. https://doi.org/10.1016/J.SCITOTENV.2014.06.051

Rogozea, E. A., Olteanu, N. L., Petcu, A. R., et al. (2016). Extension of optical properties of ZnO/SiO_2 materials induced by incorporation of Au or NiO nanoparticles. *Optical Materials, 56*, 45–48. https://doi.org/10.1016/j.optmat.2015.12.020

Sanna, V., Roggio, A. M., Pala, N., Marceddu, S., Lubinu, G., Mariani, A., et al. (2015). Effect of chitosan concentration on PLGA microcapsules for controlled release and stability of resveratrol. *International Journal of Biological Macromolecules, 72*, 531–536. https://doi.org/10.1016/j.ijbiomac.2014.08.053

Saud, T., Gautam, R., Mandal, T. K., Gadi, R., Singh, D. P., & Sharma, S. K. (2012). Emission estimates of organic and elemental carbon from household biomass fuel used over the Indo-Gangetic Plain (IGP), India. *Atmospheric Environment, 61*, 212–220. https://doi.org/10.1016/j.atmosenv.2012.07.030

Shaalan, M., Saleh, M., El-Mahdy, M., & El-Matbouli, M. (2016). Recent progress in applications of nanoparticles in fish medicine: A review. *Nanomedicine: Nanotechnology, Biology and Medicine, 12*, 701–710. https://doi.org/10.1016/j.nano.2015.11.005

Siepmann, J., Faisant, N., Akiki, J., Richard, J., & Benoit, J. P. (2004). Effect of the size of biodegradable microparticles on drug release: Experiment and theory. *Journal of Controlled Release, 96*, 123–134. https://doi.org/10.1016/j.jconrel.2004.01.011

Sigmund, W., Yuh, J., Park, H., & Maneeratana, V. (2006). Processing and structure relationships in electrospinning of ceramic fiber systems. *Journal of the American Ceramic Society, 89*, 395–407. https://doi.org/10.1111/j.1551-2916.2005.00807.x

Sikora, A., Shard, A. G., & Minelli, C. (2016). Size and f-potential measurement of silica nanoparticles in serum using tunable resistive pulse sensing. *Langmuir, 32*, 2216–2224. https://doi.org/10.1021/acs.langmuir.5b04160

Thomas, S., Harshita, B. S. P., Mishra, P., & Talegaonkar, S. (2015). Ceramic nanoparticles: Fabrication methods and applications in drug delivery. *Current Pharmaceutical Design, 21*, 6165–6188. https://doi.org/10.2174/1381612821666151027153246

Thote, A. J., & Gupta, R. B. (2005). Formation of nanoparticles of a hydrophilic drug using supercritical carbon dioxide and microencapsulation for sustained release. *Nanomedicine, 1*, 85–90. https://doi.org/10.1016/j.nano.2004.12.001

Torchynska, T. V., El Filali, B., Ballardo Rodríguez, I. C., & Shcher-byna, L. (2016). Defect related emission of ZnO and ZnO Cu nanocrystals prepared by electrochemical method. *Physica Status Solidi, 13*, 594–597. https://doi.org/10.1002/pssc.201510277

Tratnyek, P. G., & Johnson, R. L. (2006). Nanotechnologies for environ-mental cleanup. *Nano Today, 1*, 44–48. https://doi.org/10.1016/S1748-0132(06)70048-2

Ullah, H., Khan, I., Yamani, Z. H., & Qurashi, A. (2017). Sono chemical-driven ultrafast facile synthesis of SnO_2 nanoparticles: Growth mechanism structural electrical and hydrogen gas sensing proper-ties. *Ultrasonics Sonochemistry, 34*, 484–490. https://doi.org/10.1016/j.ultsonch.2016.06.025

Wan, D., Chen, H. L., Lin, Y. S., Chuang, S. Y., Shieh, J., & Chen, S. H. (2009). Using spectroscopic ellipsometry to characterize and apply the optical constants of hollow gold nanoparticles. *ACSNano, 3*, 960–970. https://doi.org/10.1021/nn8009008

Wang, J., Yang, N., Tang, H., et al. (2013). Accurate control of multi shelled Co_3O_4 hollow microspheres as high-performance anode materials in lithium-ion batteries. *Angewandte Chemie International Edition, 52*, 6417–6420. https://doi.org/10.1002/anie.201301622

Wang, Z., Pan, X., He, Y., et al. (2015). Piezoelectric nanowires in energy harvesting applications. *Advances in Materials Science and Engineering, 2015*, 1–21. https://doi.org/10.1155/2015/165631

Watson, J. T., Gayer, M., & Connolly, M. A. (2007). Epidemics after natural disasters. *Emerging Infectious Diseases, 13*, 1–5. https://doi.org/10.3201/eid1301.060779

Wilson, W. E., & Suh, H. H. (1997). Fine particles and coarse particles: Concentration relationships relevant to epidemiologic studies. *Journal of the Air & Waste Management Association (1995), 47*, 1238–1249. https://doi.org/10.1080/10473289.1997.10464074

Wong, C. Y., Al-Salami, H., & Dass, C. R. (2018). Microparticles, microcapsules and microspheres: A review of recent developments and prospects for oral delivery of insulin. *International Journal of Pharmaceutics, 537*, 223–244. https://doi.org/10.1016/j.ijpharm.2017.12.036

Yu, Z. B., Xie, Y. P., Liu, G., Lu, G. Q., Ma, X. L., & Cheng, H. M. (2013). Self-assembled CdS/Au/ZnO heterostructure induced by surface polar charges for efficient photocatalytic hydrogen evolution. *Journal of Materials Chemistry A, 1*, 2773. https://doi.org/10.1039/c3ta01476b

Yu, M., Van Der Ploeg, M., Lwanga, E. H., Yang, X., Zhang, S., Ma, X., & Geissen, V. (2019). Leaching of microplastics by preferential flow in earthworm (*Lumbricus terrestris*) burrows. *Environmental Chemistry, 16*, 31–40. https://doi.org/10.1071/EN18161

Yuan, S., Lei, F., Liu, Z., Tong, Q., Si, T., & Xu, R. X. (2015). Coaxial electrospray of curcumin-loaded microparticles for sustained drug release. *PLoS One, 10*, e0132609. https://doi.org/10.1371/journal.pone.0132609

Zhang, L., Huang, J., Si, T., & Xu, R. X. (2012). Coaxial electrospray of microparticles and nanoparticles for biomedical applications. *Expert Review of Medical Devices, 9*, 595–612. https://doi.org/10.1586/erd.12.58

Zhu, D., Bi, Q. F., Xiang, Q., Chen, Q. L., Christie, P., Ke, X., & Zhu, Y. G. (2018). Trophic predator-prey relationships promote transport of microplastics compared with the single *Hypoaspis aculeifer* and *Folsomia candida*. *Environmental Pollution, 235*, 150–154. https://doi.org/10.1016/j.envpol.2017.12.058

Chapter 3
Abundance and Distribution of MPs and NPs in Soil: A Global Scenario

Mahir Tajwar, Shamiha Shafinaz Shreya, Md. Yousuf Gazi, and Md. Bayazid Hossain

Abstract The worldwide abundance of microplastics (MP) and nanoplastics (NP) is generally identified as a persistent problem to the marine environment and is already deemed a silent threat in aquatic environments. However, their presence in agricultural soil and terrestrial environment has largely been overlooked, and our understanding of its effect on the terrestrial ecosystem is not fully understood. This chapter addressed the global accumulation and abundance of MP and NP in terrestrial ecosystems. Furthermore, the factors contributing to their distribution and widespread presence in terrestrial soil have been evaluated for better insights in microplastic studies. Based on the limited studies done on terrestrial soil, the abundance of MP and NP varies geographically with high concentrations being detected in the regions of China, Pakistan, Canada, the USA, Spain, Italy, and Australia whereas comparatively, a lower amount has been detected in France, Germany, and Antarctica. This chapter intends to (1) summarize the accumulation and distribution of MPs and NPs in the terrestrial ecosystem and (2) evaluate the factors regulating the distribution of MPs and NPs as environmental pollutants on territorial soil system. The prospects for future research include an in-depth investigation of the concentration and characterization of MPs and NPs in the terrestrial soil of various countries and analysis of different factors controlling its distribution and its potential impact.

Keywords Microplastics · Nanoplastics · Soil · Spatial Distribution · Abundance · Agriculture

M. Tajwar
Department of Geosciences, Auburn University, Auburn, AL, USA

S. S. Shreya
Department of Oceanography, University of Dhaka, Dhaka, Bangladesh

M. Y. Gazi (✉)
Department of Geology, University of Dhaka, Dhaka, Bangladesh
e-mail: yousuf.geo@du.ac.bd

M. B. Hossain
Department of Soil, Water & Environment, University of Dhaka, Dhaka, Bangladesh

N. R. Maddela et al. (eds.), *Micro and Nanoplastics in Soil*,
https://doi.org/10.1007/978-3-031-21195-9_3

3.1 Introduction

Since the mass production of plastic started, the biosphere of the earth has been confronting a steadily increasing threat. Cumulative thrust for convenient yet cheaper products and modern lifestyle has boosted the annual production of plastic materials by leaps and bounds which is estimated to be about 34,000 million tons by 2050 (Plastics Europe, 2019; Maity & Pramanick, 2020). Trends of overdependence, indiscriminate utilization, less recycling propensity (only 9%), improper management of used products, etc. are leading the global annual plastic waste production to 6300 million tons approximately; a lion's share of this finds a way to the terrestrial ecosystem (Geyer et al., 2017; Van Sebille et al., 2015). Consequently, soil possesses a far greater concentration of plastic materials than aquatic sediments as reported by Horton et al. (2017a, b) and Fischer et al. (2016). Furthermore, agro-ecosystem was reported as the most plastic contaminated terrestrial ecosystem by Nizzetto et al. (2016a, b). However, the distributive nature and interaction with components in the terrestrial ecosystem, specifically in agricultural soils, are still unclear.

Plastic products have gained universal popularity; hence widespread occurrence and distribution of plastic materials can be traced easily. Worldwide contamination of soil by smaller plastic particles has been mentioned by several studies (Fuller & Gautam, 2016; Scheurer & Bigalke, 2018; Koutnik et al., 2021). Zhang et al. (2021) identified the variation in abundance of global soil MPs which is thought to be connected with the physiographical nature, development activities, population densities, soil properties, and other features of that area (Harms et al., 2021; Kim et al., 2015). Recently, detection of plastic particles has been confirmed in distant places like the Polar Regions and the Tibetan Plateau where anthropogenic interference is severely less (Jiang et al., 2019; Peeken et al., 2018). Upon deposition on surface soil, plastics undergo disintegration and degradation processes resulting in smaller MPs and even NPs (Rocha-Santos & Duarte, 2015; Cole & Galloway, 2015; Barnes et al., 2009). Meanwhile, previously generated MPs can also directly enter the terrestrial ecosystem as primary MPs (Napper et al., 2015) (Fig. 3.1).

Previous studies have identified the sources and pathways of MP contamination in farmland soil and reported plastic mulch, application of compost and sludge, irrigation with untreated and partially treated water, plastic container, atmospheric deposition, etc. as the main sources (Corradini et al., 2019; Blasing & Amelung, 2018; Rillig, 2012; Sanchez-Hernandez et al., 2018).

Once entered the agricultural soils, MPs and NPs can be transferred both vertically and horizontally through soil pore spaces (Zhang et al., 2019; Lwanga et al., 2017a, b; Horton et al. 2017a, b; Rillig et al., 2017b) or can be bioaccumulated into the human food chain and endanger health (Zhang et al., 2019; Machado et al., 2019). Hence, a detailed understanding of the distribution and transportation of MP and NP in soil may serve as a vital factor for controlling plastic contamination. Unfortunately, the scarcity of published reports on the movement and transportation nature of MPs in the terrestrial ecosystem reveals the fact that it has gained less

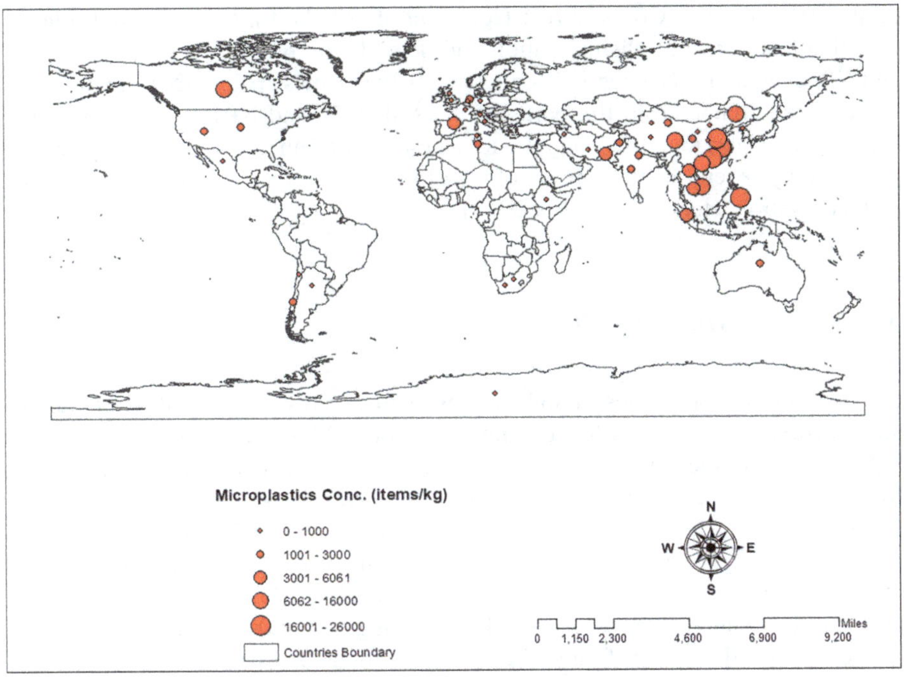

Fig. 3.1 Showing the global abundance and distribution of microplastics on the agricultural soil

attention from the researcher community, compared to aquatic systems (Möller et al., 2020; Brady & Weil, 2000). As a result, several information gaps need to be addressed. Consequently, a knowledge gap has been created.

This review study aims to gather data about the global trend of accumulation and abundance of MPs and NPs in the terrestrial ecosystem. Moreover, several factors which control the distribution behavior of MPs and NPs in farmland soils have been discussed with a view to providing a base material for the better realization of transportability of smaller plastic particles in agro-ecosystem.

3.2 Factors Controlling the Distribution of MPs and NPs in Soil

Since plastic has got an overwhelmingly ubiquitous nature, it is obvious that MPs and NPs will be found in the terrestrial ecosystem. Smaller plastic materials enter into the agro-ecosystem as manufactured MPs and NPs primarily or secondarily as produced from the bigger plastic materials (Rillig, 2012; Duis & Coors, 2016; Koelmans et al., 2015). Plastic mulch, greenhouse materials, atmospheric deposition, etc. are considered as direct sources, while indirect sources include application of organic amendments, irrigation with wastewater, application of sludge, etc. (Ng et al., 2018; Duis & Coors, 2016; Horton et al. 2017a, b). Once entered, plastic materials may undergo disintegration and degradation processes, generating MPs and NPs eventually (Napper & Thompson, 2019; Chamas et al., 2020). Having convenient size compared to the pores of the soil, MPs and NPs can be distributed spatially and horizontally. This phenomenon of movement in agro-ecosystem is governed by several factors: morphology of the plastic, precipitation, properties of soil, cultivation, etc. (Zhang et al., 2018).

3.2.1 Properties of Soil

Physicochemical properties of soil such as texture, moisture content, temperature, soil reaction, etc. directly affect the movement of MPs and NPs in soil. Soil texture straightly determines the pore space distribution which is very crucial for the translocation of smaller plastic particles within the layers of soil (Rahmatpour et al., 2018; Cey et al., 2009). Light textured soil (containing higher sand percentage) tends to have bigger pore spaces (macropores) which will enable the soil to permit more vertical movement of microplastics (Rillig et al., 2017b) than soils with higher clay and silt content, respectively. In addition, Ding et al. (2021) recently studied the abundance of MP in soils of three sites, and they found that sand soil had higher MP content than grassland and woodland.

Soils containing a greater amount of montmorillonite and other expanding clay minerals will generate cracks and fissures upon drying. These cracks can pave the way for massive transportation of plastic particles of various sizes directly at the deeper part of the soil profile very quickly (Rillig & Lehmann, 2020). Similar findings were observed in several other studies (O'Connor et al. 2019a, b; El-Farhan et al., 2000; Majdalani et al., 2008) implying the significant effects of interconnected pore space pathways and wet-dry cycles on the transport of MPs in a terrestrial ecosystem. Meanwhile, previous experiments reported the significant reduction of microplastic movement with increasing ionic attachment in a media of quartz sand (Treumann et al., 2014; Pelley & Tufenkji, 2008). Moreover, chemical properties of soil, namely, soil pH and Fe/Al content, influence the distribution of MPs and NPs as mentioned by some studies (Wu et al., 2020; Scheurer & Bigalke, 2018).

3.2.2 Morphology of MP and NP

The distribution of MP and NP in the soil is most reliant on their various morphological characteristics, such as size, density, shape, hydrophobicity, etc. (Rillig et al., 2017a). Previous several studies have enlightened on the influence of size and hydrophobicity of plastic materials upon their movement in the terrestrial ecosystem (O'Connor et al. 2019a, b; Pelley & Tufenkji, 2008). David O'Connor et al. (2019a, b) studied the mobility of five different MPs and found that mobility of smaller-sized PE-MP was greater than any other with the longest penetration. A same observation regarding the size-mobility inverse relation of MPs has also been reported by Rillig et al. (2017b). Furthermore, Liu et al. (2018) investigated arable soils around the suburbs of Shanghai and noticed relatively larger MPs in topsoil varying significantly from deeper soil (Hurley & Nizzetto, 2018).

Shape is another morphological factor that affects the fate of mobility of plastic materials in a terrestrial system. MPs and NPs are found with various shapes in soil, for instance, sphere, particle, fiber, and film mostly. Among them, sphere and particle are being widely used for recent relevant researches showing that microplastic particles of these two shapes can easily translocate to the deeper part of the soil (Lwanga et al., 2017a, b; Rillig et al., 2017b; Treumann et al., 2014; Zhuang et al., 2005). On the other hand, in a review study, Rillig et al. (2017b) assumed different distributive behavior of other shapes such as film, fiber, etc. from the sphere. They predicted that fiber and film-shaped smaller plastic particles have a possibility to be trapped in the soil matrix and can become surrounded by an eco-corona likewise aquatic system (Galloway et al., 2017). In addition, the reviewers also reported that the movement of MPs and NPs in the soil system would be substantially influenced by eco-corona. Their findings are consistent with Zhang et al. (2019). Meanwhile, low-density microplastics had fewer tendencies to move downward as mentioned by O'Connor et al. (2019a, b).

After the entrance, MPs and NPs confront several processes like attachment, sedimentation, etc. which can hinder the strolling of plastic particles in soil (Zhang & Liu, 2018; Rillig et al., 2017a). Previous experiments on movement behavior of colloids in different media found straining, attachment at the solid-liquid interface, pore exclusion, and air-water interfacial bond as the significant factors in particle movements through soil (Zhuang et al., 2005; Bradford & Torkzaban, 2012; Bradford et al., 2002).

3.2.3 *Soil Biota*

Previous experiments have mentioned that biogenic activities could instigate the transfer of smaller particles from the surface into the deeper part of the soil by creating interconnected pore space pathways (Blasing & Amelung, 2018; Zubris & Richards, 2005). In a review study, Rillig et al. (2017b) reported earthworms and roots as the most important producers of bio-macrospores in soil. They expected similar results from micro arthropods. Moreover, Huerta Lwanga et al. (2016) reported that earthworms can contribute to microplastic movement from the surface soil to the deep soil by ingestion/excretion mechanism. The capability of the conversion of primary MPs to secondary MPs and NPs through ingestion by earthworms and some other soil-dwelling organisms, for instance, burrowing mammals, collembolan, mites, etc., has been reported in some studies as well (Zhu et al., 2019; Rillig, 2012).

Previous studies found evidence of vertical and horizontal movement of LDMPs in soil facilitated by collembolan, earthworms, and other organisms (Maaß et al., 2017; Rillig et al., 2017b; Lwanga et al., 2017a, b). Besides, direct transportation of LDMPs by preferential flow through the pore space pathways was mentioned by Yu et al. (2019). Moreover, Zhang et al. (2019) conducted a study combining field investigations and laboratory simulations to examine LDMP distribution and controlling factors in agricultural soils. They found evidence of vertical as well as horizontal movement of LDMPs along with water through pore space pathways of soil. Maaß et al. (2017) conducted a study with two species of collembola and confirmed the movement and distribution of MPs by both species. In another study, Zhu et al. (2018) mentioned that mites can also disperse commercial PVC particles.

Root penetration, expansion, and water extraction create pores and channels within soils which facilitate downward translocation of smaller particles as reported by Gabet et al. (2003). In addition, the decomposition of roots produces macropores of nearly the same size, which can distribute MPs in soils (Li et al. 2019a, b). Another study revealed that fungal hyphae might facilitate MP movement by serving as preferential paths (Wick et al., 2007). Leaching contributes significant microplastic transportation vertically in soils through pore spaces created by naturally or biogenic activities (Cey et al., 2009). However, there is no comparative analysis of microplastic particles movement by earthworm or other soil biota and leaching in soil.

3.2.4 Cultivation

Agronomic activity such as plowing, mulching with plastics, application of sludge, application of organic amendments, irrigation with wastewater, etc. cause the mobilization of MP and NP in surface soil and subsurface soil. The findings of several recent pieces of research have proven the fact that cultivation activities help to spread around plastic particles in any agro-ecosystem (Nizzetto et al., 2016b; Zhang et al., 2019). Ding et al. (2020) found a significantly higher number of MPs in orchards which they thought for massive utilization of plastic for packaging. Rillig and Lehmann (2020) reviewed the effect of different tillage practices on microplastic incorporation in different depths of agricultural soil. They noted that mould board plowing brings most of the MPs present on the soil surface at the plowing depth. Besides, they noted that other tillage practices would show mixing effects throughout the tillage layer.

Mulching with plastics, a widely practiced trend of modern farming, has been established as a major source of plastic materials in agro-ecosystem by numerous studies (Sintim & Flury, 2017; Zhou et al. 2019a, b; Gao et al., 2019). Once embedded in the soil, it can be converted into MPs and NPs (Blasing & Amelung, 2018). Wang et al. (2021) studied samples of different land-use patterns from five provinces of China and found significantly higher particle abundance where plastic mulching was used. Furthermore, they noticed that the MP abundance of paddy fields was significantly higher than wheat lands. Plastic mulch affects the agro-ecosystem inversely since plastic covering can raise both soil temperature and moisture which can intensify the degradation and transportation rate of MPs (Subrahmaniyan et al., 2006). Moreover, decomposition of other agricultural plastic materials such as seed bags, packaging materials, agricultural plastic tools, etc. can add MP in farmlands (Antunes et al., 2013). Chen et al. (2020) added polytunnels, bale twines, fertilizer bags, containers, and nets to the list which can be a source factor of plastic materials in agro-ecosystem.

Application of sludge to amend the soil is reported by many studies to be a significant input pathway of smaller fractions of plastics in farmland soil (Ziajahromi et al., 2016; Zhou et al., 2020; Nizzetto et al., 2016b). An estimation made by Nizzetto et al. (2016b) revealed that about 4.4×10^4–3×10^5 tons and about 6.3×10^4–4.3×10^5 tons of MPs enter arable soil annually because of use as the amendment in North America and Europe, respectively. The findings of Chen et al. (2020) are consistent with this. Meanwhile, organic farming involves the utilization of organic fertilizers (namely composts) which are commonly produced from household waste or municipal waste. These composts may have a MP concentration of 895 items kg^{-1} as reported by Weithmann et al. (2018). On the other hand, irrigation with wastewater can serve as a source and distributor of plastic materials both spatially and vertically in agricultural soil (Scheurer & Bigalke, 2018). Mintenig et al. (2017) showed that about 20 million hectares of arable land worldwide are irrigated with untreated or partially treated sewage water on which about 10% of the world's population depends on the food. On the other hand, Rillig and Lehmann (2020) considered bio-pores created by decomposed roots after harvesting as a massive transport pathway of MPs and NPs in agricultural systems. They also reported that harvesting submerged parts of plants beneath the surface (carrots, potatoes, etc.) can also facilitate to incorporation and transportation of microplastics in farmland. Li et al. (2019a, b) also attributed that harvesting of rhizome may serve the downward movement of microplastics.

3.2.5 *Weather Pattern*

The weather pattern of an area can affect the distribution and accumulation of MPs and NPs significantly. An area with annual heavy rainfall may experience surface runoff which can mobilize plastic materials spatially over a huge area. With the flow coming from the source, plastic materials can be floated away to distant locations. Previous studies indicated that flow length and catchment size showed a positive correlation with the number of possible plastic sources and distribution (Klein et al., 2015; Tibbetts et al., 2018; Fischer et al., 2016; Ballent et al., 2016). A study on the abundance of plastics in the Swiss floodplain was carried out by Scheurer and Bigalke (2018). They identified the distribution of plastic as diffuse and found a linkage of the lateral distribution process with flood dynamics. Moreover, such flushing movement of water determined by topographical and weather factors can carry smaller plastic materials even through soil pores spaces of soil horizontally and vertically (Zhang et al., 2017; Zhang et al., 2011). Meanwhile, O'Connor et al. (2019a, b) reported upward migration of MPs if saturation prevails in soil pore spaces since MPs have relatively low specific density. On the contrary, during hot days with no precipitation, dry soils will likely have natural cracks, which can serve as entryways for microplastics to deeper soils (Li et al. 2019a, b). In addition, dry hot days will intensify MP conversion to NP with the help of UV radiation and elevated temperature (Horton et al. 2017a, b).

Ding et al. (2020) conducted a research with agricultural soils from nine sites across Shaanxi province, and they observed that in northern Shaanxi, MPs were not gone by surface runoff, causing massive accumulation while the situation appeared reverse in the case of southern Shaanxi, where they found degraded smaller MPs. They explained their findings focusing on the weather pattern difference between northern and southern Shaanxi since the northern part had the temperate monsoon climate and less rainfall for the southern Shaanxi; it was just the opposite.

Weather patterns involving subsequent cycles of rainfall events and dry periods may have an impact on the mobility of MPs into the soil (O'Connor et al. 2019a, b; McCarthy & McKay, 2004). O'Connor et al. (2019a, b) studied the mobility of five different MPs in sand soil column experiments. They found a significant positive relationship between MP penetration and wet-dry cycles. However, more research should be carried out for a better understanding of the effects of weather patterns on the distribution of MPs and in soil.

3.3 Summary of Regional MP and NP Abundance in Soil

3.3.1 Africa

The first report on the abundance and occurrence of microplastic in the African continent was done by Ryan in 1988 where the accumulation of pieces of plastic at the seafloor of the southwestern Cape Province of South Africa was studied from August 1977 to August 1978. After 27 years of the first study done in Africa, microplastic in the surface water of south-eastern major bays of South Africa was evaluated (Nel & Froneman, 2015).

In the region of South Africa, an extent of 13.3–563.8 items/kg of microplastic and nanoplastic have been detected in the river sediments of Eastern Cape Town (Nel et al., 2018), and the concentration in the region of Braamfontein Spruit, Johannesburg, has been found to be 166.8 items/kg (Dahms et al., 2020). Particles identified in the Lake Ziway sediments of Ethiopia have a range from 6.3 to 115.9 items per kg of freshwater sediments (Merga et al., 2020).

Comparatively, a high concentration of MP and NP has been found in Tunisia where an amount of more than 6920 particles has been reported in the region of

Table 3.1 Distribution of microplastics and nanoplastic in terrestrial ecosystems across Africa

Countries	Location	Concentration (items/kg)	Sample type	References
Tunisia	Bizerte	2340 ± 227.15–6920 ± 395.98	Freshwater sediment	Toumi et al. (2019)
Tunisia	South/North Lake of Tunis	316.03 ± 123	Sediment	Abidli et al. (2019)
Ethiopia	Lake Ziway	6.3–115.9	Freshwater sediment	Merga et al. (2020)
South Africa	Eastern Cape Town	13.3–563.8	River sediment	Nel et al. (2018)
South Africa	Braamfontein Spruit, Johannesburg	166.8	Stream sediment	Dahms et al. (2020)

Bizerte (Toumi et al., 2019). A much lower distribution (316.03 ± 123 items/kg) has been detected in the sediments of Tunis lake (Abidli et al., 2019) (Table 3.1).

3.3.2 *America*

Several pieces of research have been gathered in this study to give an overview of the abundance of microplastics in the soil of the Americas. These researches looked at the usage of sewage sludge and biosolids as a source of microplastics. In a study, it has been estimated that biosolid applications might provide up to 300,000 tons of MPs to farmed soils in North America each year (Nizzetto et al., 2016a, b). In Ontario, Canada, microplastic abundance is found to be at a range of 8700–14,000 MPs/kg in biosolid samples (Crossman et al., 2020).

A study had been conducted in the city of New York, USA, where samples were collected from four sites, and the findings revealed MP concentration ranged from 370 to 2060 items/kg^{-1} with a mean of 1235 items per kg. This result is comparable to the amount of sewage sludge applied to worldwide soils (Zubris & Richards, 2005).

A study (Corradini et al., 2019) of 31 agricultural fields in Chile found that the concentrations of those areas range from 0.6 to 10.4 MPs/g. But it revealed that after five sewage sludge applications, the concentration amount increased to 3600 items/kg from 1200 items/kg.

In Brazil, it has been found that the sediments near municipal dumping sites have a significant concentration of MP (Neto et al., 2019), and the study also expressed that the presence of MP was detected in 100% of the samples; however, no specific data on agricultural soil was obtained.

Another study (Lwanga et al., 2017a, b) measured plastic particles in a rural field on the Yucatán Peninsula of Mexico where a total of 870 ± 1900 items/kg of microplastics were discovered (Table 3.2).

Table 3.2 Distribution of microplastics and nanoplastic in terrestrial ecosystems across America

Region	Countries	Location	Sample type	Concentration (unit)	References
South America	Chile	–	Agricultural soil	184–306 pieces/kg	Corradini et al. (2021)
	Chile	Mellipill	Agricultural soil	2010 items/kg	Corradini et al. (2019)
	Mexico	Yucatán Peninsula	Home garden soil	870 ± 1900 items/kg	Lwanga et al. (2017a, b)
	Argentina		Farmland soil	30 ± 19 kg/ha	Ramos et al. (2015)
North America	USA	Washington, D.C.	Vegetated wetland soil	1270 pieces/kg	Helcoski et al. (2020)
		New York	Soil	1235 items/kg	Zubris and Richards (2005)
	Canada	Ontario	Biosolid samples	8700–14,000 MPs/kg	Crossman et al. (2020)

3.3.3 Asia

In Asian countries, the pervasiveness of microplastics (MP) is a severe environmental concern. According to the literature, Asia is home to seven of the top ten trash-dumping countries (Jambeck et al., 2015). To acquire an overview of the abundance and distribution of microplastics across Asia, a variety of studies have been collated and discussed.

Southern Asia

In Southern Asia, the highest abundance is found in Pakistan. A study was carried out in Pakistan to map out the regional dispersion of microplastics in the topsoil Lahore, Pakistan. The distribution of MPs in topsoil throughout the Lahore district was discovered to range from 1750 to 12,200 MPs per kg, with a mean of 4483 ± 2315 items/kg. In agricultural soil MP concentration was found in the range of 2200–6875 MPs per kg with a mean concentration of 3712 ± 2156 MPs per kg of soil (Rafique et al., 2020). In India, MPs were identified in soils collected from electronic waste-dumping sites. In Bangalore and Chennai, the average MP estimates were 302 and 1908 items/kg, respectively (Tun et al., 2022). Traditional plasticizers like dibutyl phthalate, dimethyl phthalate, and di(2-ethylhexyl) phthalate were detected in high abundance (Chakraborty et al., 2019). PE was also found to be a dominated polymer in Indian soils of dumping sites which account for 55% of total MP (Chai et al., 2020a, b).

In the case of Bangladesh, most of the recent studies have reported the abundance of microplastic in the sediments of the coastal environment (Rahman et al., 2020; Tajwar et al., 2021, 2022a, b; Rakib et al., 2021), where MP concentration had been found about 8.1 ± 2.9 particles/kg. However, the concentration of MP & NP based on the terrestrial ecosystem is yet to be evaluated (Table 3.3).

Table 3.3 Distribution of microplastics and nanoplastic in terrestrial soil across Southern Asia

Countries	Location	Sample Type	Concentration (items/kg)	References
India	Chennai	Soil of e-waste dumping sites	1908	Tun et al. (2022)
	Bangalore	Soil of e-waste dumping sites	302	
Pakistan	Lahore	Agricultural soil	3712 ± 2156	Rafique et al. (2020)
		Topsoil	4483 ± 2315	

South-eastern Asia

MPs have also been discovered in alarming figures in the countries of Southeast Asia. In Indonesia, many areas continue to operate open dumping sites, which are a potential source of MPs (The Jakarta Post, 2019). In one Indonesian sampling site, MP distribution was highest at 43,704 particles per kg, followed by 16,842 and 11,111 pieces/kg in two other sites. The dumping site's entrance revealed a significant MP abundance (16,842 pieces/kg). The median number of MPs in the soil is 6061 pieces/kg (Tun et al., 2022).

According to research in Cambodia, the median MP concentration in Cambodian soils was determined to be 4360 pieces per kilogram. It demonstrated the maximum abundance of MP (218,182 pieces/kg) in one Cambodian sampling site, with 48 MP found in only 0.22 g of soil (Tun et al., 2022). From a study on Asian countries, it is found that the average value of microplastic was highest in the Philippines counted for 24,000 pieces/kg. MP levels were found to be high in soils from two sampling sites of the Philippines, containing 31,000 and 24,000 pieces/kg, respectively (Tun et al., 2022). In Vietnam, the median of MPs in soils was about 11,337 pieces/kg, but the highest abundance is found about 83,606 pieces/kg, followed by 28,358 pieces/kg and 26,768 particles per kg. Soil samples accumulated from two e-waste recycling areas close to Ho Chi Minh City showed high abundance of MP, containing 17,568 and 26,761 items per kg (Tun et al., 2022). In Laos, the highest MP abundance was accounted for 22,222 pieces/kg from a dumping site. MPs were detected from the other three sampling sites ranging from 893 to 4651 items per kg, and the samples had been taken from a rice field near a landfill area. The median value of MPs in soils of Laos was 4651 pieces/kg (Tun et al., 2022) (Table 3.4).

Middle East

Rezaei et al. (2019) investigated the transmission of MPs by wind erosion measuring the abundance of MPs (low-density) in the soils of Iran's Fars province. 1.2 ± 0.6 and 205 ± 186 mg per kg had been identified at agriculture-based locations, but only 0.2 ± 0.1 and 38 ± 17 particles per kg were found on rangelands. The primary source of MPs in these areas was assumed to be insufficient removal of plastic mulch films (Büks & Kaupenjohann, 2020) (Table 3.5).

Table 3.4 Distribution of microplastics and nanoplastic in terrestrial soil across South-eastern Asia

Countries	Location	Sample type	Concentration (items/kg)	References
Indonesia	–	Dumping sites soils	6061	Tun et al. (2022)
Philippines	Smokey Mountain, Manila	Soils	24,000	
Vietnam	Ho Chi Minh City	Soils	11,337	
Cambodia	–	Soils	4360	
Laos	–	Soils	4651	

Table 3.5 Distribution of microplastics and nanoplastic in terrestrial soil across the Middle East

Country	Location	Sample type	Concentration	References
Iran	Fars Province	Rangeland soils	38 ± 17 items/kg	Büks and Kaupenjohann (2020)
		Agricultural soils	205 ± 186 mg / kg	

Eastern Asia

In Eastern Asia, China is the only country that performed MP measurement in more than 15 regions, and from those researches, it is found that MPs are highly distributed all over China. As significant amounts of plastics are manufactured, consumed, and discharged in China each year, soil microplastic pollution deserves special attention (Gourmelon, 2015). In China, the amount of plastic garbage that had been improperly managed was found to be the highest in 2010, at 8.82 million tons (Jambeck et al., 2015). Ding et al. (2020) examined the abundance of MPs in the agriculture-based sites of Shaanxi Province, China. They discovered that MPs were present in all agricultural soils. MP concentrations in soil ranged from 1430 to 3410 items/kg. The findings of this research validated the existence of a high concentration of MPs in farmland soil and demonstrated that activities related to agriculture could have contaminated the soil with MPs. If we look carefully, it will be observed that soils with a history of sewage sludge application, wastewater irrigation, and mulching contain more MPs (Zhang & Liu, 2018). It is estimated that sludge application in China contributes approximately 1.56×10^{14} sludge-based MPs to the natural ecosystem (Liu et al., 2018).

A distant agricultural area in China's Loess Plateau consisted only of 0.54 mg of MPs per kg of a soil sample, whereas the use of sewage sludge to agricultural soil enhanced the MPs concentration to 15,800 MPs/kg (Mahon et al., 2017; Zhang et al., 2018). Huang et al. (2020) investigated agricultural soils by examining 384 sediment samples gathered from 19 provinces across China. In places where plastic mulching was employed consistently, microplastic particle abundance rose over time, with an amount of 80.3 49.3, 308138.1, and 1075.6346.8 items per kg soil in fields with 5, 15, and 24 years of continuous mulching, respectively.

Zhang and Liu (2018) took polymer samples from one untreated afforested site close to Kunming and four farmland locations with sewage sludge and wastewater application. They discovered average concentrations of 26,070 items/kg with a minimum concentration of 13,470 items/kg and a maximum concentration of 42,960 particles per kg in farmland Gleysol, 14,440 particles per kg (min, 8180 particles/kg; max, 18,100 particles/kg) in an afforested Gleysol and 12,050 particles per kg (min, 7100 particles/kg; max, 26,630 particles /kg) in farmland Nitsol. This suggests that not only the plastic load but also the soil type are considered as factors in MP concentrations in soils. Zhou et al. (2019a, b) investigated the prevalence and quantity of MPs, as well as their interactions with heavy metals, across three different subareas in central China. The concentration of MPs ranged from 2.2×10^4 to 6.9×10^5 items/kg according to the findings. MP distribution was much greater in the forest (4.1×10^5 items/kg) compared to the vegetable land (1.6×10^5 items/kg) or barren land (1.2×10^5 items/kg) (Table 3.6).

Another study on Yunnan, China, showed MP concentrations ranged from 7100 to 42,960 particles/kg with an average concentration of 18,760 particles/kg in cropped soils (Zhang & Liu, 2018). In Shanghai, the abundance of microplastics in shallow and deep soils was 78.00 ± 12.91 and 62.50 ± 12.97 items/kg, respectively. Furthermore, topsoil had a higher amount and greater sizes of micro(meso)plastics

Table 3.6 Distribution of microplastics and nanoplastic in terrestrial soil across Eastern Asia

Countries	Location	Sample type	Concentration (items/kg)	References
China	Shanghai	Vegetable farmland	70	Liu et al. (2019)
	Nanjing and Wuxi	Agricultural land soil	855	Li et al. (2019a, b)
	Wuhan	Vegetable plots soil	16,000	Zhou et al. (2019a, b)
	Wuhan	Vegetable Farmland	2020	Chen et al. (2020)
	Hangzhou	Agricultural soils	503.3	Zhou et al. (2020)
	Shaanxi	Agricultural soils	2420	Ding et al. (2020)
	Xinjiang	Agricultural soils	308 ± 138.1	Huang et al. (2020)
	Heilongjiang	Farmland Mollisol	107	Zhang et al. (2020)
	Hebei	Beach soils	317/500 g	Zhou et al. (2016)
	Shangdong	Beach soils	1.3–14712.5	Zhou et al. (2018)
	Loess plateau	Agricultural field	<0.54 mg/kg	Zhang et al. (2018)
	Yunnan	Tree-planted soils	7100–42,960	Zhang et al. (2018)
	Dagoujian and Shangusan	Cropland soils	12,960	Zhang et al. (2018)
	Kunming	Farmland Gleysols	26,070	Zhang et al. (2018)
		Farmland Nitisols	12,050	
		Afforested Gleysols	14,440	
Korea	Yeoju City	Agricultural soils	664	Choi et al. (2021)

compared to deep soil, according to this study (Liu et al., 2018). Chai et al. (2020a, b) studied 33 samples of soil collected from an e-waste disposal region in China's Guangdong Province. MP was found in 30 soils, with a maximum distribution of 34,100 items/kg, indicating that an e-waste disposal location has developed into a hotspot for MP. In another study, MP abundance is found in soils from an e-waste recycling facility ranging from 600 to 14,200 pieces/kg (Zhang et al., 2021).

According to a Korean study, the soils of Yeoju had a mean of 700 items/kg of microplastics, with the highest amount of microplastics detected from upland soil (3440 items/kg). Though the average dispersion of microplastics in soils samples of agricultural land was 664 pieces/kg, this varied by farming type; orchard sites had the highest abundance, followed by greenhouse, upland, and paddy field sites (Choi et al., 2021). The high concentration of microplastics in agricultural soils (664 ± 83 pieces/kg) highlighted the influence of agricultural activities on soil microplastic contamination, which is likely related to the mulching and usage of vinyl films (Rodríguez-Seijo & Pereira, 2019).

3.3.4 Antarctica

Analysis of the abundance of MPs & NPs has been done along the nearshore region of Ross Sea, Antarctica, where the range has been detected to be 0.0032 to 1.18 particles/m^3 with an average amount of 0.17 ± 0.34 particles/m^3 (Cincinelli et al., 2017) (Table 3.7).

3.3.5 Australia

The concentration of MPs and NPs in soils near an industrial facility has been detected to be of the amount 2400 mg/kg on an average in Sydney of Australia. The min amount has been found to be 300 mg/kg, and the max amount has been recorded to be 67,500 mg/kg, which can be considered as a high contamination region. (Fuller & Gautam, 2016) (Table 3.8).

Table 3.7 Distribution of microplastics and nanoplastic in terrestrial ecosystems across Antarctica

Countries	Location	Concentration (items/kg)	Sample type	References
Antarctica	Ross Sea	0.17 ± 0.34	Sediment	Cincinelli et al. (2017)

Table 3.8 Distribution of microplastics and nanoplastic in terrestrial ecosystems across Australia

Countries	Location	Concentration (mg/kg)	Sample Type	References
Australia	Sydney	2400	Sediment	Fuller and Gautam (2016)

3.3.6 Europe

According to several studies, MPs and NPs have been found in sediments, seawater, and freshwater across Europe. The Table 3.9 represents the distribution of MP and NP in the terrestrial ecosystem across Europe. Italy, Spain, the Netherlands, and the UK recorded the presence of a high amount of MPs and NPs within the sediments with Spain showing the highest amount MP and NP of 3330 items/kg detected in the croplands of the rural areas of Valencia (van den Berg et al., 2020).

The highest amount of MPs and NPs has been found to be 1108 ± 983 items/m^2 in the sediments of the Lake Garda (Imhof et al., 2017). Concentration in the regions of Lake Bolsena and Lake Chiusi has been detected to be 112 and 234 items/kg (Fischer et al., 2016). The distribution of MPs and NPs in the country of Italy has been found to be directly proportionate to the presence of industries and human-induced contamination. Human activities, tourism, industrial activities, and urban development are directly related to the pollution of MPs and NPs (Frère et al., 2017; Gündoğdu & Çevik, 2017; Tubau et al., 2015; de Lucia et al., 2014; Barnes et al., 2009). MP and NP contamination in sediment is found to be higher where there's a high human settlement (Collet & Engelbert, 2013).

A higher concentration of MPs and NPs has been reported in the sediments of the Meuse River (1400 items/kg) of the Netherlands compared to the abundance detected in the sewage sludge (650 items/kg) by Leslie et al. (2017). A high concentration of MP and NP (660 and 300 items/kg) has also been recorded in the regions of River Thames Basin and Edgbaston Pool, Birmingham, of the UK (Horton et al. 2017a, b; Vaughan et al., 2017).

Table 3.9 Distribution of microplastics and nanoplastic in terrestrial ecosystems across Europe

Countries	Location	Concentration (items/kg)	Sample type	References
UK	River Thames Basin	660	Sediment	Horton et al. (2017a, b)
UK	Edgbaston Pool, Birmingham	250–300	Sediment	Vaughan et al. (2017)
Netherlands	Dutch	650	Sewage sludge	Leslie et al. (2017)
Netherlands	Meuse River	1400	Sediment	Leslie et al. (2017)
Italy	Lake Bolsena	112	Sediment	Fischer et al. (2016)
Italy	Lake Chiusi	234	Sediment	Fischer et al. (2016)
Spain	Valencia	3330	Sediment	Van den Berg et al. (2020)
Countries	**Location**	**Concentration (items/m^2)**	**Sample type**	**References**
Italy	Lake Garda	1108 ± 983	Sediment	Imhof et al. (2017)
France	Rhône River	0.06 to 1	Sediment	Schmidt et al. (2018)
Germany	Rhine	0.892777	Sediment	Mani et al. (2015)

However, the distribution of detected MP and NP has been found to be significantly low compared to the other studied regions in the terrestrial ecosystem of France and Germany where the amount has been reported to be 0.06 to 1 and 0.892777 items/kg, respectively (Schmidt et al., 2018; Mani et al., 2015).

The presence of buoyant MPs and NPs detected in the sediment serves as a testament to the existence of different types of microplastic getting trapped in sediment during the process of sedimentation (Chae et al., 2015). The deposition of these sorts of MPs within the sediments relates to the extent of maturity which modifies the item thickness, shape, and the improvement of surface biofilm (Long et al., 2015; Cózar et al., 2014).

References

Abidli, S., Lahbib, Y., & El Menif, N. T. (2019). Microplastics in commercial molluscs from the lagoon of Bizerte (Northern Tunisia). *Marine Pollution Bulletin, 142*, 243–252. https://doi.org/10.1016/j.marpolbul.2019.03.048

Antunes, J. C., Frias, J. P. G. L., Micaelo, A. C., & Sobral, P. (2013). Resin pellets from beaches of the Portuguese coast and adsorbed persistent organic pollutants. *Estuarine, Coastal and Shelf Science, 130*, 62–69.

Ballent, P. L., Corcoran, M., Odile, H., Paul, A., & Longstaffe, F. J. (2016). Sources and sinks of microplastics in Canadian Lake Ontario nearshore, tributary and beach sediments. *Marine Pollution Bulletin, 110*, 383e395. https://doi.org/10.1016/j.marpolbul.2016.06.037

Barnes, D. K., Galgani, F., Thompson, R. C., et al. (2009). Accumulation and fragmentation of plastic debris in global environments. *Philosophical Transactions of the Royal Society B: Biological Sciences, 364*, 1985–1998. https://doi.org/10.1098/rstb.2008.0205

Blasing, M., & Amelung, W. (2018). Plastics in soil: Analytical methods and possible sources. *Science of the Total Environment, 612*, 422–435.

Bradford, S. A., & Torkzaban, S. (2012). Colloid adhesive parameters for chemically heterogeneous porous media. *Langmuir, 28*, 13643–13651.

Bradford, S. A., Yates, S. R., Bettahar, M., & Simunek, J. (2002). Physical factors affecting the transport and fate of colloids in saturated porous media. *Water Resources Research, 38*, 63–61.

Brady, N. C., & Weil, R. R. (2000). *Nature and properties of soils*. Macmillan Publishing Company.

Büks, F., & Kaupenjohann, M. (2020). Global concentrations of microplastics in soils–a review. *The Soil, 6*(2), 649–662. https://doi.org/10.5194/soil-6-649-2020

Cey, E. E., Rudolph, D. L., & Passmore, J. (2009). Influence of macro porosity on preferential solute and colloid transport in unsaturated field soils. *Journal of Contaminant Hydrology, 107*, 45–57.

Chae, D. H., Kim, I. S., Kim, S. K., et al. (2015). Abundance and distribution characteristics of microplastics in surface seawaters of the Incheon/Kyeonggi coastal region. *Archives of Environmental Contamination and Toxicology, 69*, 269–278. https://doi.org/10.1007/s00244-015-0173-4

Chai, B., Li, X., Liu, H., Lu, G., Dang, Z., & Yin, H. (2020a). Bacterial communities on soil microplastic at Guiyu, an E-Waste dismantling zone of China. *Ecotoxicology and Environmental Safety, 195*, 110521. https://doi.org/10.1016/j.ecoenv.2020.110521

Chai, B. W., Wei, Q., She, Y. Z., Lu, G. N., Dang, Z., & Yin, H. (2020b). Soil microplastic pollution in an e-waste dismantling zone of China. *Waste Management, 118*, 291–301.

Chakraborty, P., Mukhopadhyay, M., Sampath, S., Ramaswamy, B. R., Katsoyiannis, A., Cincinelli, A., & Snow, D. (2019). Organic micropollutants in the surface riverine sediment along the lower stretch of the transboundary river Ganga: Occurrences, sources and ecological risk assessment. *Environmental Pollution, 249*, 1071–1080. https://doi.org/10.1016/j.envpol.2018.10.115

Chamas, A., Moon, H., Zheng, J., Qiu, Y., Tabassum, T., Jang, J. H., Abu-Omar, M., Scott, S. L., & Suh, S. (2020). Degradation rates of plastics in the environment. *ACS Sustainable Chemistry & Engineering, 8*(9), 3494–3511. https://doi.org/10.1021/acssuschemeng.9b06635

Chen, Y., Leng, Y., Liu, X., & Wang, J. (2020). Microplastic pollution in vegetable farmlands of suburb Wuhan, Central China. *Environmental Pollution, 257*, 113449. https://doi.org/10.1016/j.envpol.2019.113449

Choi, Y. R., Kim, Y. N., Yoon, J. H., Dickinson, N., & Kim, K. H. (2021). Plastic contamination of forest, urban, and agricultural soils: A case study of Yeoju City in the Republic of Korea. *Journal of Soils and Sediments, 21*(5), 1962–1973. https://doi.org/10.1007/s11368-020-02759-0

Cincinelli, A., Scopetani, C., Chelazzi, D., Lombardini, E., Martellini, T., Katsoyiannis, A., Fossi, M. C., & Corsolini, S. (2017). Microplastic in the surface waters of the Ross Sea (Antarctica): Occurrence, distribution, and characterization by FTIR. *Chemosphere, 175*, 391–400. https://doi.org/10.1016/j.chemosphere.2017.02.024

Cole, M., & Galloway, T. S. (2015). Ingestion of Nanoplastics and Microplastics by Pacific Oyster Larvae. *Environmental Science & Technology, 49*(24), 14625–14632. https://doi.org/10.1021/acs.est.5b04099

Collet, I., & Engelbert, A. (2013). Coastal regions: People living along the coastline, integration of NUTS 2010 and latest population grid. *Statistics in Focus, 30*. ISSN:2314-9647.

Corradini, F., Meza, P., Eguiluz, R., Casado, F., Huerta-Lwanga, E., & Geissen, V. (2019). Evidence of microplastic accumulation in agricultural soils from sewage sludge disposal. *Science of the Total Environment, 671*, 411–420. https://doi.org/10.1016/j.scitotenv.2019.03.368

Corradini, F., Casado, F., Leiva, V., Huerta-Lwanga, E., & Geissen, V. (2021). Microplastics occurrence and frequency in soils under different land use on a regional scale. *Science of the Total Environment, 752*, 141917. https://doi.org/10.1016/j.scitotenv.2020.141917

Cózar, A., Echevarría, F., González-Gordillo, J. I., et al. (2014). Plastic debris in the open ocean. *Proceedings of the National Academy of Sciences, 111*, 10239–10244. https://doi.org/10.1073/pnas.1314705111

Crossman, J., Hurley, R. R., Futter, M., & Nizzetto, L. (2020). Transfer and transport of microplastics from biosolids to agricultural soils and the wider environment. *Science of the Total Environment, 724*, 138334. https://doi.org/10.1016/j.scitotenv.2020.138334

Dahms, H. T. J., van Rensburg, G. J., & Greenfield, R. (2020). The microplastic profile of an urban African stream. *Science of the Total Environment, 731*, 138893. https://doi.org/10.1016/j.scitotenv.2020.138893

De Lucia, G. A., Caliani, I., Marra, S., et al. (2014). Amount and distribution of neustonic micro-plastic off the western Sardinian coast (Central-Western Mediterranean Sea). *Marine Environmental Research, 100*, 10–16. https://doi.org/10.1016/j.marenvres.2014.03.017

Ding, L., Zhang, S., Wang, X., Yang, X., Zhang, C., Qi, Y., & Guo, X. (2020). The occurrence and distribution characteristics of microplastics in the agricultural soils of Shaanxi Province, in north-western China. *Science of the Total Environment, 720*, 137525. https://doi.org/10.1016/j.scitotenv.2020.137525

Ding, L., Wang, X., Ouyang, Z., Chen, Y., Wang, X., Liu, S., Yang, X., Jia, H., & Guo, X. (2021). The occurrence of microplastic in Mu Us Sand Land soils in northwest China: Different soil types, vegetation cover and restoration years. *Journal of Hazardous Materials, 403*, 123982.

Duis, K., & Coors, A. (2016). Microplastics in the aquatic and terrestrial environment: Sources (with a specific focus on personal care products), fate and effects. *Environmental Sciences Europe, 28*, 1–25. https://doi.org/10.1186/s12302-015-0069-y

El-Farhan, Y. H., DeNovio, N. M., Herman, J. S., & Hornberger, G. M. (2000). Mobilization and transport of soil particles during infiltration experiments in an agricultural field, Shenandoah valley, Virginia. *Environmental Science & Technology, 34*, 3555e3559.

Fischer, E. K., Paglialonga, L., Czech, E., et al. (2016). Microplastic pollution in lakes and lake shoreline sediments – A case study on Lake Bolsena and Lake Chiusi (central Italy). *Environmental Pollution, 213*, 648–657. https://doi.org/10.1016/j.envpol.2016.03.012

Frère, L., Paul-Pont, I., Rinnert, E., et al. (2017). Influence of environmental and anthropogenic factors on the composition, concentration and spatial distribution of microplastics: A case study of the Bay of Brest (Brittany, France). *Environmental Pollution, 225*, 211–222. https://doi.org/10.1016/j.envpol.2017.03.023

Fuller, S., & Gautam, A. (2016). A procedure for measuring microplastics using pressurized fluid extraction. *Environmental Science & Technology, 50*(11), 5774–5780. https://doi.org/10.1021/acs.est.6b00816

Gabet, E. J., Reichman, O. J., & Seabloom, E. W. (2003). The effects of bioturbation on soil processes and sediment transport. *Annual Review of Earth and Planetary Sciences, 31*, 249–273.

Galloway, T. S., Cole, M., & Lewis, C. (2017). Interactions of microplastic debris throughout the marine ecosystem. *Nature Ecology & Evolution, 1*, 0116. https://doi.org/10.1038/s41559-017-0116

Gao, F. L., Li, J. X., Sun, C. J., Zhang, L. T., Jiang, F. H., Cao, W., & Zheng, L. (2019). Study on the capability and characteristics of heavy metals enriched on microplastics in marine environment. *Marine Pollution Bulletin, 144*, 61–67.

Geyer, R., Jambeck, J. R., & Law, K. L. (2017). Production, use, and fate of all plastics ever made. *Science Advances, 3*(7), e1700782. https://doi.org/10.1126/sciadv.1700782

Gourmelon, G. (2015). Global plastic production rises, recycling lags. *Vital Signs, 22*, 91–95.

Gündoğdu, S., & Çevik, C. (2017). Micro- and mesoplastics in Northeast Levantine coast of Turkey: The preliminary results from surface samples. *Marine Pollution Bulletin, 118*, 341–347. https://doi.org/10.1016/j.marpolbul.2017.03.002

Harms, I. K., Diekoetter, T., Troegel, S., & Lenz, M. (2021). Amount, distribution and composition of large microplastics in typical agricultural soils in Northern Germany. *Science of The Total Environment, 758*, 143615.

Helcoski, R., Yonkos, L. T., Sanchez, A., & Baldwin, A. H. (2020). Wetland soil microplastics are negatively related to vegetation cover and stem density. *Environmental Pollution, 256*, 113391. https://doi.org/10.1016/j.envpol.2019.113391

Horton, A. A., Svendsen, C., Williams, R. J., et al. (2017a). Large microplastic particles in sediments of tributaries of the River Thames, UK – Abundance, sources and methods for effective quantification. *Marine Pollution Bulletin, 114*, 218–226. https://doi.org/10.1016/j.marpolbul.2016.09.004

Horton, A. A., et al. (2017b). Microplastics in freshwater and terrestrial environments: Evaluating the current understanding to identify the knowledge gaps and future research priorities. *Science of the Total Environment, 586*, 127–141.

Huang, Y., Liu, Q., Jia, W., Yan, C., & Wang, J. (2020). Agricultural plastic mulching as a source of microplastics in the terrestrial environment. *Environmental Pollution, 260*, 114096. https://doi.org/10.1016/j.envpol.2020.114096

Huerta Lwanga, E., Gertsen, H., Gooren, H., Peters, P., Salánki, T., van der Ploeg, M., Besseling, E., Koelmans, A. A., & Geissen, V. (2016). Microplastics in the terrestrial ecosystem: Implications for *Lumbricus terrestris* (Oligochaeta, Lumbricidae). *Environmental Science & Technology, 50*(5), 2685–2691. https://doi.org/10.1021/acs.est.5b05478

Hurley, R. R., & Nizzetto, L. (2018). Fate and occurrence of micro(nano)plastics in soils: Knowledge gaps and possible risks. *Current Opinion in Environmental Science & Health, 1*, 6–11.

Imhof, H. K., Sigl, R., Brauer, E., et al. (2017). Spatial and temporal variation of macro-, meso- and microplastic abundance on a remote coral island of the Maldives, Indian Ocean. *Marine Pollution Bulletin, 116*, 340–347. https://doi.org/10.1016/j.marpolbul.2017.01.010

Jambeck, J. R., Geyer, R., Wilcox, C., Siegler, T. R., Perryman, M., Andrady, A., Narayan, R., & Law, K. L. (2015). Plastic waste inputs from land into the ocean. *Science, 347*(6223), 768–771. https://doi.org/10.1126/science.1260352

Jiang, X., Chen, H., Liao, Y., Ye, Z., Li, M., & Klobucar, G. (2019). Ecotoxicity and genotoxicity of polystyrene microplastics on higher plant Vicia faba. *Environmental Pollution, 250*, 831–838.

Kim, I. S., Chae, D. H., & Kim, S. K. (2015). Factors influencing the spatial variation of microplastics on high-tidal coastal beaches in Korea. *Archives of Environmental Contamination and Toxicology, 69*, 299–309. https://doi.org/10.1007/s00244-015-0155-6

Klein, S., Worch, E., & Knepper, T. P. (2015). Occurrence and spatial distribution of microplastics in river shore sediments of the Rhine-Main area in Germany. *Environmental Science & Technology, 49*(10), 6070–6076. https://doi.org/10.1021/acs.est.5b00492

Koelmans, A. A., Besseling, E., & Shim, W. J. (2015). Nanoplastics in the aquatic environment. In M. Bergmann, L. Gutow, & M. Klages (Eds.), *Critical review BT – Marine anthropogenic litter* (pp. 325–334). Springer International Publishing. https://doi.org/10.1007/978-3-319-16510-3_12

Koutnik, V. S., Leonard, J., Alkidim, S., Deprima, F. J., Ravi, S., Hoek, E. M. V., & Mohanty, S. K. (2021). Distribution of microplastics in soil and freshwater environments: Global analysis and framework for transport modeling. *Environmental Pollution, 274*, 116552.

Leslie, H. A., Brandsma, S. H., van Velzen, M. J. M., et al. (2017). Microplastics en route: Field measurements in the Dutch river delta and Amsterdam canals, wastewater treatment plants, North Sea sediments and biota. *Environment International, 101*, 133–142. https://doi.org/10.1016/j.envint.2017.01.018

Li, J., Song, Y., & Cai, Y. (2019a). Focus topics on microplastics in soil: Analytical methods, occurrence, transport, and ecological risks. *Environmental Pollution, 257*, 113570.

Li, Q., Wu, J., Zhao, X., Gu, X., & Ji, R. (2019b). Separation and identification of microplastics from soil and sewage sludge. *Environmental Pollution, 254*, 113076. https://doi.org/10.1016/j.envpol.2019.113076

Liu, M., Lu, S., Song, Y., Lei, L., Hu, J., Lv, W., Zhou, W., Cao, C., Shi, H., Yang, X., & He, D. (2018). Microplastic and mesoplastic pollution in farmland soils in suburbs of Shanghai, China. *Environmental Pollution, 242*, 855–862. https://doi.org/10.1016/j.envpol.2018.07.051

Liu, M., Song, Y., Lu, S., Qiu, R., Hu, J., Li, X., Bigalke, M., Shi, H., & He, D. (2019). A method for extracting soil microplastics through circulation of sodium bromide solutions. *Science of the Total Environment, 691*, 341–347. https://doi.org/10.1016/j.scitotenv.2019.07.144

Long, M., Moriceau, B., Gallinari, M., et al. (2015). Interactions between microplastics and phytoplankton aggregates: Impact on their respective fates. *Marine Chemistry, 175*, 39–46. https://doi.org/10.1016/j.marchem.2015.04.003

Lwanga, E. H., Gertsen, H., Gooren, H., Peters, P., Salanki, T., van der Ploeg, M., et al. (2017a). Incorporation of microplastics from litter into burrows of Lumbricus terrestris. *Environmental Pollution, 220*, 523–531.

Lwanga, E. H., Vega, J. M., Quej, V. K., de los Angeles Chi, J., Del Cid, L. S., Chi, C., Segura, G. E., Gertsen, H., Salánki, T., van der Ploeg, M., & Koelmans, A. A. (2017b). Field evidence for transfer of plastic debris along a terrestrial food chain. *Scientific Reports, 7*(1), 1–7. https://doi.org/10.1038/s41598-017-14588-2

Maaß, S., Daphi, D., Lehmann, A., & Rillig, M. C. (2017). Transport of microplastics by two collembolan species. *Environmental Pollution, 225*, 456–459.

Machado, A. A. D., Lau, C. W., Kloas, W., Bergmann, J., Bacheher, J. B., Faltin, E., Becker, R., Gorlich, A. S., & Rillig, M. C. (2019). Microplastics can change soil properties and affect plant performance. *Environmental Science & Technology, 53*, 6044–6052.

Mahon, A. M., O'Connell, B., Healy, M. G., O'Connor, I., Officer, R., Nash, R., & Morrison, L. (2017). Microplastics in sewage sludge: Effects of treatment. *Environmental Science & Technology, 51*(2), 810–818. https://doi.org/10.1021/acs.est.6b04048

Maity, S., & Pramanick, K. (2020). Perspectives and challenges of micro/nano plastics induced toxicity with special reference to phytotoxicity. *Global Change Biology, 26*, 3241–3250.

Majdalani, S., Michel, E., Di-Pietro, L., & Angulo-Jaramillo, R. (2008). Effects of wetting and drying cycles on in situ soil particle mobilization. *European Journal of Soil Science, 59*, 147–155.

Mani, T., Hauk, A., Walter, U., et al. (2015). Microplastics profile along the Rhine River. *Scientific Reports, 5*, 17988. https://doi.org/10.1038/srep17988

McCarthy, J. F., & McKay, L. D. (2004). Colloid transport in the subsurface. *Vadose Zone Journal, 3*, 326–337.

Merga, L. B., Redondo-Hasselerharm, P. E., Van den Brink, P. J., & Koelmans, A. A. (2020). Distribution of microplastic and small macroplastic particles across four fish species and sediment in an African lake. *Science of the Total Environment, 741*, 140527. https://doi.org/10.1016/j.scitotenv.2020.140527

Mintenig, S. M., Int-Veen, I., Löder, M. G. J., Primpke, S., & Gerdts, G. (2017). Identification of microplastic in effluents of waste water treatment plants using focal plane array-based micro-Fourier transform infrared imaging. *Water Research, 108*, 365–372.

Möller, J. N., Lö der, M. G. J., & Laforsch, C. (2020). Finding microplastics in soils: A review of analytical methods. *Environmental Science & Technology, 54*, 2078–2090.

Napper, I. E., Bakir, A., Rowland S. J., & Thompson, R. C. (2015). Characterisation quantity and sorptive properties of microplastics extracted from cosmetics. *Marine Pollution Bulletin, 99*(1–2), 178–185. S0025326X1500449X. https://doi.org/10.1016/j.marpolbul.2015.07.029

Napper, I. E., & Thompson, R. C. (2019). Environmental deterioration of biodegradable, oxo-biodegradable, compostable, and conventional plastic carrier bags in the sea, soil, and open-air over a 3-year period. *Environmental Science & Technology, 53*(9), 4775–4783. https://doi.org/10.1021/acs.est.8b06984

Nel, H. A., & Froneman, P. W. (2015). A quantitative analysis of microplastic pollution along the south-eastern coastline of South Africa. *Marine Pollution Bulletin, 101*(1), 274–279. https://doi.org/10.1016/j.marpolbul.2015.09.043

Nel, H. A., Dalu, T., & Wasserman, R. J. (2018). Sinks and sources: Assessing microplastic abundance in river sediment and deposit feeders in an Austral temperate urban river system. *Science of the Total Environment, 612*, 950–956. https://doi.org/10.1016/j.scitotenv.2017.08.298

Neto, J. A. B., de Carvalho, D. G., Medeiros, K., Drabinski, T. L., de Melo, G. V., Silva, R. C. O., Silva, D. C. P., de Sousa Batista, L., Dias, G. T. M., da Fonseca, E. M., & dos Santos Filho, J. R. (2019). The impact of sediment dumping sites on the concentrations of microplastic in the inner continental shelf of Rio de Janeiro/Brazil. *Marine Pollution Bulletin, 149*, 110558. https://doi.org/10.1016/j.marpolbul.2019.110558

Ng, E. L., Lwanga, E. H., Eldridge, S. M., Johnston, P., Hu, H. W., Geissen, V., & Chen, D. L. (2018). An overview of microplastic and nanoplastic pollution in agroecosystems. *Science of the Total Environment, 627*, 1377–1388.

Nizzetto, L., Bussi, G., Futter, M. N., Butterfield, D., & Whitehead, P. G. (2016a). A theoretical assessment of microplastic transport in river catchments and their retention by soils and river sediments. *Environmental Science: Processes & Impacts, 18*, 1050–1059.

Nizzetto, L., Futter, M., & Langaas, S. (2016b). Are agricultural soils dumps for microplastics of urban origin? *Environmental Science & Technology, 50*, 10777–10779. https://doi.org/10.1021/acs.est.6b04140

O'Connor, D., Pan, S., Shen, Z., Song, Y., Jin, Y., Wu, W., & Hou, D. (2019a). Microplastics undergo accelerated vertical migration in sand soil due to small size and wet-dry cycles. *Environmental Pollution, 249*(2019), 527–534.

O'Connor, D., Hou, D., Ok, Y. S., Mulder, J., Duan, L., Wu, Q., Wang, S., Tack, F. M. G., & Rinklebe, J. (2019b). Mercury speciation, transformation, and transportation in soils, atmospheric flux, and implications for risk management: A critical review. *Environment International, 126*, 747–761.

Peeken, I., Primpke, S., Beyer, B., Guetermann, J., Katlein, C., Krumpen, T., Bergmann, M., Hehemann, L., & Gerdts, G. (2018). Arctic sea ice is an important temporal sink and means of transport for microplastic. *Nature Communications, 9*, 1505.

Pelley, A. J., & Tufenkji, N. (2008). Effect of particle size and natural organic matter on the migration of nano- and micro scale latex particles in saturated porous media. *Journal of Colloid and Interface Science, 321*, 74–83.

Plastics Europe. (2019). *Plastics – the facts, 2019. An analysis of European plastics production, demand and waste data.*

Rafique, A., Irfan, M., Mumtaz, M., & Qadir, A. (2020). Spatial distribution of microplastics in soil with context to human activities: A case study from the urban center. *Environmental Monitoring and Assessment, 192*(11), 1–13. https://doi.org/10.1007/s10661-020-08641-3

Rahman, S. M. A., Robin, G. S., Momotaj, M., Uddin, J., & Siddique, M. A. M. (2020). Occurrence and spatial distribution of microplastics in beach sediments of Cox's Bazar, Bangladesh. *Marine Pollution Bulletin, 160*, 111587. https://doi.org/10.1016/j.marpolbul.2020.111587

Rahmatpour, S., Mosaddeghi, M. R., Shirvani, M., & Simunek, J. (2018). Transport of silver nanoparticles 794 in intact columns of calcareous soils: The role of flow conditions and soil texture. *Geoderma, 322*, 79589–79100.

Rakib, M., Jahan, R., Al Nahian, S., Alfonso, M. B., Khandaker, M. U., Enyoh, C. E., Hamid, F. S., Alsubaie, A., Almalki, A. S., Bradley, D. A., & Mohafez, H. (2021). Microplastics pollution in salt pans from the Maheshkhali Channel, Bangladesh. *Scientific Reports, 11*(1), 1–10. https://doi.org/10.1038/s41598-021-02457-y

Ramos, L., Berenstein, G., Hughes, E. A., Zalts, A., & Montserrat, J. M. (2015). Polyethylene film incorporation into the horticultural soil of small periurban production units in Argentina. *Science of the Total Environment, 523*, 74–81. https://doi.org/10.1016/j.scitotenv.2015.03.142

Rezaei, M., Riksen, M. J. P. M., Sirjani, E., Sameni, A., & Geissen, V. (2019). Wind erosion as a driver for transport of light density microplastics. *Science of the Total Environment, 669*, 273–281.

Rillig, M. C. (2012). Microplastic in terrestrial ecosystems and the soil? *Environmental Science & Technology, 46*, 6453–6454. https://doi.org/10.1021/es302011r

Rillig, M. C., & Lehmann, A. (2020). Microplastic in terrestrial ecosystems. *Science, 368*(6498), 1430–1431. https://doi.org/10.1126/science.abb5979

Rillig, M. C., Ziersch, L., & Hempel, S. (2017a). Microplastic transport in soil by earthworms. *Scientific Reports, 7*, 1362.

Rillig, M. C., Ingraffia, R., & Machado, A. A. S. (2017b). Microplastic incorporation into soil in agroecosystems. *Frontiers in Plant Science, 8*, 1805. https://doi.org/10.3389/fpls.2017.01805

Rocha-Santos, T., & Duarte, A. C. (2015). A critical overview of the analytical approaches to the occurrence, the fate and the behavior of microplastics in the environment. *TrAC – Trends in Analytical Chemistry, 65*, 47. https://doi.org/10.1016/j.trac.2014.10.011

Rodríguez-Seijo, A., & Pereira, R. (2019). Microplastics in agricultural soils: Are they a real environmental hazard? In *Bioremediation of agricultural soils* (pp. 45–60). CRC Press. eBook ISBN9781315205137.

Sanchez-Hernandez, J. C., Del Pino, J. N., Capowiez, Y., Mazzia, C., & Rault, M. J. S. O. T. T. E. (2018). Soil enzyme dynamics in chlorpyrifos-treated soils under the influence of earthworms. *Science of the Total Environment, 612*, 1407–1416. https://doi.org/10.1016/j.scitotenv.2017.09.043

Scheurer, M., & Bigalke, M. (2018). Microplastics in Swiss floodplain soils. *Environmental Science & Technology, 52*, 3591–3598.

Schmidt, N., Thibault, D., Galgani, F., et al. (2018). Occurrence of microplastics in surface waters of the Gulf of Lion (NW Mediterranean Sea). *Progress in Oceanography, 163*, 214–220. https://doi.org/10.1016/j.pocean.2017.11.010

Sintim, H. Y., & Flury, M. (2017). Is biodegradable plastic mulch the solution to agriculture's plastic problem? Environ. *Environmental Science & Technologyy, 51*, 1068–1069.

Subrahmaniyan, K., Kalaiselvan, P., Balasubramanian, T. N., & Zhou, W. (2006). Crop productivity and soil properties as affected by polyethylene film mulch and land configurations in groundnut (Arachis hypogaea L.). *Archives of Agronomy and Soil Science, 52*(1), 79–103.

Tajwar, M., Gazi, M. Y., & Saha, S. K. (2021). Characterization and spatial abundance of microplastics in the coastal regions of Cox's Bazar, Bangladesh: An integration of field, laboratory, and GIS techniques. *Soil and Sediment Contamination: An International Journal, 31*, 1–25. https://doi.org/10.1080/15320383.2021.1910622

Tajwar, M., Shreya S. S., Gazi, M. Y., Hasan, M., & Saha, S. K. (2022a). Microplastic contamination in the sediments of the Saint Martin's Island Bangladesh. *Regional Studies in Marine Science* 53102401-S2352485522001232 102401. https://doi.org/10.1016/j.rsma.2022.102401

Tajwar, M., Shreya S. S., Hasan M., Hossain M. B., Gazi M. Y., & Sakib N. (2022b). Assessment of microplastics as contaminants in a coal mining region. *Heliyon 8*(11), e11666-S2405844022029541 e11666. https://doi.org/10.1016/j.heliyon.2022.e11666

The Jakarta Post. (2019). *Inadequate landfills worsen Indonesia's waste problems*. https://www.thejakartapost.com/news/2019/03/03/inadequate-landfills-worsen-indonesias-waste-problems.html

Tibbetts, J., Krause, S., Lynch, I., & Sambrook Smith, G. (2018). Abundance, distribution, and drivers of microplastic contamination in urban river environments. *Water, 10*(11), 1597. https://doi.org/10.3390/w10111597

Toumi, H., Abidli, S., & Bejaoui, M. (2019). Microplastics in freshwater environment: The first evaluation in sediments from seven water streams surrounding the lagoon of Bizerte (Northern Tunisia). *Environmental Science and Pollution Research, 26*(14), 14673–14682. https://doi.org/10.1007/s11356-019-04695-0

Treumann, S., Torkzaban, S., Bradford, S. A., Visalakshan, R. M., & Page, D. (2014). An explanation for differences in the process of colloid adsorption in batch and column studies. *Journal of Contaminant Hydrology, 843164*, 219–229.

Tubau, X., Canals, M., Lastras, G., et al. (2015). Marine litter on the floor of deep submarine canyons of the Northwestern Mediterranean Sea: The role of hydrodynamic processes. *Progress in Oceanography, 134*, 379–403. https://doi.org/10.1016/j.pocean.2015.03.013

Tun, T. Z., Kunisue, T., Tanabe, S., Prudente, M., Subramanian, A., Sudaryanto, A., Viet, P. H., & Nakata, H. (2022). Microplastics in dumping site soils from six Asian countries as a source of plastic additives. *Science of the Total Environment, 806*, 150912. https://doi.org/10.1016/j.scitotenv.2021.150912

van den Berg, P., Huerta-Lwanga, E., Corradini, F., & Geissen, V. (2020). Sewage sludge application as a vehicle for microplastics in eastern Spanish agricultural soils. *Environmental Pollution, 261*, 114198. https://doi.org/10.1016/j.envpol.2020.114198

Van Sebille, E., Wilcox, C., Lebreton, L., Maximenko, N., Hardesty, B. D., Van Franeker, J. A., Eriksen, M., Siegel, D., Galgani, F., & Law, K. L. (2015). A global inventory of small floating plastic debris. *Environmental Research Letters, 10*(12). https://doi.org/10.1088/1748-9326/10/12/124006

Vaughan, R., Turner, S. D., & Rose, N. L. (2017). Microplastics in the sediments of a UK urban lake. *Environmental Pollution, 229*, 10–18. https://doi.org/10.1016/j.envpol.2017.05.057

Wang, J., Li, J. Y., Liu, S. T., Li, H. Y., Chen, X. C., Peng, C., Zhang, P. P., & Liu, X. H. (2021). Distinct microplastic distributions in soils of different land-use types: a case study of Chinese farmlands. *Environmental Pollution, 2021*(269), 116199.

Weithmann, N., Moller, J. N., Loder, M. G. J., Piehl, S., Laforsch, C., & Freitag, R. (2018). Organic fertilizer as a vehicle for the entry of microplastic into the environment. *Science Advances, 4*, eaap8060.

Wick, L. Y., Remer, R., Würz, B., Reichenbach, J., Braun, S., Schäfer, F., et al. (2007). Effect of fungal hyphae on the access of bacteria to phenanthrene in soil. *Environmental Science & Technology, 41*, 500–505. https://doi.org/10.1021/es061407s

Wu, X., Lyu, X., Li, Z., Gao, B., Zeng, X., Wu, J., & Sun, Y. (2020). Transport of polystyrene nanoplastics in natural soils: Effect of soil properties, ionic strength and cation type. *Science of the Total Environment, 707*, 136065.

Yu, M., van der Ploeg, M., Lwanga, E. H., Yang, X. M., Zhang, S. L., Ma, X. Y., Ritsema, C. J., & Geissen, V. (2019). Leaching of microplastics by preferential flow in earthworm (Lumbricus terrestris) burrows. *Environmental Chemistry Letters, 16*, 31–40.

Zhang, G. S., & Liu, Y. F. (2018). The distribution of microplastics in soil aggregate fractions in southwestern China. *Science of the Total Environment, 642*, 12–20. https://doi.org/10.1016/j.scitotenv.2018.06.004

Zhang, S. L., Zhang, X. Y., Huffman, T., Liu, X. B., & Yang, J. Y. (2011). Influence of topography and land management on soil nutrients variability in Northeast China. *Nutrient Cycling in Agroecosystems, 89*, 427–438.

Zhang, S. L., Huang, J., Wang, Y., Shen, Q. S., Mu, L. L., & Liu, Z. H. (2017). Spatiotemporal heterogeneity of soil available nitrogen during crop growth stages on mollisol slopes of Northeast China. *Land Degradation and Development, 28*, 856–869.

Zhang, S., Yang, X., Gertsen, H., Peters, P., Salánki, T., & Geissen, V. (2018). A simple method for the extraction and identification of light density microplastics from soil. *Science of the Total Environment, 616–617*, 1056–1065. https://doi.org/10.1016/j.scitotenv.2017.10.213

Zhang, S., Liu, X., Hao, X., Wang, J., & Zhanget, Y. (2019). Distribution of low-density microplastics in the mollisol farmlands of northeast China. *Science of the Total Environment, 708*. https://doi.org/10.1016/j.scitotenv.2019.135091

Zhang, D., Ng, E. L., Hu, W., Wang, H., Galaviz, P., Yang, H., Sun, W., Li, C., Ma, X., Fu, B., & Zhao, P. (2020). Plastic pollution in croplands threatens long-term food security. *Global Change Biology, 26*(6), 3356–3367. https://doi.org/10.1111/gcb.15043

Zhang, Z., Cui, Q., Chen, L., Zhu, X., Zhao, S., Duan, C., Zhang, X., Song, D., & Fang, L. (2021). A critical review of microplastics in the soil-plant system: Distribution, uptake, phytotoxicity and prevention. *Journal of Hazardous Materials, 424*, 127750. https://doi.org/10.1016/j.jhazmat.2021.127750

Zhou, Q., Zhang, H., Zhou, Y., Li, Y., Xue, Y., Fu, C., Tu, C., & Luo, Y. (2016). Separation of microplastics from a coastal soil and their surface microscopic features. *Chinese Science Bulletin, 61*(14), 1604–1611. https://doi.org/10.1360/N972015-01098

Zhou, B. Y., Wang, J. Q., Zhang, H. B., Shi, H. H., Fei, Y. F., Huang, S. Y., Tong, Y. Z., Wen, D., Zhou, Q., Zhang, H., Fu, C., Zhou, Y., Dai, Z., Li, Y., Tu, C., & Luo, Y. (2018). The distribution and morphology of microplastics in coastal soils adjacent to the Bohai Sea and the Yellow Sea. *Geoderma, 322*, 201–208. https://doi.org/10.1016/j.geoderma.2018.02.015

Zhou, B., Wang, J., Zhang, H., Shi, H., Fei, Y., Huang, S., Tong, Y., Wen, D., Luo, Y., & Barceló, D. (2019a). Microplastics in agricultural soils on the coastal plain of Hangzhou Bay, east China: Multiple sources other than plastic mulching film. *Journal of Hazardous Materials, 388*, 121814.

Zhou, Y., Liu, X., & Wang, J. (2019b). Characterization of microplastics and the association of heavy metals with microplastics in suburban soil of central China. *Science of the Total Environment, 694*, 133798. https://doi.org/10.1016/j.scitotenv.2019.133798

Zhou, B., Wang, J., Zhang, H., Shi, H., Fei, Y., Huang, S., Tong, Y., Wen, D., Luo, Y., & Barceló, D. (2020). Microplastics in agricultural soils on the coastal plain of Hangzhou Bay, east China: Multiple sources other than plastic mulching film. *Journal of Hazardous Materials, 388*, 121814. https://doi.org/10.1016/j.jhazmat.2019.121814

Zhu, D., Bi, Q. F., Xiang, Q., Chen, Q. L., Christie, P., Ke, X., et al. (2018). Trophic predator-prey relationships promote transport of microplastics compared with the single Hypoaspis aculeifer and Folsomia candida. *Environmental Pollution, 235*, 150–154.

Zhu, F., Zhu, C., Wang, C., & Gu, C. (2019). Occurrence and ecological impacts of microplastics in soil systems: A review. *Bulletin of Environmental Contamination and Toxicology, 102*(6), 741–749. https://doi.org/10.1007/s00128-019-02623-z

Zhuang, J., Qi, J., & Jin, Y. (2005). Retention and transport of amphiphilic colloids under unsaturated flow conditions: Effect of particle size and surface property. *Environmental Science & Technology, 39*, 7853–7859.

Ziajahromi, S., Neale, P. A., & Leusch, F. D. L. (2016). Wastewater treatment plant effluent as a source of microplastics: Review of the fate, chemical interactions and potential risks to aquatic organisms. *Water Science and Technology, 74*, 2253–2269.

Zubris, K. A. V., & Richards, B. K. (2005). Synthetic fibers as an indicator of land application of sludge. *Environmental Pollution, 138*(2), 201–211. https://doi.org/10.1016/j.envpol.2005.04.013

Chapter 4
Methodology of Assessing Microplastics and Nanoplastics in the Environment: Recent Advances in the Practical Approaches

Dayakar Govindu, Radhika Tippani, Mahendar Porika, and Syam Prasad Sura

Abstract Microplastics (MPs) and nanoplastics (NLPs) have emerged as emergent particle anthropogenic contaminants, quickly gaining scientific and popular attention. These microscopic plastic particles have been identified in the nature, portable water, and foodstuff all around the world, raising worries about their effects on the environment and human health. Reliable information on MP and NLP concentrations in the environment is required to fully address these challenges. MP and NLP particles, on the other side, vary greatly in shape, density, size, polymer type, surface characteristics, and other factors. While particle concentrations in various mediums may range by up to ten orders of magnitude, analysing these complex samples might seem like looking for a needle in a haystack. This emphasises the vital need of using the right methodologies to identify, quantify, and characterise MP and NLPs with an emphasis on sensitivity and detection limits.

Keywords Microplastics (MPs) · Anthropogenic contaminants · Nanoplastics (NLPs) · Polymer type

---

D. Govindu · R. Tippani · M. Porika (✉)
Department of Biotechnology, Kakatiya University, Warangal, Telangana, India

S. P. Sura
Department of Botany, Government Degree College, Bhupalpally, India

N. R. Maddela et al. (eds.), *Micro and Nanoplastics in Soil*,
https://doi.org/10.1007/978-3-031-21195-9_4

4.1 Introduction

Environmental plastic pollution has a significant influence on the development, growth, and longevity of a variety of living species, including humans, prompting scientists to design novel monitoring and purification methods. Despite the many benefits of plastic materials in everyday life, their limited biodegradability, incorrect usage, and ineffective disposal contribute to increased environmental contamination. Plastic derivatives are exposed to the environment, which stimulates chemical, physical, and biological degradation processes, resulting in the accretion of tiny plastic particles in both aquatic and terrestrial environments, such as soil (Li et al., 2020), freshwater, air (Prata, 2018), foodstuff, and sediments.

Microplastics (MPs, 5 mm) and nanoplastics (NLPs, 1 nm to 1 mm) are small particles of synthetic polymers that can be discharged into the environment (soils, water (sea, fresh, and drinking), biota, food, air, and sediments) and are thus recognised as evolving particulate anthropogenic contaminants (Dehaut et al., 2019; Hale et al., 2020; Delgado-Gallardo et al., 2021). Thompson et al. (2004) used the term microplastics to describe microscopic plastic pieces found in the ocean. Arthur et al. (2010) recommended a microplastics size limit of 5 mm in 2009. NP and MP are now defined as plastic particles and fibres smaller than 1 μm and in the size range of 1 μm to 1 mm, respectively (Gigault et al., 2018). Large microplastics are defined as fragments with a diameter of 15 mm or more. In the following, we'll use the acronyms MPs for microplastics, NPLs for nanoplastics (rather than NPs to prevent misperception with nanoparticles), and NMPs for both nanoplastics and microplastics when discussing microscopic plastic particles and fibres in general. In complex environmental samples, MPs and NPLs are found. This study does not cover micro- and nanoplastic analyses in full; however a few essential topics are worth mentioning (Fig. 4.1).

Aside from the fact that plastic materials increase the standard of lifespan for millions of individuals throughout the world wide by rendering it easier, harmless, and additional pleasurable. They are lightweight, adaptable, durable, formable, corrosion- and flame-resistant, and so on. On the other hand, as soon as plastics end up in nature or in food, it raises global worries. While plastic output in Europe has decreased somewhat (61.8 Mt in 2018 and 59.7 Mt in 2019), worldwide production has increased year after year, reaching 368 Mt in 2019. Production of thermoplastics like polyethylene (PE) of high density (HDPE) and low density (LDPE), polypropylene (PP), polystyrene (PS), polyvinyl chloride (PVC), and polyethylene terephthalate (PET) is an indicator of the extent to which MP pollution is present in freshwater and drinking water, for example (Koelmans et al., 2019).

Bioplastics are manufactured in addition to traditional polymers such as those described above, as well as poly(methyl methacrylate) (PMMA), polyurethane (PUR), and polyamide (PA). Food packaging (e.g., polylactide, PLA) and agriculture are increasingly using the latter (e.g., polybutylene adipate-co-terephthalate, PBAT). Tire wear particles (TWP), which comprise 40–60% synthetic polymers (e.g., styrene-butadiene rubber, SBR) and paint particles/surface coatings, are also

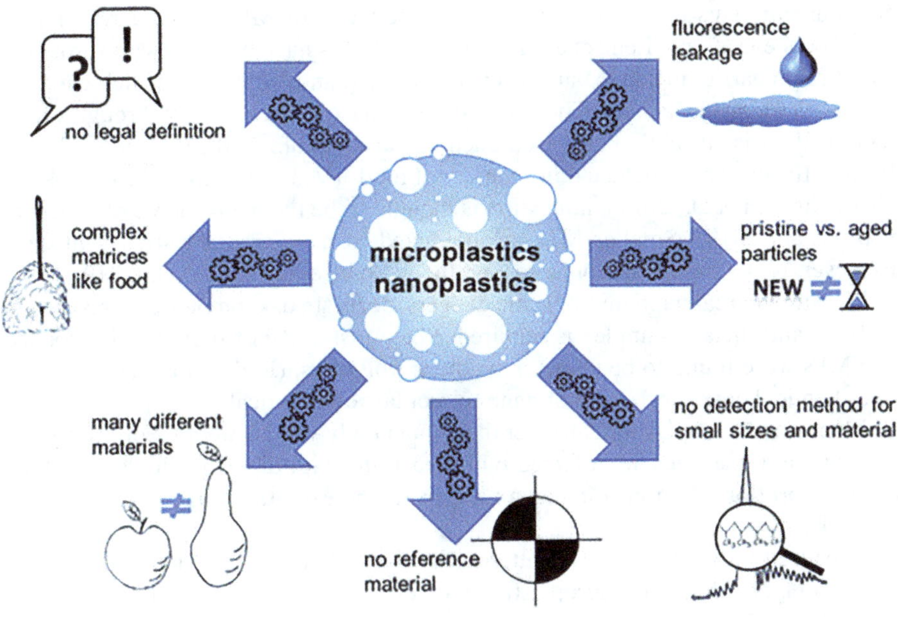

Fig. 4.1 Obstacles and opportunities in the realm of MP and NPLs study. (Adapted from Paul et al., 2020)

attributed to MPs, according to a new definition (Hartmann et al., 2019). In these systems, synthetic polymers act as film formers and are mixed with binders and fillers (Hartmann et al., 2019).

"Primary" and "secondary" origin nanomicroplastics are distinguishable. "Primary": it is possible to use nano microplastics in diverse ways (e.g., pellets for industrial production, industrial cleaners, and nano- and microbeads for personal care products). Plastic waste in the environment is fragmented and degraded by mechanical, UV, and microbiological degradation (Frias & Nash, 2019) to create secondary nano microplastic particles and fibres.

Across the globe, MPs are discovered in deep-sea sediments (Cunningham et al., 2020) and even on Mount Everest (PES fibres were identified at 8440 m) (Napper et al., 2020). The many reports of MP occurrence globally sparked several issues about MP consequences on biota. Leaching of monomers and additives may have undesirable effects, some of which are poisonous, carcinogenic, or endocrine disruptive (Tian et al., 2021). Also, oxidative photodegradation of plastic trash may produce toxic volatile organic compounds (VOCs) such as acrolein and benzene (Lomonaco et al., 2020). The MPs may also function as a vector for harmful and/or antibiotic-resistant microbes (Bakir et al., 2014; Brennecke et al., 2016; Bank et al., 2020).

There is a wide range of MP effects on biota recorded thus far: negative (including deadly), neutral, or even detoxifying. Many of these investigations used MP particle concentrations that were 102–107 times higher than those found in the environment. This observation emphasises the value of MP exposure research at actual concentrations (Lenz et al., 2016). While NPLs have been shown to cross the blood-brain barrier in fish (Mattsson et al., 2017), quantifiable data on their environmental incidence is lacking. The extent of human exposures to MPs through water, food, and air is currently being examined (Catarino et al., 2018; Paul et al., 2020). While MPs have been detected in a variety of foods (Van Cauwenberghe & Janssen, 2014; Kirstein et al., 2021), inhalation is thought to be the main source of exposure (Cox et al., 2019). Smaller MPs are predicted to have greater harmful impacts. However, NPLs have been found to pass the gut barrier (Lehner et al., 2019).

To estimate real nano microplastic dangers, accurate data on particle presence in ambient and dietary samples is required. Since just 1.4% of particles that looked like MPs were found to be made of synthetic polymers, (Löder et al., 2015), accurate chemical characterization of nano microplastics is critical.

Micro- and nanoplastics can enter the human body via the mouth. Following oral consumption, particles are affected by interactions with digestive fluids, intestinal cells, absorption and transit in the gut and liver, and excretion. Figure 4.2 shows an example.

Because plastic sources, use patterns, emission channels, and material qualities vary widely, so do nano microplastic particles (Koelmans et al., 2019; Hale, 201; Zarfl, 2019). This analyte is one among the most difficult to identify, quantify, and characterise in the environment and food.

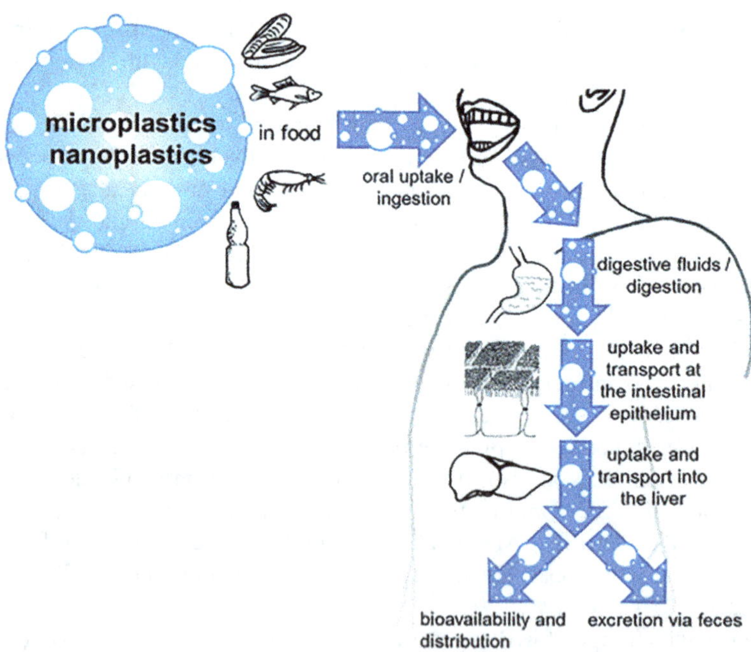

Fig. 4.2 Human exposure and micro- and nanoplastic particle pathways. (Adapted from Paul et al., 2020)

The main goals of this chapter include: (i) draw attention to the difficulties in studying micro and nanoplastics, (ii) that may be utilised to perform a trustworthy and exact chemical evaluation of particles, and (iii) to share viewpoints both inside and beyond the area of nano microplastic research. As a result, the benefits and drawbacks of both mass-based and particle-based methodologies for identifying and quantifying MPs are explored first, with an emphasis on sensitivity and lower detection size limits, and also automation and high-throughput analysis. New and promising approaches are given alongside well-evolved techniques for the examination of model and actual samples, as well as their applications. The complementary nature of several analytical approaches for thorough MP characterisation is next emphasised. A special section of the study is dedicated to the rapidly evolving subject of nanoplastic research, with an emphasis on NPLs with tiny masses and sizes. Finally, attempts to validate, harmonise, and standardise nano microplastic investigations are discussed, as well as the future uses of sophisticated technologies for the examination of plastic and non-plastic micro nanoparticles.

4.2 Analysis of Microplastics

4.2.1 Mass-Based Analysis

(a) *Thermal Degradation/GC/MS Combination*

Thermal degradation approaches have been revealed to be particularly successful in identifying and quantifying plastic pollution in food stuff and environment. These technologies depend on breakdown products created at certain temperatures in the absence of oxygen. After gas chromatographic (GC) separation, the pyrogram displays the fingerprint of the specific polymer. The volatile breakdown products may then be detected on a molecular level using mass spectrometry (MS). The measurement of polymer mass can be done based on particular pyrolysis products, allowing for concurrent detection and quantification of distinct MP in complicated environmental samples. For mass balances and modelling, and also future regulation, this knowledge for various polymers is essential. Moreover, such approaches enable the identification of plastic-associated additives and also degradation by products and so provide the information required for a meaningful risk analysis of MP for the habitat and human safety. These mass-related data, on the other hand, must be regarded as large quantities of a specific plastic type, such as PS, regardless of whether it is a pure polymer or a component of a copolymer, and are unaffected by particle properties like size, shape, form, and so on (Primpke et al., 2020).

De Leeuw et al. (1986) were the first to disclose the presence of PS as an anthropogenic contaminant in sediment and soil studied by Py-GC/MS. In another study, the Py-GC/MS examination of PS and PVC in coastal sediments, and also polybutadiene (PB), poly(vinyl acetate) (PVA), block (SBS) copolymers, poly(acrylonitrile-costyrene-co-butadiene) (ABS), and styrenebutadiene random (SBR) (Fabbri et al.,

2000; Fabbri, 2001). These approaches are commonly used in marine and freshwater environments (sediments, (Fries et al., 2013; Hermabessiere et al., 2018; Käppler et al., 2018; Dierkes et al., 2019; Ceccarini et al., 2018); Hermabessiere et al., 2018) water (Primpke et al., 2020; McCormick et al., 2016; Ravit et al., 2019; Dümichen et al., 2017; Hendrickson et al., 2018) biota (Hermabessiere et al., 2018; Dehaut et al., 2016; Dümichen et al., 2015), sewage sludge, (Dierkes et al., 2019; El Hayany et al., 2020) airborne emissions from laundry dryers, soil, (Watteau et al., 2018; Steinmetz et al., 2020) and commercial sea salt, (Fischer et al., 2019). Py-GC/MS has recently been shown to be capable of evaluating nanoplastics in both model and actual samples (Sullivan et al., 2020).

There are two kinds of pyrolysis units and their connection with gas chromatographs in terms of applicable instrumentation: I Py-GC/MS and (ii) TED (thermo-extraction and desorption) GC/MS.

(i) *Pyrolysis-Based Methods*

Py-GC/MS can be used in a variety of ways, including (I) single-shot analysis, (ii) double-shot (or "multi-shot") analysis, (iii) evolved gas analysis (EGA-MS), and (iv) reactive or thermochemolysis Py-GC/MS (Picó & Barceló, 2020). Pyrolysis is carried out in "single-shot" mode at a certain temperature, generally over 500 °C. The sample temperature is quickly raised from ambient to pyrolysis temperature (<20 ms for contemporary systems). The macromolecules are virtually instantaneously broken in the pyrolyzer, and the pyrolysis products are separated in the GC column and utilised for MS-based polymer(s) and additive identification (Primpke et al., 2020; Käppler et al., 2018).

Double-shot mode, also known as thermal desorption (TD) Py-GC/MS, is a way to analyse various kinds of compounds at different times. For example, volatile compounds that are released at low temperatures during a thermal desorption step can be analysed in the same way as the decomposition fragments of the larger macromolecules that are formed at high temperatures during pyrolysis. This is a good way to find out about the different types of polymer additives that are both volatile and nonvolatile (Herrera et al., 2003; Jansson et al., 2007) and even sorbed organic compounds (Reichel et al., 2020) pooled with the detection of polymer(s) based on the investigation of pyrolysis products (Fries et al., 2013; Dekiff et al., 2014). Also, before pyrolysis, the "double-shot" mode may be employed to eliminate organic molecules that can impede with identifying and quantifying MP breakdown products from complicated organic-rich materials (Okoffo et al., 2020).

EGA-MS is a type of chromatography that doesn't use a chromatographic column. Instead, it uses a short and narrow (2.5 m, 0.15 mm i.d.) deactivated silica capillary tube without a stationary phase to connect the GC injector and the MS detector directly (Picó & Barceló, 2020). Thermal chemistry Py-GC/MS involves adding a derivatization agent, like tetramethylammonium hydroxide (TMAH) solution, which causes an ester and ether to be broken down and then methylated (Primpke et al., 2020; Picó & Barceló, 2020). Individual plastic particles or a small amount of a sample are put into a pyrolyzer target for the Py-GC/MS method of identifying polymers, which is how it works. Each pyrolyzer has a different size and can process a different number of samples at a time (Fries et al., 2013; Nuelle et al., 2014; Funck et al., 2020) placed in a platinum coil. CP pyrolyzers utilize semi-closed ferromagnetic targets (typical dimensions ⌀ 2 mm, 8 mm height) (Fischer & Scholz-Böttcher, 2017). MF pyrolyzers use stainless steel cups (typical dimensions approximately ⌀ 4 mm, 8 mm height (Fischer et al., 2019). Heating the sample to a certain temperature in an inert gas (generally He or N2) is done in each case. This gas is also used as a carrier gas for GC separation (Primpke et al., 2020). A quadrupole mass spectrometer is usually used to make sure that the polymers made from decomposition products can be identified and counted with a high level of accuracy. A new study by Sullivan et al. shows that GC time-of-flight mass spectrometry (Py-GC/ToF) can significantly increase the amount of information that can be found (Sullivan et al., 2020).

Based on the pyrolysis product complexity and how quickly they break down, the pyrograms of different types of polymers look very different. The pyrogram of a single polymer can be very complicated (e.g., PE, PP, PET) over moderate (e.g., PS) to simple (e.g., PMMA) (Primpke et al., 2020; Fotopoulou & Karapanagioti, 2017).

Identification of Particles and Additives

Py-GC/MS has been used by a lot of different groups to find individual (plastic) particles that have been isolated from marine and river sediments, surface water, and biological samples (Hermabessiere et al., 2018; Hendrickson et al., 2018; Dehaut et al., 2016). Using "double-shot" mode (TD-Py-GC/MS), Fries and coworkers investigated marine microplastic particles down to 100 nm. These researchers could identify many plastics (PE, PP, PS), as well as numerous additives (benzophenone, 1,2-benzenedicarboxylic acid, dimethyl phthalate, diethylhexyl phthalate, dibutoxyphthalate, dibutyl phthalate, phenol, and 2,4-di-tert-butylphenol) (Fries et al., 2013; Dekiff et al., 2014). Py-GC/MS can be used to correctly determine particles and get more useful details about additives and copolymers, but it takes a long time. Each GC-MS run can take up to half an hour or more. FTIR and Raman microspectroscopy data were recently compared with Py-GC/MS findings by Käppler et al. (2018) and Hermabessiere et al. (2018). Thus, Py-GC/MS and spectroscopic approaches may be used together to analyse individual particles. EGA-MS may also be used to quickly identify additives and polymers. While Py-GC/MS can analyse individual particles, its true ability for simultaneous MP identification and quantification in complicated materials has lately been discovered (Kirstein et al., 2021; Fischer et al., 2019; Dibke et al., 2021).

Simultaneous Identification and Quantification of Polymers from Complex Samples

Analysing samples immediately (Funck et al., 2020) or after one or more sample preparation steps may involve chemical and enzymatic digestion of organic matrix and elimination of inorganic matrix by density separation (Kirstein et al., 2021; Primpke et al., 2020). Soluble extraction (Dierkes et al., 2019; Okoffo et al., 2020) and pressurised liquid extraction (PLE) may also be used to preconcentrate polymers from complicated matrices prior to Py-GC/MS. CPE has also been used to preconcentrate nanoplastics from water samples (Zhou et al., 2018). Watteau et al. (2018) found MPs in bulk soil and soil factions. Using a 0.5–1 mg sample permits identification of plastic characteristics (e.g., typical for PS) that vary from soil organic matter. Funk et al. used Py-GC/MS to identify and quantify MPs in wastewater after cascade filtering with no sample preparation other than extraction and drying. PS and PE LOQs were 0.03 μg and 1 μg absolute, respectively (Funck et al., 2020).

The elimination of organic and inorganic matrices has shown to be effective in increasing the sensitivity of Py-GC/MS for the measurement of various polymers in complicated samples. MPs deposited on a filter are pyrolyzed after sample preparation and drying; for this, fragments or even the complete glass fibre filter with a diameter of 15 mm may be immediately introduced into the pyrolyzer for analysis (Kirstein et al., 2021).

Before Py-GC/MS analysis, liquid extraction of soluble polymers may be used to preconcentrate polymers from complicated matrices (Okoffo et al., 2020). For MP measurement in environmental samples, Dierkes et al. (2019) devised a technique that combines PLE with Py-GC/MS. A pre-extraction process with methanol is used to decrease matrix effects before a PLE with tetrahydrofuran is used (THF). LOQs as low as 0.007 mg/g have been obtained for the most commonly used synthetic polymers (PE, PP, and PS). For the detection and enumeration of PP, PET, PS, PVC, PE, PMMA, and PC in biosolids, Okoffo et al. coupled PLE (using dichloromethane, DCM) with "double-shot" Py-GC/MS (treated sewage sludge). For MP analysis in complex organic-rich samples, thermal desorption of possibly interfering coextracted chemicals before pyrolysis has been shown to be particularly effective. The method's validation demonstrated a linear range of polymer absolute between 0.01 μg and 2 μg, with MP pollution of biosolids ranging from 0.1 mg/g to 4.1 mg/g dry weight across samples. Moreover, the use of 1,2,4-trichlorobenzene (TCB) for dissolving PE, PP, and PS in soil for Py-GC/MS analysis has been shown, with the technique detection limits being 1–86 g/g, while the instrumental detection limits are 186 ng absolute (Steinmetz et al., 2020). A combination of solvent extraction (using DCM) followed by gel permeation chromatography (GPC) to distinguish higher and lower molecular weight fractions was described by Ceccarini et al. (2018) as a method for characterising MPs and their degradation products in coastal sediments. In one kilogramme of sand, the researchers discovered up to 30 mg MPs.

Cloud point extraction in conjunction with Py-GC/MS has recently been proved to be effective for the study of nanoplastics in ambient waters. PS (about 65 nm) and PMMA (roughly 85 nm) Nanoplastics were enriched factored up to 500 utilizing Triton X-45-based CPE, without affecting their original shape or sizes (Zhou et al., 2018).

While adequate sample preparation may help Py-GC/MS analysis perform better, the accurate identification and quantification of polymers in complicated mixtures remains a difficult but necessary element. Specific pyrograms of distinctive and selective breakdown products representing various polymers may be used to identify them. The relative strength of indicator chemicals varies from polymer to polymer, which has a big impact on detection sensitivity. Ion chromatograms are often used for the identification of polymers in complicated samples because they may increase detection sensitivity (Dibke et al., 2021). The ion chromatograms show the ion current over time as calculated from mass spectrometric data for a specific fragment ion of an indicator molecule. In this scenario, choosing typical indicator products and their corresponding ions for each polymer based on their intensity and specificity becomes critical, with the latter being crucial for appropriate polymer identification and quantification (Primpke et al., 2020). PS has two preferred indicator compounds, styrene and its trimer, which are distinct in terms of specificity and quantity. The former is plentiful but generic, while the latter is the polar opposite. As a result, styrene is an excellent PS indicator chemical in matrix-free samples. It may be made from a variety of artificial polymers and natural chemicals, such as chitin, in natural matrices. As a result, in this scenario, using a less intensive styrene trimer is more dependable since its creation can be traced back to the presence of PS in the sample without a doubt (Primpke et al., 2020). Matrix interfered n-alkanes and n-alkenes are good choices for PE identification and quantification. Furthermore, when the carbon number, n-alkenes, rises, the interferences diminish dramatically. Polymer identification is often ensured by the presence of additional polymer-specific breakdown products (Primpke et al., 2020). The rider is directed to a recent article by Primpke et al. for an expanded list of indicator chemicals and related indicator ions that allow for the simultaneous identification and measurement of various kinds of plastics utilising thermal breakdown procedures (Primpke et al., 2020). The area beneath the signals of the indicator ions coincides with the mass of the polymer contained in the sample vessel, resulting in an ion chromatogram when pyrolysis is done under repeatable circumstances. This relationship is linear throughout a system-dependent concentration range and may be utilised for external polymer calibration. The implementation of an internal standard (or standards) will enhance data quality even more. Deuterated molecules (styrene (Funck et al., 2020) polystyrene (Eisentraut et al., 2018) and chlorobenzene (Sullivan et al., 2020) or a combination of 9-dodecyl-1,2,3,4,5,6,7,8-octahydro anthracene, anthracene-d10, androstane, and cholanic acid may be (Fischer et al., 2019; Gobmann et al., 2021).

Retrospective analysis of the pyrograms generated may detect new polymer indicator ions despite thermal degradation methods'destructive character, which prevents any further observations. Even semiquantitative information on these novel polymers may be gained by using internal standards (Primpke et al., 2020). The Py-GC/MS data for complex environmental samples like freshwater, marine sediments, road dust, blue mussels, and marine salts were recently analysed retrospectively by Gobmann et al. (2021) to determine the pollution with tyre wear particles, which are assumed to be the major source of habitat MPs. The authors discovered that in all samples studied, automobile tyre wear mass loads dominated truck tyre wear mass loads (ratios of car to truck tyre wear were up to 16 to 1). There was a substantial difference in TWP concentrations in road dust and thermoplastic (PE) MP (about 5 g of TWP vs 0.3 g of MP per kg road dust, dry weight), while TWP contamination was reduced or absent in samples collected further away from TWP sources. Nevertheless, thermoplastic polymers were still widely used (Gobmann et al., 2021).

The sensitivity of various polymer detection in complex mixtures is determined by indicator product relative intensities. A broad backdrop and any interference are quantified as well. Also, the solubility of the polymers directly affects the calibration range. Calibration of the soluble PS, for example, may be done down to 0.01 m. The LOD (S/N-ratio > 3) for the conspicuous, but unspecific pyrolysis product styrene monomer and the extremely specific but substantially weaker styrene-trimer, respectively, was derived from this calibration. The comparable LOQ values are 16 and 282 ng, respectively (S/N-ratio > 10). Lower calibration points for PP and PA 6 have been reported as 0.3 μg and 0.5 μg, respectively (Fischer et al., 2019). When working with solid standards, the LOQ for Py-GC/MS may be determined by the available balance and varies between 0.7 and 1 μg absolute, depending on the polymer type (Primpke et al., 2020). Consequently, depending on the polymer type and pyrolysis unit, Py-GC/MS analysis may be carried out with a LOQ of 0.01–1 μg. (Braun et al., 2020). It should be noted that the direct pyrolytic products of various polymers (e.g., PET and PC) exhibit a significant degree of variety and polarity, resulting in poor chromatography and limited sensitivity. Thermochemolysis, for example, by adding TMAH, may be used to boost the method's sensitivity for these polymers. The latter causes an ester and ether cleavage process, which is then followed by methylation. PET and PC have more specific thermochemolysis products, resulting in increased sensitivity for these polymers. Simultaneously, most other polymers' pyrolytic behaviour was unchanged. Using online pyrolytic derivatization, it is possible to effectively quantify PE, PP, PS, PET, PVC, PMMA, PC, PA 6, and methylene-diphenyldiisocyanate-PUR (Primpke et al., 2020; Fischer & Scholz-Böttcher, 2017). The initial sample volume needs to be adjusted for MP quantification with Py-GC-MS using the preconcentration process, taking into consideration the pyrolyzer's maximum sample capacity (g-range), as well as the predicted content of MPs and the appropriate calibration range (s).

The use of thermoextraction and desorption (TED) GC/MS allows for the analysis of much higher sample amounts (mgrange).

(ii) *TED-GC/MS*

In the TED-GC/MS method, the material is pyrolyzed using a thermogravimetric analyzer (TGA) under inert gas (typically N2) and temperature-ramped conditions up to 600 °C. The breakdown products are purged from the TGA and transported to a solid-phase adsorber bar (containing, for example, polydimethylsiloxane, PDMS), which is only linked to the decomposition product gas flow in a certain temperature range (s). The trapped gases' temperature range may be set ahead of time, for example, 25–650 °C or 350–600 °C. All volatile pyrolysis products are represented in the first range. The second is typical of most common polymers with degradation temperatures over 350 °C; however it leaves out a significant portion of pyrolysis products produced by thermo-labile organic matrix components (Primpke et al., 2020; Eisentraut et al., 2018). The adsorber is moved to a thermal desorption unit (TDU) of the GC/MS instrument after the solid phase is filled with an extract of the breakdown products (Eisentraut et al., 2018; Dümichen et al., 2019). The breakdown products are thermally desorbed and mobilised in the TDU unit and then cryo-focused in a chilled injection system, separated using a GC column, and quantified using MS (Eisentraut et al., 2018).

Dümichen et al. (2015) used the TED-GC/MS for the first time to analyse environmental samples spiked with PE and identify this polymer down to 1 wt %. Meanwhile, TEDGC/MS has been shown to be a viable method for quantifying various polymers in complicated environmental matrices. PE, PP, and PS had equivalent LOQs of roughly 10, 1, and 0.2 μg, respectively (Dümichen et al., 2019). Moreover, the use of TED-GC/MS for the study of tyre wear content in environmental samples has recently been shown. (Eisentraut et al., 2018; Klöckner et al., 2019; Klöckner et al., 2020). Amounts of styrene butadiene rubber (SBR, primary component of passenger automobile tyres) discovered in highway runoff samples ranged from 3.9 to 9.3 mg/g (Fig. 4.3; Eisentraut et al., 2018). Braun et al. (2021) recently established the usefulness of TED-GC/MS for determining the MP mass content in drinks packed in plastic bottles. The scientists created a smart filter crucible as a sampling and identification instrument, allowing MPs to be filtered down to 5 μm. MP levels as low as 0.01 g/L and as high as 2 g/L were recorded, depending on the kind of beverage bottle. TED-GC/MS has a far bigger sample capacity than Py-GC/MS, up to 100 mg (approximately 200 times more than Py-GC/MS). This is beneficial for both sensitivity (albeit the LOD and LOQ for TEDGC/MS are lower than for Pyr-GC/MS) and representativity of environmental sample analysis. As a result, MP investigation of highly contaminated samples (containing more than 0.5%–1% wt % of each kind of polymer examined) may be accomplished without sample pretreatment, which might be either insufficient or difficult (e.g., for PET and PA) (Castelvetro et al., 2021). However, when using the complete temperature range of 25–650 °C, sample-dependent organic matrix may still interfere with polymer quantification, and an adsorption cutoff below 350 °C leads to losses of highly thermolabile polymers like PVC (Primpke et al., 2020).

(b) *Further Thermoanalytical Methods*

Materić et al. (2020) have suggested a novel approach for chemical characterisation of NMPs based on thermal desorption proton transfer reaction-mass spectrometry (TD-PTR/MS, where hydronium ions produced from water vapour are employed for analyte ionisation). As of now, the approach is extensively employed in environmental studies, including real-time monitoring of volatile organic compounds, semivolatiles, and organic aerosols in the air and dissolved organic matter (DOM) in water and ice (Materić et al., 2020; Peacock et al., 2018). The studies revealed a LOD of <1 ng for PS compounds found in a sample and used this approach to (semi)quantify NMPs in Alpine snow. Because of the method's great sensitivity, it was possible to utilise tiny amounts of samples (1 mL) and conduct tests without any preconcentration processes. Even when samples include mixes of other organic compounds, unique characteristics in the high-resolution mass spectrum of distinct synthetic polymers were shown to be acceptable for fingerprinting, e.g., a valid fingerprint was obtained when just 10 ng of PS was contained inside the DOM of snow. The PET, PVC, and polypropylene carbonate (PPC) were discovered in melded cores, but only PET was found after 0.2 μm filtering, showing that PET

Peak Assignment	Structure	Substance	Characteristic fragment ions	Retention time in min	Parental Elastomer
S	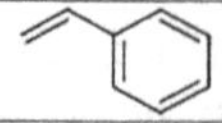	Styrene	104, 78, 51	7.2	SBR
MeS	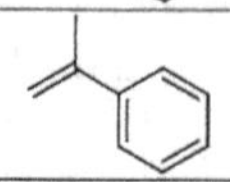	Methylstyrene	118, 103, 78	9.6	
	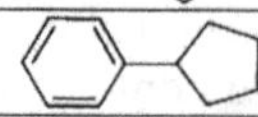	Cyclopentylbenzene	117, 104, 146, 91	16.6	
	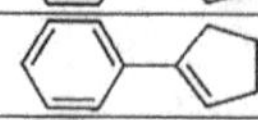	Cyclopentenylbenzene	144, 129, 115	17.8	
SB	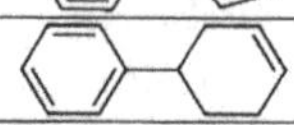	Cyclohexenylbenzene	104, 158, 129, 115	19.6	
SBB	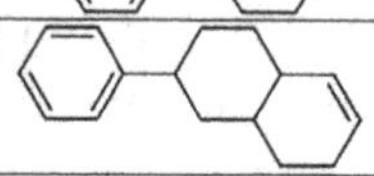	Phenyl-[4.4.0]bicyclodecene	104, 91, 156, 212	26.8	
B2	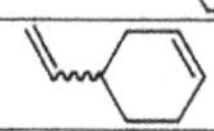	Vinylcyclohexene	79, 54, 93, 108	5.6-6.4	SBR, BR
B3	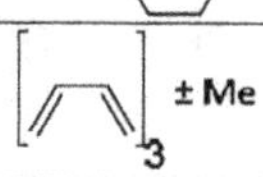	Trimers of Butadiene & Homologues	91, 148, 162, 176	13.1-21.7	BR
I2	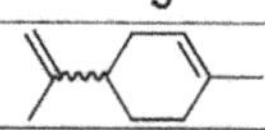	Dipentene	68, 93, 121, 136	11.0	NR
I3	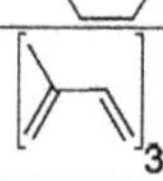	Trimers of Isoprene	119, 162, 189, 204	21.5-23.3	
I4	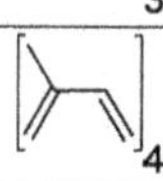	Tetramers of Isoprene	121, 93, 134, 272	31.0-36.0	

Fig. 4.3 TED-GC-MS discovered breakdown products for elastomers and tyres. (Adapted with permission from Eisentraut et al., 2018)

fibres are the most common component of airborne pollution. Despite the low recovery rates for PS (Materić et al., 2020) and the need to account for interference from even modest contaminants emanating from various sources, the TD-PTR/MS approach seems to offer promise for sensitive NMP analysis. TGA-MS may also be used to provide a quantitative investigation of MP in complicated matrices. David et al. devised a non-pretreatment technique for quantifying PET in soil samples. The mass loss and MS signal intensity of typical PET pyrolysis products were measured, while sample mixes (ca. 50 mg) were pyrolyzed at a 5 °C/min ramp (40–1000 °C). The LOD and LOQ, respectively, were 0.07 and 1.72 wt % PET (David et al., 2018). TGAFTIR may also be used to do spectroscopic examination of the gas that is produced during TGA (Dittmann et al., 2018; Becker et al., 2020).

TGA and differential scanning calorimetry (DSC) are two thermoanalytical methods for determining MP content in complex materials (Braun et al., 2020; Castañeda et al., 2014), where endothermic phase transition temperatures may be used for polymer identification and quantification. This approach can be used to test polymers having crystalline components (PE, PP, PA, and PET), but it cannot be used to study polymers without crystalline components (e.g., PS). Majewsky et al. (2016) investigated endothermic phase transition heat fluxes and peak temperatures of LDPE, PP, PET, PES, and PA by heating them from 20 to 800 °C at a rate of 5 °C/min under N2. According to the literature, LDPE, PP, and PA have low melting points, peaking at roughly 101 ± 2 °C, 164 ± 1 °C, and 216 °C, respectively, with no overlap with the other examined plastic kinds. The other polymers' peak temperatures range from 250 to 261 °C, and they generally overlap. The authors concentrated on the determination of PE and PP since unambiguous polymer identification in the presence of numerous polymers with phase transition temperatures >200 °C is difficult. Using specific polymer combinations and a total sample weight of 10 mg, they reported LOD of 2.5 and 5 wt.% for PE and PP, respectively.

Rodríguez Chialanza et al. (2018) studied at the performance of DSC for the analysis of LDPE, HDPE, PP, and PET, as well as the effect of particle size on the DSC signal for polymer mixtures. They employed size fractions of 23–256, 256–645, and 645–1000 μm and discovered that using a 10 °C/min heating rate, the signals of four polymers were easily distinct. However, particle size had a significant impact on both polymer identification and mass quantification. As a result, the authors advocated correct sample treatment, including sieving of suspended particles for MP measurement using DSC, and evaluated this method on seawater samples spiked with polymers (Rodríguez Chialanza et al., 2018).

Bitter and Lackner recently published an expanded research for quantifying semicrystalline MP in industrial wastewaters (Bitter & Lackner, 2020). They were able to analyse the samples treated with H_2O_2 in the size ranges for small (10–1000 μm) and large (1000–5000 μm) MPs by using a modified DSC protocol proposed by Majewsky et al. (2016) (which included three steps in a N2 atmosphere: first heat-up step from 30 to 290 °C at 20 °C/min heating rate, subsequent cooling step from 290 to 0 °C at 10 °C/min cooling rate). The most prevalent polymers were PE and PP, although PA and PET were also discovered. Low mass concentrations of MPs ranging from 0.5 to 35.5 μg/L were identified at all three industrial locations, which are equivalent to the amount of organic micropollutants in municipal WWTP effluents. The removal capacity of one example industrial WWTP was found to be >99.99 percent when both influent and effluent were analysed (Bitter & Lackner, 2020).

4.3 Particle-Based Methods for Nondestructive Analysis of Microplastics

4.3.1 Vibrational Spectroscopy

It is possible to examine micro- and nanoplastic particles using infrared (IR) and Raman spectroscopy, both of which use radiation interaction with molecular vibrations. FTIR spectroscopy and Raman microspectroscopy are now frequently used because they allow for the identification of polymer type as well as the number, size/ size distribution, and form of particles. The presence of plastic particles and fibres in the sample preparation and detection procedures necessitates plastic-free (or limited) working environments (Braun et al., 2020; Koelmans et al., 2019). Plastic products should be avoided throughout the procedure, and samples should be prepared in a (MP)-particle/fibre-free or MP-poor environment. In addition, protocol and experimental negative controls must be determined. Pollutions while obtaining sample, preparation of sample, and identification are all taken into account by procedural blanks values. When case preparation and detection are done in different labs, laboratory blank values become essential. These are used to identify internal MP pollutants and assist in locating and removing their origins. LOD and LOQ data for the laboratory may be derived using lab blank values (Johnson et al., 2020). At least three (ideally ten) laboratory blank values are suggested for the LOD and LOQ determinations.

(a) *IR Spectroscopy*

The detection of MPs from marine specimens was achieved using FTIR spectroscopy in a work by Thompson et al. (2004), which was the first time the term "microplastics" was used. Since then, this method has proven to be effective for the identification, quantification, and characterization of MP pollutions in aquatic habitats (Cincinelli et al., 2017; Napper et al., 2020; Piehl et al., 2018), as well as in influents, effluents, and sludge of WWTPs (Horton et al., 2021) and ambient air (Catarino et al., 2018; Trainic et al., 2020; Pivokonsky et al., 2018; Johnson et al., 2020). IR spectroscopy is a nondestructive method that analyses molecular vibrations caused by the absorption of light in the mid-infrared (MIR) region of the electromagnetic spectrum (4000–400 cm^{-1}). Using spectrum databases (Cowger et al., 2020) or other chemometric approaches, the distinctive vibrational fingerprint spectra may be used to accurately identify the polymer type for MP as well as to assign nonplastic particles (Renner et al., 2017; Hufnagl et al., 2019). Water has extremely strong and wide IR bands, which may partly or totally overlap the spectral signature of plastic and nonplastic particles of interest; hence the samples must be dried before examination. The method's vulnerability to water is seen to be its most serious flaw. IR spectroscopy outperforms many other approaches due to the wide range of measuring options available. IR analysis may be carried out in either reflectance or transmission mode (Löder et al., 2015). IR radiation that has entered the sample is measured in transmission mode. Infrared transparent substrates or filters (e.g., aluminium oxide (Anodisc) membranes), (Primpke et al., 2018; Löder et al., 2015), silicon filters, (Käppler et al., 2016), or zinc selenide windows are necessary for this mode. Despite the fact that high-quality data is often collected and the resultant spectrum is representative for the full sample thickness (or complete particle) and hence advantageous for MP identification, this mode may be influenced by total absorption (Primpke et al., 2020). The IR beam may be partly or fully obstructed for colourful, dark, or opaque particles, resulting in low-quality spectra. This disadvantage may be overcome by using the reflectance mode, which measures the IR beam reflected by the sample (Cabernard et al., 2018; Tagg et al.,

2020). Reflective surfaces, such as metal-coated (Au, Ag, Al) (Horton et al., 2021; Cabernard et al., 2018), are necessary for this sort of study. Although this mode is excellent for analysing the (aged) surface of a sample or particle, light scattering might cause the data to be distorted. Attenuated total reflection (ATR) may be used to analyse MPs efficiently. This approach has been adopted in 58% of IR research, particularly for bigger particles, since it is the most cost-effective (Primpke et al., 2020). It also doesn't need any sample preparation or complicated mathematical adjustments (which are required for transmission and (pure and diffuse) reflection modes, respectively). An ATR crystal with a high refractive index is pushed onto the sample surface for the measurement (e.g., diamond, zinc selenide, or germanium). The IR light penetrates the sample to a depth of a few micrometres (evanescent wave) after reflection at the crystal/sample contact, and the sample's IR data is acquired. ATR-FTIR is often utilised for the identification of visibly presorted particles (sizes bigger than 200–500 μm) and the characterisation of weathered MPs because information on changing particle surfaces owing to ageing may be readily collected (Primpke et al., 2020; Cabernard et al., 2018). ATR-FTIR may also be used to differentiate between natural and synthetic (micro) fibres (Dris et al., 2018). Tiny particles and fibres may be analysed directly on filters or windows using μ-ATR objective that comes into touch with the sample (Vianello et al., 2013). However, the sample might be damaged or destroyed as a result of the applied pressure, which is needed to produce the essential contact between the crystal and particle surface. Furthermore, the close contact between crystal and stiff inorganic particles (such as MPs) may cause costly μ-ATR instruments to be damaged. Furthermore, since particles must be examined one by one, the ATR-FTIR analysis takes a long time (Ivleva et al., 2017; Braun et al., 2020).

The most widely used FTIR-based approach for analysing (MP) particles is 500 mis micro-FTIR spectroscopy (μ-FTIR), in which an FTIR spectrometer is linked to an optical microscope. The spatial resolution of the study is restricted by diffraction (theoretically ca. 1.7 μm at 4000 cm^{-1} to 13 μm at 500 cm^{-1}), although particles bigger than 10 μm (Cabernard et al., 2018) or 20 μm may be easily recognised and quantified by μ-FTIR. The removal of inorganic and organic matrices is critical for IR analysis, which includes tiny particles. Density separation (Pico et al., 2019; Imhof et al., 2012) and chemical (Pico et al., 2019) or enzymatic (Löder et al., 2017) digestion are often used for this purpose.

μ-FTIR analysis may be carried out (i) for pre-selected particles or (ii) for the full filter area. Prior to IR observations, particles may be manually selected (Cunningham et al., 2020; Ziajahromi et al., 2017) or automatically selected (Brandt et al., 2020) using optical images. For Raman microspectroscopic investigation of MP particles, the preselection technique is often used. Furthermore, using staining methods for preselection and subsequent chemical identification by IR has been demonstrated to improve the identification rate and eliminate researcher bias (Shim et al., 2016). The removal of organic matrix before staining has been advised to eliminate or decrease "false-positive" identification caused by costaining of part of the natural organic material (Zarfl, 2019; Shim et al., 2016).

FTIR imaging (chemical) measures all particles in the studied region, allowing for a more detailed examination of the chemical composition of overlapping and agglomerated particles than the particle preselection option. The number of spectra that must be measured and processed, on the other hand, is much larger (Primpke et al., 2020). Chemical imaging by mercury cadmium telluride (MCT) detectors is conceivable; however measuring vast regions takes a long time. As a result, spectra are often obtained from filter subareas, such as 0.17% (3 mm^2 areas on 47 mm diameter PC membranes) (Harrison et al., 2012) or 5.6% (12 sampling unit areas of 4.5 mm^2 each on 47 mm diameter fibreglass filters) (Vianello et al., 2013). According to Johnson et al. (2020) and Horton et al. (2021), 92% of the filtration area of 11.6 mm × 11.6 mm can be studied by optimising a measuring methodology.

Recent technological developments in μ-FTIR analysis have resulted in multiple high-detail investigations on MP contamination of various ecosystems, waste management systems, (Tagg et al., 2020), and drinking water (Lorenz et al., 2019; Johnson et al., 2020). Aluminium oxide (Anodisc) filters are often employed for transmission mode measurements in FPA μ-FTIR for high-throughput analysis. Silicon filter substrates (e.g., with a pore size of 10 μm) that offer adequate transparency for the wide mid-infrared region of 4000–600 cm^{-1} may be utilised if a greater spectral range has to be monitored (Käppler et al., 2016). Au-coated PC filters were found to be adequate for measurements in reflection mode (Cabernard et al., 2018). Complementary Raman analysis is possible with both kinds of filters (Si and Au-coated PC) (Cabernard et al., 2018; Von der Esch et al., 2020). Large data sets occur from imaging-based analysis, particularly when the (automatic) FPA-FTIR option is used. These data sets must be processed to provide information on particle identification as well as other attributes (particle number, size, and shape), which are essential for the comprehensive quantitative analysis. As a result, automated data analysis procedures, such as spectrum preprocessing (baseline correction, smoothing, and so on) and assessment, are essential (Primpke et al., 2020; Renner et al., 2019). Library search is often used for spectra assignment, with search algorithms employed to build a hit quality index (Renner et al., 2019) (HQI). The HQI is a comparison of the query and reference spectrums. Despite the fact that various HQI levels (e.g., 0.7) (Yang et al., 2015) have been recommended as a threshold in different research, the HQI values created using different methods and software may not be comparable. These numbers are also heavily influenced by the database's spectral quality and sample type. It's also worth noting that reference spectra taken with various kinds of apparatus (ATR and μ-ATR, (FPA) μ-FTIR in transmittance or reflectance, detector type) and with various parameter settings (number of scans, spectral resolution) and spectral range might vary greatly (Andrade et al., 2020). As a result, the database(s) in use must be customised for specific applications, and the HQI index level utilised must be verified. Furthermore, databases must include not only the reference spectra of pure synthetic polymers (as in commercial libraries) but also the spectra of various plastic products (including additives) and weathered plastic particles (as in custom-made libraries) (Cowger et al., 2020).

Model-based classification for the automated assessment of FTIR imaging data, where labelled training data are used to predict the class affiliations of unknown data, seems to be a promising alternative to the conventional instance-based spectrum library search (Hufnagl et al., 2019; Xu et al., 2019). The main difference between traditional database searches and model-based categorization is that instead of utilising reference data to determine class membership, the latter employs a multivariate model of the actual data. Hufnagl et al. (2019) introduced a system that uses random decision forest (RDF) classifiers to discriminate between distinct polymer types and assess their abundance and size distributions with good accuracy. The approach was used to identify five different polymer kinds (i.e., PE, PP, PMMA, PS, and polyacrylonitrile, PAN). The expanded RDF technique was recently utilised to effectively identify 11 polymer classes in mussel samples evaluated by FPA-μFTIR imaging (Kumar et al., 2021). Da Silva et al. (2020) devised a model-based technique for the assessment of FPA-μFTIR hyperspectral imaging data using partial least-squares discriminant analysis (PLS-DA) and soft independent modelling of class analogy (SIMCA) models. The method worked well for classifying and quantifying MPs of <100 μm in nine of the most commonly manufactured polymers in the globe (PA, PC, PE, PET, PMMA, PP, PS, PU, PVC). PLS-DA had greater analytical performance than SIMCA models, according to the authors, and was characterised by higher sensitivity, sensibility, and reduced misclassification error. PLS-DA, on the other hand, was less affected by spectral edge effects and poorly focused particle areas (da Silva et al., 2020). It should be highlighted that the creation of classifiers (training data sets) takes time and needs expert operators; also, more work is needed to expand the number of polymer kinds (and include nonplastic analytes). However, given the fast development of hyperspectral imaging technologies, model-based approaches are becoming more appealing because they can reliably evaluate large data sets that often include spectra with poor signal-to-noise ratios.

An exploratory investigation of FPA-μFTIR imaging data acquired from environmental microplastic samples was recently published. The multivariate similarity of spectra is used in this method to identify species or particles without the need for previous information. The dimensionality reduction using PCA and uniform manifold approximation and projection (UMAP) were used as a key idea, which increased data visual accessibility and provided a chemical two-dimensional picture of the sample. Particle spectra were isolated from blank spectra (substantially lowering the quantity of data) and analysed using PCA and UMAP. Cluster analysis utilising k-means and density-based and interactive manual clustering revealed groups of similar spectra, which were then assigned to chemical species based on reference spectra. While the acquired findings were in excellent agreement with a focused study based on automated library search, exploratory analysis highlights a set of unidentified spectra that persisted and would otherwise be disregarded (Wander et al., 2020).

Aside from FPA-μFTIR systems, Scircle et al. recently stated that an alternate technique, laser direct infrared (LDIR) analysis, seems to have a great potential for the quick and automated detection and quantification of MP particles (Scircle et al.,

2020). In the aquatic environment and soil (Scircle et al., 2020; Mughini-Gras et al., 2021; Hildebrandt et al., 2020), LDIR has been used to analyse MP particles >20 μm (Scircle et al., 2020; Mughini-Gras et al., 2021; Hildebrandt et al., 2020). (Ng et al., 2021; Li et al., 2021). The light source, a customised quantum cascade laser, is the most innovative feature (QCL). The quantum cascade laser (QCL) is a semiconductor-based laser in which electrons cascade (tunnel) through a succession of quantum wells produced by thin semiconductor layers. The thickness and distribution of semiconductor layers, not the semiconductor materials, influence the wavelength of photons. The LDIR systems have been used to examine particles >20 μm, yet it is believed that in the automated mode, the size limit for studied particles might be reduced to about 10 μm (Hildebrandt et al., 2020). However, it has been noted that for tiny particles (<30 μm), the system may need to automatically refocus in order to get the best spectrum. The per-particle analysis time in this example may be as long as 8 s. The use of this approach for MP analysis of water samples revealed that LDIR detects more particles than the fluorescence-based method (Nile Red staining) (Scircle et al., 2020), albeit a more extensive comparison will be required in the future to confirm this trend. In addition, the performance of LDIR in contrast to FTIR-based approaches has yet to be determined.

(b) *Near-IR Spectroscopy*

Apart from the mid-infrared region (MIR) (4000–400 cm^{-1}) area of fundamental molecular vibrations, which is most often employed for MP identification, quantification, and characterisation, the near-infrared (NIR) region (12800–4000 cm^{-1} or 780–2500 nm) may also be used, despite the fact that NIR spectroscopy was already utilised for decades for online food quality verification and online plastic packaging sorting in recycling (Braun et al., 2020; Moroni et al., 2015). This technique has recently been identified for MP testing in various environmental materials, such as seawater (Karlsson et al., 2016; Shan et al., 2019) and surface water (Schmidt et al., 2018), biota (Zhang et al., 2019), and soil (Paul et al., 2019) has been recognized.

NIR spectra are defined by vast overlapping bands of overtone and combination vibrations for a small number of chemical vibrations, commonly of type X-H, e.g., C-H, O-H, and N-H. For NIR applications, automated statistical approaches from the area of chemometrics, as well as relevant databases, are needed. In comparison to MIR, however, using the NIR area for MP analysis has significant benefits. NIR radiation may penetrate deeper than MIR because higher overtones have lower absorption coefficients than fundamental vibrations, allowing it to handle greater sample volumes and providing fingerprints. Furthermore, the NIR area has a decreased sensitivity to water and pollutants biofilms.

In addition, the ability to use quartz materials for fibres and optical elements in NIR spectroscopy (Paul et al., 2019) leads to a wide range of instrumentation arrangements, ranging from hand-held spectrometers appropriate for in-field investigation (Crocombe, 2018) to laboratory equipment commonly used for hyperspectral imaging, as explained below. The applied equipment and analysed materials have a substantial influence on the spectral range employed, and MP particles have a smaller size limit. The lower spatial resolution of NIR analysis compared to MIR might be explained by the larger sample volume required to provide acceptable signal for weak overtone and combined vibrations. For hyperspectral investigation of MP pollution in seawater filtrates, Karlsson et al. (2016) examined three diverse imaging systems with wavelength ranges of 375–970, 960–1662, and 1000–2500 nm. They discovered that the wavelength range 1000–2500 nm, along with the PCA model technique, is the best suitable for this sort of sample, allowing them to analyse preselected MP particles down to 300 μm. (Karlsson et al., 2016). Schmidt et al. (2018) described a semiautomated approach for detecting MP particles bigger than 450 μms in surface water samples filtered via glass fibre filters. 10 complete filters with a diameter of 47 mm could be scanned in around 20 min (measurement speed, 52048 mm^2 per hour). Counting MP particles, classifying plastic kinds, and estimating particle sizes are all possible using hyperspectral pictures with a pixel size of $280 \times 280\ \mu m^2$ and a spectral signature consisting of 256 spectral bands within the wavelength range of 968–2498 nm. Schmidt et al. (2018) and Paul et al. (2019) used NIR analysis in conjunction with chemometrics models such as support vector machine regression (SVR) and PLS-DA to achieve high-throughput MP identification in soil. Artificial MP/soil mixes comprising prescribed ratios of PE, PET, PP, and PS < 125 μm were utilised for calibration. Without any chemical pretreatment, accurate detection and categorization of MP at levels exceeding 0.5 to 1.0 wt%, depending on the polymer, has been shown (Paul et al., 2019). Zhang et al. (2019) showed the potential of analysing MP particles in fish digestive tracts quickly and efficiently without using any reagents (reagent-free). For the detection, identification, and characterisation of five kinds of MPs > 200 μm, the scientists used HSI in conjunction with a support vector machine classification model (Zhang et al., 2019). As a result, NIR-based technologies, particularly when combined with HSI and chemometric methods, may be highly effective for detecting MP contamination without the need for sample preparation. NIR-based monitoring might be utilised as a first step in MP prescreening (e.g., using the traffic-light approach) before doing a full study of particles <500 μm for all or just suspect samples using μ-(FT)IR or μ-Raman spectroscopy.

(c) *Raman Spectroscopy*

Raman spectroscopy is a nondestructive analytical approach that is becoming more popular, particularly for the study of tiny microplastics in a variety of environmental materials, including marine and freshwater (Cabernard et al., 2018; Trainic et al., 2020), sediments (Imhof et al., 2016; Enders et al., 2019), biota (Missawi et al., 2020; Collard et al., 2015), compost (El Hayany et al., 2020), and ambient particulate matter (Trainic et al., 2020; Levermore et al., 2020), and also in urban wastewater treatment plant effluent (Wolff et al., 2019) and in drinking (tap and bottled) water, (Weber et al., 2021; Shruti et al., 2020) beverages, and food (Karami et al., 2017). Using handmade and commercial spectrum databases, it is possible to properly identify plastic particles and certain additives and also other (in)organic and (micro)biological chemicals (Enders et al., 2015). Raman spectroscopy can analyse MP particles as well as synthetic and natural fibres (Remy et al., 2015; Wiesheu et al., 2016).

By combining Raman spectroscopy with confocal optical microscopy and using visible excitation lasers, the spatial resolution may be improved to 1 μm and even lower (~300 nm). As a result, Raman spectroscopy is recommended for examining plastic particles with sizes of 10–20 μm (Anger et al., 2018).

Fluorescence intervention caused by inorganic (clay minerals, dust particles) and organic (humic compounds) contaminants in the matrix (Ivleva et al., 2017; Anger et al., 2018), as well as some additives (pigments), is a main drawback of Raman spectroscopy, particularly when analysing MPs in habitat samples (Araujo et al., 2018; Lenz et al., 2015). Before Raman analysis, inorganic and organic nonplastic particles (Enders et al., 2020) must frequently be removed by density separation (Imhof et al., 2012; Coppock et al., 2017) and chemical (Ivleva et al., 2017; Al-Azzawi et al., 2020) or enzymatic (Löder et al., 2017). The removal of the matrix will also result in a large rise in the plastic/nonplastic particle ratio, which will enhance the MP analyses' representativity and statistical certainty. Additionally, agglomeration and overlapping of MP with natural particles may be reduced, resulting in an over- or underestimate of particle number and size (Primpke et al., 2020). To limit or prevent interferences generated by intense fluorescence, it's also crucial to choose the right measurement settings (laser wavelength and power, photo bleaching, and acquisition time, as well as objective magnification and confocal mode). It's critical to choose the right laser wavelength (Anger et al., 2018).

In general, excessively high laser power (e.g., more than 10 mW for 532 nm) should be avoided since it might induce the thermal disintegration of plastic particles and, more often, organic contaminants, as well as the formation of characteristic soot bands in Raman spectra. Additionally, photobleaching before or during Raman measurements (by using longer collection periods) may be highly effective for reducing fluorescence and therefore improving the signal-to-noise ratio. Longer acquisition durations, in particular, may assist in the accurate detection of coloured plastic and paint particles (Anger et al., 2018). The latter include pigments, film formers, curing coating systems, and physically drying systems (acryl and vinyl(co) polymers) that are generated from surface coatings (such as paints) (Hartmann et al., 2019). Because of the comparatively high (pre)resonant Raman signals of pigments (e.g., Cu phthalocyanine), spectra recorded at little acquisition periods (about 1 s) might be misattributed to paint particles, but extending acquisition time can assist to get more Raman signals of polymers. Pigments (in the absence of a strong fluorescence signal) do not generally obstruct the detection of the polymer type of MPs since pigments and polymers usually have crisp signals.

(d) *Nonconventional Raman Techniques*

Nonlinear Raman methods like coherent anti-Stokes Raman scattering (CARS) and stimulated Raman scattering may increase the sensitivity of Raman analysis (SRS). Only the molecular vibrational modes of curiosity give a significant signal in CARS and SRS. Thus, if fluorescing pollutants emit no light in the frequency range of interest, by removing the (in)organic and biological matrix, the fluorescence problem may be completely eliminated.

Cole et al. (2013) were the first to demonstrate the use of CARS for detecting and photographing MPs consumed by zooplankton. The scientists employed Raman bands at 2845 and 3050 cm^{-1} (aliphatic and aromatic C-H str., respectively) to see PS beads (0.4–3.8 μm) in 2D pictures scanned from 2775 to 3100 cm^{-1}. Meanwhile, CARS has been used to visualise 8 μm amino-coated and carboxylated PS beads in shore crab gills (Watts et al., 2016). CARS can therefore do 2D analysis of microscopic microplastics and even nanoplastics down to 80 nm in environmental samples. The CARS application, on the other hand, requires a complicated and costly instrumental setup as well as user skill. Furthermore, CARS may be impacted by an electronic, nonchemically specific background, like those created by solvents, making interpretation difficult (Riberio et al., 2017; Goodhead et al., 2015).

SRS microscopy is the next potential method for imaging MPs at a high speed. SRS microscopy is based on the coherent interaction of two laser beams with vibrational levels of the sample molecules. The SRS signal is created when the photon energy difference between the beams matches a molecule's vibrational state. Normally, the modulation transfer imposed on the other beam is identified by amplitude modulating one of the beams before the sample. The resultant SRS signature at various wavenumbers mimics the target analyte's spontaneous Raman spectrum (Zada et al., 2018). Zada et al. (2018) showed that this method may be used to image MPs from five different polymers, including nylon, PET, PS, PP, and PE. For the examination of particles with a spatial resolution limit of 12 μm, the spectral range from 950 to 1850 cm^{-1} was employed, and 1 cm^2 of the filter was scanned in less than 5 h (Zada et al., 2018). Laptenok et al. (2020) recently shown the usefulness of SRS for determining natural and manufactured microfibres from environmental samples (i.e., fish gastrointestinal tract, deep-sea and coastal sediments, surface seawater, and drinking water). The majority of the studied ambient fibres are of natural origin, as predicted.

As proven by Liao et al. (2017), a fibre delivered hand-held SRS microscope may provide quick in situ imaging of MP. By temporally splitting the two ultrafast pulses travelling through the fibre and then overlaying them on a sample via a highly dispersive material, a stimulated Raman signal of PS and PMMA beads (both 5 μm in diameter) has been recorded (e.g., paper). The described system, which enables for imaging in the areas 2800–3100 cm^{-1} (CH str. vibrations) and 1550–1800 cm^{-1} with a spatial resolution of 1.4 μm, seems to be extremely promising for chemical investigation of plastic and nonplastic microparticles.

According to Zhang et al. (2017), SRS offers the potential for high-throughput single particle analysis. The newly established 32-channel multiplex stimulated Raman scattering flow cytometry (SRS-FC) technology enables for the chemical analysis of single particles (e.g., 10 μm PS and PMMA beads, and polycaprolactone, PCL) at a rate of 5 s per Raman spectra. At 0.4 m/s flow speed and a throughput of up to 11,000 particles per second, the spectral range from roughly 2800 to 3100 cm^{-1} (CH str. vibrations) was employed to discriminate between distinct particles in suspension.

Thus, SRS-based approaches offer a lot of promise for quick and sensitive MP study, but the complicated arrangement and the essential for a lot of user experience (like to CARS) are still restrictive issues for SRS's widespread application in MP investigations.

(e) *Combination of (FT) IR and Raman Analysis*

The complementary nature of (FT)IR and Raman analysis should not be overlooked. If the absorption of IR radiation results in a change in the dipole moment of the molecule during the vibration process, the molecular vibrations are said to be IR active. If the polarizability of the whole molecule's atomic electron shell changes, the vibrations become Raman active. IR and Raman spectroscopy produce various spectra with regard to vibration activities and intensities for diverse functional groups due to different selection procedures. As a result, further information on polymers and (in)organic additives may be gathered (Käppler et al., 2016; Xu et al., 2019).

Käppler et al. (2016) who analysed habitat materials using both Raman and FTIR spectroscopy offered a rigorous comparison and validation of both spectroscopic approaches with regard to MPs. The authors conclude that both approaches are acceptable for detecting MP particles in the environment in theory. However, in other situations, particularly for coloured particles, a combination of both spectroscopic approaches was required for comprehensive and consistent chemical composition determination. While acrylic resin can be identified better using FTIR spectroscopy and also characterization of particles with a high fluorescence background, μ-Raman spectroscopy can offer comprehensive pigment information. Furthermore, the scientists discovered a substantial underestimating (approximately 35%) of MP by FTIR imaging compared to Raman for particles placed on Si filter substrate (fraction <400 μm), notably in the size range <20 μm. Raman imaging, however, has shown to be much more time-consuming. As a result, the authors recommended size split of samples into two fractions at 50 μm and the use of quick FTIR imaging for particle analysis on filters (MP < 500 μm) (Käppler et al., 2016).

Kumar et al. (2021) recently utilised the suggested size split at 50 μm to the investigation of MP down to a size of 3 μm in economically relevant mussels (Kumar et al., 2021). The number of MP particles per sample discovered using FPA-FTIR imaging in the size fraction >50 m ranged from 0.13 to 2.45/g wet weight (g ww) of mussel samples, with an average of 0.63 ± 0.59 MP particles/g ww. PP (39% ± 6.3%), PET (32% ± 2.8%), PAN (8.2% ± 1.4%), and PE (7.2% ± 0.6%) were the most prevalent synthetic polymer types found. PA (40.2%), PP (16.5%), PE (14.6%), and PAN were the most prevalent synthetic polymer types in the fraction <50 μm, where 211 MP particles were discovered by Raman spectroscopy (13.2%). The findings imply that various polymer types may dominate different size fractions of MP particles, such as PP and PET or PA for particle fractions >50 μm and <50 μm, respectively (Kumar et al., 2021).

Cabernard et al. (2018) evaluated the quantification of MP particles from the aquatic environment deposited on Au-coated PC filters using FPA μ-FTIR (reflection mode) and μ-Raman coupled with automatic particle identification (Cabernard et al., 2018). They discovered that for MPs ≤ 500 μm, μ-Raman analysis quantified two times greater MP counts but took four times as long as FTIR imaging.

Furthermore, compared to the ten polymer types recognised by FTIR imaging, the μ-Raman technique allowed the identification of 19 distinct polymer kinds. Based on these findings, the authors believe that the ambient concentration of MPs ≤ 500 μm has been underestimated up to this point, which they ascribe to the unusual rise in concentration with declining MP size (Cabernard et al., 2018).

The findings show that using a combination of (FT)IR and Raman analysis to analyse MPs may yield complimentary results and allow for accurate size easily resolved chemical analysis. The evaluation of MP-related threats to the environment and human health requires extensive and reliable information on MP contamination. Small MP particles must be identified and quantified with special care, since their quantity is unknown or most likely overestimated, despite the fact that this MP fraction is the most important in terms of ecotoxicity (Cabernard et al., 2018).

Until recently, the only way to identify and measure MP fractions smaller than 10 μm was to use μ-Raman spectroscopy. The development of optical photothermal (O-PT) IR spectroscopy, on the other hand, means that noncontact IR analysis with submicrometre resolution is now possible (Hale et al., 2020). Probes for visible lasers are used to measure the photothermal response of particles that have been absorbed by a pulsed laser in the MIR range (532 nm). Furthermore, the setups allow for simultaneous IR and Raman investigation at the same location and with the same spatial resolution by detecting inelastic light scattering induced by the visible probe laser (Li et al., 2019; Marcott et al., 2020). This offers up new opportunities for future submicrometre-resolution complementary IR and Raman investigation of (plastic) particles (Hale et al., 2020; Marcott et al., 2020).

4.4 Analysis of NPLs

4.4.1 Mass-Based Methods

The knowledge on the existence of NPLs and associated mass in separate size fractions (e.g., <1 μm) can be adequate, depending on the analytical query, for example, for monitoring and modelling. The Py-GC/MS technique is still the most popular (Zhou et al., 2018). The identification of nanoplastics in the North Atlantic Subtropical Gyre (NASG) was originally published by Ter Halle et al. (2017) using Py-GC/MS. A commercial polymer database was integrated with a chemometric approach that used PCA to detect polymers. The presence of PVC, PET, PS, and PE was detected in the colloidal fraction <1.2 μm after filtering, with 70, 17, 9, and 4 percent of their anthropogenic pyrolytic fingerprints, respectively. The relative abundance of PVC and PET NPLs compared to PE, PS, and PP indicates the relative abundance of PVC and PET NPLs likened to PE, PS, and PP (Ter Halle et al., 2017). An analysis of sand water extracts from cost subjected to NASG using Py-GC/MS has revealed the presence of NPLs (PS and PVC) (Davranche et al., 2020). Citing Blancho et al. (2021), finding NPLs in complex environmental matrices remains problematic due to low concentrations of NPLs compared to NOM. The authors identified PP and PS and investigated possible environmental matrices interventions by spiking NPLs in various organic matter suspensions. Two complementary approaches were devised based on plastic composition and NOM concentration. PS NPLs must be handled first, unlike PP NPLs. H_2O_2 and UV light were employed to specifically destroy NOM and not damaging NPLs for this purpose.

Mintenig et al. (2018) devised a method that combines cross-flow ultrafiltration, AF4, and Py-GC/MS to analyse NPLs in aqueous environmental samples. The scientists utilised PS NPLs (50, 100, 200, 500, and 1000 nm in size) as the model particles to spike several drinking and surface water samples and obtained LODs and LOQs ranging from 50 to 250 ng. The LOD and LOQ of 4 mg/L and 410 mg/L were determined using the provided conditions and pyrolyzed quantities of 25 L. When the initial concentration of PS in watery sample was >20 g/L, it was possible to identify it by preconcentrating NPLs using cross-flow ultrafiltration. Wahl et al. (2021) recently established the viability of coupling AF4 to Py-GC/MS for the detection of NPLs in NOM-reach environmental samples, like in soil modified by plastic trash. Prior to chemical analysis, AF4-based size fractionation of aquatic extracts (0.8 m fractions) may be used to avoid the influence of organic matter on NPL detection. For the first time, PP, PS, and PVC NPLs by diameters varying from 20 to 150 nm were discovered in soil using this method. Py-GC/ToF spectrometry can be used to do sensitive examination of distinct NPL particles with a size limit of 100 nm (Sullivan et al., 2020).

Furthermore, based on thermal desorption-proton transfer reaction-mass spectrometry (TD-PTR/MS), a potential approach for highly sensitive detection and quantification of NMPs has recently been described by Materić et al. (2020). PS has an estimated LOD of less than 1 ng, and it could be detected in complex samples down to 10 ng. Following polymer extraction and depolymerization, an alternate method based on HPLC was developed (Castelvetro et al., 2021). The method with LOD and LOQ of 15.3 and 51.1 g/L for PET has been found to be appropriate for the detection and quantification of PET and PA NMPs in complicated samples. As a result, several methods for identifying and (semi)quantifying NPLs in various environmental samples have been developed and tested. Optimisation and validation of detection methods and also efficient preconcentration and enrichment for NPLs will be required to improve analytical dependability.

4.4.2 Nondestructive Spectroscopic Methods

The diffraction limit of light limits the spatial resolution of spectroscopic methods used to study microplastics. This limit is around 10 m for (FT) IR and 300 nm for Raman, which allows for the study of (almost) the entire size range of MPs. Although μ-Raman appears to be suitable for nanoscale particle analysis, recognising particles smaller than 500–1000 μm is problematic. As a consequence, SEM and Raman spectroscopy for high-resolution images and particle identification have been created and employed to examine microscopic MPs and, more recently, NPLs (Sobhani et al., 2020; Sobhani et al., 2020). Sobhani et al. (2020) showed that Raman imaging can observe and identify NPLs down to 100 nm by differentiating the laser spot, pixel size/image resolution, NPL size/position (inside a laser spot), Raman signal strength, and sample preparation. It was used to examine dust samples collected from a driveway after a vehicle's clear polyacrylic finish was hand shined. By hand-polishing an engine hood, the scientists calculated that billions of trillions of NMPs with sizes as small as 200 nm were produced (Sobhani et al., 2020). As a result of their work, the scientists have identified NPLs of sizes between 30 and 600 nm. It is possible to visualise and observe individual nanoplastics by decreasing the mapping pixel size and offsetting the colour to capture just the high-intensity component of the Raman signal generated by the laser point. It was feasible to image particles in the 30–80 nm range, but it was difficult since the Raman signal becomes extremely faint and difficult to separate from noise. Despite this, the SEM-Raman combo has shown to have a great deal of promise. Furthermore, commercially accessible devices provide correlative Raman imaging and SEM, offering up new opportunities for optimising and detailing NPL particle morphology and chemical analyses.

A recent study by Zhang et al. (2019) employed Raman imaging and SEM to directly examine NPL release from commercially recycled plastics. Several challenges must be addressed when combining SEM with Raman, such as particle carbon accumulation and destruction by the electron beam during SEM, particle switching concerns, and the need for vacuum in the SEM chamber (Primpke et al., 2020; Cardell & Guerra, 2016). A comprehensive morphological and chemical analysis of NPLs looks promising for studying small plastic particles.

The use of surface-enhanced Raman scattering (SERS) to overcome the problem of small NPL particles with weak Raman signals has recently been researched (Lv et al., 2020; Zhou et al., 2021; Lê et al., 2021). Colloids or rough surfaces that are near to or connected to nanometre-sized metallic objects (Ag or Au) have stronger Raman emissions. Electromagnetic ("localised surface plasmon resonance, LSPR") and chemical enhancement effects may produce amplification factors of 103–1011. Lv et al. (2020) have recently shown that by utilising Ag colloid as SERS medium, the Raman signal of PS beads with diameters of 100 and 500 nm may be greatly improved (up to 5 × 102 and 4 × 104, respectively). PE and PP MPs did not get as much of a boost as PS NPLs. The authors demonstrated how SERS can be used to identify NPLs in both pure water and saltwater (Lv et al., 2020). Zhou et al. (2021) publised on SERS enrichment for PS beads with a size of 50 nm using Ag colloid and used the approach to analyse model NPLs in river water, virtually simultaneously. Lê et al. (2021) created unique nanostructured Raman substrates for sensitive identification of NPLs in water in this light. They made anisotropic nanostar dimer-embedded nanopore substrates and successfully evaluated the approach for sensitive identification of PS beads with a size of 400 nm, but no substantial improvement was shown for PS NMPs with diameters of 800 nm, 2.3 m, or 4.8 m. (Lê et al., 2021).

(a) *Scanning Probe Microscopy Coupled to Spectroscopy*

While the previously stated vibrational spectroscopic approaches have been demonstrated to be effective in identifying, quantifying, and characterising MP and NPL particles, they can't overcome the diffraction limit of light's spatial resolution (Verma, 2017). Scanning probe technologies for chemical analysis at the nanoscale (Verma, 2017; Dazzi & Prater, 2017; Xiao & Schultz, 2018) shows great potential in NPL research. AFM-IR, nano-FTIR, and tip-enhanced Raman spectroscopy are some of the nanoscale methods used (TERS). It is focused on a sample at the AFM tip and tuned to an absorption band in AFM-IR. The absorbed light causes local photothermal expansion of the material, which the AFM tip detects. AFM cantilever oscillation amplitude monitoring as a function of wavelength produces local absorption spectrum with nanoscale spatial resolution (Dazzi & Prater, 2017; Xiao & Schultz, 2018; Hermann & Gordon, 2018). Felts et al. (2012) used the AFM-IR approach to successfully identify and image polymer nanostructures at the nanometre scale. They used the total internal reflection mode to examine PE and PS nanowires placed on an IR-transparent ZnSe prism, a spatial resolution of almost 100 nm. Pancani et al. (2018) claim that AFM-IR can quickly locate and chemically characterise NPLs inside a cell without any labeling. They studied macrophages treated with PLA NPLs smaller than 200 nm, which are often employed in drug delivery.

Analysing broadband IR absorption spectra of surfaces with spatial resolution of 10–20 nm is possible using nano-FTIR and scattering-type scanning near-field optical microscopy (s-SNOM). The IR beam is attentive on the near-field probe, usually a metal coated tip, in nano-FTIR investigations, and a local antenna effect provides a nanoscaled focus with the tip's dimension. During scanning of the surface with the tip, the near-field interactions between tip and sample alter. An asymmetric Michelson interferometer is then used to monitor the ensuing variations in local scattering intensity. The sample's local IR absorption bands may be connected to the amplitude and phase of scattered light, and the resulting spectra correlate well with bulk FTIR data for a broad variety of materials (Hermann & Gordon, 2018). The nano-FTIR technique has been used to analyse NPLs with success. Brehm et al. (2006) published a paper on the detection of PMMA beads with a diameter of 30–70 nm.

Huth et al. (2012) showed that PMMA samples can be chemically analysed with a spatial precision of 20 nm in this light (Huth et al., 2012). Meyns et al. (2019) examined the suitability of library-based search for the identification of distinct polymers identified by nano-FTIR using commercial and open source analytic software (siMPle). It was discovered that this technology can accurately identify polymer samples that have weathered in the environment without the need for preliminary cleaning, opening up a broad range of applications for the identification and characterisation of various polymer samples.

(b) *Optical Tweezers for Raman Analysis of Nanoplastics*

Raman analysis of NPLs and micro MPs can be done under aqueous settings since water is a feeble Raman scatterer. For this type of research, optical tweezers might be utilised, which retain the particles in the laser beam's focus and allow for spectroscopic identification. Gillibert et al. (2019) demonstrated the utility of a method that combines optical tweezers and μ-Raman spectroscopy for trapping and chemical detection of NMPs. Using 633 and 785 nm excitation lasers, plastic particles dispersed in saltwater (PE, PP, PS, PET, PVC, PMMA, and PA 6) with sizes ranging from 20 m to 50 nm were investigated. The researchers were able to distinguish plastics from organic matter and mineral deposits at the single-particle level, as well as analyse the size and shape of NMPs (beads, pieces, and fibres), with only diffraction limiting spatial resolution. The approach was evaluated on model particles as well as naturally aged environmental samples, demonstrating its ability to characterise real-world samples Gillibert et al. (2019).

Schwaferts et al. (2020) presented online connection of field-flow fractionation and Raman microspectroscopy for the investigation of NPLs using optical tweezers in this area. The authors coupled particle separation and characterisation with chemical identification using online μ-Raman spectroscopy in a flow cell. It was possible to identify particles of various reference materials (polymers and inorganic, specifically PS, PMMA, and SiO_2 at concentrations of 1 mg/L (109 particles/L) using 2D optical tweezers for particle entrapment. The new approach has a wide range of applications in nanomaterial characterisation, including NPL analysis (Schwaferts et al., 2020). It is necessary to create appropriate preconcentration and enrichment of NPLs from environmental and dietary matrices in order to get accurate and representative results. Actual reference materials are necessary for the proper development, optimization, and validation of methods for NPL detection.

4.5 Conclusions and Future Perspectives

Plastic has become the most damaging manmade trash in the environment in recent decades due to its increased production and usage by humans. However, there is a scarcity of data on the prevalence of MPs in various environmental matrices, as well as their impact on human health and the ability to detect them quickly.

There is currently no one-size-fits-all approach of identification and characterisation that can be applied to all of the cases investigated. We have described the most commonly used strategies for detecting and characterising MPs in environmental samples, as well as their benefits and drawbacks, in this study.

References

Al-Azzawi, M. S., Kefer, S., Weißer, J., Reichel, J., Schwaller, C., Glas, K., Knoop, O., & Drewes, J. E. (2020). Validation of sample preparation methods for microplastic analysis in wastewater matrices—Reproducibility and standardization. *Water., 12*(9), 2445.

Andrade, J. M., Ferreiro, B., López-Mahía, P., & Muniategui-Lorenzo, S. (2020). Standardization of the minimum information for publication of infrared-related data when microplastics are characterized. *Marine Pollution Bulletin, 154*, 111035.

Anger, P. M., von der Esch, E., Baumann, T., Elsner, M., Niessner, R., & Ivleva, N. P. (2018). Raman microspectroscopy as a tool for microplastic particle analysis. *TrAC Trends in Analytical Chemistry, 109*, 214–226.

Araujo, C. F., Nolasco, M. M., Ribeiro, A. M., & Ribeiro-Claro, P. J. (2018). Identification of microplastics using Raman spectroscopy: Latest developments and future prospects. *Water Research, 142*, 426–440.

Arthur, C., Baker, J., Bamford, H. (2010). Proceedings of the second research workshop on microplastic Debris, November 5–6, 2010; NOAA Technical Memorandum NOS-OR&R-39; Marine Debris Division, Office of Response and Restoration, Ocean Service, NOAA, 2011; https://marinedebris.noaa.gov/proceedings-second-researchworkshop-microplastic-marine-debris.

Bakir, A., Rowland, S. J., & Thompson, R. C. (2014). Enhanced desorption of persistent organic pollutants from microplastics under simulated physiological conditions. *Environmental Pollution, 185*, 16–23.

Bank, M. S, Ok, Y. S., & Swarzenski, P. W. (2020). Microplastic's role in antibiotic resistance. *Science, 369*(6509), 1315.

Becker, R., Altmann, K., Sommerfeld, T., & Braun, U. (2020). Quantification of microplastics in a freshwater suspended organic matter using different thermoanalytical methods–outcome of an interlaboratory comparison. *Journal of Analytical and Applied Pyrolysis, 1*(148), 104829.

Bitter, H., & Lackner, S. (2020). First quantification of semi-crystalline microplastics in industrial wastewaters. *Chemosphere, 258*, 127388.

Blancho, F., Davranche, M., Hadri, H. E., Grassl, B., & Gigault, J. (2021). Nanoplastics identification in complex environmental matrices: Strategies for polystyrene and polypropylene. *Environmental Science & Technology, 55*(13), 8753–8759.

Brandt, J., Bittrich, L., Fischer, F., Kanaki, E., Tagg, A., Lenz, R., Labrenz, M., Brandes, E., Fischer, D., & Eichhorn, K. J. (2020). High-Throughput analyses of microplastic samples using Fourier transform infrared and Raman spectrometry. *Applied Spectroscopy, 74*(9), 1185–1197.

Braun, U., Altmann, K., Bannick, C. G., Becker, R., Bitter, H., Bochow, M., Dierkes, G., Enders, K., Eslahian, K. A., Fischer, D. et al. (2020). Status report: Analysis of microplastics sampling, preparation and detection methods within the scope of the Bmbf research focus plastics in the environment: Sources, sinks, solutions, 2020.

Braun, U., Altmann, K., Herper, D., Knefel, M., Bednarz, M., & Bannick, C. G. (2021). Smart filters for the analysis of microplastic in beverages filled in plastic bottles. *Food Additives & Contaminants: Part A, 38*(4), 691–700.

Brehm, M., Taubner, T., Hillenbrand, R., & Keilmann, F. (2006). Infrared spectroscopic mapping of single nanoparticles and viruses at nanoscale resolution. *Nano Letters, 6*(7), 1307–1310.

Brennecke, D., Duarte, B., Paiva, F., Caçador, I., & Canning-Clode, J. (2016). Microplastics as vector for heavy metal contamination from the marine environment. *Estuarine, Coastal and Shelf Science, 178*, 189–195.

Cabernard, L., Roscher, L., Lorenz, C., Gerdts, G., & Primpke, S. (2018). Comparison of Raman and Fourier transform infrared spectroscopy for the quantification of microplastics in the aquatic environment. *Environmental Science & Technology, 52*(22), 13279–13288.

Cardell, C., & Guerra, I. (2016). An overview of emerging hyphenated SEM-EDX and Raman spectroscopy systems: Applications in life, environmental and materials sciences. *TrAC Trends in Analytical Chemistry., 77*, 156–166.

Castañeda, R. A., Avlijas, S., Simard, M. A., & Ricciardi, A. (2014). Microplastic pollution in St. Lawrence river sediments. *Canadian Journal of Fisheries and Aquatic Sciences, 71*(12), 1767–1771.

Castelvetro, V., Corti, A., Ceccarini, A., Petri, A., & Vinciguerra, V. (2021). Nylon 6 and nylon 6, 6 micro-and nanoplastics: A first example of their accurate quantification, along with polyester (PET), in wastewater treatment plant sludges. *Journal of Hazardous Materials, 407*, 124364.

Catarino, A. I., Macchia, V., Sanderson, W. G., Thompson, R. C., & Henry, T. B. (2018). Low levels of microplastics (MP) in wild mussels indicate that MP ingestion by humans is minimal compared to exposure via household fibres fallout during a meal. *Environmental Pollution, 237*, 675–684.

Ceccarini, A., Corti, A., Erba, F., Modugno, F., La Nasa, J., Bianchi, S., & Castelvetro, V. (2018). The hidden microplastics: New insights and figures from the thorough separation and characterization of microplastics and of their degradation byproducts in coastal sediments. *Environmental Science & Technology, 52*(10), 5634–5643.

Cincinelli, A., Scopetani, C., Chelazzi, D., Lombardini, E., Martellini, T., Katsoyiannis, A., Fossi, M. C., & Corsolini, S. (2017). Microplastic in the surface waters of the Ross Sea (Antarctica): Occurrence, distribution and characterization by FTIR. *Chemosphere, 175*, 391–400.

Cole, M., Lindeque, P., Fileman, E., Halsband, C., Goodhead, R., Moger, J., & Galloway, T. S. (2013). Microplastic ingestion by zooplankton. *Environmental Science & Technology, 47*(12), 6646–6655.

Collard, F., Gilbert, B., Eppe, G., Parmentier, E., & Das, K. (2015). Detection of anthropogenic particles in fish stomachs: An isolation method adapted to identification by Raman spectroscopy. *Archives of Environmental Contamination and Toxicology, 69*(3), 331–339.

Coppock, R. L., Cole, M., Lindeque, P. K., Queirós, A. M., & Galloway, T. S. (2017). A small-scale, portable method for extracting microplastics from marine sediments. *Environmental Pollution, 230*, 829–837.

Cowger, W., Gray, A., Christiansen, S. H., DeFrond, H., Deshpande, A. D., Hemabessiere, L., Lee, E., Mill, L., Munno, K., Ossmann, B. E., & Pittroff, M. (2020). Critical review of processing and classification techniques for images and spectra in microplastic research. *Applied Spectroscopy., 74*(9), 989–1010.

Cox, K. D., Covernton, G. A., Davies, H. L., Dower, J. F., Juanes, F., & Dudas, S. E. (2019). Human consumption of microplastics. *Environmental Science & Technology, 53*(12), 7068–7074.

Crocombe, R. A. (2018). Portable spectroscopy. *Applied Spectroscopy, 72*(12), 1701–1751.

Cunningham, E. M., Ehlers, S. M., Dick, J. T., Sigwart, J. D., Linse, K., Dick, J. J., & Kiriakoulakis, K. (2020). High abundances of microplastic pollution in deep-sea sediments: Evidence from Antarctica and the Southern Ocean. *Environmental Science & Technology, 54*(21), 13661–13671.

da Silva, V. H., Murphy, F., Amigo, J. M., Stedmon, C., & Strand, J. (2020). Classification and quantification of microplastics (< 100 μm) using a focal plane Array–Fourier transform infrared imaging system and machine learning. *Analytical Chemistry, 92*(20), 13724–13733.

David, J., Steinmetz, Z., Kučerík, J., & Schaumann, G. E. (2018). Quantitative analysis of poly (ethylene terephthalate) microplastics in soil via thermogravimetry–mass spectrometry. *Analytical Chemistry, 90*(15), 8793–8799.

Davranche, M., Lory, C., Le Juge, C., Blancho, F., Dia, A., Grassl, B., El Hadri, H., Pascal, P. Y., & Gigault, J. (2020). Nanoplastics on the coast exposed to the North Atlantic Gyre: Evidence and traceability. *NanoImpact, 20*, 100262.

Dazzi, A., & Prater, C. B. (2017). AFM-IR: Technology and applications in nanoscale infrared spectroscopy and chemical imaging. *Chemical Reviews, 117*(7), 5146–5173.

De Leeuw, J. W., De Leer, E. W., Damste, J. S., & Schuyl, P. J. (1986). Screening of anthropogenic compounds in polluted sediments and soils by flash evaporation/pyrolysis gas chromatography-mass spectrometry. *Analytical Chemistry, 58*(8), 1852–1857.

Dehaut A, Cassone AL, Frère L, Hermabessiere L, Himber C, Rinnert E, Rivière G, Lambert C, (2016) Soudant P, Huvet A, Duflos G. Microplastics in seafood: Benchmark protocol for their extraction and characterization. Environmental Pollution. ;215:223–233.

Dehaut, A., Hermabessiere, L., & Duflos, G. (2019). Current frontiers and recommendations for the study of microplastics in seafood. *TrAC Trends in Analytical Chemistry, 116*, 346–359.

Dekiff, J. H., Remy, D., Klasmeier, J., & Fries, E. (2014). Occurrence and spatial distribution of microplastics in sediments from Norderney. *Environmental Pollution., 186*, 248–256.

Delgado-Gallardo, J., Sullivan, G. L., Esteban, P., Wang, Z., Arar, O., Li, Z., Watson, T. M., & Sarp, S. (2021). From sampling to analysis: A critical review of techniques used in the detection of micro-and nanoplastics in aquatic environments. *ACS ES&T Water, 1*(4), 748–764.

Dibke, C., Fischer, M., & Scholz-Böttcher, B. M. (2021). Microplastic mass concentrations and distribution in German bight waters by pyrolysis–gas chromatography–mass spectrometry/

thermochemolysis reveal potential impact of marine coatings: Do ships leave skid marks? *Environmental Science & Technology, 55*(4), 2285–2295.

Dierkes, G., Lauschke, T., Becher, S., Schumacher, H., Földi, C., & Ternes, T. (2019). Quantification of microplastics in environmental samples via pressurized liquid extraction and pyrolysis-gas chromatography. *Analytical and Bioanalytical Chemistry, 411*(26), 6959–6968.

Dittmann, D., Braun, U., Jekel, M., & Ruhl, A. S. (2018). Quantification and characterisation of activated carbon in activated sludge by thermogravimetric and evolved gas analyses. *Journal of Environmental Chemical Engineering, 6*(2), 2222–2231.

Dris, R., Gasperi, J., Rocher, V., & Tassin, B. (2018). Synthetic and non-synthetic anthropogenic fibers in a river under the impact of Paris Megacity: Sampling methodological aspects and flux estimations. *Science of the Total Environment, 618*, 157–164.

Dümichen, E., Barthel, A. K., Braun, U., Bannick, C. G., Brand, K., Jekel, M., & Senz, R. (2015). Analysis of polyethylene microplastics in environmental samples, using a thermal decomposition method. *Water Research, 85*, 451–457.

Dümichen, E., Eisentraut, P., Bannick, C. G., Barthel, A. K., Senz, R., & Braun, U. (2017). Fast identification of microplastics in complex environmental samples by a thermal degradation method. *Chemosphere, 174*, 572–584.

Dümichen, E., Eisentraut, P., Celina, M., & Braun, U. (2019). Automated thermal extraction-desorption gas chromatography mass spectrometry: A multifunctional tool for comprehensive characterization of polymers and their degradation products. *Journal of Chromatography, 1592*, 133–142.

Eisentraut, P., Dümichen, E., Ruhl, A. S., Jekel, M., Albrecht, M., Gehde, M., & Braun, U. (2018). Two birds with one stone—fast and simultaneous analysis of microplastics: Microparticles derived from thermoplastics and tire wear. *Environmental Science & Technology Letters., 5*(10), 608–613.

El Hayany, B., El Fels, L., Quénéa, K., Dignac, M. F., Rumpel, C., Gupta, V. K., & Hafidi, M. (2020). Microplastics from lagooning sludge to composts as revealed by fluorescent staining-image analysis, Raman spectroscopy and pyrolysis-GC/MS. *Journal of Environmental Management., 275*, 111249.

Enders, K., Lenz, R., Stedmon, C. A., & Nielsen, T. G. (2015). Abundance, size and polymer composition of marine microplastics≥ 10 μm in the Atlantic Ocean and their modelled vertical distribution. *Marine Pollution Bulletin, 100*(1), 70–81.

Enders, K., Käppler, A., Biniasch, O., Feldens, P., Stollberg, N., Lange, X., Fischer, D., Eichhorn, K. J., Pollehne, F., Oberbeckmann, S., & Labrenz, M. (2019). Tracing microplastics in aquatic environments based on sediment analogies. *Scientific Reports, 9*(1), 1–5.

Enders, K., Lenz, R., do Sul, J. A., Tagg, A. S., & Labrenz, M. (2020). When every particle matters: A QuEChERS approach to extract microplastics from environmental samples. *MethodsX., 7*, 100784.

Fabbri, D. (2001). Use of pyrolysis-gas chromatography/mass spectrometry to study environmental pollution caused by synthetic polymers: A case study: The Ravenna Lagoon. *Journal of Analytical and Applied Pyrolysis., 58*, 361–370.

Fabbri, D., Tartari, D., & Trombini, C. (2000). Analysis of poly (vinyl chloride) and other polymers in sediments and suspended matter of a coastal lagoon by pyrolysis-gas chromatography-mass spectrometry. *Analytica Chimica Acta, 413*(1–2), 3–11.

Felts, J. R., Kjoller, K., Lo, M., Prater, C. B., & King, W. P. (2012). Nanometer-scale infrared spectroscopy of heterogeneous polymer nanostructures fabricated by tip-based nanofabrication. *ACS Nano, 6*(9), 8015–8021.

Fischer, M., & Scholz-Böttcher, B. M. (2017). Simultaneous trace identification and quantification of common types of microplastics in environmental samples by pyrolysis-gas chromatography–mass spectrometry. *Environmental Science & Technology, 51*(9), 5052–5060.

Fischer, M., Gobmann, I., & Scholz-Böttcher, B. M. (2019). Fleur de Sel—An interregional monitor for microplastics mass load and composition in European coastal waters? *Journal of Analytical and Applied Pyrolysis, 144*, 104711.

Fotopoulou, K. N., & Karapanagioti, H. K. (2017). Degradation of various plastics in the environment. In *Hazardous chemicals associated with plastics in the marine environment* (pp. 71–92).

Frias, J. P., & Nash, R. (2019). Microplastics: Finding a consensus on the definition. *Marine Pollution Bulletin, 138*, 145–147.

Fries, E., Dekiff, J. H., Willmeyer, J., Nuelle, M. T., Ebert, M., & Remy, D. (2013). Identification of polymer types and additives in marine microplastic particles using pyrolysis-GC/MS and scanning electron microscopy. *Environmental Science: Processes & Impacts, 15*(10), 1949–1956.

Funck, M., Yildirim, A., Nickel, C., Schram, J., Schmidt, T. C., & Tuerk, J. (2020). Identification of microplastics in wastewater after cascade filtration using Pyrolysis-GC–MS. *MethodsX., 7*, 100778.

Gigault, J., Ter Halle, A., Baudrimont, M., Pascal, P. Y., Gauffre, F., Phi, T. L., El Hadri, H., Grassl, B., & Reynaud, S. (2018). Current opinion: What is a nanoplastic? *Environmental Pollution., 235*, 1030–1034.

Gillibert, R., Balakrishnan, G., Deshoules, Q., Tardivel, M., Magazzù, A., Donato, M. G., Maragò, O. M., Lamy de La Chapelle, M., Colas, F., Lagarde, F., & Gucciardi, P. G. (2019). Raman tweezers for small microplastics and nanoplastics identification in seawater. *Environmental Science & Technology., 53*(15), 9003–9013.

Gobmann, I., Halbach, M., & Scholz-Böttcher, B. M. (2021). Car and truck tire wear particles in complex environmental samples - a quantitative comparison with "Traditional" microplastic polymer mass loads. *Science of the Total Environment, 773*, 145667.

Goodhead, R. M., Moger, J., Galloway, T. S., & Tyler, C. R. (2015). Tracing engineered nanomaterials in biological tissues using coherent anti-Stokes Raman scattering (CARS) microscopy–a critical review. *Nanotoxicology, 9*(7), 928–939.

Hale, R. C., Seeley, M. E., La Guardia, M. J., Mai, L., & Zeng, E. Y. (2020). A global perspective on microplastics. *Journal of Geophysical Research: Oceans, 125*(1), e2018JC014719.

Harrison, J. P., Ojeda, J. J., & Romero-González, M. E. (2012). The applicability of reflectance micro-Fourier-transform infrared spectroscopy for the detection of synthetic microplastics in marine sediments. *Science of the Total Environment, 416*, 455–463.

Hartmann, N. B., Huffer, T., Thompson, R. C., Hassellöv, M., Verschoor, A., Daugaard, A. E., Rist, S., Karlsson, T., Brennholt, N., Cole, M., & Herrling, M. P. (2019). Are we speaking the same language? Recommendations for a definition and categorization framework for plastic debris. *Environmental Science & Technology, 2019*(53), 1039–1047.

Hendrickson, E., Minor, E. C., & Schreiner, K. (2018). Microplastic abundance and composition in western Lake Superior as determined via microscopy, Pyr-GC/MS, and FTIR. *Environmental Science & Technology, 52*(4), 1787–1796.

Hermabessiere, L., Himber, C., Boricaud, B., Kazour, M., Amara, R., Cassone, A. L., Laurentie, M., Paul-Pont, I., Soudant, P., Dehaut, A., & Duflos, G. (2018). Optimization, performance, and application of a pyrolysis-GC/MS method for the identification of microplastics. *Analytical and Bioanalytical Chemistry, 410*(25), 6663–6676.

Hermann, R. J., & Gordon, M. J. (2018). Nanoscale optical microscopy and spectroscopy using near-field probes. *Annual Review of Chemical and Biomolecular Engineering, 9*, 365–387.

Herrera, M., Matuschek, G., & Kettrup, A. (2003). Fast identification of polymer additives by pyrolysis-gas chromatography/mass spectrometry. *Journal of Analytical and Applied Pyrolysis, 70*(1), 35–42.

Hildebrandt, L., El Gareb, F., Zimmermann, T., Klein, O., Emeis, K., Proefrock, D., & Kerstan, A. (2020). Fast, automated microplastics analysis using laser direct chemical imaging: Characterizing and quantifying microplastics in water samples from marine environments.

Horton, A. A., Cross, R. K., Read, D. S., Jürgens, M. D., Ball, H. L., Svendsen, C., Vollertsen, J., & Johnson, A. C. (2021). Semi-automated analysis of microplastics in complex wastewater samples. *Environmental Pollution, 268*, 115841.

Hufnagl, B., Steiner, D., Renner, E., Löder, M. G., Laforsch, C., & Lohninger, H. (2019). A methodology for the fast identification and monitoring of microplastics in environmental samples using random decision forest classifiers. *Analytical Methods, 11*(17), 2277–2285.

Huth, F., Govyadinov, A., Amarie, S., Nuansing, W., Keilmann, F., & Hillenbrand, R. (2012). Nano-FTIR absorption spectroscopy of molecular fingerprints at 20 nm spatial resolution. *Nano Letters, 12*(8), 3973–3978.

Imhof, H. K., Schmid, J., Niessner, R., Ivleva, N. P., & Laforsch, C. (2012). A novel, highly efficient method for the separation and quantification of plastic particles in sediments of aquatic environments. *Limnology and Oceanography: Methods, 10*(7), 524–537.

Imhof, H. K., Laforsch, C., Wiesheu, A. C., Schmid, J., Anger, P. M., Niessner, R., & Ivleva, N. P. (2016). Pigments and plastic in limnetic ecosystems: A qualitative and quantitative study on microparticles of different size classes. *Water Research, 98*, 64–74.

Ivleva, N. P., Wiesheu, A. C., & Niessner, R. (2017). Microplastic in aquatic ecosystems. *Angewandte Chemie, International Edition, 56*(7), 1720–1739.

Jansson, K. D., Zawodny, C. P., & Wampler, T. P. (2007). Determination of polymer additives using analytical pyrolysis. *Journal of Analytical and Applied Pyrolysis, 79*(1–2), 353–361.

Johnson, A. C., Ball, H., Cross, R., Horton, A. A., Jurgens, M. D., Read, D. S., Vollertsen, J., & Svendsen, C. (2020). Identification and quantification of microplastics in potable water and their sources within water treatment works in England and Wales. *Environmental Science & Technology, 54*(19), 12326–12334.

Käppler, A., Fischer, D., Oberbeckmann, S., Schernewski, G., Labrenz, M., Eichhorn, K. J., & Voit, B. (2016). Analysis of environmental microplastics by vibrational microspectroscopy: FTIR, Raman or both? *Analytical and Bioanalytical Chemistry, 408*(29), 8377–8391.

Käppler, A., Fischer, M., Scholz-Böttcher, B. M., Oberbeckmann, S., Labrenz, M., Fischer, D., Eichhorn, K. J., & Voit, B. (2018). Comparison of μ-ATR-FTIR spectroscopy and py-GCMS as identification tools for microplastic particles and fibers isolated from river sediments. *Analytical and Bioanalytical Chemistry, 410*(21), 5313–5327.

Karami, A., Golieskardi, A., Ho, Y. B., Larat, V., & Salamatinia, B. (2017). Microplastics in eviscerated flesh and excised organs of dried fish. *Scientific Reports, 7*(1), 1–9.

Karlsson, T. M., Grahn, H., van Bavel, B., & Geladi, P. (2016). Hyperspectral imaging and data analysis for detecting and determining plastic contamination in seawater filtrates. *Journal of Near Infrared Spectroscopy, 24*(2), 141–149.

Kirstein, I. V., Hensel, F., Gomiero, A., Iordachescu, L., Vianello, A., Wittgren, H. B., & Vollertsen, J. (2021). Drinking plastics?–Quantification and qualification of microplastics in drinking water distribution systems by μFTIR and Py-GCMS. *Water Research, 188*, 116519.

Klöckner, P., Reemtsma, T., Eisentraut, P., Braun, U., Ruhl, A. S., & Wagner, S. (2019). Tire and road wear particles in road environment–Quantification and assessment of particle dynamics by Zn determination after density separation. *Chemosphere, 222*, 714–721.

Klöckner, P., Seiwert, B., Eisentraut, P., Braun, U., Reemtsma, T., & Wagner, S. (2020). Characterization of tire and road wear particles from road runoff indicates highly dynamic particle properties. *Water Research, 185*, 116262.

Koelmans, A. A., Nor, N. H., Hermsen, E., Kooi, M., Mintenig, S. M., & De France, J. (2019). Microplastics in freshwaters and drinking water: Critical review and assessment of data quality. *Water Research, 155*, 410–422.

Kumar, B. V., Löschel, L. A., Imhof, H. K., Löder, M. G., & Laforsch, C. (2021). Analysis of microplastics of a broad size range in commercially important mussels by combining FTIR and Raman spectroscopy approaches. *Environmental Pollution, 269*, 116147.

Laptenok, S. P., Martin, C., Genchi, L., Duarte, C. M., & Liberale, C. (2020). Stimulated Raman microspectroscopy as a new method to classify microfibers from environmental samples. *Environmental Pollution, 267*, 115640.

Lê, Q. T., Ly, N. H., Kim, M. K., Lim, S. H., Son, S. J., Zoh, K. D., & Joo, S. W. (2021). Nanostructured Raman substrates for the sensitive detection of submicrometer-sized plastic pollutants in water. *Journal of Hazardous Materials, 402*, 123499.

Lehner, R., Weder, C., Petri-Fink, A., & Rothen-Rutishauser, B. (2019). Emergence of nanoplastic in the environment and possible impact on human health. *Environmental Science & Technology, 53*(4), 1748–1765.

Lenz, R., Enders, K., Stedmon, C. A., Mackenzie, D. M., & Nielsen, T. G. (2015). A critical assessment of visual identification of marine microplastic using Raman spectroscopy for analysis improvement. *Marine Pollution Bulletin, 100*(1), 82–91.

Lenz, R., Enders, K., & Nielsen, T. G. (2016). Microplastic exposure studies should be environmentally realistic. *Proceedings of the National Academy of Sciences of the United States of America, 113*(29), E4121–E4122.

Levermore, J. M., Smith, T. E., Kelly, F. J., & Wright, S. L. (2020). Detection of microplastics in ambient particulate matter using Raman spectral imaging and chemometric analysis. *Analytical Chemistry, 92*(13), 8732–8740.

Li, X., Zhang, D., Bai, Y., Wang, W., Liang, J., & Cheng, J. X. (2019). Fingerprinting a living cell by raman integrated mid-infrared photothermal microscopy. *ACS Photonics, 91*(16), 10750–11076.

Li, D., Shi, Y., Yang, L., Xiao, L., Kehoe, D. K., Gunko, Y. K., Boland, J. J., & Wang, J. J. (2020). Microplastic release from the degradation of polypropylene feeding bottles during infant formula preparation. *Nature Food, 1*(11), 746–754.

Li, Q., Zeng, A., Jiang, X., & Gu, X. (2021). Are microplastics correlated to phthalates in facility agriculture soil? *Journal of Hazardous Materials, 412*, 125164.

Liao, C. S., Wang, P., Huang, C. Y., Lin, P., Eakins, G., Bentley, R. T., Liang, R., & Cheng, J. X. (2017). In vivo and in situ spectroscopic imaging by a handheld stimulated Raman scattering microscope. *ACS Photonics, 5*(3), 947–954.

Löder, M. G., Kuczera, M., Mintenig, S., Lorenz, C., & Gerdts, G. (2015). Focal plane array detector-based micro-Fourier-transform infrared imaging for the analysis of microplastics in environmental samples. *Environment and Chemistry, 12*(5), 563–581.

Löder, M. G., Imhof, H. K., Ladehoff, M., Löschel, L. A., Lorenz, C., Mintenig, S., Piehl, S., Primpke, S., Schrank, I., Laforsch, C., & Gerdts, G. (2017). Enzymatic purification of microplastics in environmental samples. *Environmental Science & Technology, 51*(24), 14283–11492.

Lomonaco, T., Manco, E., Corti, A., La Nasa, J., Ghimenti, S., Biagini, D., Di Francesco, F., Modugno, F., Ceccarini, A., Fuoco, R., & Castelvetro, V. (2020). Release of harmful volatile organic compounds (VOCs) from photo-degraded plastic debris: A neglected source of environmental pollution. *Journal of Hazardous Materials, 394*, 122596.

Lorenz, C., Roscher, L., Meyer, M. S., Hildebrandt, L., Prume, J., Löder, M. G., Primpke, S., & Gerdts, G. (2019). Spatial distribution of microplastics in sediments and surface waters of the southern North Sea. *Environmental Pollution, 252*, 1719–1729.

Lv, L., He, L., Jiang, S., Chen, J., Zhou, C., Qu, J., Lu, Y., Hong, P., Sun, S., & Li, C. (2020). In situ surface-enhanced Raman spectroscopy for detecting microplastics and nanoplastics in aquatic environments. *Science of Total Environment, 728*, 138449.

Majewsky, M., Bitter, H., Eiche, E., & Horn, H. (2016). Determination of microplastic polyethylene (PE) and polypropylene (PP) in environmental samples using thermal analysis (TGA-DSC). *Science of Total Environment, 568*, 507–511.

Marcott, C., Kansiz, M., Dillon, E., Cook, D., Mang, M. N., & Noda, I. (2020). Two-dimensional correlation analysis of highly spatially resolved simultaneous IR and Raman spectral imaging of bioplastics composite using optical photothermal Infrared and Raman spectroscopy. *Journal of Molecular Structure, 1210*, 128045.

Materić, D., Kasper-Giebl, A., Kau, D., Anten, M., Greilinger, M., Ludewig, E., Van Sebille, E., Röckmann, T., & Holzinger, R. (2020). Micro-and nanoplastics in alpine snow: A new method for chemical identification and (semi) quantification in the nanogram range. *Environmental Science & Technology, 54*(4), 2353–2359.

Mattsson, K., Johnson, E. V., Malmendal, A., Linse, S., Hansson, L. A., & Cedervall, T. (2017). Brain damage and behavioural disorders in fish induced by plastic nanoparticles delivered through the food chain. *Scientific Reports, 7*(1), 1–7.

McCormick, A. R., Hoellein, T. J., London, M. G., Hittie, J., Scott, J. W., & Kelly, J. J. (2016). Microplastic in surface waters of urban rivers: Concentration, sources, and associated bacterial assemblages. *Ecosphere., 7*(11), e01556.

Meyns, M., Primpke, S., & Gerdts, G. (2019). Library based identification and characterisation of polymers with nano-FTIR and IR-sSNOM imaging. *Analytical Methods, 11*(40), 5195–5202.

Mintenig, S. M., Bäuerlein, P. S., Koelmans, A. A., Dekker, S. C., & Van Wezel, A. P. (2018). Closing the gap between small and smaller: Towards a framework to analyse nano-and micro-plastics in aqueous environmental samples. *Environmental Science: Nano, 5*(7), 1640–1649.

Missawi, O., Bousserrhine, N., Belbekhouche, S., Zitouni, N., Alphonse, V., Boughattas, I., & Banni, M. (2020). Abundance and distribution of small microplastics ($\leq$ 3 μm) in sediments and seaworms from the Southern Mediterranean coasts and characterisation of their potential harmful effects. *Environmental Pollution, 263*, 114634.

Moroni, M., Mei, A., Leonardi, A., Lupo, E., & Marca, F. L. (2015). PET and PVC separation with hyperspectral imagery. *Sensors, 15*(1), 2205–2227.

Mughini-Gras, L., van der Plaats, R. Q., van der Wielen, P. W., Bauerlein, P. S., & de Roda Husman, A. M. (2021). Riverine microplastic and microbial community compositions: A field study in the Netherlands. *Water Research, 192*, 116852.

Napper, I. E., Davies, B. F., Clifford, H., Elvin, S., Koldewey, H. J., Mayewski, P. A., Miner, K. R., Potocki, M., Elmore, A. C., Gajurel, A. P., & Thompson, R. C. (2020). Reaching new heights in plastic pollution—preliminary findings of microplastics on Mount Everest. *One Earth, 3*(5), 621–630.

Ng, E. L., Lin, S. Y., Dungan, A. M., Colwell, J. M., Ede, S., Lwanga, E. H., Meng, K., Geissen, V., Blackall, L. L., & Chen, D. (2021). Microplastic pollution alters forest soil microbiome. *Journal of Hazardous Materials, 409*, 124606.

Nuelle, M. T., Dekiff, J. H., Remy, D., & Fries, E. (2014). A new analytical approach for monitoring microplastics in marine sediments. *Environmental Pollution, 184*, 161–169.

Okoffo, E. D., Ribeiro, F., O'Brien, J. W., O'Brien, S., Tscharke, B. J., Gallen, M., Samanipour, S., Mueller, J. F., & Thomas, K. V. (2020). Identification and quantification of selected plastics in biosolids by pressurized liquid extraction combined with double-shot pyrolysis gas chromatography–mass spectrometry. *Science of the Total Environment, 715*, 136924.

Pancani, E., Mathurin, J., Bilent, S., Bernet-Camard, M. F., Dazzi, A., Deniset-Besseau, A., & Gref, R. (2018). High-resolution label-free detection of biocompatible polymeric nanoparticles in cells. *Particle & Particle Systems Characterization., 35*(3), 1700457.

Paul, A., Wander, L., Becker, R., Goedecke, C., & Braun, U. (2019). High-throughput NIR spectroscopic (NIRS) detection of microplastics in soil. *Environmental Science and Pollution Research, 26*(8), 7364–7374.

Paul, M. B., Stock, V., Cara-Carmona, J., Lisicki, E., Shopova, S., Fessard, V., Braeuning, A., Sieg, H., & Böhmert, L. (2020). Micro-and nanoplastics–current state of knowledge with the focus on oral uptake and toxicity. *Nanoscale Advances, 2*(10), 4350–4367.

Peacock, M., Materić, D., Kothawala, D. N., Holzinger, R., & Futter, M. N. (2018). Understanding dissolved organic matter reactivity and composition in lakes and streams using proton-transfer-reaction mass spectrometry (PTR-MS). *Environmental Science & Technology Letters., 5*(12), 739–744.

Picó, Y., & Barceló, D. (2020). Pyrolysis gas chromatography-mass spectrometry in environmental analysis: Focus on organic matter and microplastics. *Trends in Analytical Chemistry, 130*, 115964.

Pico, Y., Alfarhan, A., & Barcelo, D. (2019). Nano-and microplastic analysis: Focus on their occurrence in freshwater ecosystems and remediation technologies. *Trends in Analytical Chemistry, 113*, 409–425.

Piehl, S., Leibner, A., Löder, M. G., Dris, R., Bogner, C., & Laforsch, C. (2018). Identification and quantification of macro-and microplastics on an agricultural farmland. *Scientific Reports, 8*(1), 1–9.

Pivokonsky, M., Cermakova, L., Novotna, K., Peer, P., Cajthaml, T., & Janda, V. (2018). Occurrence of microplastics in raw and treated drinking water. *Science of Total Environment, 643*, 1644–1651.

Prata, J. C. (2018). Airborne microplastics: Consequences to human health? *Environmental Pollution., 234*, 115–126.

Primpke, S., Wirth, M., Lorenz, C., & Gerdts, G. (2018). Reference database design for the automated analysis of microplastic samples based on Fourier transform infrared (FTIR) spectroscopy. *Analytical and Bioanalytical Chemistry, 410*(21), 5131–5141.

Primpke, S., Fischer, M., Lorenz, C., Gerdts, G., & Scholz-Böttcher, B. M. (2020). Comparison of pyrolysis gas chromatography/mass spectrometry and hyperspectral FTIR imaging spectroscopy for the analysis of microplastics. *Analytical and Bioanalytical Chemistry, 412*(30), 8283–8298.

Ravit, B., Cooper, K., Buckley, B., Yang, I., & Deshpande, A. (2019). Organic compounds associated with microplastic pollutants in New Jersey, USA surface waters. AIMS. *Environmental Sciences, 6*(6), 445–459.

Reichel, J., Graßmann, J., Letzel, T., & Drewes, J. E. (2020). Systematic development of a simultaneous determination of plastic particle identity and adsorbed organic compounds by thermodesorption–pyrolysis GC/MS (TD-Pyr-GC/MS). *Molecules, 25*(21), 4985.

Remy, F., Collard, F., Gilbert, B., Compère, P., Eppe, G., & Lepoint, G. (2015). When microplastic is not plastic: The ingestion of artificial cellulose fibers by macrofauna living in seagrass macrophytodetritus. *Environmental Science & Technology, 49*(18), 11158–11166.

Renner, G., Schmidt, T. C., & Schram, J. (2017). A new chemometric approach for automatic identification of microplastics from environmental compartments based on FT-IR spectroscopy. *Analytical Chemistry, 89*(22), 12045–12053.

Renner, G., Nellessen, A., Schwiers, A., Wenzel, M., Schmidt, T. C., & Schram, J. (2019). Data preprocessing & evaluation used in the microplastics identification process: A critical review & practical guide. *TrAC Trends in Analytical Chemistry, 111*, 229–238.

Ribeiro-Claro, P., Nolasco, M. M., & Araújo, C. (2017). In T. A. P. Rocha-Santos & A. C. Duarte (Eds.), *Comprehensive analytical chemistry: Characterization and analysis of microplastics*. Elsevier.

Rodríguez Chialanza, M., Sierra, I., Pérez Parada, A., & Fornaro, L. (2018). Identification and quantitation of semi-crystalline microplastics using image analysis and differential scanning calorimetry. *Environmental Science and Pollution Research, 25*(17), 16767–16775.

Schmidt, L. K., Bochow, M., Imhof, H. K., & Oswald, S. E. (2018). Multi-temporal surveys for microplastic particles enabled by a novel and fast application of SWIR imaging spectroscopy–study of an urban watercourse traversing the city of Berlin, Germany. *Environmental Pollution, 239*, 579–589.

Schwaferts, C., Sogne, V., Welz, R., Meier, F., Klein, T., Niessner, R., Elsner, M., & Ivleva, N. P. (2020). Nanoplastic analysis by online coupling of Raman microscopy and field-flow fractionation enabled by optical tweezers. *Analytical Chemistry, 92*(8), 5813–5820.

Scircle, A., Cizdziel, J. V., Tisinger, L., Anumol, T., & Robey, D. (2020). Occurrence of microplastic pollution at oyster reefs and other coastal sites in the Mississippi Sound, USA: Impacts of freshwater inflows from flooding. *Toxics., 8*(2), 35.

Shan, J., Zhao, J., Zhang, Y., Liu, L., Wu, F., & Wang, X. (2019). Simple and rapid detection of microplastics in seawater using hyperspectral imaging technology. *Analytica Chimica Acta, 1050*, 161–168.

Shim, W. J., Song, Y. K., Hong, S. H., & Jang, M. (2016). Identification and quantification of microplastics using Nile Red staining. *Marine Pollution Bulletin, 113*(1–2), 469–476.

Shruti, V. C., Pérez-Guevara, F., & Kutralam-Muniasamy, G. (2020). Metro station free drinking water fountain-A potential "microplastics hotspot" for human consumption. *Environmental Pollution, 261*, 114227.

Sobhani, Z., Zhang, X., Gibson, C., Naidu, R., Megharaj, M., & Fang, C. (2020). Identification and visualisation of microplastics/nanoplastics by Raman imaging (i): Down to 100 nm. *Water Research, 174*, 115658.

Steinmetz, Z., Kintzi, A., Muñoz, K., & Schaumann, G. E. (2020). A simple method for the selective quantification of polyethylene, polypropylene, and polystyrene plastic debris in soil by pyrolysis-gas chromatography/mass spectrometry. *Journal of Analytical and Applied Pyrolysis, 147*, 104803.

Sullivan, G. L., Gallardo, J. D., Jones, E. W., Hollliman, P. J., Watson, T. M., & Sarp, S. (2020). Detection of trace sub-micron (nano) plastics in water samples using pyrolysis-gas chromatography time of flight mass spectrometry (PY-GCToF). *Chemosphere, 249*, 126179.

Tagg, A. S., Sapp, M., Harrison, J. P., Sinclair, C. J., Bradley, E., Ju-Nam, Y., & Ojeda, J. J. (2020). Microplastic monitoring at different stages in a wastewater treatment plant using reflectance micro-FTIR imaging. *Frontiers in Environmental Science, 8*, 145.

Ter Halle, A., Jeanneau, L., Martignac, M., Jardé, E., Pedrono, B., Brach, L., & Gigault, J. (2017). Nanoplastic in the North Atlantic subtropical gyre. *Environmental Science & Technology, 51*(23), 13689–13697.

Thompson, R. C., Olsen, Y., Mitchell, R. P., Davis, A., Rowland, S. J., John, A. W., McGonigle, D., & Russell, A. E. (2004). Lost at sea: Where is all the plastic? *Science, 304*(5672), 838.

Tian, Z., Zhao, H., Peter, K. T., Gonzalez, M., Wetzel, J., Wu, C., Hu, X., Prat, J., Mudrock, E., Hettinger, R., & Cortina, A. E. (2021). A ubiquitous tire rubber–derived chemical induces acute mortality in coho salmon. *Science, 371*(6525), 185–189.

Trainic, M., Flores, J. M., Pinkas, I., Pedrotti, M. L., Lombard, F., Bourdin, G., Gorsky, G., Boss, E., Rudich, Y., Vardi, A., & Koren, I. (2020). Airborne microplastic particles detected in the remote marine atmosphere. *Communication Earth and Environment, 1*(1), 1–9.

Van Cauwenberghe, L., & Janssen, C. R. (2014). Microplastics in bivalves cultured for human consumption. *Environmental Pollution, 193*, 65–70.

Verma, P. (2017). Tip-enhanced Raman spectroscopy: Technique and recent advances. *Chemical Reviews, 117*(9), 6447–6466.

Vianello, A., Boldrin, A., Guerriero, P., Moschino, V., Rella, R., Sturaro, A., & Da Ros, L. (2013). Microplastic particles in sediments of Lagoon of Venice, Italy: First observations on occurrence, spatial patterns and identification. *Estuarine, Coastal and Shelf Science, 30*, 54–61.

Von der Esch, E., Lanzinger, M., Kohles, A. J., Schwaferts, C., Weisser, J., Hofmann, T., Glas, K., Elsner, M., & Ivleva, N. P. (2020). Simple generation of suspensible secondary microplastic reference particles via ultrasound treatment. *Frontiers in Chemistry, 8*, 169.

Wahl, A., Le Juge, C., Davranche, M., El Hadri, H., Grassl, B., Reynaud, S., & Gigault, J. (2021). Nanoplastic occurrence in a soil amended with plastic debris. *Chemosphere, 262*, 127784.

Wander, L., Vianello, A., Vollertsen, J., Westad, F., Braun, U., & Paul, A. (2020). Exploratory analysis of hyperspectral FTIR data obtained from environmental microplastics samples. *Analytical Methods, 12*(6), 781–791.

Watteau, F., Dignac, M. F., Bouchard, A., Revallier, A., & Houot, S. (2018). Microplastic detection in soil amended with municipal solid waste composts as revealed by transmission electronic microscopy and pyrolysis/GC/MS. *Frontiers in Sustainable Food System, 2*(Article 81), 14.

Watts, A. J., Urbina, M. A., Goodhead, R., Moger, J., & Lewis, C. (2016). Galloway TS. Effect of microplastic on the gills of the shore crab Carcinus maenas. *Environmental Science & Technology, 50*(10), 5364–5369.

Weber, F., Kerpen, J., Wolff, S., Langer, R., & Eschweiler, V. (2021). Investigation of microplastics contamination in drinking water of a German city. *Science of the Total Environment, 755*, 143421.

Wiesheu, A. C., Anger, P. M., Baumann, T., Niessner, R., & Ivleva, N. P. (2016). Raman microspectroscopic analysis of fibers in beverages. *Analytical Methods, 8*(28), 5722–5725.

Wolff, S., Kerpen, J., Prediger, J., Barkmann, L., & Müller, L. (2019). Determination of the microplastics emission in the effluent of a municipal waste water treatment plant using Raman microspectroscopy. *Water Research, 2*, 100014.

Xiao, L., & Schultz, Z. D. (2018). Spectroscopic imaging at the nanoscale: Technologies and recent applications. *Analytical Chemistry, 90*(1), 440.

Xu, J. L., Thomas, K. V., Luo, Z., & Gowen, A. A. (2019). FTIR and Raman imaging for microplastics analysis: State of the art, challenges and prospects. *TrAC Trends in Analytical Chemistry, 119*, 115629.

Yang, D., Shi, H., Li, L., Li, J., Jabeen, K., & Kolandhasamy, P. (2015). Microplastic pollution in table salts from China. *Environmental Science & Technology, 49*(22), 13622–13627.

Zada, L., Leslie, H. A., Vethaak, A. D., Tinnevelt, G. H., Jansen, J. J., de Boer, J. F., & Ariese, F. (2018). Fast microplastics identification with stimulated Raman scattering microscopy. *Journal of Raman Specroscopy, 49*(7), 1136–1144.

Zarfl, C. (2019). Promising techniques and open challenges for microplastic identification and quantification in environmental matrices. *Analytical and Bioanalytical Chemistry, 411*(17), 3743–3756.

Zhang, C., Huang, K. C., Rajwa, B., Li, J., Yang, S., Lin, H., Liao, C. S., Eakins, G., Kuang, S., Patsekin, V., & Robinson, J. P. (2017). Stimulated Raman scattering flow cytometry for label-free single-particle analysis. *Optica., 4*(1), 103–109.

Zhang, Y., Wang, X., Shan, J., Zhao, J., Zhang, W., Liu, L., & Wu, F. (2019). Hyperspectral imaging based method for rapid detection of microplastics in the intestinal tracts of fish. *Environmental Science & Technology, 53*(9), 5151–5158.

Zhou, X. X., Hao, L. T., Wang, H. Y., Li, Y. J., & Liu, J. F. (2018). Cloud-point extraction combined with thermal degradation for nanoplastic analysis using pyrolysis gas chromatography–mass spectrometry. *Analytical Chemistry, 91*(3), 1785–1790.

Zhou, X. X., Liu, R., Hao, L. T., & Liu, J. F. (2021). Identification of polystyrene nanoplastics using surface enhanced Raman spectrosc. *Talanta, 221*, 121552.

Ziajahromi, S., Neale, P. A., Rintoul, L., & Leusch, F. D. (2017). Wastewater treatment plants as a pathway for microplastics: Development of a new approach to sample wastewater-based microplastics. *Water Research, 112*, 93–99.

Chapter 5
Persistence of Micro- and Nanoplastics in Soil

N. Chaitanya, Suresh Babu Bastipati, and D. Bhagawan

Abstract Plastic pollution has been a heavy drawback for a number of years; these plastics (MNPs) have gathered notice from researchers all over the world. The fate, determination, and properties of microplastics (MPs) and nanoplastics (NPs) in soil are not well-known. In fact, yearly three hundred million plastics are created within the environment, and due this plastic trash, the soil acts as a logterm sink. In soil, the fate of MPs and NPs is powerfully determined by plastic physical properties, considering negligible impact is applied by their chemical structures. The derivative of plastic, termed deteriorate, other than generating micro- and nano-size waste, can produce marked changes in their properties (chemical and physical) with applicable impact on their reactivity. Further, these processes might cause the discharge of harmful monomeric and oligomeric components from plastics, likewise as cyanogenic additives, which can enter within the food chain, constituting a potential harm to human health and affecting the flora and fauna within the environment. In relevance their persistence in soil, soil inhabiting list, plastic uptake bacterium, fungi and insect are increasing daily. One among the most ecological functions due to MPs is expounded to their aim as vectors for microorganisms through the soil. However, the most ecological effect of NPs (limited to the fraction size <50 nm) is their capability to suffer the membrane of each being and organism cells. Soil biota, significantly earthworms and order Collembola, are often each MPs and NPs carriers through profile. The utilization of molecular techniques, particularly omics approaches, will gain insights into the results of MPs and NPs on composition and activity of microorganism communities inhabiting the soil and into those living on MP surface and within the gut of the soil plastic-ingesting fauna.

N. Chaitanya · D. Bhagawan
Centre for Environment, IST, JNTUH, Hyderabad, India

S. B. Bastipati (✉)
Centre for Biotechnology, IST, JNTUH, Hyderabad, India

N. R. Maddela et al. (eds.), *Micro and Nanoplastics in Soil*,
https://doi.org/10.1007/978-3-031-21195-9_5

Keywords Nanoplastics · Microplastics · Soil microbes · Environment

5.1 Introduction

Plastics, easy to supply, strong, and flexible, were exponentially growing in production and consumption over the previous few decades worldwide (Jambeck et al., 2015). More than 300 million tons of plastics are manufactured every year wherein 50% of which are primarily single use (Chen et al., 2020). In maximum parts of the world, this plastic waste was mismanaged because the decomposition takes place more than 1000 years because it's more chemically stable and corrosion-resistant and degradation is very hard (Shen et al., 2019). About 6300 million metric tons of global plastic waste had exceeded by 2015, and the current production and waste management trends will attain 12,000 million metric tons by 2050 (Geyer et al., 2017). Due to the ubiquity, persistence, and less management, plastic wastes are scarcely recycled (Barnes et al., 2009; Dris et al., 2015; Nizzetto et al., 2016). Due to this the accumulation of plastic was covered inside the aquatic, terrestrial, and atmospheric environments or all types of environmental media, inclusive of aquatic ecosystems, landfills, and agricultural ecosystems together with mulching and biofilm in addition to air (Shen et al., 2019; Kumar et al., 2021; Wang et al., 2021; Adam et al., 2021). Plastic pollution contributes to a rebellion problem faced by the sector today. The plastic waste can break into smaller particles that can affect the terrestrial and aquatic surroundings significantly. These particles are known as microplastics (MPs) (0.1 μm–5 mm) and nanoplastics (NPs) (0.001–0.1 μm) (Gigault et al., 2018).

Microplastic pollution in soil became first addressed by Rillig (2012), and there have been an increasing number of studies centered in this crucial subject matter as soil and environmental scientists alike have realized its gravity (Chae & An, 2017; De Souza Machado et al., 2018a; Hurley & Nizzetto, 2018; Mai et al., 2018; Ng et al., 2018; Rillig et al., 2017). On World Environment Day in 2018, the United Nations Environment Programme (UNEP) referred to more research on the effects of microplastic pollutants in the soil environment (UNEP, 2018). One critical factor influencing the UNEP's decision was the fact that soil is probably a greater essential sink for microplastics than marine environments. It's far envisioned that plastic released yearly to the terrestrial environment is 4–23-fold greater than that released to the marine environment (Horton et al., 2017).

In general, the researchers of the studies make a special focus on MPs, ignoring the reality that NPs, the much less-explored plastic fragments, can also have an effect on the soil system (Revel et al., 2018) due to the fact NPs can act as the carriers for pathogens and serve as habitat for microorganism and virus. NPs are more dangerous than MPs as they could penetrate through the biological membranes (Ng et al., 2018). Notwithstanding the significance of MNPs in terrestrial environments, maximum of the MNPs research has focused on the marine environment (Horton et al., 2017; Mofijur et al., 2021; Soares et al., 2020). The enormous amount of MNPs present in the soil leads to unveiling pathways to the microorganisms and human health (Hurley & Nizzetto, 2018).

MNPs impact the soil plant system extensively because they accumulate in the soil and interact with PTEs and organic pollutants, concentrating the soil (Chai et al., 2020). MNPs can have an effect on the increase of plant life and its chlorophyll content material. Besides, MNPs can have interaction with soil organic particles and survive in soil for 100 years which may additionally contaminate the soil properties (physicochemical) and groundwater (Wahl et al., 2021). The crop yield of agricultural soil contaminated with MNPs could be affected negatively due to low pH levels and much smaller number of productive earthworms. Further, MNPs can act as the organic pollutant and pathogens carriers because of their excessive surface area-to-quantity ratio and hydrophobicity (Atugoda et al., 2021; Tiwari et al., 2020).

Microorganisms connected to MNPs can pose threats to the surroundings as they are able to act as the medium in transferring MNPs from the soil to plants and eventually to other living things through the food chain (Chai et al., 2020). The consuming of MNPs by human have a higher threat to several problems, such as reproductive harm, cancer, developmental postpone or organs issues. Visual exam, vibrational spectroscopy, and mass spectrometry are the analysis techniques of MNPs in the soil samples; however these analytical methods are most applicable to identify MPs in place of NPs because of the particles size of NPs that make the detection process tough. Moreover, those analytical methods commonly use samples from marine surroundings, creating the doubt whether that analytical equipment can be utilized in soil samples. Prior to the identification of MNPs, pretreatment and isolation of plastic particles from the samples, which includes soil matrices, are essential to avoid confusion. Nevertheless, there is loss of strong separation and pretreatment techniques from complex environmental samples and this may be a chief impediment (Hurley & Nizzetto, 2018). MNPs can interact with PTEs in soil, inflicting soil toxicity (Antoniadis et al., 2017, 2019; Palansooriya et al., 2020). The control or removal of MNPs particles in the soil is very difficult and hard (Oke et al., 2020; Hou et al., 2020). Currently, the soil and environmental scientists are trying to adjust and invent new techniques to assess the presence of MPs and NPs in soil (Qi et al., 2018). To reduce the impacts of MPs and NPs within the soil systems, there is a need to manipulate MNPs which includes microbial remediation, biotechnology use of biodegradable bioplastics, and public education.

5.2 Scenario of Plastic and Its Waste in the World and in India

The petrochemical sector is seemed as the backbone of plastic production; it is also considered a yardstick for measuring global monetary growth, wherein plastic processing and manufacturing are important. Only 2 million tons/year (worldwide) of plastic was produced in 1950. Since then, it was increased nearly 200-fold and

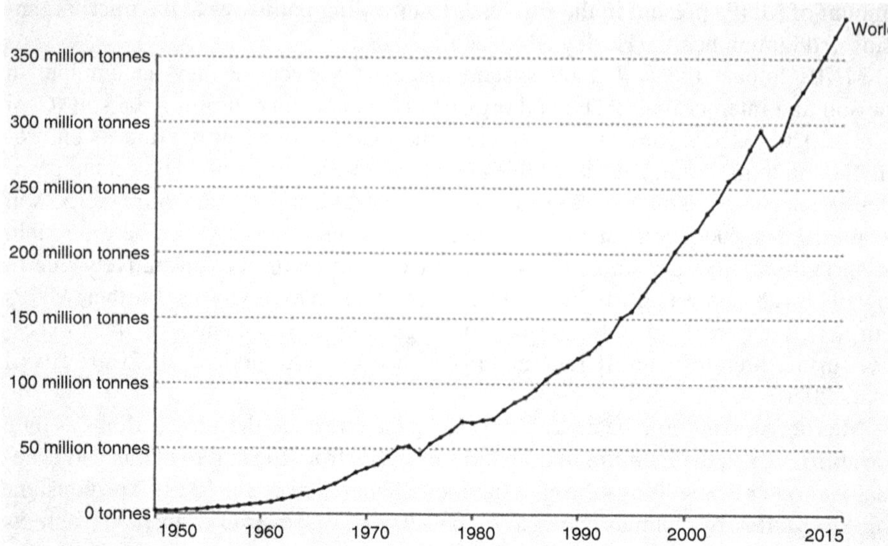

Fig. 5.1 Global plastic production, 1950 to 2015. (Modified from Geyer et al., 2017)

reached to 381 million tons/year in 2015 (Fig. 5.1). It's far predicted that inside the current financial year 2020, exports would cross 8 billion USD with an extended increase of 9.5% within the first half of FY 2018 as compared to the past year. Over 6.3 billion tons of plastic waste were generated globally up to now. That level is alarmingly high, and there are fears that if this situation isn't addressed, the world will turn out to be "drowning" in plastic. Plastic use has appreciably increased through the years; particularly, as it's a reasonably a cheap form of material, it could effortlessly be molded, and unlike paper, plastic keeps ingredients sparkling for longer intervals. Of late, there was a growing trend of making much less long-lasting plastic materials which makes it hard to reuse. But, at a matching rate, the quantity of plastic waste has also grown through the years, no longer just in India, but globally. The overall plastic produced almost 79% entering into the environment as wastes. Only 9% is recycled from the total global plastic waste. The Central Pollution Control Board (CPCB) reports (2018–2019) that 3.3 million metric tons/year of plastic wastes were generated in India. Even this information, horrifying as it is, might be sarcasm. While India's plastic waste trouble isn't always as massive as that of other countries, it's really growing. Some rich states like Delhi and Goa produce 37 grams and 60 grams/capita/day, respectively, against a national average of 8 grams/capita/day. These plastics are referred to as single-use plastics and are said to account for 40% of all plastics manufactured. Statistics also show that from the entire plastics produced globally, a measure 9% has been recycled. Prior to 1980, recycling and incineration of plastic was negligible; 100% was therefore discarded. For incineration in 1980 and recycling in 1990, rates increased on average by about 0.7 percent per year. An estimated 55% of global plastic waste was discarded in 2015, 25% was incinerated, and 20% was recycled (Fig. 5.2).

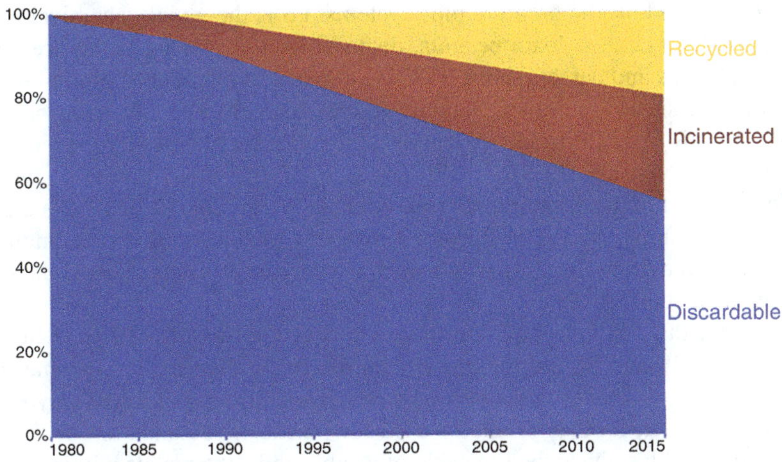

Fig. 5.2 Global plastic waste by disposal, 1950 to 2015. (Modified from Geyer et al., 2017)

The predominant task, however, is segregation and re-aggregation of plastic waste streams together with packaging waste, consisting of laminated plastic. As a lot as 79% of the plastic manufactures in the world ends up in landfills or in the environment in our oceans and our water bodies. We want to be aware of the plastic usage and need to find other alternative energy supplies for the recovery of the material immediately in order to save the environment.

5.3 Different Categories and Uses of Plastic

A main classification of plastics is based on the durability or non-durability of their shapes, or whether they are thermosets or thermoplastics. Thermosets consist of polyurethane, epoxy, and alkyd, and they are regularly used as insulators, adhesives, and plywood. The thermosetting procedure is based on heat-prompted go linking to form new and irreversible covalent bonds, which makes the thermosets strong and not easy to decompose (Rudyak, et al., 2018). On the contrary, thermoplastics have no newly shaped chemical bonds and may be recycled and remolded, making them extra broadly used than thermosets in patron goods (Mattsson et al., 2015; Battulga et al., 2019; Xu et al., 2019; Raddadi & Fava 2019). There are different types of thermoplastics, which are polyethylene (PE), polypropylene (PP), polystyrene (PS), and polyvinyl chloride (PVC). PE is used in a wide range of less expensive plastic products, consisting of plastic bags and bottles. There are two typically used subtypes of PE: (1) the high-density polyethylene (HDPE), which is generally used in milk cans, detergent bottles, and molded plastic cases; and (2) the low-density polyethylene (LDPE) used in outside furniture, siding, ground tiles, clamshell packaging, and bath curtains. PP is mainly used to prepare ingesting straws, bottle cap appliances, yogurt containers, car bumpers, fishing traces, and plastic pressure pipe systems. PS is the mainly chemical used to provide food containers, foam peanuts, disposable cups, plastic tableware, plates, cutlery, CD discs and cassette bins. PVC is an essential element of plumbing pipes and guttering, shower curtains, window frames, and flooring. In addition to the everyday plastic classifications listed above, microplastic fibers (MFs), which might be made of polyester (PES) or PP, are one of the most common types of MPs located within the environment (Hü-er et al., 2018; Cole, 2016). MFs are generally used in an expansion of fibrous substances, inclusive of textile, industrial, agricultural, and household semi-finished or ancillary products used in different fields (Liu et al., 2019).

Generally, PE, PS, and PVC are three important forms of MPs used in scientific research. PE and PS are the popular plastic materials used in customer products, and they have shorter carrier lives than different types of plastics. Moreover, PVC is primarily used for plastic wire insulation or the cable jacket of data cables. Once the life cycle of a cable ends, the metals within the cable could be recycled, but the plastic elements containing PVC are commonly discarded into the environments because of the expensive for separation and recycling value. It was reported that 82% of PVC waste is discarded in landfills, 15% incinerated, and most effective 3% recycled (Suresh et al., 2017). This courting between the large output, brief life cycle and plentiful environmental discharge of these plastics makes them the primary recognition of scientific research (Deng et al., 2015; Yang et al., 2019; Deng et al., 2018; Wu et al., 2019).

5.4 Sources and Formation of MNPs

There are many viable routes through which microplastics can enter soil, along with soil amendment with compost and sewage sludge, plastic mulching, irrigation and flooding, littering, and atmospheric deposition (Bleäsing & Amelung, 2018). Compost and sewage sludge are extensively used as fertilizers, which may serve as a significant source of soil plastic pollutants in these areas. For example, in Ireland, 79.3% of sewage sludge became reused in agricultural lands (EPA, 2015). In Asia, China South Korea, and Japan accounted for 80% of world plastic mulching (Espı et al., 2006). Similarly, in China, plastic mulching increased fourfold from 0.64 million tons in 1991 to 2.60 million tons in 2015 (NBSC, 2019). Fragmentation of massive plastics result in the formation of MPs (>100 nm and <5 mm) Or NPs (<100 nm) (Horton et al., 2017). Larger plastics can be degraded in two mechanisms which are abiotic (mechanical method, solar irradiation, warmth, chemical substances) and biotic degradation (Chamas et al., 2020; Zhang et al., 2021). Plastics can be broken down mechanically through external forces, for instance, freezing and thawing of plastics in the marine environment. Wind and waves can cause plastics to collide with rocks and sands, ensuing in mechanical degradation (Fiend et al., 2018). The impact of external forces is determined by the plastic mechanical properties. Moreover, photodegradation mainly by UV radiation (UV-A and UV-B) is one of the vital procedures to degrade most of the plastics that require free radical-mediated reactions and are brought on by way of solar irradiation (Liu et al., 2019). There are few techniques to assess plastic degradation charge including the mass loss, evolution of carbon dioxide, chemical analysis, thermal analysis, surface analysis, and so on. The MPs in the environment can be leached from the agriculture practices, runoff, unintentional direct disposal, and fragmentation of larger plastic waste. These sources are then split into direct and indirect sources. Microplastics and nanoplastics input agro ecosystems either as primary (synthetic) micro and nano materials (e.g. in medical applications, waterborne paints, coatings, electronics, adhesives), or secondary microplastics and nanoplastics generated by the breakdown indirectly of large plastic debris (Koelmans et al., 2015; Duis & Coors, 2016; Rillig, 2012). Later, it turned into photodegradation of recovered marine microplastic debris (Lambert & Wagner, 2016), 1-cm2-portions of disposable polystyrene espresso cup lid (Lambert & Wagner, 2016) generated nanoplastics. Direct resources in agriculture consist of plastic mulches and soil conditioners and greenhouse materials (e.g., polyurethane foam and polystyrene flakes). Oblique assets consist of fashionable littering and the usage of dealt with wastewater and biosolids (Duis & Coors, 2016; Horton et al., 2017). Microplastic and nanoplastic emissions per capita vary greatly between regions because of populace affluence, size, presence, and efficacy of waste control practices (Ziajahromi et al., 2016; Nizzetto et al., 2016). However, we consciousness on plastics that become in agro ecosystems. Using present statistics and estimates, we have derived capability annual and maximum plastic loadings in agro ecosystems for Europe, United States of America and Australia to illustrate the capability scale of the plastic hassle. MNPs are broadly used within the electronics, plastic goods, automotive, textile, 3D printing, personal care, and cosmetics sectors (particularly toothpaste and exfoliating merchandise) (Tiwari et al., 2020; Carr et al., 2016) at which can be discharged (MNPs) into soil, affecting the soil environment. MNPs may be categorized into two companies which are primary and secondary MNPs (Fig. 5.3). Primary MNPs are processed plastics which are commonly delivered to personal care products (Auta et al., 2017; Anderson et al., 2016; Godoy et al., 2019; Guerranti et al., 2019; Praveena et al., 2018). These PE microbeads are extensively used as exfoliants in detergents, cosmetics, toothpastes, scrub facial cleansers, and drug carriers. Due to this, the primary MNPs especially cleansers serve as stimulus (physical) and after use, it was discharged into the environment (Cheg et al., 2018; Yurtsever, 2019). In addition, a recent study also suggested that glitters that are usually utilized in crafts, cosmetics, and textiles are some other essential supplies of plastic infection caused by primary MNPs (Yurtsever, 2019). The secondary MNPs are plastic debris that degrades from the massive portions of plastics because of UV radiation, physical wear, and biodegradation within the environment (Sharma & Chatterjee, 2017; Lehner et al., 2019; Adawi et al., 2018; Auta et al., 2017; Deng et al., 2017; Yurtsever, 2019).

Once the plastic enters into the environment, these plastics become brittle because of UV radiation that catalyzes the photooxidation. Similarly due to the wind, waves, and other abrasive interactions, the structural integrity of the

Fig. 5.3 Different categories of plastic

plastics further break and form MNPs (Sharma & Chatterjee, 2017; Lehner et al., 2019; Adawi et al., 2018; Auta et al., 2017; Deng et al., 2017; Yurtsever, 2019; Au et al., 2017; Hebner & Maurer-Jones, 2020; Weinstein et al., 2016; Wright & Kelly, 2017). Those outcomes suggest that both MPs and NPs can be produced within the degradation process of disposable plastic waste and accumulate over time (Lambert & Wagner, 2016; Sommer et al., 2018; Kole et al., 2017).

5.5 General Soil Properties

The properties of soil are determined by the structure of the soil, depending on different amounts of biotic and abiotic components. The combinations of these components determine the properties (physical and chemical) of soil.

5.5.1 Physical Properties

The following physical properties of soil are:

(a) *Soil Texture*

- Soil texture depends on the soil particle size that is based on the related components of minerals like silt, sand, and clay.
- Soil texture is an addition to the soil infiltration, porosity, and water retention capacity.
- The texture of soil differs with soil type; silt feels smooth, sandy soil feels gritty, and clay is sticky and moldable.

(b) *Soil Structure*

- The components of soil structure, together with sand, silt, and clay, might develop in aggregates because of their clumping and after together to form peds.
- This soil structure gives information on the matter content, soil texture, and biological activity.
- Soil structure is formed by physical processes that are probably improved or destroyed by the farming practices.

(c) *Soil Density*

- The common soil density ranges from 2.60 to 2.75 grams/cm^3, which generally remains unchanged for a given soil.
- The soil particle density is lower for soils with high organic matter content and higher for soil with higher mineral content.
- Soil particle density varies from soil bulk density which is less than soil particle density.
- Soil density usually depends on the soil structure and texture of the composition of the soil.

(d) *Soil Porosity*

- Soil porosity means the number of pores present within the soil.
- The determination of air and water movement within the soil is known as porosity.
- Usually, more number of pores between and within soil aggregates mean this soil is healthy, whereas few pores or cracks soil mean poor quality soils.
- The structure and texture of soil influence the soil porosity. Based on these pore sizes, the food substances like water, oxygen, and other gases/minerals will enter into the plants and other organisms.

(e) *Soil Consistency*

- Soil consistency refers to the ability of the soil to stick to itself or other objects and to resist deformation and rupture.
- Three moisture conditions define soil consistency: air-dry, moist, and wet.
- The consistency of dry soil ranges from loose to hard, whereas that of wet soil ranges from non-sticky to sticky.
- Soil consistency is an important property that determines the ability of soil to support buildings and roads.

(f) *Soil Color*

- Soil color is determined primarily by the organic composition of the soil.
- Soil color is one of the factors that help in the prediction of other soil characteristics within a soil profile.
- The quality of soil can be identified by the color; it means measuring organic matter, iron oxide, and the clay contents of the soil.
- Besides, soil color is also influenced by the mineral content of the soil as the color might change as a result of oxidation of degradation.

5.5.2 Chemical Properties

(a) *Cation Exchange Capacity (CEC)*

- At particular pH, the maximum holding capacity of total cations that a soil sample is known as cation exchange capacity.
- In soil, the cation exchange capacity is taken as an indicator of nutrient retention, soil fertility, and the ability of soil to protect groundwater from cation contamination.

(b) *Soil pH*

- The reactivity of soil is expressed in terms of the soil pH, which determines the acidity and alkalinity of the soil.
- It is the measure of the concentration of hydrogen ion in the aqueous solution of soil which ranges between 3.5 and 9.5.
- Usually, the higher amounts of manganese and aluminum present in the soil have high acidity, and higher concentration of sodium carbonate has higher alkalinity.
- In terms of soil fertility, agricultural production tends to be more in acidic soil.

(c) *Soil salinity*

- Salts in the soil are transported from salt tables in water resources that then accumulate due to evaporation.
- During irrigation processes the salinization of soil also occurs from drainages. The salt accumulation affects the degradation of organic matter in soil and the vegetation on the soil.
- The most common salts that are present in soil include magnesium sulfate, potassium sulfate, and carbonates.

5.6 Impact on Soil Properties

Once in soil, MNPs accumulation may cause a series of adverse effects in soil ecosystems. MNPs may alter soil physical properties, decrease soil fertility, and disrupt resident microbial groups, therefore affecting soil quality and nutrient cycling (Awet et al., 2018; De Souza Machado et al., 2018b; Wan et al., 2019; Zhu et al., 2018). Moreover, these MNPs ingested by way of earthworms had been observed to be transferred through the food chain, which may also pose a potential threat to terrestrial predators and even human beings (Huerta Lwanga et al., 2017). Several research have simulated the abiotic transport of microplastics, especially nanoplastics, in soil using column experiments packed with porous media (Alimi et al., 2018; He et al., 2018; O'Connor et al., 2019). The movement of earthworms also can provide a pathway for downward microplastic transport, potentially into groundwater systems (Yu et al., 2019). Similarly, microplastic generally tends to adsorb various classes of potentially toxic chemicals, which include polycyclic aromatic hydrocarbons (PAHs), polychlorinated biphenyls (PCBs), dichlorodiphenyltrichloroethane (DDT), hexachlorocyclohexanes (HCHs), pharmaceuticals and personal care products (PPCPs), pesticides, perfluoroalkyl substances (PFASs), and heavy metals (Engler, 2012; Holmes et al., 2012; Lee et al., 2014; Liu et al., 2017; Llorca et al., 2018; Seidensticker et al., 2018; Velzeboer et al., 2014; Wu et al., 2019; Xu et al., 2019). Significantly, due to the sorption of contaminants (e.g., organic and inorganic compounds) onto microplastics (in particular nanoplastics), the migration of plastic debris likely enables the transport of absorbed contaminants throughout soil and contributes to a more ecological chance (Liu et al., 2018; Liu et al., 2019).

5.6.1 Soil Chemical-Physical Properties

The presence of MPs and NPs in soil aggregates can alter biological, chemical, and bodily properties of soil (Rillig et al., 2017; Moreno & Moreno, 2008; Atuanya et al., 2012; Kasirajan & Ngouajio, 2013) and have an effect on the estimation of soil carbon sequestration (Santos et al., 2017). The effect of plastics on soil combination formation and humic acid properties are shown in Fig. 5.4. Certainly, Atuanya et al. (2012) observed that the addition of plastic granules to soil improved the whole natural carbon content of soil because the modern-day methods used to quantify the soil organic carbon also determine the invisible MP fraction of soil aggregates (e.g., polystyrene or polyethylene incorporate almost 90% carbon) (Atuanya et al., 2012). Consequently, Riling advocated re-comparing what's the "actual" soil carbon storage in plastic-polluted soils. De Souza Machado et al. (2018a, 2018b) studied the results of MPs with shape, density and chemical composition on the bulk density, water conserving capability, water strong aggregates, and microbial sports of the soil.

As previously mentioned, the interactions of MNPs with soil reactive components and with the main extracellular biological molecules may affect soil functionality, which could directly affect the soil fertility and consequently yield and quality of crops. But, only a few reviews confirm the environmental relevance of MPs in soil. MPs and NPs can interact with various practical of the dissolved fraction (dissolved organic count number, DOM), which can be associated with inorganic pollution (Chen et al., 2014, 2017; Sun et al., 2017). The end result would be the formation of natural complexes being probably very toxic and shifting thru the soil profile (Chen et al., 2018a). Certainly, humic substances interacted with MPs (phenanthrene) and heavy metals (Pd and Cd) while being adsorbed on a clay mineral (bentonite) (Zhang et al., 2015). MPs can have interaction with the fragrant DOM structure via conjugation and then with carboxyl organizations and C=O bonds (Chen et al., 2018a). Microplastics assume to act as electron donors to humic materials main to highly conjugated co-polymers with an improved electron density (Fig. 5.4). Those consequences depend on the plastic size and the pH of the

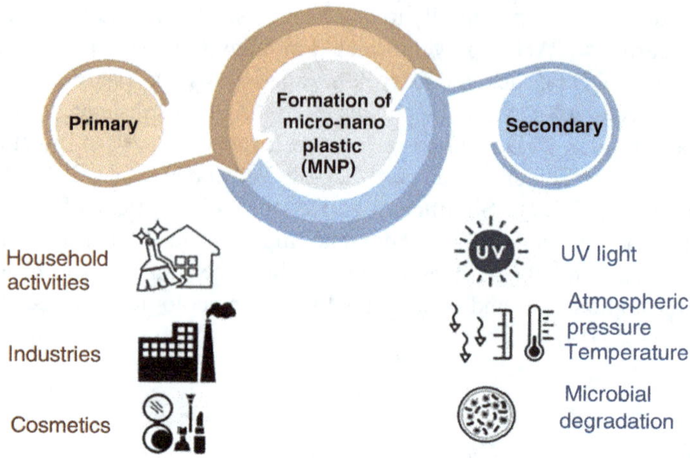

Fig. 5.4 Formations of MNPs. Primary MNPs are produced for a specific purpose, for instance, microbeads in cosmetics, microfibers in textiles, etc. On the other hand, secondary MNPs are produced due to "environmental actions": sunlight, wave actions, and microbial degradation

soil answer. On the other hand, NPs can boost up the kinetic assembly rate of DOM through susceptible electrostatic interactions and hydrophobic interactions as a consequence forming the particulate natural count (Chen et al., 2018b). The interaction of MPs with natural compounds also relies upon at the plastic's age, but contradictory variations in reactivity among the elderly and the relative unique plastics were observed (Hü-er et al., 2018; Liu et al., 2019). Interactions of MPs with natural matter can also affect the nutrient availability to biota in soil, as an instance, with the aid of reducing the dissolved organic N and dissolved natural P forms (Liu et al., 2017).

5.6.2 *Soil Microbial and Enzyme Activities*

Recent studies have focused on the potential of microplastics to affect or disturb soil microbial communities and enzymatic activities, but the findings of these studies disagree on the magnitude of the impact of microplastic pollution on these endpoints. Polyacrylic and polyester microfibers (0.1% w/w) significantly decreased soil microbial activities (De Souza Machado et al., 2018b). Polystyrene (PS) nanoplastics at 100 and 1000 ng g^{-1} significantly decreased soil microbial biomass and increased basal respiration, respectively (Awet et al., 2018). An intensely enhance the rate (28% w/w) of microplastics PP expanded up to 3 times of respiration rate of soil basal (Yang et al., 2019). And the presence of PS nanoplastics (100 ng g^{-1}) in silt loam soils may reduce the enzyme activity which are involved in the C, N, and P cycles (Awet et al., 2018), while PP microplastics (28% w/ w) mainly stimulated the activity of fluorescein diacetate hydrolase in sandy loam soils and hence increased nutrient availability to plants by enhancing microbial hydrolytic activity on SOM (Liu et al., 2017; Yang et al., 2019). Likely the effect of microplastic on soil physical properties and soil microbial and enzyme activities also depend on size, shape (linear vs. nonlinear), addition rate, soil texture, and polymer composition. Plastic film residues (67.5 kg ha^{-1}) can cause significant declines in soil microbial community diversity and decrease soil microbial C and N, as well as decrease the activity of fluorescein diacetate hydrolase and dehydrogenase by 10% and 20%, respectively (Wang et al., 2016). However, it determined that these negative effects might be related to concomitant phthalate pollution and not to the occurrence of plastic film residues (Wang et al., 2016). Overall, anthropogenic microplastic pollution may alter soil microbial community diversity, as well as the activity of enzymes and microbiota in the soil and thus disturbs microbial ecosystems and affects soil nutrient cycles. Due to the environmental stresses exerted on soil microbial communities, the adaptation and evolutionary response of soil microorganisms to microplastics in soil should be considered and addressed in future research (Rillig et al., 2018).

5.6.3 Soil Microbial Community

The transport of microbial species through plastic waste, specifically the function of MNPs in transporting microorganisms, was poorly identified because of methodological issues. Pesticides can pass through the soil with MNPs as hypothesized (Horton et al., 2017). The concern on those problems should suggest destiny studies with the intention to aware on MNPs role not only as vectors of microorganisms however also as vector of pollutants. MPs can affect a few microbial residences along with MNP-associated microorganism, which confirmed higher plasmid transfer costs than free-living microorganism (McCormick et al., 2014). Because the elimination of biofilm through bacterial community can collect a large antibiotic resistance potential, it is important to hypothesize that MPs might also act as vector for antibiotic resistance (McCormick et al., 2014). In the biofilm, the DNA transfer is involved by both transformation and conjugation (Li et al., 2001). The cell function is affected by the NPs through the cell's lipid membranes (Rossi et al., 2014). But the prevention of NPs from the cells through microorganisms by means of adopting exceptional self-protective mechanisms, which include secretion of molecules, changes of intracellular membrane and neutralizing the contaminants (Henriques & Love, 2007; Leriche et al., 2003) and any biofilm matrix and other the bacterial cell wall barriers (Jing et al., 2014; Tang et al., 2018). Perhaps the motion of cells from a biofilm state which is in a planktonic state will contact with NPs (Jing et al., 2014). The biofilm action for the protective towards toxicity of NPs is exerted with the aid of active and passive mechanisms. The passive safety is because of the biofilm matrix properties (physical-chemical), consisting of practical agencies and the high density at the extracellular polymeric materials that are efficient of binding and entrapping NPs on the surface of biofilm layer, respectively (Feng et al., 2018). Further, NPs separate from biofilms and after they penetrate on the surface of biofilm (Jing et al., 2014), which is probably because of tendency NPs to form hetero-aggregates within the presence of natural organic matter and inorganic colloids the bacterial pastime (Jing et al., 2014). But the NP presence may result in resilience and useful redundancy houses within the bacterial network inhabiting the biofilm (Tang et al., 2018). There is no exact information about the soil MNPs because of the soil microbial composition. So, further researches should fill those gaps by way of monitoring the relative consequences via figuring out microbial diversity via metagenomics or amplicon sequencing (Schöler et al., 2017; Vestergaard et al., 2017).

5.7 Soil Fauna

Soil biota plays a significant role in soil ecosystem techniques and gives a number of environment services such as decomposition of natural count, nutrient biking, bioturbation, and suppression of soil-borne illnesses and pests (Brussaard, 1997). According to the body width of soil fauna, they're categorized into three categories, i.e., microfauna (<0.1 mm), mesofauna (0.1–2 mm), and macrofauna (>2 mm) (Orgiazzi et al., 2016; Wardle, 2002).

5.8 Soil Quality

5.8.1 Soil Physical Environment

Numerous studies have supplied restricted statistics concerning the have an effect on of MNPs on soil physical homes. The prevailing proof has revealed that the impact of MNPs on soil residences depends on plastic type. Those with a shape and size extra just like soil debris have a much less pronounced effect on soil structure and water cycle. as an example, polyester fibers (0.4%,w/w) appreciably reduced soil bulk density and water-strong aggregates, and extended water-conserving capacity and evapotranspiration, at the same time as other kinds of microplastics, together with PE fragments or polyamide (PA) beads (2.0%w/w) prompted results with a lower significance (De Souza Machado et al., 2018a, 2018b, 2019; Lau et al., 2018). Furthermore, polyester fibers had been found to significantly decrease the bulk density of sand soils (0.4% w/w), however not clay loam soils (0.3% w/w) (De Souza Machado et al., 2018b; Zhang et al., 2021).

Consequently, it is speculated that soil texture is an important aspect to decide the impact of MNPs on soil properties; however extra research are required to test this speculation. Polyester fibers (0.3% w/w) have additionally been observed to affect the pore shape of a clay loam soil (Zhang et al., 2019). PE movie (1% w/w) has been observed to increase the rate of water evaporation in a clay soil to a full-size diploma (Wan et al., 2019). In addition, plastic-film residues (15 kg ha^{-1}) instigated significant changes to preliminary gravimetric water content, bulk density, total porosity, and soil water distribution (Jiang et al., 2019). These deviations from a natural state suggest that the presence of microplastics should pose a capacity hazard to soil ecosystems.

5.8.2 *Soil Chemical Fertility*

Microplastics may also affect the soil chemical properties. An amendment rate of 28% (w/w) microplastics significantly increased the concentrations of dissolved organic C, inorganic N, and total P in a sandy loam soil, but these changes were not observed at a reduced amendment rate of 7% (w/w) over 30 days (Liu et al., 2017). Soils with a range of textures and long-term effects need to be considered in future research to gain a better understanding of these phenomena in the field. Due to the presence of plastic residues, the soil organic matter (SOM) and alkali-hydrolysable N at 500 kg ha^{-1} and available N and P by 55% and 60%, respectively, at 2000 kg ha^{-1} (Dong et al., 2015). Those findings recommend that plastic-mulching residues certainly decrease soil fertility and possibly result in decreased plant growth while the impact of microplastic under environmental conditions isn't properly understood (Fig. 5.5).

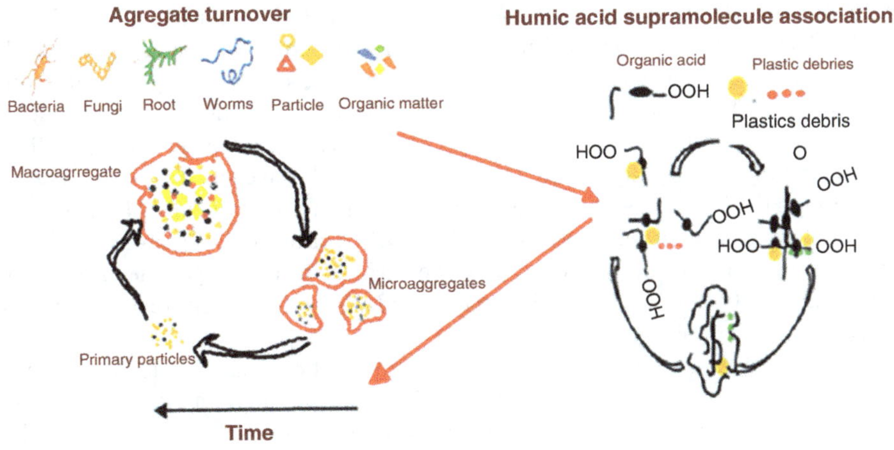

Fig. 5.5 Plastics effects on humic acid properties and soil aggregate formation (OM Organic Matter)

5.9 Soil Pedogenesis

An interesting issue mentioned above is the effect of MNPs on the soil properties, due to an extended residence time and excessive reactivity along with the soil pedological procedures. It's far feasible to hypothesize the presence of MNPs characterizing thing to categories the soil horizons (surface and subsurface). Furthermore, how should this debris alter the pedological processes isn't discussed within the research study to our expertise. This feasibility is preeminent and harbinger of thrilling trends. In this context, it's far crucial to don't forget these days observed pyro-plastics originating from the regularly burning waste particles (Ehlers & Ellrich, 2020; Turner et al., 2019). These plastics may also emerge as part of the soil's geological cycle due to their recalcitrance to degradation.

5.10 Plants

The contamination of soil plastics could apply a direct and indirect influence developed plants as an outcome of their root take-up or impacts on soil substance physical-chemical and organic characteristics individually. Concerning the direct impacts over the most recent 2 years, there were various studies of MNPs' effect in plants (Huerta Lwanga et al., 2017); the pollutants in plant metabolism or their stockpiling of headstrong pollutants could be the primary issue (Sandermann, 1992). The plant take-up of MNPs depends upon the physiological and anatomical characters of the plant on plastics properties, particularly those of the eco-corona and environmental ageing affecting surface chemistry and behavior (Ng et al., 2018). Li et al. (2019) have established that plastic material with size up to 0.2 μm can be absorbed by roots (lettuce) intercellular presence of plastics as a "grape-like" group. As of late, Sun et al. (2020) revealed that take-up of NPs altogether relies on their surface. The researchers showed that collection and take-up of adversely charged NPs was a lot higher in attaches contrasted with emphatically charged NPs, due to their high restricting proclivity with extremist adhesive and their size expanding through the hetero-conglomeration acceptance by root's exudates. Nonetheless, these outcomes do not bar the danger of the MP section into the human pecking order for food as a result of their grip to the surfaces of root vegetables and salad (Sun et al., 2020). Smith et al. (2018) studied on MPs from modern created fertilizer could aggregate in plants compost.

The immoderate accumulation of plastic in the root can cause various problems for the plants which include the disruption of the transport system (nutrient) via the interference of the cell connection and/or the pores in the cell wall (Jiang et al., 2019) and the extreme production of reactive oxygen species (ROS) (Jiang et al., 2019), and the plant disease resistance can reduce by inducing down regulation of disease resistance genes (Sun et al., 2020). Further these plastic particles enter into the roots, stems, and leaves through the vascular system following the transpiration and finally detected at intercellular level as "string-like" cluster and dispersed forms, respectively (Li et al., 2019). The NPs (<0.1 um) can normally enter in the cell membrane, but the MPs cannot enter inter the cell membrane because of their particle size (from 0.1 μm to 5 mm).

NPs can take up by the plant cell endocytosis, through ion transport channels along with the aquaporins or carrier proteins. The particle size of NPs is important for plant uptake due to nano-beads (polystyrene) with a diameter of 20 to 40 nm were entered into tobacco cells, but not those of 100 nm (Bandmann et al., 2012; Kettler et al. (2014) advised that the threshold value is 50 nm. And it also depends on physiological and anatomical properties of the plant; plant species; properties of nanoparticle; plant parts, especially the eco-corona; environmental ageing; and activity of NPs, affecting the surface chemistry. More investigations is need to well known MNPs translocation, toxicity and storage on plants and the defense mechanisms of plants against NPs (Wang et al., 2013).

In indirect MP and NP effect on cultivated plants, the primary form to be considered is their effects on nutrient immobilization, soil structure, impurities in diffusion and adsorption, soil microbial community root-associated microbiome, and root symbionts. The toxicity of plastic waste could cause degradation in soil; and this may evidence the plants' high enantioselectivity, which is as recently investigated for hexabromocyclododecanes (HBCDs) and pentabromocyclododecenes (PBCDEs) monomers (Huang et al., 2018).

5.11 Biological Indicators in Soil

Plant growth and seed germination (ISO 11269-2 Soil Quality–Part2, 2012), focusing on the inhibition of nitrification and toxicity to earthworms (Bandmann et al., 2012), act like biological indicators of plastic toxicity. As previously indicated, these soil filtering earthworms can accumulate both MPs and NPs into the soil, and thus these earthworms can extract plastics from the soil (Zhu et al., 2018). The microbial degradation of MPs can produce volatile compounds (VOCs), includes ethylene and methane, MPs indicators presence in the analyzed sample (Huerta & Wanga et al., 2018; Kyaw et al., 2012; Royer et al., 2018) . However, the application of this process is aimed by the lack of awareness by (such as those due to microbe-microbe and plant-microbe interactions) producing VOC mechanism during the degradation of plastic and by various factors, such as moisture, pH, clay minerals and organic carbon content, and different microbial diversities affecting this production in soil (Serrano et al., 2006; Heribert, 2014).

5.12 Environmental Risk of MNPs in Soil

MNPs entering the soil pose a potential environmental hazard to terrestrial ecosystems. Those anthropogenic substances may also pressure environmental changes in soil that cause pressure to soil fauna (De Souza Machado et al., 2018a; Rillig et al.,

2017). But few research have documented the environmental effect of MNPs on the soil environment. It is critical to conduct further studies on the risk of these classes of pollutants so one can direct efforts to address their presence inside the environment.

5.13 Environmental Management of MNPs

Micro-nano plastics (MNPs) are very hard to degrade within the soil. So, we must mitigate or minimize the impacts of MNPs at the soil environment and human beings, which are legislation (regulation on the plastic waste management and plastic manufacturing in various industries), technical (biodegradable bioplastics and microbial biotechnology) and social (public education on reducing using single-use plastics or disposable plastics, adopting recycling habits, use of biodegradable products).

First, biological technology using organic retailers (bacteria and fungus) and metabolic enzymes has been advised and explored due to their competencies to degrade natural and synthetic polymers. Microbes will adhere to the plastic surface, which bring about the formation of microbial biofilm. Following that, microorganisms secrete extracellular enzymes and exopolysaccharides to adhesion of biofilm on the plastic surfaces, biodeterioration triggering, and breakdown of the plastic substances into monomers, dimers, and oligomers. Ultimately, mineralization takes place with microbial biomass, carbon dioxide, and water as the end products (Ganesh Kumar et al., 2020). From the past research, microorganisms that have proven to degrade plastics were *Streptomyces setonii, Pseudomonas aeruginosa, Rhodococcus ruber*, *Pseudomonas stutzeri, Streptomyces badius, Aspergillus niger, Aspergillus flavus,* and *Fusarium lini* (Pathak & Navneet, 2017). Liang et al. (2016) isolated a bacterial strain *Pseudomonas tamsuii* TKU0155 from Taiwanese soil, to reduce PLA, fibrinogen, and tributyrin successfully, except casein, triolein, and poly (β-hydroxybutyrate). Gajendiran et al. (2016) demonstrated the biodegradable capability LDPE *(Aspergillus clavatus)* of an aqueous medium. The degradation of LDP E became observed through the burden and morphological changes through microscopy and CO_2 evolution test (Sturm test). Mor and Sivan (2008) proved the potential of biofilm-producing *Rhodococcus ruber* in inducing partial degradation of PS. The biodegradation of plastic substances can be influenced by the microorganism species involved, resources of carbon, and size and types of plastic substances. Microorganism can utilize the plastic substances for the carbon sources through biodegradable procedure, but the high stability of MNPs in the environment is making them hard to be used as carbon resources. Consequently, biodegradation of MNPs necessitates the suitable conditions, which are not always possible in field conditions (Shen et al., 2019). Biotechnology, for example, enzyme engineering, stress engineering, and metagenomics, can be used to enhance microbial generation or boost up enzyme activity within the biodegradation of plastic materials or MNPs. Wei et al. (2016) confirmed the residual exchange of amino acid of polyester hydrolases, Tf Cut2 from *Thermo bifidafusca* KW3 with LC-cutinase

(LCC) elevated the PET degradation. Islam et al. (2019) investigated the cutinases from *Thermomonospora curvata* through the combination peptide tachystatin A2 which facilitated the enzyme kinetics and biodegradation rate of polyester-polyurethane nanoparticles (NPs). Huang et al. (2018) utilized a strain by means of inactivating dual arginine translocation complexes and expanded the secretion of PETase by 3.8-fold in *Bacillus subtilis* strain to increase the MP/NP degradation, specifically PET. Moog et al. (2019) used *Phaeodactylum tricornutum* (a photosynthetic microalga) as a chassis to produce engineered model of PETase within the bacterium *Ideonella sakaiensis* to degrade PET. Apart from the microbial generation, physicochemical-biological treatments also may be used to degrade MNPs. Siipola et al. (2020) studied on the production of low-price biochar from the bark of Scots pine *(Pinus sylvestris) and spruce (Picea sp.) trees in the remediation of chemical contaminant complexes from MPs or immobilize larger MPs particles. Cunha et al. (*2020) determined the bioflocculant *from Cyanothece* sp, which can EPS to mixture NPs and MPs. MNPs induced a negative impact on the microalgal growth up to 47%. Similarly, to remediate MNPs, few chemical substances can also be used. Ramirez Arenas et al. (2020) assessed the capability of nanoparticles (TiO_2/CeO_2) collectively with chemical coagulants, polyaluminum chloride, or iron chloride to remove PS NPs through water treatment method. The findings suggest that polyaluminum chloride became greater green in comparison with iron chloride seeing that all nanoparticles are coagulated at low dosage at the turn aspect, NPs coagulation was found much less efficient compared with TiO_2 and CeO_2 NPs, indicating that NPs was more solid and harder to dispose of. Additionally, the usage of bioplastics (biodegradable materials) can assist to reduce the MNP effect in the soil environment; these biopolymers biodegrade completely into carbon dioxide without producing any other hazardous products. Biodegradable bioplastics (PLA, polyhydroxyalkanoates) that made from distinctive biomass feed stocks, for instance, microalgae, could help resolve the problem of plastic pollution within the environment (Chia et al., 2020; Mofijur et al., 2021; Zhang et al., 2019). The legislation or policies have to be focused to reduce the plastic waste (MNPs) in the environment, for instance, reusable and recyclable plastic manufacturing products, with restriction of MPs in personal care products and cosmetics in some European countries (Boyle & Ormeci, 2020). Under REACH regulation in 2019, ECHA reported that the limitation of MPs to products and look forward to reduce and release of 500,000 tonnes of MPs. Besides, for the products containing MPs, there may be a necessity for labeling to reduce MPs release and enhance their right disposal as well as pre considered necessary for monitoring and reporting in order to increase data collection and discover feasible future risks.

Some countries, for example, Malaysian authorities, have implemented "No Plastic Bag" campaign in any state imposing the minimal fee of RM0.20 per plastic bag to customers in all commercial enterprise premises. This has recommended the citizens to carry their own buying bags and decrease the plastic utilization, adapting the "green" lifestyle. The implementation of government law can help to reduce and control using plastic, especially MNPs in numerous industries. Lastly, public schooling is likewise critical in dealing with and decreasing MNPs within the environment.

Public training also aids in increasing consciousness to the public about the terrible effects of MNPs to the environment. The public community wants to be educated to practice 3 "R" in their life, for instance, reusing plastic baggage; reducing using single-use disposable plastics, plastic bottles, or plastic straws; and adopting recycling habits. While grocery buying, it's really helpful to use personal buying bags, use recycle bag, and buy bins for bottles to lessen using plastic. When buying foods for take-out at the café or restaurants, use personal food boxes to reduce the usage of Styrofoam.

References

Adam, M. G., Tran, P. T. M., Bolan, N., & Balasubramanian, R. (2021). Biomass burning-derived airborne particulate matter in Southeast Asia: A critical review. *Journal of Hazardous Materials, 407, 124760.* https://doi.org/10.1016/j.jhazmat.2020.124760

Adawi, H. I., Newbold, M. A., Reed, J. M., Vance, M. E., Feitshans, I. L., Bickford, L. R., et al. (2018). Nano-enabled personal care products: Current developments in consumer safety. *Nano, 11*, 170–179. https://doi.org/10.1016/j.impact.2018.08.002

Alimi, O. S., Farner Budarz, J., Hernandez, L. M., & Tufenkji, N. (2018). Microplastics and nanoplastics in aquatic environments: Aggregation, deposition, and enhanced contaminant transport. *Environmental Science & Technology, 52*(4), 1704–1724. https://doi.org/10.1021/acs.est.7b05559

Anderson, A. G., Grose, J., Pahl, S., Thompson, R. C., & Wyles, K. J. (2016). Microplastics in personal care products: Exploring perceptions of environmentalists, beauticians and students. *Marine Pollution Bulletin, 113*, 454–60. https://doi.org/10.1016/j.marpolbul.2016.10.048

Antoniadis, V., Levizou, E., Shaheen, S. M., Ok, Y. S., Sebastian, A., Baum, C., et al. (2017). Trace elements in the soil-plant interface: Phytoavail ability, translocation, and phytoremediation– A review. *Earth Science Reviews, 171*, 621–645. https://doi.org/10.1016/j.earscirev.2017.06.005

Antoniadis, V., Shaheen, S. M., Levizou, E., Shahid, M., Niazi, N. K., Vithanage, M., et al. (2019). A critical prospective analysis of the potential toxicity of trace element regulation limits in soils worldwide: Are they protective concerning health risk assessment? A review. *Environment International, 127*, 819–847. https://doi.org/10.1016/j.envint.2019.03.039

Atuanya, E. I., Aborisade, W. T., & Nwogu, N. A. (2012). Impact of plastic enriched composting on soil structure, fertility and growth of maize plants. *European Journal of Applied Science, 4*, 105–109.

Atugoda, T., Vithanage, M., Wijesekara, H., Bolan, N., Sarmah, A. K., Bank, M.S., et al. (2021). Interactions betweenmicroplastics, pharmaceuticals and personal care products: Implications for vector transport. *Environment International, 149*, 106367. https://doi.org/10.1016/j.envint.2020.106367

Au, S. Y., Lee, C. M., Weinstein, J. E., van den Hurk, P., & Klaine, S. J. (2017). Trophic transfer of microplastics in aquatic ecosystems: Identifying critical research needs. *Integrated Environmental Assessment and Management, 13*, 505–509. https://doi.org/10.1002/ieam.1907

Auta, H. S., Emenike, C. U., & Fauziah, S. H. (2017). Distribution and importance of microplastics in the marine environment: A review of the sources, fate, effects and potential solutions. *Environment International, 102*, 165–176. https://doi.org/10.1016/j.envint.2017.02.013

Awet, T. T., Kohl, Y., Meier, F., Straskraba, S., Grün, A. L., Ruf, T., et al. (2018). Effects of polystyrene nanoparticles on the microbiota and functional diversity of enzymes in soil. *Environmental Sciences Europe, 30*, 11. https://doi.org/10.1186/s12302-018-0140-6

Bandmann, V., Müller, J. D., Köhler, T., & Homann, U. (2012). Uptake of fluorescent nano beads into BY2-cells involves clathrin-dependent and clathrin-independent endocytosis. *FEBS Letters, 586*, 3626–3632.

Barnes, D. K. A., Galgani, F., Thompson, R. C., & Barlaz, M. (2009). Accumulation and fragmentation of plastic debris in global environments. *Philosophical Transactions of the Royal Society B, 364*(1526), 1985–1998.

Battulga, B., Kawahigashi, M., & Oyuntsetseg, B. (2019). Distribution and composition of plastic debris along the river shore in the Selenga River basin in Mongolia. *Environmental Science and Pollution Research International, 26*, 14059–14072. https://doi.org/10.1007/s11356-019-04632-1

Bleäsing, M., & Amelung, W. (2018). Plastics in soil: Analytical methods and possible sources. *Science of the Total Environment, 612*, 422–435.

Boyle, K., & Örmeci, B. (2020). Microplastics and nanoplastics in the freshwater and terrestrial environment: A review. *Water, 12*. https://doi.org/10.3390/w12092633

Brussaard, L. (1997). Biodiversity and ecosystem functioning in soil. *Ambio, 26*(8), 563–570. Retrieved from https://www.jstor.org/stable/4314670

Chae, Y., & An, Y. J. (2017). Effects of micro- and nanoplastics on aquatic ecosystems: Current research trends and perspectives. *Marine Pollution Bulletin, 124*, 624–632. https://doi.org/10.1016/j.marpolbul.2017.01.070

Chai, B., Wei, Q., She, Y., Lu, G., Dang, Z., & Yin, H. (2020). Soil microplastic pollution in an E-waste dismantling zone of China. *Waste Management, 118*, 291–301. https://doi.org/10.1016/j.wasman.2020.08.048

Chia, W. Y., Ying Tang, D. Y., Khoo, K. S., Kay Lup, A. N., Chew, K. W. (2020). Nature's fight against plastic pollution: algae for plastic biodegradation and bioplastics production. *Environmental Science & Technology, 4*, 100065. https://doi.org/10.1016/j.ese.2020.100065

Chamas, A., Moon, H., Zheng, J., Qiu, Y., Tabassum, T., Jang, J. H., et al. (2020). Degradation rates of plastics in the environment. *ACS Sustainable Chemistry & Engineering, 8*, 3494–3511. https://doi.org/10.1021/acssuschemeng.9b06635

Carr, S. A., Liu, J., & Tesoro, A. G. (2016). Transport and fate of microplastic particles in wastewater treatment plants. *Water Research, 91*, 174–182. https://doi.org/10.1016/j.watres.2016.01.002

Cheg P. K, Ouyang, Z.-Y., Qian, C., & Yu, H.-Q. (2018). Induced structural changes of humic acid by exposure of polystyrene microplastics: A spectroscopic insight. *Environmental Pollution, 233*, 1–7. https://doi.org/10.1016/j.envpol.2017.10.027

Chen, W., Qian, C., Liu, X. Y., & Yu, H. Q. (2014). Two-dimensional correlation spectroscopic analysis on the interaction between humic acids and TiO2 nanoparticles. *Environmental Science & Technology, 48*, 11119–11126.

Chen, W., Liu, X. Y., & Yu, H. Q. (2017). Temperature–dependent conformational variation of chromophoric dissolved organic matter and its consequent interaction with phenanthrene. *Environmental Pollution, 222*, 23–31.

Chen, W., Ouyang, Z. Y., Qian, C., & Yu, H. Q. (2018a). Induced structural changes of humic acid by exposure of polystyrene microplastics: A spectroscopic insight. *Environmental Pollution, 233*, 1–7.

Chen, C. S., Le, C., Chiu, M. H., & Chin, W. C. (2018b). The impact of nanoplastics on marine dissolved organic matter assembly. *Science of the Total Environmental, 634*, 316–320.

Chen, Q., Li, Y., & Li, B. (2020). Is color a matter of concern during microplastic exposure to Scenedesmus obliquus and Daphnia magna? *Journal of Hazardous Materials, 383*, 121224.

Cole, M. (2016). A novel method for preparing microplastic fibers. *Scientific Reports, 6*, 34519. https://doi.org/10.1038/srep34519

Cunha, C., Silva, L., Paulo, J., Faria, M., Nogueira, N., & Cordeiro, N. (2020). Microalgalbased biopolymer for nanoand microplastic removal: A possible biosolution for wastewater treatment. *Environmental Pollution, 263*, 114385. https://doi.org/10.1016/j.envpol.2020.114385

Deng, Y., Zhang, Y., Lemos, B., Ren, H. (2017). Tissue accumulation of microplastics in mice and biomarker responses suggest widespread health risks of exposure. *Scientific Reports, 7*, 46687. https://doi.org/10.1038/srep46687

Deng, Y. F., Zhang, Y., Qiao, R. X., Bonila, M. M., Yang, X. L., Ren, H. Q., et al. (2018). Evidence that microplastics aggravate the toxicity of organophosphorus flame retardants in mice (Mus musculus). *Journal of Hazardous Materials, 357*, 348–354. https://doi.org/10.1016/j.jhazmat.2018.06.017

De Souza Machado, A. A., Kloas, W., Zarfl, C., Hempel, S., & Rillig, M. C. (2018a). Microplastics as an emerging threat to terrestrial ecosystems. *Global Change Biology, 24*(4), 1405–1416. https://doi.org/10.1111/gcb.14020

De Souza Machado, A. A., Lau, C. W., Till, J., Kloas, W., Lehmann, A., Becker, R., & Rillig, M. C. (2018b). Impacts of microplastics on the soil biophysical environment. *Environmental Science & Technology, 52*(17), 9656–9665. https://doi.org/10.1021/acs.est.8b02212

De Souza Machado, A. A., Lau, C. W., Kloas, W., Bergmann, J., Bachelier, J. B., Faltin, E., & Rillig, M. C. (2019). Microplastics can change soil properties and affect plant performance. *Environmental Science & Technology, 53*(10), 6044–6052. https://doi.org/10.1021/acs.est.9b01339

Dong, H., Liu, T., Han, Z., Sun, Q., & Li, R. (2015). Determining time limits of continuous film mulching and examining residual effects on cotton yield and soil properties. *Journal of Environmental Biology, 36*(3), 677–684. Retrieved from http://www.jeb.co.in/journal_issues/201505_may15/paper_25.pdf

Dris, R., Imhof, H., Sanchez, W., Gasperi, J., Galgani, F. O., Tassin, B., & Laforsch, C. (2015). Beyond the ocean: Contamination of freshwater ecosystems with (micro-) plastic particles. *Environment and Chemistry, 12*(5), 539–550.

Duis, K., & Coors, A. (2016). Microplastics in the aquatic and terrestrial environment: Sources (with a specific focus on personal care products), fate and ects. *Environmental Sciences Europe, 28*, 2.

Ehlers, S. M., & Ellrich, J. A. (2020). First record of 'plasticrusts' and 'pyroplastic' from the Mediterranean Sea. *Marine Pollution Bulletin, 151*, 110845.

Engler, R. E. (2012). The complex interaction between marine debris and toxic chemicals in the ocean. *Environmental Science & Technology, 46*(22), 12302–12315. https://doi.org/10.1021/es3027105

Environmental Protection Agency (EPA). (2015). *Urban waste water treatment in 2014: A report for the year 2014*. Retrieved from http://www.epa.ie/pubs/reports/water/wastewater/2014%20waste%20water%20report_web.pdf

Espı, E., Salmeron, A., Fontecha, A., Garcıa, Y., & Real, A. I. (2006). Plastic films for agricultural applications. *Journal of Plastic Film & Sheeting, 22*(2), 85–102. https://doi.org/10.1177/8756087906064220

Feng, L. J., Wang, J. J., Liu, S. C., Sun, X. D., Yuan, X. Z., & Wang, S. G. (2018). Role of extracellular polymeric substances in the acute inhibition of activated sludge by polystyrene nanoparticles. *Environmental Pollution, 179*, 122–131.

Gajendiran, A., Krishnamoorthy, S., & Abraham, J. (2016). Microbial degradation of low-density polyethylene (LDPE) by Aspergillus clavatus strain JASK1 isolated from landfill soil. *3 Biotech, 6*, 52. https://doi.org/10.1007/s13205-016-0394-x

Ganesh Kumar, A., Anjana, K., Hinduja, M., & Sujitha, K. G. D. (2020). Review on plastic wastes in marine environment – Biodegradation and biotechnological solutions. *Marine Pollution Bulletin, 150*, 110733. https://doi.org/10.1016/j.marpolbul.2019.110733

Geyer, R., Jambeck, J. R., & Law, K. L. (2017). Production, use, and fate of all plastics ever made. *Science Advances, 3*(7), e1700782.

Gigault, J., ter Halle, A., Baudrimont, M., Pascal, P. Y., Gauffre, F., Phi, T. L., El Hadri, H., Grassl, B., & Reynaud, S. (2018). Current opinion: What is a nanoplastic? *Environmental Pollution, 235*, 1030–1034.

Godoy, V., Martin-Lara, M. A., Calero, M., & Blazquez, G. (2019). Physical-chemical characterization of microplastics present in some exfoliating products from Spain. *Marine Pollution Bulletin, 139*, 91–99. https://doi.org/10.1016/j.marpolbul.2018.12.026

Guerranti, C., Martellini, T., Perra, G., Scopetani, C., & Cincinelli, A. (2019). Microplastics in cosmetics: Environmental issues and needs for global bans. *Environmental Toxicology and Pharmacology, 68*, 75–79. https://doi.org/10.1016/j.etap.2019.03.007

He, D., Luo, Y., Lu, S., Liu, M., Song, Y., & Lei, L. (2018). Microplastics in soils: Analytical methods, pollution characteristics and ecological risks. *Trends in Analytical Chemistry, 109*, 163–172. https://doi.org/10.1016/j.trac.2018.10.006

Hebner, T. S., & Maurer-Jones, M. A. (2020). Characterizing microplastic size and morphology of photodegraded polymers placed in simulated moving water conditions. *Environmental Science. Processes & Impacts, 22*, 398–407. https://doi.org/10.1039/c9em00475k

Henriques, I. D. S., & Love, N. G. (2007). The role of extracellular polymeric substances in the toxicity response of activated sludge bacteria to chemical toxins. *Water Research, 41*, 4177–4185.

Heribert, I. (2014) Soil volatile organic compounds as tracers for microbial activities in soils. In P., Nannipieri, G., Pietramellara, G., Renella (Eds.), *Omics in soil science* (pp. 127–138). Caister Academic Press: Norfolk, UK.

Holmes, L. A., Turner, A., & Thompson, R. C. (2012). Adsorption of trace metals to plastic resin pellets in the marine environment. *Environmental Pollution, 160*(1), 42–48. https://doi.org/10.1016/j.envpol.2011.08.052

Horton, A. A., Walton, A., Spurgeon, D. J., Lahive, E., & Svendsen, C. (2017). Microplastics in freshwater and terrestrial environments: Evaluating the current understanding to identify the knowledge gaps and future research priorities. *Science of the Total Environment, 586*, 127–141. https://doi.org/10.1016/j.scitotenv.2017.01.190

Hou, D., O'Connor, D., Igalavithana, A. D., Alessi, D. S., Luo, J., Tsang, D. C. W., et al. (2020). Metal contamination and bioremediation of agricultural soils for food safety and sustainability. *Nature Reviews Earth and Environment, 1*, 366–381. https://doi.org/10.1038/s43017-020-0061-y

Hü-er, T., Weniger, A. K., & Hofmann, T. (2018). Sorption of organic compounds by aged polystyrene microplastic particles. *Environmental Pollution, 236*, 218–225.

Huang, X., Cao, L., Qin, Z., Li, S., Kong, W., & Liu, Y. (2018). Tat-independent secretion of polyethylene terephthalate hydrolase PETase in Bacillus subtilis 168mediated by its native signal peptide. *Journal of Agricultural and Food Chemistry, 66*, 13217–13227. https://doi.org/10.1021/acs.jafc.8b05038

Huang, H., Wang, D., Wan, W., Wen, B. (2017). Hexabromocyclododecanes in soils and plants from a plastic waste treatment area in North China: Occurrence, diastereomer- and enantiomer-specific profiles, and metabolization. *Environmental Science Pollution Research 24*, 21625–21635.

Hurley, R. R., & Nizzetto, L. (2018). Fate and occurrence of micro(nano)plastics in soils: Knowledge gaps and possible risks. *Current Opinion in Environmental Science and Health, 1*, 6–11. https://doi.org/10.1016/j.coesh.2017.10.006

Huerta Lwanga, E., Mendoza Vega, J., Ku Quej, V., JdlA, C., Cid, L., Sanchez del Chi, C., et al. (2017). Field evidence for transfer of plastic debris along a terrestrial food chain. *Scientific Reports, 7*, 14071. https://doi.org/10.1038/s41598-017-14588-2

Islam, S., Apitius, L., Jakob, F., & Schwaneberg, U. (2019). Targeting microplastic particles in the void of diluted suspensions. *Environment International, 123*, 428–435. https://doi.org/10.1016/j.envint.2018.12.029

Jambeck, J. R., Geyer, R., Wilcox, C., Siegler, T. R., Perryman, M., Andrady, A., Narayan, R., & Law, K. L. (2015). Plastic waste inputs from land into the ocean. *Science, 347*(6223), 768–771.

Jiang, X., Chen, H., Liao, Y., Ye, Z., Li, M., & Klobuˇcar, G. (2019). Ecotoxicity and genotoxicity of polystyrene microplastics on higher plant Vicia faba. *Environmental Pollution, 250*, 831–838.

Jing, H., Mezgebe, B., Aly Hassan, A., Sahle-Demessie, E., Sorial, G. A., & Bennett-Stamper, C. (2014). Experimental and modeling studies of sorption of ceria nanoparticle on microbial biofilms. *Bioresource Technology, 161*, 109–117.

Kasirajan, S., & Ngouajio, M. (2013). Erratum: Polyethylene and biodegradable mulches for agricultural applications: A review. *Agronomy for Sustainable Development, 33*, 501–539.

Kettler, K., Veltman, K., van de Meent, D., van Wezel, A., & Hendriks, A. J. (2014). Cellular uptake of nanoparticles as determined by particle properties, experimental conditions, and cell type. *Environmental Toxicology and Chemistry, 33*, 481–492. [CrossRef] [PubMed].

Kole, P. J., Lohr, A. J., Van Belleghem, F., & Ragas, A. M. J. (2017). Wear and tear of tyres: A stealthy source of microplastics in the environment. *International Journal of Environmental Research and Public Health*, 14. https://doi.org/10.3390/ijerph14101265

Koelmans, A. A., Besseling, E., Shim, W. J. (2015). Nanoplastics in the aquatic environment. Critical review. In *Marine anthropogenic litter* (pp. 325–340). Springer: Cham, Switzerland.

Kumar, M., Chen, H., Sarsaiya, S., Qin, S., Liu, H., Awasthi, M. K., et al. (2021). Current research trends on microand nano-plastics as an emerging threat to global environment: a review. *Journal of Hazardous Materials, 409, 124967.* https://doi.org/10.1016/j.jhazmat.2020.124967

Lambert, S., & Wagner, M. (2016). Characterisation of nanoplastics during the degradation of polystyrene. *Chemosphere, 145*, 265–268. https://doi.org/10.1016/j.chemosphere.2015.11.078

Lau, C. W., Till, J., Kloas, W., Lehmann, A., Becker, R., & Rillig, M. C. (2018). Impacts of microplastics on the soil biophysical environment. *Environmental Science & Technology, 52*, 9656–9665. https://doi.org/10.1021/acs.est.8b02212

Lee, H., Shim, W. J., & Kwon, J. H. (2014). Sorption capacity of plastic debris for hydrophobic organic chemicals. *Science of the Total Environment, 470–471*, 1545–1552. https://doi.org/10.1016/j.scitotenv.2013.08.023

Lehner, R., Weder, C., Petri-Fink, A., & Rothen-Rutishauser, B. (2019). Emergence of nanoplastic in the environment and possible impact on human health. *Environmental Science & Technology, 53*, 1748–1765. https://doi.org/10.1021/acs.est.8b05512

Leriche, V., Briandet, R., & Carpentier, B. (2003). Ecology of mixed biofilms subjected daily to a chlorinated alkaline solution: Spatial distribution of bacterial species suggests a protective effect of one species to another. *Environmental Microbiology, 5*, 64–71.

Li, Y. H., Lau, P. C. Y., Lee, J. H., Ellen, R. P., & Cvitkovitch, D. G. (2001). Natural genetic transformation of streptococcus mutans growing in biofilms. *Journal of Bacteriology, 183*, 897–908.

Li, L., Zhou, Q., Yin, N., Tu, C., & Luo, Y. (2019). Uptake and accumulation of microplastics in an edible plant. KexueTongbao/Chinese. *Scientific Bulletin, 64*, 928–934. [CrossRef].

Liang, T.-W., Jen, S.-N., Nguyen, A. D., & Wang, S.-L. (2016). Application of chitinous materials in production and purification of a poly(l-lactic acid) depolymerase from Pseudomonas tamsuii TKU015. *Polymers, 8*, 98. https://doi.org/10.3390/polym8030098

Liu, J., Lezama, N., Gasper, J., Kawata, J., Morley, S., Helmer, D., et al. (2016). Burn pit emissions exposure and respiratory and cardiovascular conditions among airborne hazards and open burn pit registry participants. *Journal of Occupational Environmental Medicine, 58*, e249–e255. https://doi.org/10.1097/JOM.0000000000000776

Liu, H., Yang, X., Liu, G., Liang, C., Xue, S., Chen, H., Ritsema, C. J., & Geissen, V. (2017). Response of soil dissolved organic matter to microplastic addition in Chinese loess soil. *Chemosphere, 185*, 907–917.

Liu, J., Ma, Y., Zhu, D., Xia, T., Qi, Y., Yao, Y., … Chen, W. (2018). Polystyrene nanoplastics- enhanced contaminant transport: Role of irreversible adsorption in glassy polymeric domain. *Environmental Science & Technology, 52*(5), 2677–2685. https://doi.org/10.1021/acs.est.7b05211

Liu, G., Zhu, Z., Yang, Y., Sun, Y., Yu, F., & Ma, J. (2019). Sorption behavior and mechanism of hydrophilic organic chemicals to virgin and aged microplastics in freshwater and seawater. *Environmental Pollution, 246*, 26–33.

Llorca, M., Schirinzi, G., Martınez, M., Barcelo, D., & Farre, M. (2018). Adsorption of perfluoroalkyl substances on microplastics under environmental conditions. *Environmental Pollution, 235*, 680–691. https://doi.org/10.1016/j.envpol.2017.12.075

Mai, L., Bao, L.-J., Wong, C. S., & Zeng, E. Y. (2018). Microplastics in the terrestrial environment. In E. Y. Zeng (Ed.), *Microplastic contamination in aquatic environments: An emerging matter of environmental urgency* (pp. 365–378). Elsevier. https://doi.org/10.1016/B978-0-12-813747-5.00012-6

Mattsson, K., Hansson, L. A., & Cedervall, T. (2015). Nano-plastics in the aquatic environment. *Environmental Science. Processes & Impacts, 17*, 1712–1721. https://doi.org/10.1039/c5em00227c

McCormick, A., Hoellein, T. J., Mason, S. A., Schluep, J., & Kelly, J. J. (2014). Microplastic is an abundant and distinct microbial habitat in an urban river. *Environmental Science & Technology, 48*, 11863–11871.

Mofijur, M., Ahmed, S. F., Rahman, S. M. A., Arafat Siddiki, S. K. Y., Islam, A. B. M. S., Shahabuddin, M., et al. (2021). Source, distribution and emerging threat of micro and nanoplastics to marine organism and human health: Socioeconomic impact and management strategies. *Environmental Research, 195*, 110857. https://doi.org/10.1016/j.envres.2021.110857

Moog, D., Schmitt, J., Senger, J., Zarzycki, J., Rexer, K.-H., Linne, U., et al. (2019). Using a marine microalga as a chassis for polyethylene terephthalate (PET) degradation. *Microbial Cell Factories, 18*, 171. https://doi.org/10.1186/s12934-019-1220-z

Mor, R., & Sivan, A. (2008). Biofilm formation and partial biodegradation of polystyrene by the actinomycete Rhodococcus ruber. *Biodegradation, 19*, 851–858. https://doi.org/10.1007/s10532-008-9188-0

Moreno, M. M., & Moreno, A. (2008). E_ect of di_erent biodegradable and polyethylene mulches on soil properties and production in a tomato crop. *Scientia Horticulture (Amst.), 116*, 256–263.

National Bureau of Statistics of China (NBSC). (2019). *National data*. Retrieved from http://data.stats.gov.cn.

Ng, E.-L., Huerta Lwanga, E., Eldridge, S. M., Johnston, P., Hu, H.-W., Geissen, V., & Chen, D. (2018). An overview of microplastic and nanoplastic pollution in agroecosystems. *Science of the Total Environment, 627*, 1377–1388. https://doi.org/10.1016/j.scitotenv.2018.01.341

Nizzetto, L., Futter, M., & Langaas, S. (2016). Are agricultural soils dumps for microplastics of urban origin? *Environmental Science & Technology, 50*(20), 10777–10779.

Oke, Y. S., Rinklebe, J., Hou, D., Tsang, D. C. W., & Tack, F. M. G. (2020). *Soil and groundwater remediation. A practical guide*. CRC Press, Technologies.

O'Connor, D., Pan, S., Shen, Z., Song, Y., Jin, Y., Wu, W.-M., & Hou, D. (2019). Microplastics undergo accelerated vertical migration in sand soil due to small size and wet-dry cycles. *Environmental Pollution, 249*, 527–534. https://doi.org/10.1016/j.envpol.2019.03.092

Orgiazzi, A., Bardgett, R. D., Barrios, E., Behan-Pelletier, V., Briones, M. J. I., Chotte, J.-L., Wall, D. H. (2016). *Global soil biodiversity atlas*. Luxembourg: European Commission, Publications Office of the European Union. https://doi.org/10.2788/2613

Palansooriya, K. N., Shaheen, S. M., Chen, S. S., Tsang, D. C. W., Hashimoto, Y., Hou, D., et al. (2020). Soil amendments for immobilization of potentially toxic elements in contaminated soils: A critical review. *Environment International, 134*, 105046. https://doi.org/10.1016/j.envint.2019.105046

Pathak, V. M., & Navneet. (2017). Review on the current status of polymer degradation: A microbial approach. *Bioresources and Bioprocessing, 4*, 15. https://doi.org/10.1186/s40643-017-0145-9

Praveena, S. M., Shaifuddin, S. N. M., & Akizuki, S. (2018). Exploration of microplastics from personal care and cosmetic products and its estimated emissions to marine environment: An evidence from Malaysia. *Marine Pollution Bulletin, 136*, 135–140. https://doi.org/10.1016/j.marpolbul.2018.09.012

Qi, Y., Yang, X., Pelaez, A. M., Huerta Lwanga, E., Beriot, N., Gertsen, H., et al. (2018). Macro and micro- plastics in soil-plant system: Effects of plastic mulch film residues on wheat (Triticum aestivum) growth. *Science of the Total Environment, 645*, 1048–1056. https://doi.org/10.1016/j.scitotenv.2018.07.229

Raddadi, N., & Fava, F. (2019). Biodegradation of oil-based plastics in the environment: Existing knowledge and needs of research and innovation. *Science of the Total Environment, 679*, 148–158. https://doi.org/10.1016/j.scitotenv.2019.04.419

Ramirez Arenas, L., Ramseier Gentile, S., Zimmermann, S., & Stoll, S. (2020). Coagulation of TiO2, CeO2 nanoparticles, and polystyrene nanoplastics in bottled mineral and surface waters. Effect of water properties, coagulant type, and dosage. *Water Environment Research, 92*, 1184–1194. https://doi.org/10.1002/wer.1313

Revel, M., Châtel, A., & Mouneyrac, C. (2018). Micro(nano)plastics: A threat to human health? *Current Opinion in Environmental Science and Health, 1*, 17–23. https://doi.org/10.1016/j.coesh.2017.10.003

Rillig, M. C. (2012). Microplastic in terrestrial ecosystems and the soil? *Environmental Science & Technology, 46*(12), 6453–6454. https://doi.org/10.1021/es302011r

Rillig, M. C., Ziersch, L., & Hempel, S. (2017). Microplastic transport in soil by earthworms. *Scientific Reports, 7*, 1362.

Rillig, M. C., & Bonkowski, M. (2018). Microplastic and soil protists: A call for research. *Environmental Pollution, 241*, 1128–1131. https://doi.org/10.1016/j.envpol.2018.04.147

Rossi, G., Barnoud, J., & Monticelli, L. (2014). Polystyrene nanoparticles perturb lipid membranes. *Journal of Physical Chemistry Letters, 5*, 241–246.

Rudyak, V. Y., Efimova, E. A., Guseva, D. V., & Chertovich, A. V. (2018). Thermoset polymer matrix structure and properties: Coarse-grained simulations. *Polymers (Basel), 11*. https://doi.org/10.3390/polym11010036

Sandermann Jr., H., (1992). Plant metabolism of xenobiotics. *Trends in Biochemical Sciences, 17*, 82–84. https://doi.org/10.1016/0968-0004(92)90507-6

Santos, F. C. F., Gomes, S. I. L., Scott-Fordsmand, J. J., & Amorim, M. J. B. (2017). Hazard assessment of nickel nanoparticles in soil—The use of a full life cycle test with Enchytraeus crypticus. *Environmental Toxicology and Chemistry, 36*, 2934–2941.

Schöler, A., Jacquiod, S., Vestergaard, G., Schulz, S., & Schloter, M. (2017). Analysis of soil microbial communities based on amplicon sequencing of marker genes. *Biology and Fertility of Soils, 53*, 485–489.

Seidensticker, S., Grathwohl, P., Lamprecht, J., & Zarfl, C. (2018). A combined experimental and modeling study to evaluate pH-dependent sorption of polar and non-polar compounds to polyethylene and polystyrene microplastics. *Environmental Sciences Europe, 30*(1), 1–12.

Sharma, S., & Chatterjee, S. (2017). Microplastic pollution, a threat to marine ecosystem and human health: A short review. *Environmental Science and Pollution Research International, 24*, 21530–21547. https://doi.org/10.1007/s11356-017-9910-8

Shen, M., Zhu, Y., Zhang, Y., Zeng, G., Wen, X., Yi, H., et al. (2019). Micro(nano) plastics: unignorable vectors for organisms. *Marine Pollution Bulletin, 139, 328–331.* https://doi.org/10.1016/j.marpolbul.2019.01.004

Siipola, V., Pflugmacher, S., Romar, H., Wendling, L., Koukkari, P., (2020). Low-cost biochar adsorbents for water purification including microplastics removal. *Applied Sciences, 10.* https://doi.org/10.3390/app10030788

Smith, M., Love, D. C., Rochman, C. M., & Neff, R. A. (2018). Microplastics in seafood and the implications for human health. *Current Environmental Health Reports, 5*, 375–386.

Soares, J., Miguel, I., Venâncio, C., Lopes, I., & Oliveira, M. (2020). Perspectives on micro(nano) plastics in the marine environment: Biological and societal considerations. *Water, 12*. https://doi.org/10.3390/w12113208

Sommer, F., Dietze, V., Baum, A., Sauer, J., Gilge, S., Maschowski, C., et al. (2018). Tire Abrasion as a major source of microplastics in the environment. *Aerosol and Air Quality Research, 18*, 2014–2028. https://doi.org/10.4209/aaqr.2018.03.0099

Sun, F., Polizzotto, M. L., Guan, D., Wu, J., Shen, Q., Ran, W., Wang, B., & Yu, G. (2017). Exploring the interactions and binding sites between Cd and functional groups in soil using two-dimensional correlation spectroscopy and synchrotron radiation based spectromicroscopies. *Journal of Hazardous Materials, 326*, 18–25.

Sun, X. D., Yuan, X. Z., Jia, Y., Feng, L. J., Zhu, F. P., Dong, S. S., Liu, J., Kong, X., Tian, H., Duan, J. L., et al. (2020). Differentially charged nanoplastics demonstrate distinct accumulation in Arabidopsis thaliana. *Nature Nanotechnology*. Available online: https://www.nature.com/articles/s41565-020-0707-4. Accessed on 29 Aug 2020

Suresh, S. S., Mohanty, S., & Nayak, S. K. (2017). Composition analysis and characterization of waste polyvinyl chloride (PVC) recovered from data cables. *Waste Management, 60*, 100–111. https://doi.org/10.1016/j.wasman.2016.08.033

Tang, J., Wu, Y., Esquivel-Elizondo, S., Sørensen, S. J., & Rittmann, B. E. (2018). How microbial aggregates protect against nanoparticle toxicity. *Trends in Biotechnology, 36*, 1171–1182.

Tiwari, N., Santhiya, D., & Sharma, J. G. (2020). Microbial remediation of micro-nano plastics: Current knowledge and future trends. *Environmental Pollution, 265*, 115044. https://doi.org/10.1016/j.envpol.2020.115044

Turner, A., Wallerstein, C., Arnold, R., & Webb, D. (2019). Marine pollution from pyroplastics. *Science of the Total Environment, 694*, 133610.

UNEP. (2018). *Plastic planet: How tiny plastic particles are polluting our soil*. Retrieved from https://www.unenvironment.org/news-and-stories/story/plastic-planet-how-tiny-plastic-particlesare-polluting-oursoil.

Velzeboer, I., Quik, J. T. K., van de Meent, D., & Koelmans, A. A. (2014). Rapid settling of nanoparticles due to heteroaggregation with suspended sediment. *Environmental Toxicology and Chemistry, 33*, 1766–1773.

Vestergaard, G., Schulz, S., Schöler, A., & Schloter, M. (2017). Making big data smart—How to use metagenomics to understand soil quality. *Biology and Fertility of Soils, 53*, 479–484.

Wahl, A., Le Juge, C., Davranche, M., El Hadri, H., Grassl, B., Reynaud, S., et al. (2021). Nanoplastic occurrence in a soil amended with plastic debris. *Chemosphere, 262*, 127784. https://doi.org/10.1016/j.chemosphere.2020.127784

Wan, Y., Wu, C., Xue, Q., & Hui, X. (2019). Effects of plastic contamination on water evaporation and desiccation cracking in soil. *Science of the Total Environment, 654*, 576–582. https://doi.org/10.1016/j.scitotenv.2018.11.123

Wang, J., Luo, Y., Teng, Y., Ma, W., Christie, P., & Li, Z. (2013). Soil contamination by phthalate esters in Chinese intensive vegetable production systems with different modes of use of plastic film. *Environmental Pollution, 180*, 265–273.

Wang, J., Lv, S., Zhang, M., Chen, G., Zhu, T., Zhang, S., … Luo, Y. (2016). Effects of plastic film residues on occurrence of phthalates and microbial activity in soils. *Chemosphere, 151*, 171–177. https://doi.org/10.1016/j.chemosphere.2016.02.076

Wang, L., Wu, W. -M., Bolan, N. S., Tsang, D. C. W., Li, Y., Qin, M., et al. (2021). Environmental fate, toxicity and risk management strategies of nano plastics in the environment: current status and future perspectives. *Journal of Hazardous Materials, 401, 123415.* https://doi.org/10.1016/j.jhazmat.2020.123415

Wardle, D. A. (2002). Communities and ecosystems. Linking the aboveground and belowground components. Princeton, NJ: Princeton University Press. Sandermann, H. Plant metabolism of xenobiotics. *Trends in Biochemical Sciences, 2*, 82–84.

Weinstein, J. E., Crocker, B. K., & Gray, A. D. (2016). From macroplastic to microplastic: Degradation of high-density polyethylene, polypropylene, and polystyrene in a salt marsh habitat. *Environmental Toxicology and Chemistry, 35*, 1632–1640. https://doi.org/10.1002/etc.3432

Wei, R., Oeser, T., Schmidt, J., Meier, R., Barth, M., Then, J., et al., (2016). Engineered bacterial polyester hydrolases efficiently degrade polyethylene terephthalate due to relieved product inhibition. *Biotechnololgy and Bioengineering, 113*, 1658–1665. https://doi.org/10.1002/bit.25941

Wright, S. L., & Kelly, F. J. (2017). Plastic and human health: A micro issue? *Environmental Science & Technology, 51*, 6634–6647. https://doi.org/10.1021/acs.est.7b00423

Wu, C., Zhang, K., Huang, X., & Liu, J. (2016). Sorption of pharmaceuticals and personal care products to polyethylene debris. *Environmental Science and Pollution Research, 23*(9), 8819–8826. https://doi.org/10.1007/s11356-016-6121-7

Wu, P., Cai, Z., Jin, H., & Tang, Y. (2019). Adsorption mechanisms of five bisphenol analogues on PVC microplastics. *Science of the Total Environment, 650*, 671–678. https://doi.org/10.1016/j.scitotenv.2018.09.049

Xu, Y., He, Q., Liu, C., & Huangfu, X. (2019). Are micro- or nanoplastics leached from drinking water distribution systems? *Environmental Science & Technology, 53*, 9339–9340. https://doi.org/10.1021/acs.est.9b03673

Yang, Y. F., Chen, C. Y., Lu, T. H., & Liao, C. M. (2019). Toxicity-based toxicokinetic/ toxicodynamic assessment for bioaccumulation of polystyrene microplastics in mice. *Journal of Hazardous Materials, 366*, 703–713. https://doi.org/10.1016/j.jhazmat.2018.12.048

Yu, M., van der Ploeg, M., Lwanga, E. H., Yang, X., Zhang, S., Ma, X., et al. (2019). Leaching of microplastics by preferential flow in earthworm (Lumbricus terrestris) burrows. *Environment and Chemistry, 16*, 31–40. https://doi.org/10.1071/EN18161

Yurtsever, M. (2019). Tiny, shiny, and colorful microplastics: Are regular glitters a significant source of microplastics? *Marine Pollution Bulletin, 146*, 678–682. https://doi.org/10.1016/j.marpolbul.2019.07.009

Zhang, K., Hamidian, A. H., Tubić, A., Zhang, Y., Fang, J. K. H., Wu, C., et al. (2021). Understanding plastic degradation and microplastic formation in the environment: A review. *Environmental Pollution, 274*, 116554. https://doi.org/10.1016/j.envpol.2021.116554

Zhang, W., Zheng, J., Zheng, P., Tsang, D. C. W., & Qiu, R. (2015). The roles of humic substances in the interactions of phenanthrene and heavy metals on the bentonite surface. *Journal of Soils and Sediments, 15*, 1463–1472.

Zhang, C., Show, P.-L., Ho, S.-H. (2019). Progress and perspective on algal plastics – a critical review. *Bioresource Technology, 289*, 121700. https://doi.org/10.1016/j.biortech.2019.121700

Zhu, B.-K., Fang, Y.-M., Zhu, D., Christie, P., Ke, X., & Zhu, Y.-G. (2018). Exposure to nanoplastics disturbs the gut microbiome in the soil oligochaete Enchytraeus crypticus. *Environmental Pollution, 239, 408–415.* https://doi.org/10.1016/j.envpol.2018.04.017

Ziajahromi, S., Neale, P. A., Leusch, D. D. L. (2016). Wastewater treatment plant effluent as a source of microplastics: Review of the fate, chemical interactions and potential risks to aquatic organisms. *Water Science and Technology, 74*, 2253–2269.

Chapter 6
Micro- and Nanoplastics as Carriers for Other Soil Pollutants

Nahid Khoshnamvand

Abstract Annual releases of plastic to the terrestrial environment are 4–23 times as high as releases to the marine environment. Microplastics can enter the soil in many routes, for example, compost and sewage sludge as fertilizer, plastic mulching, irrigation and flooding, and atmospheric deposition. The process of top-down irrigation into the soil causes MP/NPs to be transported downwards along with soil cavities and eventually possibly into groundwater. Contact of toxic and harmful metal pollutants with M&NPs will inevitably occur during the migration process in the environment. Various factors are considered in their transportation such as microplastic properties, pore water forms, and properties of packing materials to influence microplastic transport that can indicate the environmental chance of microplastics in soil conditions. Among the important roles in the environmental behavior of M&Ms are absorption and migration. Microplastics or nanoplastic particles as a carrier, adsorb contaminants and increase or decrease their transportation. The transfer of microplastics in the soil environment occurs in the form of vertical and horizontal migration and nonliving transport. Microplastics are known to adsorb toxic chemicals such as PCBs, PAHs, DDTs, PFASs, PPCPs, and heavy metals.

Keywords Microplastic · Nanoplastic · Soil pollutants · Transportation · Carriers

6.1 Introduction

Microplastics soil pollution was first studied by Rillig (2012), and subsequent research has focused on this important issue. On World Environment Day 2018, the United Nations Environment Programme (UNEP) called for a more in-depth consideration of the impacts of microplastic pollution on the soil environment (Schnurr et al., 2018). The fact that soil is a more important sink for microplastics than marine

N. Khoshnamvand (✉)
Department of Environmental Health Engineering, School of Public Health, Tehran University of Medical Sciences, Tehran, Iran

N. R. Maddela et al. (eds.), *Micro and Nanoplastics in Soil*,
https://doi.org/10.1007/978-3-031-21195-9_6

environments has been a critical factor influencing UNEP's decision. Annual releases of plastic to the terrestrial environment are 4–23 times as high as releases to the marine environment (Horton et al., 2017). Plastics are categorized into two groups: primary plastics produced in the size range, such as MPs/NPs in pharmaceutical and personal care products (PPCPs) (Rochman et al., 2015). Secondary plastics are generated by crushing major plastics, such as agricultural plastic mulch or car tires (Huang et al., 2020). Therefore, it is necessary to investigate the physical structure and surface properties of microplastics to facilitate a comprehensive understanding of the factors influencing the environmental fate of microplastics in soil. Inputs from agricultural practices, the influence of runoff and deposits, and degradation or fragmentation of plastic debris are the major sources of entry into the soil. The shape and composition of microplastics are almost identical to their sources. Microplastics can enter the soil in many routes, for example, compost and sewage sludge as fertilizer, plastic mulching, irrigation and flooding, and atmospheric deposition (Bläsing & Amelung, 2018). The degradation by ultraviolet (UV) radiation also may serve as a significant source of soil plastic pollution. Plastics can be degraded through various processes, such as hydrolysis, oxidation, photodegradation, mechanical corrosion, and biological degradation (Alimi et al., 2018). Usage of membranes in modern agriculture led to the discharge of MPs/NPs on the soil. The process of top-down irrigation into the soil causes MP/NPs to be transported downwards along with soil cavities and eventually possibly into groundwater (Zeng et al. 2020a, b). Microplastic accumulation may also occur in the soil, causing a number of adverse effects on the soil ecosystem, including alterations in the chemical, physical, and fertility properties of the soil, leading to disruption of the microbial population living in the soil (Zhang et al., 2017).

6.1.1 Micro- and Nanoplastic Transportation to Soil

Based on columnar experiments conducted in laboratory environments, the transfer of microplastics and nanoplastics has been investigated. Simulation is one of the main methods to analyze the transport behavior of pollutants in soil and groundwater. Glass beads and quartz sand are used as model porous media to simulate the soil environment. Polystyrene microspheres were used as research objects for the transport study of colloids (Bradford et al., 2002). Generally, to date investigations on the transfer of M&NPs have been conducted mainly on the basis of colloidal transfer, especially similar to that of engineered nanomaterials (Hüffer et al., 2017). Coexisting with other substances affected the stability of microplastics in a porous medium. Also, the interaction between M&NPs and collectors and thus changed transport and deposition in porous media are influenced. Other properties of porous media, solution environment, and the characteristics of M&NPs exhibit an effect on transport behavior. Quartz sands because of their homogeneous texture and negative charge are used as environment porous in transport investigations. Here, too, the simultaneous presence of other materials may occupy sedimentary sites in porous media for M&NPs. For example, the transportation of microplastics under positively charged surfactants was great than those under negatively charged surfactants due to competitive adsorption sites (Pelley & Tufenkji, 2008). Additional deposition sites could be provided by coexisting pollutants. The effect of dissolved black carbon (DBC) on the transport of polystyrene nanoplastics was explored with dissipation monitoring equipment (QCM-D) techniques (Gul et al., 2021). Two main items in the study of transport in porous media are natural and artificial colloids (Si et al., 2019). Another nanomaterial such as graphene oxide (GO) also supplied more sites to nanoplastic, especially at relatively high ionic power in $CaCl_2$ (Xia et al., 2021). In another study by O'Connor et al. (2019), a sand column was used to study the penetration process of polypropylene (PP) and polyethylene (PE) microplastics. Polypropylene (PP) and polyethylene (PE) microplastics with various sizes and densities were used in the experiment, and they found that the microplastics all more moved downwards in the range of 1.5–7.5 cm (O'Connor et al., 2019). Although the simulation experiment helps us to better understand the mechanism of transmission of M&NPs in the soil, the results of laboratory studies are not generalizable to the actual soil environment. Generally, results showed that the existence of microplastics alters the bulk density, water holding capacity, structure, and hydrodynamics of the soil (de Souza Machado et al., 2018, 2019). Conversely, that may change the transfer of M&NPs in soil (Xia et al., 2021). M&NPs affect the soil media and, in turn, affect the movement of M&NPs. Since soil is a complex media with intricate pore systems and rich biological residents, therefore, the results of research in the actual natural environment are different from the laboratory.

6.1.1.1 Factors Involving the Transportation of M&NPs

Various factors are considered in laboratories such as microplastic properties, pore water forms, and properties of packing materials to influence microplastic transport that can indicate the environmental chance of microplastics in real soil conditions (Alimi et al., 2018). The co-occurrence of contaminants in porous media alters the transmission of M&NPs. Also, particle size and specific surface area characteristics are desirable for M&NPs to affect the fate of other contaminants such as metal and some organic contaminants (Bradford & Bettahar, 2006; Pelley & Tufenkji, 2008).

6.1.1.2 Micro- and Nanoplastic Movement Model in Soil

The transfer of microplastics in the soil environment occurs in the form of vertical and horizontal migration and nonliving transport. The transportation downwards of microplastics, especially nano-sheets, may pose a potential risk of groundwater contamination. The penetration of microplastics into the soil inevitably affects a series of biological processes that affect their carrier and fate (de Souza Machado et al., 2018).

6.1.1.3 Microplastic Transportation Through Porous Media

In previous studies, the PS microsphere transfer model (less than 5 mm) as a colloid has been examined due to its inertia and fluorescent traceability (Tong et al., 2005; Peng et al. 2017a, b, c). Generally, microplastics because of their similar properties and behaviors have a similar transmission. Compared to hydrophilic plastics, PS hydrophobic plastics had higher colloidal retention in unsaturated columns of sand (Wan & Wilson, 1994). Increasing the surface negative charge of PS nanoplastics by UV or O_3 aging processes can lead to a significant increase in the movability of spherical PS nanoplastics (Zeng et al. 2020a, b). Factors such as flow velocity, water content, ionic strength, and natural organic matter significantly affect the transport of microplastics. This means that the microplastic transportation increases with the high velocity of water pores. The results showed that increasing the flow velocity reduces the shelf life of PS microplastics in quartz sand columns (Wu et al., 2012; Zhang et al., 2015). A significant correlation between substantiated and accelerated deposition and reduced mobility of PS microplastics with increasing ionic strength had been approved by most researchers (Elimelech & O'Melia, 1990; Kobayashi et al., 2009). Sedimented microplastics with lower ionic strength were more prone to re-entrain from glass surfaces into the bulk solutions. (Franchi & O'Melia, 2003). In addition, the wet-dry cycle affected the vertical migration of microplastics in sand columns (O'Connor et al., 2019). Besides mentioned factors, also, natural organic matter (NOM) contributes to the movability of microplastics. The existence of NOM not only led to the resistance of PS particles (Deshiikan et al., 1998) but also the larger the molecular size of NOM, the lower the particle size in absorbent media owing to the increase in spatial share. (Amirbahman & Olson, 1995).

6.1.1.4 Microplastic Migration in Soil Media

During the tillage functions, plastic fibers and mulching parts were observed in the deeper layers of the soil (>20 cm), which is clear proof of the transportation of microplastic downwards in the soil environment (Huang et al., 2020). Besides, soil fauna is counted as a carrier of microplastics in soil. Following the migration of earthworms (*L. terrestris*) and providing a downwards path for the vertical transport of microplastics, the microplastics of polyethylene could be transported from layers near the ground to the deeper layer (>50 cm) (Huerta Lwanga et al., 2017). In addition to vertical transfer, digestion and adhesion to the exterior body of the earthworm were also involved in microplastic transfer (Rillig et al., 2017). In horizontal transport, microplastics are moved and distributed by collembolans. Interestingly, *F. candida*, due to its larger size, plays a larger role in the displacement of large particles than *Proisotoma minuta* (Maaß et al., 2017). The presence of feeding and nutrition relationships between various species such as collembola (*F. candida*) and mite (*Hypoaspis aculeifer*) in soil facilitated the movement of micro- and nanoplastics up to 40% compared with the presence of a homogenous species (Zhu et al., 2018). Totally, biogenic actions can elevate the forwarding of microplastics, particularly vertical transport, which led to a likely risk for groundwater and biotas' existence in soil.

6.2 M&NPs as Carriers for Other Soil Pollutants

Understanding the interaction and simultaneous transmission of M&NPs with contaminants has been designed and popularized by many researchers. Because as a carrier, M&NPs are likely to carry certain dangerous pollutants over long distances and pose risks to the ecosystem and human health. During the interaction with micro- and nanoplastics, the transport of symbiotic materials may be facilitated or inhibited. In addition to their roles, microplastics are known to adsorb toxic chemicals such as PCBs, PAHs, DDTs, PFASs, PPCPs, and heavy metals (Peng et al. 2017a, b, c). The properties of plastics such as hydrophobicity of surfaces led to the condensation of PCBs and dichlorodiphenyldichloroethylene in PP pellets up to 10^5–10^6 times higher than concentrations in seawater (Mato et al., 2001). Microplastics and nanoplastics can attach to toxic emerging contaminants, such as hormones and pharmaceutical and personal care products, and bioaccumulate and would remain in the body of humans and animals, persistently (Zhou et al., 2022). Microplastics are able to carry bisphenol A and then release it as a source of environmental pollution (Zhou et al., 2022). Microplastics can also act as an effective sink for tetracycline by increasing deposition sites. The results of previous studies exhibited that polyester fibers (0.3% by weight) can influence the configuration of clay loam soils (Zhang et al., 2019). Besides, polyethylene film (1% w/w) can significantly raise the speed of water vaporization in clay (Wan et al., 2019). Polystyrene nanoplastics (PS) in different concentrations lead to modifications in the soil media, for example, polystyrene nanoplastics at 100 and 1000 ng g^{-1} which seriously reduced the soil microbial communities and raised basal aspiration, respectively (Awet et al., 2018). In extremely high modification rates, PP microplastics can increase soil basal aspiration rate about three times (Yang et al., 2018). PS nanoplastics (100 ng g^{-1}) also reduced the activities of enzymes involved in C, N, and P cycles in soils with silt loam textures (Awet et al., 2018). PP microplastics greatly promoted the activity of fluorescein diacetate hydrolase in sandy loam soils and consequently improved the availability of nutrients for plants by enhancing microbial hydrolytic action on soil organic matter (SOM) (Yang et al., 2018). Plastic fragment remains (67.5 kg ha^{-1}) can lead to considerable decreases in soil microbial biota type and reduce soil microbial C and N, plus decreased the activity of fluorescein diacetate hydrolase and dehydrogenase by 10% and 20%, respectively (Wang et al., 2016). The shape of microplastic (linear versus nonlinear), size, rate of growth, polymer design, and soil texture affect the enzyme activity and microbial biota of the soil. There is a possibility that the negative effects are probably related to concomitant phthalate contamination and not to the existence of plastic layers alone (Wang et al., 2016). Earthworms, one of the most well-known animals in the soil, have a layer of viscous body fluid on their surface that M&NPs may attach to, and the earthworms' motion causes the spatial transfer of plastics (Rillig et al., 2017). Although their role is still unclear, Rong et al. lately recommended that the bacterial community could slow the transport of plastic particles, as biofilms narrow the pathway and increase the surface roughness, as well as the OH and NH groups at cellular levels. Rate probability of plastic particles forming hydrogen bonds was very high (He et al., 2020). Due to the large specific surface area and powerful hydrophobicity, micro- and nanoplastics can easily absorb contaminants and act as a carrier.

6.2.1 Adsorption and Migration

Among the important roles in the environmental behavior of M&Ms are absorption and migration. Microplastics or nanoplastic particles as a carrier adsorb contaminants and increase or decrease their transportation. The adsorption action of pollutants into M&NPs occurs with mechanisms such as hydrophobic action, electrostatic action, pore filling, van der Waals forces, hydrogen bond, and π-π interaction (Torres et al., 2021). Hydrophobic activity is the fundamental mechanism in describing the M&NP and pollutants. Polymer type, surface functional groups, and material structure of M&Ms define the relations between microplastics and related substances (Tourinho et al., 2019). Hydrophobic organic materials have a high tendency to adsorb on non-polar microplastics, due to their hydrophobic properties (Tourinho et al., 2019).

Owing to adsorption on the plastic, the transformation of organic contaminants from water to organisms could be increased. For example, by transferring polybrominated diphenyl ethers (PBDEs) on the surface, microplastics would persist in the body of the fish for many years (Wardrop et al., 2016). The type and size of microplastics impact the adsorption capacity. For example, the higher adsorption capacity of PS microplastics than PE and PP on tetracycline is mostly due to their polar properties (Xu et al., 2018). In addition to knowledge about the environmental manners of M&NPs as vectors for other pollutants, also discussion of the mutual effect available cotransport studies of M&NPs and coexisting pollutants is important, while there are rare investigations such as a number of academic reviews on cotransport of M&NPs and coexisting pollutants compared with adsorption studies. More researchers focused on individual transport behavior than cotransport of M&NPs which cannot explain the entire transport behavior completely (Alimi et al., 2018). The results of studies analyzing the interaction of M & NPs with hydrophobic organic matter have shown that they, especially certain persistent organic pollutantshave a great affinity for absorbing to surfaces of plastic particles. For example, microplastics and nanoplastics can absorb PCBs. The adsorption of PCBs to nano-PS was 1–2 times stronger than that to micro-PE due to higher aromaticity and surface-area-to-volume ratio (Velzeboer et al., 2014). Compared to adsorption examinations, investigations of the simultaneous transport of M&NPs with numerous pollutants in porous media is not adequate. According to previous studies,

M&NPs as noteworthy carriers can improve or hinder the transfer of biocloids and non-biological colloids. Oppositely, the transport and storage of M&NPs may be altered by symbiotic substances.

6.3 Adsorption of Various Toxic Chemicals into M&NPs

One of the most common types of microplastics that are examined by researchers is polyethylene, which is probably due to the high consumption in industry and the successive pollution levels of plastics (Andrady & Neal, 2009). The types of organic compounds that are adsorbed by micro- and nanoplastics have various adsorption amounts based on LogK_{MP} values. Organic chemicals which supposed nonpolar and have a LogK_{ow} more than two important positive linear correlations were observed between LogK_{ow} and LogK_{MP} for PE ($p < .001$), PP ($p < .001$), PS ($p < .001$), and PVC ($p < .01$) microplastics. This indicates that the hydrophobic property has a great role in the sorption of HOCs (Log$K_{ow} > 2$), i.e., the higher the hydrophobic property, the higher the sorption tendency; this result is consistent with the results of previous studies (Wang et al., 2018). The distribution coefficients (LogK_{MP}) of HOCs (e.g., PAHs and HCHs) on PE, PP, and PS microplastics showed fit linear correlations with their LogK_{ow} values, with R^2 values of 0.92, 0.94, and 0.84, respectively.

Hydrophobic distribution coefficients are considered a predominant sorption process for nonpolar organic compounds onto microplastics, compared with other processes such as electrostatic interchange and hydrogen bonding (Wang et al., 2015). When microplastic particles are exposed to organic compounds, non-polar organic compounds have stronger sorption due to their higher hydrophobic properties than polar compounds. The important correlation between distribution factor and hydrophobicity of HOCs shows that hydrophobic relations with microplastics are noteworthy (Seidensticker et al., 2018), in addition to the hydrophobicity properties of polar organic compounds, with Log$K_{ow} < 2$ and various acid dissociation constants (pKa), which are probably not the only adsorption regulators. Some types of polar organic compounds, under the influence of pH, can wildly change their adsorption to microplastics. The type of charge of the contaminants adsorbed on the micro- and nanoplastics also affects the adsorption power. Charged species are usually weaker adsorbed by microplastics than charge neutral species (Seidensticker et al., 2018). For example, due to the similar charge of tetracycline and microplastics such as PE, PP, and PS, the adsorption of tetracycline on the surface of microplastic particles is strongly hindered due to the incidence of electrostatic repulsion (Zhu et al., 2018). The adsorption of charged oxytetracycline on microplastics also had this fate (Wang et al., 2018). In contrast, because tylosin has a positive charge in acidic conditions that is opposite to the charge of microplastics, it is more easily absorbed by PS and PVC microplastics than neutral species. The extended role of organic matters is also proved in the sorption of organic compounds to microplastics. Fluorescence measurements confirmed a minor exchange between original microplastics and soluble organic matter (DOM; e.g., humic acid) (Seidensticker et al., 2018), which creates competition in the adsorption of HOCs between micro- and nanoplastics and DOM. For example, in the presence of DOM, the partitioning of phenanthrene and tonalide among microplastics and water was extremely altered (150–1000 mg L^{-1}). In the attendance of DOM that easily affected polar organic compounds, even at little concentrations of DOM (e.g., 20 mg L^{-1}), the adsorption of polar compounds was remarkably slow (Shen et al., 2018). According to Wang et al. (2018) the sorption of oxytetracycline on weathered PS microplastics was improved with the existence of DOM; it's probably because of the complexation of humic acid with weathered PS (Wang et al., 2018). Also, when the size of origin microplastic is decreased to the nanometer size, these particles can interact with the aromatic structure of DOM among p-p conjugation (Chen et al., 2018).

6.3.1 Microplastic Properties

One of the effective parameters in the adsorption of organic compounds is the microplastic properties, plastic polymers usually have amorphous and crystalline regions. For example, PE and PP have crystalline and amorphous parts in their structure and are considered semi-crystalline polymers, their amorphous regions being desirable sites for the adsorption of organic chemicals (Endo & Koelmans, 2016). Study results of Guo and Wang (2019) demonstrate that the crystallinity of microplastics has an important impact on the adsorption of pollutants; they noted that the adsorption coefficients of phenanthrene, naphthalene, and lindane to PE declined with raising crystallinity of PE. Other researchers (Lu et al., 2019) report a positive relationship between the adsorption of 17b-estradiol and the crystallinity of microplastics (Velzeboer et al., 2014).

6.3.1.1 Transition Temperature

The glass transition temperature (Tg) of microplastics that can affect chemical sorption is different between species plastic polymers. Polymers that have Tg amounts more than ambient temperature are known as glassy polymers (e.g., PVC and PS), while Tg amounts lower than ambient temperature are called rubbery polymers (e.g., PE and PP). Following the pore-filling process of organic chemicals into glassy polymers, nonlinear isotherms are observed on glassy polymers, while linear sorption isotherms were usually observed on rubbery polymers (Endo & Koelmans, 2016; Seidensticker et al., 2018).

6.3.1.2 Size of Microplastics

Stronger adsorption occurs between micro- and nanoplastics and organic compounds (e.g., phenanthrene) as the size decreases (Chen et al., 2018). Nevertheless, for some microplastics, the size of the microplastics and the related surface area has no important role in the adsorption of organic compounds, while the properties of the microplastics (e.g., chemical structure and composition of the microplastic polymers) have a significant role (Hüffer & Hofmann, 2016).

6.3.1.3 Environmental Conditions

Typically, polymers of microplastic are exposed to abiotic and biological aging processes in environmental situation, which have the potential to change the nature of interactions with chemical contaminants. Result of studies showed the aging process can shift surface physical and chemical properties by raising the existence of oxygen-containing functional groups (e.g., carbonyl), reducing molecular weight, and producing a rough surfaces (Song et al., 2020). In laboratory conditions acceleration of the aging process led to the formation of polar functional groups on the surface of micro- and nanoplastics, which influences the sorption of organic compounds. Results of Huffer's research, in which PS microplastics were treated with UV and 10% H_2O_2, showed that sorption coefficients of nonpolar organic compounds on older PS were one-time extent lower than those on origin PS (Hüffer & Hofmann, 2016).

Also, lower adsorption of benzene, toluene, ethylbenzene, and xylene on PS microplastics (MPs) after the aging process has been confirmed by researchers (Müller et al., 2018). On the contrary, after aging by UV, the sorption tendencies of ciprofloxacin to PS and PVC microplastics for hydrophilic organic compounds were increased by 123% and 20%, respectively, which may be the main reason for the formation of oxygen-containing functional groups (e.g., hydroxyl and carbonyl) on the surface of aged microplastics (Zhu et al., 2018). Of course, the confirmation of weathering in laboratory and field conditions requires further research; one of the factors that should be evaluated is the environmentally relevant concentration of organics and the aging process.

6.4 Co-transport of Microplastics with Colloids

In actual soil media, colloids and manmade products released certainly meet with micro nanoparticles and affect microplastic transportation. For instance, the co-presence of graphene oxide had a significant result on the movement of PS microplastics, and the effect depended on the ionic strength of the solution (Peng et al. 2017a, b, c). This interaction between graphene oxide and microplastics and nanoplastics can influence the transport and deposition of engineered nanoparticles. In one study microplastics raised the transport and lowered the deposition of $nTiO_2$ in quartz sand at pH 7; also nanoplastics had a substantial effect on fullerene (C60) transportation (Dong et al., 2019).

6.4.1 Transfer of Microplastics Attached to Contaminants

Because of the size exclusion effect which is important in screening larger particles from small pores, the movement of colloids or nanoparticles could be quicker than that of the pore water. PS nanoplastics at low concentrations greatly increase the transfer of non-polar and weakly polar pollutants but have little effect on the transfer of polar pollutants (Liu et al., 2018). And also, the aging process increased the dynamism of PS nanoplastics, thereby greatly enhancing the ability to transport contaminants (e.g., non-polar pyrene and polar 4-nonylphenol) through slow-release kinetics and immutable adsorption of pollutants (Wang et al. 2019a, b, c). The chemical configurations and compounds of plastics definitely influence the sorption of organic pollutants and consequently affect transport by nanoplastics and microplastics (Guo et al., 2012).

6.4.2 Soil Fauna's Role in Pollutant Transport by Microplastics

After devouring microplastics by present organisms in the soil media and then transferring them to humans along the food chain, various toxic chemicals are able to enter the human body.

6.4.3 Organic Pollutants

Results of previous studies displayed that microplastics may transport HOCs to the aquatic amphipod *Allorchestes compressa* (Chua et al., 2014), while current analyses exhibited that in real soil media, microplastics have a narrow role in the expansion of HOCs in sea organisms such as the deposit-dwelling lugworm (Endo & Koelmans, 2016). In the soil which was rich with organic compounds (e.g., PAHs and PCBs) and despite the ingestion of microplastics by earthworms (*E. fetida*), a minor impact of microplastics on the bioaccumulation of HOCs in *E. fetida* with 10% (w/w) microplastics in agrarian soil was observed (Wang et al. 2019a, b, c). This contrast results of studies indicate that microplastics can hardly be the carrier of HOCs to earthworms, or facilitate the bioaccumulation of HOCs in earthworms, so further investigations are needed to clarify the role of microplastics as carriers of HOCs in other soil organisms.

6.4.4 Inorganic Contaminants

There is no conclusive evidence that microplastics increase or decrease the risk associated with trace elements (e.g., As and Zn) for earthworms. Although zinc is an element with high bioavailability on microplastics (greater desorption) than soil, no detectable effect of zinc-contaminated microplastics on zinc accumulation, fatality, and increase or decrease in weight of earthworms (*L. terrestris*) faced with earthworms to zinc-contaminated microplastics in arable soil was observed (Hodson et al., 2017). Nevertheless, one study stated that the presence of microplastics reduced As accumulation and prevented As (V) accumulation in earthworms (*Metaphire californica*), resulting in less toxicity to *M. californica* (Wang et al. 2019a, b, c).

6.4.5 Antibiotics

Combination of microplastics in soil with antibiotics can contribute to increased biological resistance. The presence of tetracycline and microplastic significantly disturbs the microbial residents in soil (Ma et al., 2020). On a laboratory scale, disturbance between the microbial communities by microplastics combined with tetracycline has been proved. Moreover, the diversity of antibiotic resistance genes (ARGs) was increased (Ma et al., 2020). As we know ARGs is recognized as one of the most important emerging pollutants which is a severe threat to the ecosystem (Sanderson et al., 2016). Recent research has confirmed that microplastics can act as a vector for ARGs in landfill leachate. As mentioned above antibiotics exhibited intense disorder to ecology and include disrupting the endocrine system and chronic toxicity (Ma et al., 2020).

The interaction between microplastics and antibiotics remains to be explored because antibiotics are among the widely used pharmaceutical and personal care products that are resistant to biodegradation, and their adsorption on microplastics played a significant function during cotransport (Li et al., 2021).

For example, the results of studies showed that the sorption of oxytetracycline on the surface of polyamide (PA) microplastics takes place weaker than sorption on soil. The aging process of PP and PE microplastic is a factor that improved enrichment for ARGs due to changes in MP surface properties and oxygen-containing active groups. Interaction between microplastic and PPCPs is intrinsically related to characteristics of the sorbate and environmental factors such as solution pH. This is because PPCPS are usually hydrophilic. Inverse hydrophobic substances and other mechanisms also control the adsorption of PPCPs on M&NPs including electrostatic interaction, not merely hydrophobicity. In current studies, the sorption of three different nonsteroidal anti-inflammatory drugs (NSAIDs) to microplastics is examined. They found an apparent pH dependency. This mechanism could be defined by changes in the surface charge of drugs and microplastics (Elizalde-Velazquez et al., 2020).

6.4.6 Heavy Metals

Contact of toxic and harmful metal pollutants with M&NPs will inevitably occur during the migration process in the environment. The adsorbed pollutants on M&NPs undoubtedly make complex situations that interfere with the growth and survival of organisms (Wang et al. 2019a, b, c). For example, metals accumulated on the microplastics may pose a higher risk to aquatic organisms. Mixed contaminants related to heavy metals can enter the food chain and affect the human body indirectly (Dobaradaran et al., 2018). A previous study demonstrated that micro- and nanoplastics acting as vectors then could transport metal into organisms. Higher desorption of Zn is enriched in fragmented plastic bags (particle size was approximately 1.32 ± 0.72 mm and 0.71 ± 0.43 mm) than in soil. The results confirm the carrier role of microplastics and intensify their effect when exposed to metals (Hodson et al., 2017). The surface of the microplastics when exposed to UV forms more holes making it easier to absorb metals. Typically environmental factors such as the aging process showed dominant effects on the sorption capacity of heavy metals on M&NPs. Sorption of heavy metal on microplastic increased with increasing aging treatment time (Mao et al., 2020). Solution chemistry also influences the aging process. Adsorption capacity of cadmium onto the MPs first increased and then decreased when solution pH increased from 2.0 to 9.0, reaching highest at pH 6 (Zhou et al., 2020). The ionic strength of the solution also affects the adsorption behavior strongly (Ren et al., 2021). For example, along with the raising NaCl concentration, lead (II) adsorption to aged nylon microplastics is reduced likely due to competing for adsorption areas and decreased electrostatic potential of microplastic. Seidensticker et al. (2018) reported that the partition coefficient (KMP) of Cr on aged PE microplastics was one degree of extent higher than that on original microplastic particles, which was also comparable with or higher than that of some PPCPs in a similar concentration range (Seidensticker et al., 2018). By other studies by Turner and Holmes (Turner & Holmes, 2015), the improved sorption of heavy metals on aged microplastics has been documented. A recent study also indicated that nanoplastics had a high sorption capacity for Pb (II), with a removal rate of up to 79–97% (Wang et al., 2019a). Generally, microplastics with aged surfaces and smaller sizes have the highest chance, to carry heavy metals. Nevertheless, the role of organic matter in this sorption process needs more elucidation, i.e., it is ambiguous whether sorption to microplastics and organic matter will be synergistic or competitive in nature.

6.5 Competition Microplastics and Soils in Sorption of Toxic Chemicals

If the amount of LogK_{ow} of organic compounds is more than two, there is a powerful linear relevance between LogK_{ow} and LogK_{oc}. Doucette et al. (Doucette, 2003) suggested that LogK_{ow} is the suitable predictor for the sorption of neutral HOC$_s$ onto soils media and deposits but is not suitable for highly polar or ionizable organic chemicals. This might explain the relatively low linear correlation (R2 ¼ 0.425) and the scattered points with LogK_{ow} < 2, because many organic compounds investigated are polar and ionizable (e.g., some PPCPs).

6.5.1 Biodegradable Plastics

In addition to conventional plastics, the application of biodegradable plastics has also increased environmental concerns. With the advent of biodegradable plastics, their entry into the environment, especially the soil system, is inevitable (Liao & Yang, 2020). In various studies, considering the interaction of degradable microplastics with contaminants, the ability of biodegradable plastics to absorb certain organic compounds compared to ordinary plastics has been proven. For example, the role of biodegradable plastics in the absorption of antibiotics and drugs was greater than that of traditional plastics (Fan et al., 2021). However, investigation on the environmental behavior of degradable plastics, particularly its transport behavior in soil and groundwater, is limited and requires further research. The interaction of M&NPs and pollutants in the meningeal environment coincides with contamination. The transfer of auxiliary pollutants generally depends on the characteristics of the M&NPs and external environmental factors. Compared to microplastics, the simultaneous transport of nanoplastics deserves more attention. Because, as mentioned earlier, due to the size of the nano and the large specific surface area of the nanoplastics, especially in porous media, it may be a threat to human safety and, in addition, they are easier to transport and move. To reverse the migration of symbiotic substances. The impact of M&NPs on the transmission of organic pollutants, natural or synthetic colloids, and bacteria still needs to be investigated.

Different adsorption tendencies played a key part. Polar compounds are just absorbed on the surface of polystyrene, whereas nonpolar compounds were entrapped in the inner matrices due to the glassy polymeric structure. A similar study showed the cotransport behavior of polystyrene nanoplastics and naphthalene in different ionic strengths. As a nonpolar organic pollutant, naphthalene quickly contacted the sorption sites. Transport of nonpolar naphthalene was enhanced by nanoplastics since the binding strength of naphthalene to PSNP was stronger than that of porous media (Liu et al., 2018).

6.5.2 M&NPs and Natural Colloid

As soil pollution intensifies, the impact of natural soil minerals on the fate of pollutants becomes more important and complicates interactions between M&NPs and natural colloids. According to a column experiment, researchers have found interactions between natural colloids and M&NPs. Their results showed that smaller plastic particles (0.02 and 0.2 μm) improved the transfer of goethite and hematite, while larger plastic particles (2 μm) did not. Competitive deposition and spatial repulsion help increase the transfer of iron oxides, while the adsorption of 0.2 μm megapixels on iron oxides had a great function in increasing the transfer of iron oxides. In the case of a 2 μm transfer reduction, this was because iron oxides, due to their attractive interaction, preferred to be deposited on quartz sand rather than plastic particles. Consequence, the interaction between microplastics and iron oxides showed low impact on the transfer of iron oxides. At the same time, the surface properties of M&NPs affect equity. Carboxyl microplastic modification can increase the transport of clay particles due to the microplastic rival adsorption sites, while amino-modified microplastics did not show an important influence on kaolinite transport behavior (Li et al., 2021). However, following the electrostatic attraction, the exchange between positively charged amino microplastics and negatively charged quartz sand was very strong. The mobility of kaolinite was limited by the formation of aggregates with amino microplastics.

6.5.3 Engineered Nanomaterials

As we know natural or synthesized fabrics are continually utilized to adsorb toxic and harmful substances to reach the objective of treating contaminated soil. Nanomaterials are commonly well-known as adsorbents to coexisting pollutants in the soil. When exposed to microplastics, both of their surface characteristics may change, thus changing their transport and sediment in porous media. Graphene oxide (GO) is widely used as a soil remediation. Studies were tested cotransport of difference-sized microplastic and GO (Peng et al. 2017a, b, c). The results recommended that cotransport of both two components was mostly controlled by GO, due to the adsorption of microplastic on GO. Parallel analysis on cotransport of GO, reduced-GO (RGO), and polystyrene nanoplastic delivered that the presence of nanoplastics declined transport of (R)GO due to further retention sites supplied by plastic particles. The sediment of nanoplastics on the sand surface restricted pore throats and hindered transport of (R) GO (Xia et al., 2021). Examination of metal nanomaterials and microplastics revealed that the surface charge of titanium dioxide ($nTiO_2$) in acidic solution was positive. Though positively charged $nTiO_2$ easily attracted negatively charged microplastics forming aggregates, transport of $nTiO_2$ did not change owing to attractive electrostatic interaction between $nTiO_2$-microplastic cluster and porous media, whereas microplastic enhanced transport of nanomaterial at pH 7. The plastic particles preferred to be adsorbed on sand rather than on $nTiO_2$. This competition on the deposition site of quartz sand led to increased transport of $nTiO_2$. The effect of soluble chemical conditions on joint transport cannot be underestimated. Weathering processes also affect the interaction between microplastics and metal nanomaterials. According to stated researchers, weathered XPs adsorbed CeO2 nanoparticles more easily than pristine ones (Singh et al., 2021).

6.6 Conclusions

Indeed, co-contamination in the actual environment is not a simple binary system. The transmission route and fate of combined pollutants containing multiple pollutants has always been a concern (Zhang et al., 2017). Zhao et al. investigated the effect of simultaneous transfer of metal composite contaminants and antibiotics with graphene oxide (Yu et al., 2020). Taken together, existing research on transport of M&NPs and coexisting pollutants focus more on such binary compound system as mentioned above, and there are still few studies on the cotransport of M&NPs with a diversity of contaminants. Microorganisms are everywhere in the realistic conditions and can perform self-migrate. Microplastics were potential carriers for colonizing or forming biofilms in marine environments and then may take microorganisms such as bacteria transport to faraway areas (Deshiikan et al., 1998). He et al. found that the low ionic strength of the solution had little effect on the transfer of M&NPs, whereas increased bacterial transfer occurs under conditions of high ionic strength. The adsorption mechanisms of M&NPs under experimental conditions varied with different particle sizes during common transport (He et al., 2018). Research on interaction and cotransport of microorganisms and M&NPs is far from adequate. Because the laboratory environment is a simple environment compared to real soil, more research is needed to better analyze the interaction of M&NPs and microorganisms in the porous medium. Laboratory research on M&NPs verified their great transportability, but field studies or pilot-scale experiments of M&NPs are still necessary to more comprehend the true state of transport in soil and groundwater.

6.7 Perspective

Generally, the transfer of pollutants by micro-nanoplastics is influenced by factors such as the existence of natural organic environments, heterogeneous porosity, and type variety of microplastics and the impact of coupling environmental factors on the transfer of chemicals and requires further investigation. The interchange of microplastics or nanoplastics with other symbiotic materials must be carefully considered. The properties of M&NPs, including surface properties and density, may change with the coexistence of other materials, resulting in different environmental behaviors. Nonetheless, there is yet a gap between the experimental environment and the actual conditions. Biofilms are always present on the surface of M&NPs in the real environment, whereas this effect has always been mostly forgotten in previous laboratory studies. The physical and chemical properties of the micro- and nanoplastics will change by coating with biofilm (Rummel et al., 2017). The difference in properties such as surface roughness, zeta potential, hydrophobicity, and surface energy will impact adsorption capacity and then affect their interaction with other pollutants in the cotransport system. For example, mass recovery of polystyrene nanoplastic (PSNP) declined from 77.60% to 62.48% because of naphthalene. This mechanism could be explained by the charge shielding effect. Because naphthalene is non-polar, it may trap some negative charges on the PSNP surface and increase neoplastic retention on the sand (Hu et al., 2020). Differences in electrostatic repulsion force and hydrophobicity would well clarify the simultaneous transport behavior (Yu et al., 2020). Also, the existence of NOM (natural organic matter) in the environment has a significant function in the transport of M&NPs (Pelley & Tufenkji, 2008). The adsorption of pollutants and aggregation of microplastics are greatly affected by the existence of NOM (Rong et al., 2021). Although organic matter is everywhere in the background, it should be regarded as an important element in the simultaneous transfer of micro- and nanoplastic with other materials. According to the present examinations on cotransport of M&NPs with contaminants, solution chemical conditions such as ionic strength and ion species were the principal influencing parts. The effect of some coexisting substances on microplastics migration cannot be considered from a single perspective mentioned above. For example, two kinds of bacteria, *E. coli* (Gram (−)) and *B. subtilis* (Gram (+)), decreased transport of negatively charged MPs (carboxylate-modified, CMPs), yet raised positively charged microplastic (amine-modified, AMPs). For example, in 5 mM NaCl, 73.1% CMPs, and 0.94% AMPs without bacteria passed through the columns. Meantime, the bulk recovery of CMPs and AMPs with *E. coli* is 58.6% and 18.8%, and the corresponding percentage with B. subtilis was 38.9% and 30%. Changing surface charge, extra deposition sites, and aggregation all contributed to such results (He et al., 2021).

References

Alimi, O. S., Farner Budarz, J., Hernandez, L. M., & Tufenkji, N. (2018). Microplastics and nanoplastics in aquatic environments: Aggregation, deposition, and enhanced contaminant transport. *Environmental Science & Technology, 52*(4), 1704–1724.

Amirbahman, A., & Olson, T. M. (1995). Deposition kinetics of humic matter-coated hematite in porous media in the presence of Ca2+. *Colloids and Surfaces A: Physicochemical and Engineering Aspects, 99*(1), 1–10.

Andrady, A. L., & Neal, M. A. (2009). Applications and societal benefits of plastics. *Philosophical Transactions of the Royal Society B: Biological Sciences, 364*(1526), 1977–1984.

Awet, T., Kohl, Y., Meier, F., Straskraba, S., Grün, A.-L., Ruf, T., Jost, C., Drexel, R., Tunc, E., & Emmerling, C. (2018). Effects of polystyrene nanoparticles on the microbiota and functional diversity of enzymes in soil. *Environmental Sciences Europe, 30*(1), 1–10.

Bläsing, M., & Amelung, W. (2018). Plastics in soil: Analytical methods and possible sources. *Science of the Total Environment, 612*, 422–435.

Bradford, S. A., & Bettahar, M. (2006). Concentration dependent transport of colloids in saturated porous media. *Journal of Contaminant Hydrology, 82*(1–2), 99–117.

Bradford, S. A., Yates, S. R., Bettahar, M., & Simunek, J. (2002). Physical factors affecting the transport and fate of colloids in saturated porous media. *Water Resources Research, 38*(12), 63-61-63-12.

Chen, W., Ouyang, Z.-Y., Qian, C., & Yu, H.-Q. (2018). Induced structural changes of humic acid by exposure of polystyrene microplastics: A spectroscopic insight. *Environmental Pollution, 233*, 1–7.

Chua, E. M., Shimeta, J., Nugegoda, D., Morrison, P. D., & Clarke, B. O. (2014). Assimilation of polybrominated diphenyl ethers from microplastics by the marine amphipod, Allorchestes compressa. *Environmental Science & Technology, 48*(14), 8127–8134.

de Souza Machado, A. A., Lau, C. W., Till, J., Kloas, W., Lehmann, A., Becker, R., & Rillig, M. C. (2018). Impacts of microplastics on the soil biophysical environment. *Environmental Science & Technology, 52*(17), 9656–9665.

de Souza Machado, A. A., Lau, C. W., Kloas, W., Bergmann, J., Bachelier, J. B., Faltin, E., Becker, R., Görlich, A. S., & Rillig, M. C. (2019). Microplastics can change soil properties and affect plant performance. *Environmental Science & Technology, 53*(10), 6044–6052.

Deshiikan, S. R., Eschenazi, E., & Papadopoulos, K. D. (1998). Transport of colloids through porous beds in the presence of natural organic matter. *Colloids and Surfaces A: Physicochemical and Engineering Aspects, 145*(1–3), 93–100.

Dobaradaran, S., Schmidt, T. C., Nabipour, I., Khajeahmadi, N., Tajbakhsh, S., Saeedi, R., Mohammadi, M. J., Keshtkar, M., Khorsand, M., & Ghasemi, F. F. (2018). Characterization of plastic debris and association of metals with microplastics in coastline sediment along the Persian Gulf. *Waste Management, 78*, 649–658.

Dong, Z., Zhang, W., Qiu, Y., Yang, Z., Wang, J., & Zhang, Y. (2019). Cotransport of nanoplastics (NPs) with fullerene (C60) in saturated sand: Effect of NPs/C60 ratio and seawater salinity. *Water Research, 148*, 469–478.

Doucette, W. J. (2003). Quantitative structure-activity relationships for predicting soil-sediment sorption coefficients for organic chemicals. *Environmental Toxicology and Chemistry: An International Journal, 22*(8), 1771–1788.

Elimelech, M., & O'Melia, C. R. (1990). Kinetics of deposition of colloidal particles in porous media. *Environmental Science & Technology, 24*(10), 1528–1536.

Elizalde-Velazquez, A., Subbiah, S., Anderson, T. A., Green, M. J., Zhao, X., & Cañas-Carrell, J. E. (2020). Sorption of three common nonsteroidal anti-inflammatory drugs (NSAIDs) to microplastics. *Science of the Total Environment, 715*, 136974.

Endo, S., & Koelmans, A. A. (2016). *Sorption of hydrophobic organic compounds to plastics in the marine environment: Equilibrium. Hazardous chemicals associated with plastics in the marine environment* (pp. 185–204). Springer.

Fan, X., Zou, Y., Geng, N., Liu, J., Hou, J., Li, D., Yang, C., & Li, Y. (2021). Investigation on the adsorption and desorption behaviors of antibiotics by degradable MPs with or without UV ageing process. *Journal of Hazardous Materials, 401*, 123363.

Franchi, A., & O'Melia, C. R. (2003). Effects of natural organic matter and solution chemistry on the deposition and reentrainment of colloids in porous media. *Environmental Science & Technology, 37*(6), 1122–1129.

Gul, C., Mahapatra, P. S., Kang, S., Singh, P. K., Wu, X., He, C., Kumar, R., Rai, M., Xu, Y., & Puppala, S. P. (2021). Black carbon concentration in the central Himalayas: Impact on glacier melt and potential source contribution. *Environmental Pollution, 275*, 116544.

Guo, X., & Wang, J. (2019). The chemical behaviors of microplastics in marine environment: A review. *Marine Pollution Bulletin, 142*, 1–14.

Guo, X., Wang, X., Zhou, X., Kong, X., Tao, S., & Xing, B. (2012). Sorption of four hydrophobic organic compounds by three chemically distinct polymers: Role of chemical and physical composition. *Environmental Science & Technology, 46*(13), 7252–7259.

He, L., Wu, D., Rong, H., Li, M., Tong, M., & Kim, H. (2018). Influence of nano-and microplastic particles on the transport and deposition behaviors of bacteria in quartz sand. *Environmental Science & Technology, 52*(20), 11555–11563.

He, L., Rong, H., Wu, D., Li, M., Wang, C., & Tong, M. (2020). Influence of biofilm on the transport and deposition behaviors of nano-and micro-plastic particles in quartz sand. *Water Research, 178*, 115808.

He, L., Rong, H., Li, M., Zhang, M., Liu, S., Yang, M., & Tong, M. (2021). Bacteria have different effects on the transport behaviors of positively and negatively charged microplastics in porous media. *Journal of Hazardous Materials, 415*, 125550.

Hodson, M. E., Duffus-Hodson, C. A., Clark, A., Prendergast-Miller, M. T., & Thorpe, K. L. (2017). Plastic bag derived-microplastics as a vector for metal exposure in terrestrial invertebrates. *Environmental Science & Technology, 51*(8), 4714–4721.

Horton, A. A., Walton, A., Spurgeon, D. J., Lahive, E., & Svendsen, C. (2017). Microplastics in freshwater and terrestrial environments: Evaluating the current understanding to identify the knowledge gaps and future research priorities. *Science of the Total Environment, 586*, 127–141.

Hu, D., Zhang, Y., & Shen, M. (2020). Investigation on microplastic pollution of Dongting Lake and its affiliated rivers. *Marine Pollution Bulletin, 160*, 111555.

Huang, Y., Liu, Q., Jia, W., Yan, C., & Wang, J. (2020). Agricultural plastic mulching as a source of microplastics in the terrestrial environment. *Environmental Pollution, 260*, 114096.

Huerta Lwanga, E., Mendoza Vega, J., Ku Quej, V., Chi, J. D. L. A., Sanchez del Cid, L., Chi, C., Escalona Segura, G., Gertsen, H., Salánki, T., & van der Ploeg, M. (2017). Field evidence for transfer of plastic debris along a terrestrial food chain. *Scientific Reports, 7*(1), 1–7.

Hüffer, T., & Hofmann, T. (2016). Sorption of non-polar organic compounds by micro-sized plastic particles in aqueous solution. *Environmental Pollution, 214*, 194–201.

Hüffer, T., Praetorius, A., Wagner, S., Von der Kammer, F., & Hofmann, T. (2017). *Microplastic exposure assessment in aquatic environments: Learning from similarities and differences to engineered nanoparticles*. ACS Publications.

Kobayashi, M., Nanaumi, H., & Muto, Y. (2009). Initial deposition rate of latex particles in the packed bed of zirconia beads. *Colloids and Surfaces A: Physicochemical and Engineering Aspects, 347*(1–3), 2–7.

Li, J., Guo, K., Cao, Y., Wang, S., Song, Y., & Zhang, H. (2021). Enhance in mobility of oxytetracycline in a sandy loamy soil caused by the presence of microplastics. *Environmental Pollution, 269*, 116151.

Liao, Y.-L., & Yang, J.-Y. (2020). Microplastic serves as a potential vector for Cr in an in-vitro human digestive model. *Science of the Total Environment, 703*, 134805.

Liu, M., Lu, S., Song, Y., Lei, L., Hu, J., Lv, W., Zhou, W., Cao, C., Shi, H., & Yang, X. (2018). Microplastic and mesoplastic pollution in farmland soils in suburbs of Shanghai, China. *Environmental Pollution, 242*, 855–862.

Lu, L., Luo, T., Zhao, Y., Cai, C., Fu, Z., & Jin, Y. (2019). Interaction between microplastics and microorganism as well as gut microbiota: A consideration on environmental animal and human health. *Science of the Total Environment, 667*, 94–100.

Ma, H., Pu, S., Liu, S., Bai, Y., Mandal, S., & Xing, B. (2020). Microplastics in aquatic environments: Toxicity to trigger ecological consequences. *Environmental Pollution, 261*, 114089.

Maaß, S., Daphi, D., Lehmann, A., & Rillig, M. C. (2017). Transport of microplastics by two collembolan species. *Environmental Pollution, 225*, 456–459.

Mao, S., Gu, W., Bai, J., Dong, B., Huang, Q., Zhao, J., Zhuang, X., Zhang, C., Yuan, W., & Wang, J. (2020). Migration of heavy metal in electronic waste plastics during simulated recycling on a laboratory scale. *Chemosphere, 245*, 125645.

Mato, Y., Isobe, T., Takada, H., Kanehiro, H., Ohtake, C., & Kaminuma, T. (2001). Plastic resin pellets as a transport medium for toxic chemicals in the marine environment. *Environmental Science & Technology, 35*(2), 318–324.

Müller, A., Becker, R., Dorgerloh, U., Simon, F.-G., & Braun, U. (2018). The effect of polymer aging on the uptake of fuel aromatics and ethers by microplastics. *Environmental Pollution, 240*, 639–646.

O'Connor, D., Pan, S., Shen, Z., Song, Y., Jin, Y., Wu, W.-M., & Hou, D. (2019). Microplastics undergo accelerated vertical migration in sand soil due to small size and wet-dry cycles. *Environmental Pollution, 249*, 527–534.

Pelley, A. J., & Tufenkji, N. (2008). Effect of particle size and natural organic matter on the migration of nano-and microscale latex particles in saturated porous media. *Journal of Colloid and Interface Science, 321*(1), 74–83.

Peng, G., Zhu, B., Yang, D., Su, L., Shi, H., & Li, D. (2017a). Microplastics in sediments of the Changjiang Estuary, China. *Environmental Pollution, 225*, 283–290.

Peng, J., Wang, J., & Cai, L. (2017b). Current understanding of microplastics in the environment: Occurrence, fate, risks, and what we should do. *Integrated Environmental Assessment and Management, 13*(3), 476–482.

Peng, S., Wu, D., Ge, Z., Tong, M., & Kim, H. (2017c). Influence of graphene oxide on the transport and deposition behaviors of colloids in saturated porous media. *Environmental Pollution, 225*, 141–149.

Ren, Z., Gui, X., Wei, Y., Chen, X., Xu, X., Zhao, L., Qiu, H., & Cao, X. (2021). Chemical and photo-initiated aging enhances transport risk of microplastics in saturated soils: Key factors, mechanisms, and modeling. *Water Research, 202*, 117407.

Rillig, M. C. (2012). *Microplastic in terrestrial ecosystems and the soil?* ACS Publications.

Rillig, M. C., Ziersch, L., & Hempel, S. (2017). Microplastic transport in soil by earthworms. *Scientific Reports, 7*(1), 1–6.

Rochman, C. M., Tahir, A., Williams, S. L., Baxa, D. V., Lam, R., Miller, J. T., Teh, F.-C., Werorilangi, S., & Teh, S. J. (2015). Anthropogenic debris in seafood: Plastic debris and fibers from textiles in fish and bivalves sold for human consumption. *Scientific Reports, 5*(1), 1–10.

Rong, L., Zhao, L., Zhao, L., Cheng, Z., Yao, Y., Yuan, C., Wang, L., & Sun, H. (2021). LDPE microplastics affect soil microbial communities and nitrogen cycling. *Science of the Total Environment, 773*, 145640.

Rummel, C. D., Jahnke, A., Gorokhova, E., Kühnel, D., & Schmitt-Jansen, M. (2017). Impacts of biofilm formation on the fate and potential effects of microplastic in the aquatic environment. *Environmental Science & Technology Letters, 4*(7), 258–267.

Sanderson, H., Fricker, C., Brown, R. S., Majury, A., & Liss, S. N. (2016). Antibiotic resistance genes as an emerging environmental contaminant. *Environmental Reviews, 24*(2), 205–218.

Schnurr, R. E., Alboiu, V., Chaudhary, M., Corbett, R. A., Quanz, M. E., Sankar, K., Srain, H. S., Thavarajah, V., Xanthos, D., & Walker, T. R. (2018). Reducing marine pollution from single-use plastics (SUPs): A review. *Marine Pollution Bulletin, 137*, 157–171.

Seidensticker, S., Grathwohl, P., Lamprecht, J., & Zarfl, C. (2018). A combined experimental and modeling study to evaluate pH-dependent sorption of polar and non-polar compounds to polyethylene and polystyrene microplastics. *Environmental Sciences Europe, 30*(1), 1–12.

Shen, X.-C., Li, D.-C., Sima, X.-F., Cheng, H.-Y., & Jiang, H. (2018). The effects of environmental conditions on the enrichment of antibiotics on microplastics in simulated natural water column. *Environmental Research, 166*, 377–383.

Si, W., Cai, Y., Liu, J., Shen, J., Chen, Q., Chen, C., & Ning, L. (2019). Investigating the role of colloids on the distribution of bisphenol analogues in surface water from an ecological demonstration area, China. *Science of the Total Environment, 673*, 699–707.

Singh, N., Khandelwal, N., Tiwari, E., Naskar, N., Lahiri, S., Lützenkirchen, J., & Darbha, G. K. (2021). Interaction of metal oxide nanoparticles with microplastics: Impact of weathering under riverine conditions. *Water Research, 189*, 116622.

Song, C., Liu, Z., Wang, C., Li, S., & Kitamura, Y. (2020). Different interaction performance between microplastics and microalgae: The bio-elimination potential of Chlorella sp. L38 and Phaeodactylum tricornutum MASCC-0025. *Science of the Total Environment, 723*, 138146.

Tong, M., Li, X., Brow, C. N., & Johnson, W. P. (2005). Detachment-influenced transport of an adhesion-deficient bacterial strain within water-reactive porous media. *Environmental Science & Technology, 39*(8), 2500–2508.

Torres, F. G., Dioses-Salinas, D. C., Pizarro-Ortega, C. I., & De-la-Torre, G. E. (2021). Sorption of chemical contaminants on degradable and non-degradable microplastics: Recent progress and research trends. *Science of the Total Environment, 757*, 143875.

Tourinho, P. S., Kočí, V., Loureiro, S., & van Gestel, C. A. (2019). Partitioning of chemical contaminants to microplastics: Sorption mechanisms, environmental distribution and effects on toxicity and bioaccumulation. *Environmental Pollution, 252*, 1246–1256.

Turner, A., & Holmes, L. A. (2015). Adsorption of trace metals by microplastic pellets in fresh water. *Environmental Chemistry, 12*(5), 600–610.

Velzeboer, I., Kwadijk, C., & Koelmans, A. (2014). Strong sorption of PCBs to nanoplastics, microplastics, carbon nanotubes, and fullerenes. *Environmental Science & Technology, 48*(9), 4869–4876.

Wan, J., & Wilson, J. L. (1994). Colloid transport in unsaturated porous media. *Water Resources Research, 30*(4), 857–864.

Wan, Y., Wu, C., Xue, Q., & Hui, X. (2019). Effects of plastic contamination on water evaporation and desiccation cracking in soil. *Science of the Total Environment, 654*, 576–582.

Wang, F., Shih, K. M., & Li, X. Y. (2015). The partition behavior of perfluorooctanesulfonate (PFOS) and perfluorooctanesulfonamide (FOSA) on microplastics. *Chemosphere, 119*, 841–847.

Wang, J., Lv, S., Zhang, M., Chen, G., Zhu, T., Zhang, S., Teng, Y., Christie, P., & Luo, Y. (2016). Effects of plastic film residues on occurrence of phthalates and microbial activity in soils. *Chemosphere, 151*, 171–177.

Wang, F., Wong, C. S., Chen, D., Lu, X., Wang, F., & Zeng, E. Y. (2018). Interaction of toxic chemicals with microplastics: A critical review. *Water Research, 139*, 208–219.

Wang, H.-T., Ding, J., Xiong, C., Zhu, D., Li, G., Jia, X.-Y., Zhu, Y.-G., & Xue, X.-M. (2019a). Exposure to microplastics lowers arsenic accumulation and alters gut bacterial communities of earthworm Metaphire californica. *Environmental Pollution, 251*, 110–116.

Wang, J., Coffin, S., Sun, C., Schlenk, D., & Gan, J. (2019b). Negligible effects of microplastics on animal fitness and HOC bioaccumulation in earthworm Eisenia fetida in soil. *Environmental Pollution, 249*, 776–784.

Wang, J., Liu, X., Li, Y., Powell, T., Wang, X., Wang, G., & Zhang, P. (2019c). Microplastics as contaminants in the soil environment: A mini-review. *Science of the Total Environment, 691*, 848–857.

Wardrop, P., Shimeta, J., Nugegoda, D., Morrison, P. D., Miranda, A., Tang, M., & Clarke, B. O. (2016). Chemical pollutants sorbed to ingested microbeads from personal care products accumulate in fish. *Environmental Science & Technology, 50*(7), 4037–4044.

Wu, L., Gao, B., Muñoz-Carpena, R., & Pachepsky, Y. A. (2012). Single collector attachment efficiency of colloid capture by a cylindrical collector in laminar overland flow. *Environmental Science & Technology, 46*(16), 8878–8886.

Xia, T., Lin, Y., Li, S., Yan, N., Xie, Y., He, M., Guo, X., & Zhu, L. (2021). Co-transport of negatively charged nanoparticles in saturated porous media: Impacts of hydrophobicity and surface O-functional groups. *Journal of Hazardous Materials, 409*, 124477.

Xu, B., Liu, F., Brookes, P. C., & Xu, J. (2018). Microplastics play a minor role in tetracycline sorption in the presence of dissolved organic matter. *Environmental Pollution, 240*, 87–94.

Yang, X., Bento, C. P., Chen, H., Zhang, H., Xue, S., Lwanga, E. H., Zomer, P., Ritsema, C. J., & Geissen, V. (2018). Influence of microplastic addition on glyphosate decay and soil microbial activities in Chinese loess soil. *Environmental Pollution, 242*, 338–347.

Yu, F., Li, Y., Huang, G., Yang, C., Chen, C., Zhou, T., Zhao, Y., & Ma, J. (2020). Adsorption behavior of the antibiotic levofloxacin on microplastics in the presence of different heavy metals in an aqueous solution. *Chemosphere, 260*, 127650.

Zeng, L.-J., Huang, Y.-H., Chen, X.-T., Chen, X.-H., Mo, C.-H., Feng, Y.-X., Lü, H., Xiang, L., Li, Y.-W., & Li, H. (2020a). Prevalent phthalates in air-soil-vegetable systems of plastic greenhouses in a subtropical city and health risk assessments. *Science of the Total Environment, 743*, 140755.

Zeng, Q.-L., Yu, Z.-J., Gou, J.-J., Li, G.-M., Ma, S.-H., Zhang, G.-F., Xu, J.-H., Lin, W.-B., Cui, G.-L., & Zhang, M.-M. (2020b). Effect of convalescent plasma therapy on viral shedding and survival in patients with coronavirus disease 2019. *The Journal of Infectious Diseases, 222*(1), 38–43.

Zhang, Q., Raoof, A., & Hassanizadeh, S. (2015). Pore-scale study of flow rate on colloid attachment and remobilization in a saturated micromodel. *Journal of Environmental Quality, 44*(5), 1376–1383.

Zhang, K., Xiong, X., Hu, H., Wu, C., Bi, Y., Wu, Y., Zhou, B., Lam, P. K., & Liu, J. (2017). Occurrence and characteristics of microplastic pollution in Xiangxi Bay of Three Gorges Reservoir, China. *Environmental Science & Technology, 51*(7), 3794–3801.

Zhang, Y., Xu, X., Li, Z., Liu, M., Xu, C., Zhang, R., & Luo, W. (2019). Effects of vegetation restoration on soil quality in degraded karst landscapes of southwest China. *Science of the Total Environment, 650*, 2657–2665.

Zhou, Y., Wang, J., Zou, M., Jia, Z., Zhou, S., & Li, Y. (2020). Microplastics in soils: A review of methods, occurrence, fate, transport, ecological and environmental risks. *Science of the Total Environment, 748*, 141368.

Zhou, D., Cai, Y., & Yang, Z. (2022). Key factors controlling transport of micro-and nanoplastic in porous media and its effect on coexisting pollutants. *Environmental Pollution, 293*, 118503.

Zhu, D., Bi, Q.-F., Xiang, Q., Chen, Q.-L., Christie, P., Ke, X., Wu, L.-H., & Zhu, Y.-G. (2018). Trophic predator-prey relationships promote transport of microplastics compared with the single Hypoaspis aculeifer and Folsomia candida. *Environmental Pollution, 235*, 150–154.

Chapter 7
Microplastics as a Carrier of Antibiotic Resistance Genes: A Revision of Literature

Amira Bouaziz, Aicha Asma Houfani, Mounia Arab, and Hafida Baoune

Abstract Antibiotic resistance genes (ARGs) and microplastics (MPs) are one of the major threats, representing a pressing concern to the environment and human health. MPs have been shown to influence bacterial growth and community composition, which could have an impact on the spread of ARGs in the environment. Antibiotic use has resulted in mutated genes in bacteria, and as a result, they can reduce the efficacy of antibiotic therapies, necessitating the use of more of them. Humans produce a significant amount of waste, and global quantities are constantly increasing, resulting in the colonization of many antibiotic resistance bacteria in MPs, which have been turned into a reservoir for ARGs. This chapter highlights the main sources of MPs in the environment, mechanisms of transfer, relationships, and interactions between MPs and ARGs. This revision of the literature aims more specifically to provide the potential implications of the antibiotic resistome of the plastisphere for human health.Graphical Abstract

A. Bouaziz (✉)
Laboratoire de Biotechnologie des Molécules Bioactives et de la Physiopathologie Cellulaire (LBMBPC), Faculté Des Sciences de la Nature et de la Vie, Université de Batna 2, Batna, Algeria

A. A. Houfani
Michael Smith Laboratories and Department of Biochemistry & Molecular Biology, University of British Columbia, Vancouver, BC, Canada

M. Arab
Laboratory of Applied Microbiology in Food, Biomedical and Environment, Department of Biology, Faculty of Nature Sciences and Life, University of Tlemcen, Tlemcen, Algeria

Faculty of Biological Sciences, University of Sciences and Technology Houari Boumediene, Algiers, Algeria

H. Baoune
Faculty of Natural and Life Sciences, Université Kasdi Merbah Ouragla, Ouragla, Algeria

N. R. Maddela et al. (eds.), *Micro and Nanoplastics in Soil*,
https://doi.org/10.1007/978-3-031-21195-9_7

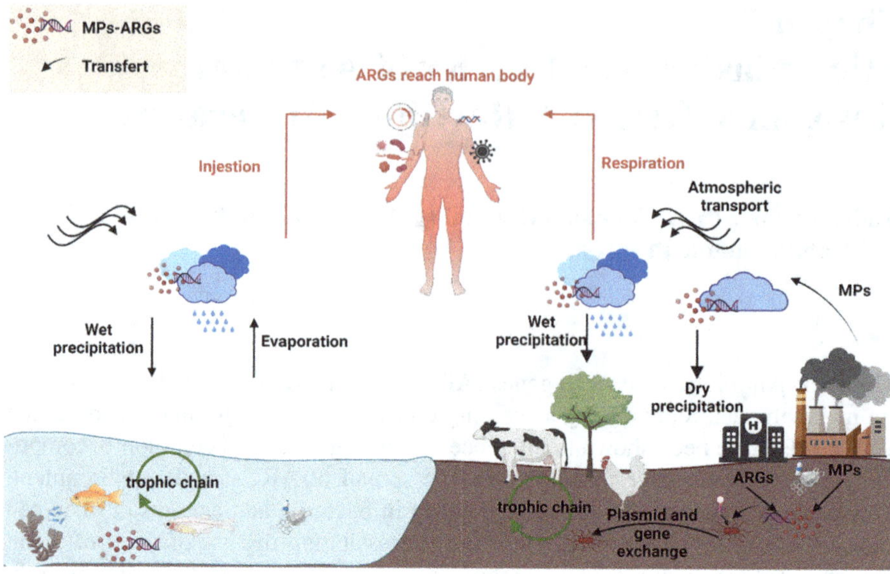

Created with BioRender.com

Keywords Microplastics · Antibiotic resistance genes · Environment · Human health

7.1 Introduction

Plastics are large molecular weight polymeric compounds, which have emerged as a global environmental issue (Sathicq et al., 2021; Zhu et al., 2021). Polystyrene (a strong plastic made from ethylene and benzene) was initially invented in 1839 by Eduard Simon as a polymer containing complex combinations of mostly synthetic organic components that are joined together during polymerization (Alimba & Faggio, 2019). The scientific community provides evidence on a daily basis of the harmful consequences of macroplastics and their generated microplastics and nanoplastics waste on both aquatic and terrestrial ecosystems (Caruso, 2019).

Plastic particles are typically classified based on their size and origin. Microplastics (MPs) are a very diverse group, differing in diameters (<5 mm), shape, color, chemical composition, density, and other properties (Bergmann et al., 2015). Based on their source, they can also be classified as primary or secondary MPs. The latter are derived from textiles, medications, personal care items, pellets, microfibers, and microbeads, which are described as "primary MPs," whereas "secondary MPs" come from the fragmentation of large plastic pieces into small ones as a result of physical and chemical reactions over time (Sathicq et al., 2021; Syranidou & Kalogerakis, 2022). Biotic activity, such as microbial degradation or animal activity, can contribute to the fragmentation, through photodegradation fragments of plastic particles (Burns & Boxall, 2018). The colonization of the material by fouling organisms (including microbial biofilms) increases the density of the particles and forces them to sink, facilitating the transport of small particles to the seafloor and their deposition in the benthic environment (Amaral-Zettler et al., 2021; Bergmann et al., 2015).

MPs quickly absorb pollutants, such as heavy metals, endocrine-disrupting compounds, and antibiotics, due to their huge specific surface area (Shi et al., 2020). MPs are known as a distinct microhabitat, offering a perfect support to stimulate the growth of biofilm and may create a protected niche capable of maintaining a diverse range of microorganisms, referred to as "Plastisphere" (Keswani et al., 2016; Pham et al., 2021).

MPs' potential as "hotspots" of antibiotic resistance genes has received much attention lately, due to their areas of increased nutrient availability and high microbial cell densities, favoring intense interactions (Arias-Andres et al., 2018). Molecular investigation of microplastic biofilms reveals a diverse set of antibiotic resistance genes (ARGs), a serious threat that has been recognized by the World Health Organization (Arias-Andres et al., 2018; Pham et al., 2021). Because of their durability, MPs could serve as a long-term reservoir for ARG and an additional ecological habitat for resistant bacteria (Sathicq et al., 2021).

This chapter briefly summarizes the main sources of MPs in the ecological systems, and the following sections focus on the role of MPs as reservoirs of ARGs and their accumulation, dissemination, and potential risks to human health.

7.1.1 Sources, Distributions, and Behavior Characteristics of MPs Harboring ARG in the Environment

Plastic litter is deposited along shorelines from terrestrial, marine, and atmospheric sources. The pollution caused by microplastics varies geographically and has been found in quite diverse media where the human activities (e.g., domestic wastewater systems, industry, agriculture, and fisheries) are directly responsible for their pollution.

7.2 Aquatic Environment

Human activities place constant and intense pressure on coastal and marine areas. In recent years, it has become increasingly clear that plastics and microplastics are contaminating and affecting the marine environment (Barrows et al., 2018; Coyle et al., 2020; He et al., 2019). Originally described as plastic particles in 1970 by Carpenter and Smith (1972) on the Sargasso Sea surface, MPs have been studied since then by researchers all over the world for their abundance in marine ecosystems (Rezania et al., 2018), the term "microplastics" was coined in 2004 (Thompson et al., 2004).

MPs pollution becomes a severe concern for the marine biota because of their interactions with pollutants such as heavy metals and hydrophobic pollutants (Rezania et al., 2018). In marine environments, microplastics' properties, such as their small size and low density, result in long-distance transport and determine their distribution in the water column (Guo & Wang, 2019; Li et al., 2020). MPs can be transported over long distances (horizontally) and through the water column (vertically) after changes in biofouling, which represents a type of ecological succession that may appear in the aquatic plastisphere and includes microorganisms, aquatic plants, and macrofauna (Rogers et al., 2020). Furthermore, the biofouling affects particle density while serving as vectors for pathogenic bacteria, harmful algae, and invasive species in all types of aquatic systems (Arias-Andres et al., 2018).

Big plastic trash has a density less than that of the seawater and floats at fast speeds across relatively long distances due to the surface current and the wind forces, whereas large, non-buoyant plastics primarily deposit on close shore sediments (Zhang, 2017). Microplastic abundance has been discovered in regions as distant as Antarctica, as well as high concentrations of these microplastics have been detected close to the local wastewater treatment facility, ship traffic, and coastal research operations (Shahul Hamid et al., 2018).

MPs may be consumed by a variety of marine organisms (invertebrates, fishes, birds, etc.) that lead to their morbidity or mortality (Alimba & Faggio, 2019). The side effects of their consumption include blockage of the digestive system, reduction of stomach enzyme release, diminishing of eating stimulation, lowering of steroid hormone levels, ovulation delays, and inability to reproduce (Li et al., 2016). MPs may have the ability to alter population structure, which might have an influence on ecosystem dynamics, including bacteria and viruses (GESAMP, 2015).

Recently, MPs have been discovered to affect microbial community evolution and to promote exchange of ARGs. To date, there are only a very few investigations into a profusion and variety of antibiotic resistance genes in bacterial community on MPs in the marine environment (Hu et al., 2021; Yang et al., 2019). The study of Wang et al. (2020) indicates potentially a greater horizontal gene transfer (HGT) for ARGs between bacterial communities on microplastics and water samples collected from two urban rivers in Jiaxing City, Zhejiang Province, China. In the North Pacific Gyre, the presence of ARGs on microplastics was 5.69 times that of ARGs in the water, indicating the enrichment of ARG on microplastics (Yang et al., 2019).

7.3 Terrestrial Environment

Soils can be contaminated by microplastic debris due to various anthropogenic activities, like wastewater effluent, agricultural plastic mulch, and sludge landfills (Vázquez & Rahman, 2021). Despite the abundance of MP emission in soils

(>40,000 particles/kg) (de Souza Machado et al., 2019), the distribution of terrestrial microplastics is less understood and scientifically poorly studied in comparison to marine microplastic contamination and transport (de Souza Machado et al., 2018; Laermanns et al., 2021; The Royal Society, 2019). MPs are an emerging soil pollutant that is causing concern worldwide, probably because their degradation is slow and incremental (Rillig & Bonkowski, 2018). In fact, there have been some suggestions that terrestrial systems may even be more susceptible to microplastic pollution than oceans (Helmberger et al., 2020). Furthermore, plastic incorporated into or emitted from associated biofilms has been implicated in plastic transfers from marine to terrestrial ecosystems. For instance, bears and wolves were also observed eating plastic marine debris on Alaskan shores, resulting in feces containing MPs (Hale et al., 2020).

Microplastics in soils can be a source of antibiotic-resistant bacteria and have an adverse effect on biodiversity and soil ecosystems (Baho et al., 2021; Corradini et al., 2021). However, it might come as a surprise, but some reports indicate that the impact on soil microbes and ARGs may be positive on soil microbial activity depending on microplastics particle type and concentration (Verla et al., 2019). Thereby, the addition of microplastics to soil has been experimented with and found to enhance or flaw microbial activity (Lin et al., 2020). By analyzing soil enzymatic activity, Zhao et al. (2021) demonstrated that MPs, depending on how they were shaped, negatively affected some enzyme activities related to cellulose and chitin bioconversion. However, to gain more insights, this needs to be investigated further, especially as few data exist regarding microplastic concentrations in the terrestrial environment (Duis & Coors, 2016).

7.4 Atmosphere

MPs have been detected not only in water and soil but also in the atmosphere: an underestimated but a potential source of respiratory and oral exposure (Amato-Lourenço et al., 2020; Gasperi et al., 2018; Prata, 2018), as well as a source of ARGs (Syranidou & Kalogerakis, 2022).

ARGs may reach the atmosphere via biological aerosols generated by wind-borne dust, wastewater treatment plants, or biomass burning, as well as via jet streams (Li et al., 2018; McEachran et al., 2015; Xie et al., 2018). These aerosols can subsequently be deposited on the Earth's surface as snow and rain, creating a relationship between the atmosphere and the Earth's surface and establishing a global ARG cycle, facilitating their spread over great distances.

ARGs are thoroughly mixed in the air under the influence of long-term atmospheric circulation, and their abundance and profile may achieve a relatively stable and homogeneous state. Due to the differences in their mode of dissemination and mechanism of action, several ARGs exhibit a rise or a decrease in abundance in response to air pollution (Allen et al., 2010; Li et al., 2018; McEachran et al., 2015; Xie et al., 2018) Thus, they can be used as indicators of local air pollution.

Zhu et al. (2020) have reported that air pollution amplified the precipitation of widespread ARGs in fresh snow, as it has a considerable impact on the variety and the abundance of ARGs and MGEs. Furthermore, they have found a high positive correlation between integrons, ARGs, and poor air quality but no correlation with good air quality, indicating that particulate pollution may enhance both ARG spread and horizontal transfer.

7.4.1 The Role of MPs as Reservoirs for Microbes, ARGs, and Their Accumulation and Dissemination

MPs are ubiquitous since they have been found in the majority of the components of the environment (soil, freshwater, wastewater, sea, atmosphere, etc.), mediated by a variety of microorganisms, animals, plants, wind, rainfall, as well as anthropogenic activities (Chen et al., 2021; Sun et al., 2021; Yang et al., 2020b; Zhang et al., 2020a). In addition to food processing and packaging as significant sources of microplastics, reports claim microplastics can travel across ecosystems, such as soil to vegetables, and become part of our food or ingested by other organisms (Hirt & Body-Malapel, 2020; Li et al., 2021; Liu et al., 2021).

Due to their large specific surface area, MPs are able to absorb different types of pollutants such as heavy metals, antibiotics, microorganisms, and organic pollutants (Verla et al., 2019). In fact, most of commercial MPs have hydrophobic surface, presenting the ideal environment for microbial colonization by providing a protective shelter (Zettler et al., 2013). Some microbes are able to form a complex matrix of biopolymers called biofilm (Bank & Hansson, 2022). The one formed on MPs is known as "plastisphere" (He et al., 2018) which might include pathogenic bacteria such as *Aeromonas* spp., *Vibrio* spp., and the human opportunistic bacteria *E. coli* (Marathe & Bank, 2022). Plastisphere communities can also be formed by antibiotic-resistant bacteria (ARB). Their inactivation or damage releases antibiotic resistance genes (ARGs) that can be adsorbed by MPs (Syranidou & Kalogerakis, 2022). Thus, MPs serve as a habitat for the spreading of antibiotic resistance (Pham et al., 2021).

Generally, MPs stimulate horizontal transfer of ARGs, supporting gene exchange between bacteria, therefore leading to pathogenicity augmentation and antibiotic resistance dissemination in the environment (Imran et al., 2019). Moreover, because MPs increase cell membrane permeability, ARGs might be more readily available to bacteria. As a result, the transfer of MGEs may lead to an increase in bacteria that could become receptors for the ARGs (Shi et al., 2020). The transfer occurs through ARG's spread in the environment through vertical gene transfer by transmitting genetic information during prokaryotic cell division and HGT driven by mobile genetic elements (MGEs), involving bacteriophages, insertion sequences, integrons, membrane vesicles, plasmids, and transposons (Abe et al., 2020; Lu et al., 2021; Wang et al. 2021a, b; Wu et al., 2019; Yuan et al., 2022; Zhang et al., 2021a). HGT is known to occur in Prokaryotes (Bacteria and Archaea) as well as among fungi

(Boto et al., 2019). These HGT mechanisms occur during conjugation, transformation of genes from one cell to another, or transduction by phages (Abe et al., 2020; Lu et al., 2021; Wang et al. 2021a, b; Wu et al., 2019; Yuan et al., 2022; Zhang et al., 2021a).

In addition, ARG abundance itself may also be affected by changes in the structure of the microbial community (Zhang et al., 2021b). A positive association was found between MPs and ARGs in freshwater and seawater, notably between ARGs, microbial community on MPs and MPs pollution (Hu et al., 2021; Yang et al., 2019). Recently, Arias-Andres et al. (2018) have found that MPs control the microbial community's evolution and increase the ARGs exchange between different bacterial taxa. However, little is known about the relation between ARGs, microbial communities, and MPs. In fact, the horizontal gene transfer is considered as the primary mechanism for gene transfer (Dong et al., 2021). It can happen through three mechanisms: transformation, transduction, or conjugation (Allen et al., 2008). It was demonstrated that the ARGs located on the chromosomes or MGEs are transferred through horizontal gene transfer, mostly by conjugation (Partridge et al., 2018), while extracellular DNA contains free ARGs and spreads through transformation to cells (Zarei-Baygi & Smith, 2021).

In plastisphere, HGT is one of the strategies bacteria use to maximize community dispersal and survival (Lear et al., 2021). In fact, in microplastic-plastisphere, plasmid transfer has been observed to be more prevalent than in planktonic forms or the non-attached free-living bacteria (Arias-Andres et al., 2018; Pham et al., 2021). Nevertheless, the genomic and physiological structure and composition of these types of biofilms differ from those of a typical biofilm forming on a natural substrate. This means that in every situation, specific bacteria with different evolutionary characteristics are involved. However, some cases of marine microplastics forming biofilm communities have been demonstrated to be similar to those found on other types of surfaces, such as glass (Parthasarathy et al., 2019).

Biofilms alter metabolic processes, resulting in pathogenicity and antibiotic resistance in microbial communities as well as metabolic pathways that lead to MP degradation (Okelly et al., 2021), hence, these microorganisms could be explored in soil bioremediation (Guo et al., 2020). The biodegradation process is slower, however, compared to the ultraviolet solar radiation (Oliveira et al., 2020). In a different scenario, MPs may also prevent ARGs transfer by inhibiting antibiotic-resistant bacteria growth and reducing antibiotic adsorption by large MPs (Wang et al., 2021c). Many factors contribute to the spreading and the proliferation of ARGs on MPs surface. An example of these factors can be surface substrate chemistry and the extent weathering of MPs. Besides, internal biofilm processes may also shape the plastisphere community (Syranidou & Kalogerakis, 2022). In addition, the presence of antibiotics in the environment increases the spreading and the persistence of ARGs. Otherwise, antibiotics can be introduced to microbial communities and cause mutations, leading to the development of resistance against a particular antibiotic, which can then be passed on to other bacterial strains (Atugoda et al., 2021). Besides antibiotics, other antimicrobial agents such as biocides and heavy metals can also promote the spread of HGT (Marathe & Bank, 2022). In such cases, heavy metals and antibiotics are absorbed by microplastic surfaces, which is increased by biofilms, resulting in high levels of exposure (Niegowska et al., 2021; Redhead et al., 2020). The adsorption of antibiotics on MPs could be controlled by many mechanisms, such as the hydrophobicity, electrostatic, Van der Waals forces, and π-π interactions (Syranidou & Kalogerakis, 2022). The hydrophobic antibiotics with high logKow (Octanol/Water partition coefficient) showed higher adsorption to MPs (Razanajatovo et al., 2018). Moreover, it was reported in the study of Razanajatovo et al. (2018) that the adsorption of sulfamethoxazole to MPs is associated positively to hydrophobicity.

7.5 ARGs in the Plastisphere and Potential Risks to Human Health

The risks associated with plastispheres can be categorized into four types: pathogens retained within the plastisphere, potential exchange of ARGs via HGT, enhanced antibiotic tolerance, and trophic transmission via the food web.

Several bacteria have been isolated from plastisphere (water, soil, and air) and they have been found to be invasive, virulent, and antibiotic-resistant strains (Bank et al., 2020; Liu et al., 2021). Yang et al. (2020a) identified some bacterial pathogen strains in plastisphere isolated from high-density polyethylene food bags. These bacterial strains were related to human diseases, where the ratio of pathogens to overall bacteria in the plastisphere was higher than that in water obtained from the surface of Xinglin Bay, suggesting that the plastisphere may provide a greater health risk than water. Moreover, they also identified 12 MGEs in the plastisphere, including integrase and transposase genes. Given the aggregated microbial cluster on plastics, the plastisphere provides an excellent environment for infections to gain resistance to antibiotics through horizontal gene transfer, which may be facilitated further by the adsorbed organic and inorganic compounds on plastics (Yang et al., 2020a).

Potentially pathogenic *Vibrio* spp. were detected on microplastics found in various locations (Dussud et al., 2018; Kesy et al., 2019; Kirstein et al., 2016; Rodrigues et al., 2019; Sun et al., 2020). Other putative pathogens were also found selectively enriched on MPs such as *Pseudomonas*, *E. coli*, and *Arcobacter* (Curren & Leong, 2018; Rodrigues et al., 2019; Silva et al., 2019; Wu et al., 2019). Laganà et al. (2019) have reported multiple antibiotic resistances to beta-lactams, cephalosporins, and quinolones in selected strains belonging to Pseudoalteromonas and Shewanella genera.

Potentially pathogenic *Vibrio* spp. and *E. coli* strains carrying virulence genes (EaeA, stx. and lt) on MPs samples were also detected (Silva et al., 2019). Zhang et al. (2019), in their study carried out on the surface of microplastics in mariculture system, have reported the prevalence of pathogenic Vibrio alginolyticus, as well as 160 multi-antibiotic resistant bacteria (MRAB) isolated, for the majority, from the microplastics. Most multiple antibiotic resistance patterns were tetracycline, sulfisoxazole, erythromycin, and penicillin. In addition, five different types of class 1 integrons (intI1) associated gene cassette arrays and seven different types of gene cassettes were identified (Zhang et al., 2020b). In another study, Lu et al. (2019) have reported 25 bacterial strains to be positively associated with intI1, sul1, sul2, tetG, ermF, and qnrS genes on MPs. Members of Pseudomonas, Desulfovibrio, and Flavobacteriaceae were also defined as important hosts for ARGs and MRGs on MPs biofilms (Sun et al., 2021; Wu et al., 2019; Yang et al., 2019). MPs act as hotspots for the co-selection of metal-driven, multi-antibiotic resistant human pathogens (Imran et al., 2019). It was demonstrated that plastispheres are a reservoir for both metal and antibiotic resistance genes, indicating that the selection of metal or antibiotic resistance is an important factor that influences the resistome of microbiota colonizing microplastics (Yang et al., 2019).

Multi-metal resistance genes, multi-drug resistance genes, aminoglycoside-resistance genes, and sulfonamide resistance genes are the most often encountered ARGs found on MPs (Lu et al., 2019; Sun et al., 2021; Wu et al., 2019; Yang et al., 2019). Whereas fosmidomycin, chloramphenicol, rifamycin, kasugamycin, and vancomycin resistance genes appear to be among the least often found ARGs in plastisphere (Yang et al., 2019). ARGs such as beta-lactam resistance gene (blaVEB-9), aminoglycoside resistance genes (aadA13, APH(9)-Ia, APH(3″)-VI, aadA16), and multidrug resistance genes (smeE, mdsC) have been identified as selectively abundant on microplastics (Wu et al., 2019). Moreover, MRGs such as mdsB, tolC, and mexW have been identified on plastisphere samples from marine environment (Sun et al., 2021).

The food chain plays a significant role in the animal ecosystem in terms of delivering and reserving nutritious elements. In addition, the food cycle can transmit microplastics and substances attached to them (Founou et al., 2016; Lehel & Murphy, 2021). Recent research has also shown that ARGs can spread through the trophic level and into the food chain (Zhang et al., 2019; Zhu et al., 2019). As a result, ARGs present on plastics may infiltrate the food chain via food products, posing a risk to human health. Danopoulos et al. (2020) have found the maximum annual human MPs uptake can reach 55,000 MPs particles.

The significant dissemination of antibiotic resistance has the inevitable effect of reducing the efficacy of antibiotic therapies used nowadays in human medicine, with potentially disastrous consequences by the spread of antibiotic-resistant bacterial diseases (Lewis, 2013). The rising issue of antibiotic resistance requires a holistic and multi-sectoral approach, known as the one-health paradigm, which views human, animal, and environmental health as intertwined and interdependent (Robinson et al., 2016). As a result, it's critical to look at the role of clinical and veterinary settings in the spread of ARGs, as well as varied contexts, from the least to the most human-affected.

7.6 Conclusion

The continual discharge of MPs in the biosphere is now a worldwide concern that needs immediate management solutions to prevent worsening possible dangers to organism life and human health, as well as declining aesthetic environmental qualities. MPs act as reservoirs for ARGs in the environment, as a result, more studies are required to further investigate their composition, distribution, transfer, and evolution to elucidate the best ways of dealing with the current MPs reservoir status. The plastics industry should be held accountable for the end-of-life of their products. One of the most pressing current issues that scientists are worried about is the spread of MGEs on microplastics in the environment, urging the burning need to develop effective techniques for tracing the smallest plastic particles.

References

Abe, K., Nomura, N., & Suzuki, S. (2020). Biofilms: Hot spots of horizontal gene transfer (HGT) in aquatic environments, with a focus on a new HGT mechanism. *FEMS Microbiology Ecology, 96*(5), 31.

Alimba, C. G., & Faggio, C. (2019). Microplastics in the marine environment: Current trends in environmental pollution and mechanisms of toxicological profile. *Environmental Toxicology and Pharmacology, 68*, 61–74.

Allen, H. K., Moe, L. A., Rodbumrer, J., Gaarder, A., & Handelsman, J. (2008). Functional metagenomics reveals diverse β-lactamases in a remote Alaskan soil. *The ISME Journal, 3*(2), 243–251.

Allen, H. K., Donato, J., Wang, H. H., Cloud-Hansen, K. A., Davies, J., & Handelsman, J. (2010). Call of the wild: Antibiotic resistance genes in natural environments. *Nature Reviews Microbiology, 8*(4), 251–259.

Amaral-Zettler, L. A., Zettler, E. R., Mincer, T. J., Klaassen, M. A., & Gallager, S. M. (2021). Biofouling impacts on polyethylene density and sinking in coastal waters: A macro/micro tipping point? *Water Research, 201*, 117289.

Amato-Lourenço, L. F., dos Santos, G. L., de Weger, L. A., Hiemstra, P. S., Vijver, M. G., & Mauad, T. (2020). An emerging class of air pollutants: Potential effects of microplastics to respiratory human health? *Science of the Total Environment, 749*, 141676.

Arias-Andres, M., Klümper, U., Rojas-Jimenez, K., & Grossart, H. P. (2018). Microplastic pollution increases gene exchange in aquatic ecosystems. *Environmental Pollution, 237*, 253–261.

Atugoda, T., Vithanage, M., Wijesekara, H., Bolan, N., Sarmah, A. K., et al. (2021). Interactions between microplastics, pharmaceuticals and personal care products: Implications for vector transport. *Environment International, 149*, 106367.

Baho, D. L., Bundschuh, M., & Futter, M. N. (2021). Microplastics in terrestrial ecosystems: Moving beyond the state of the art to minimize the risk of ecological surprise. *Global Change Biology, 27*(17), 3969–3986.

Bank, M. S., & Hansson, S. V. (2022). The microplastic cycle: An introduction to a complex issue. In M. S. Bank (Ed.), *Microplastic in the environment: Pattern and process* (pp. 1–16). Springer.

Bank, M. S., Ok, Y. S., & Swarzenski, P. W. (2020). Microplastic's role in antibiotic resistance. *Science, 369*(6509), 1315.

Barrows, A. P. W., Cathey, S. E., & Petersen, C. W. (2018). Marine environment microfiber contamination: Global patterns and the diversity of microparticle origins. *Environmental Pollution, 237*, 275–284.

Bergmann, M., Gutow, L., & Klages, M. (2015). *Marine anthropogenic litter* (pp. 1–447). Springer.

Boto, L., Pineda, M., & Pineda, R. (2019). Potential impacts of horizontal gene transfer on human health and physiology and how anthropogenic activity can affect it. *The FEBS Journal, 286*(20), 3959–3967.

Burns, E. E., & Boxall, A. B. A. (2018). Microplastics in the aquatic environment: Evidence for or against adverse impacts and major knowledge gaps. *Environmental Toxicology and Chemistry, 37*(11), 2776–2796.

Carpenter, E. J., & Smith Jr, K. L. (1972). Plastics on the Sargasso Sea surface. *Science, 175*(4027), 1240–1241.

Caruso, G. (2019). Microplastics as vectors of contaminants. *Marine Pollution Bulletin, 146*, 921–924.

Chen, J., Wang, W., Liu, H., Xu, X., & Xia, J. (2021). A review on the occurrence, distribution, characteristics, and analysis methods of microplastic pollution in ecosystem s. *Environmental Pollutants and Bioavailability, 33*(1), 227–246. https://doi.org/10.1080/26395940.2021.1960198

Corradini, F., Casado, F., Leiva, V., Huerta-Lwanga, E., & Geissen, V. (2021). Microplastics occurrence and frequency in soils under different land uses on a regional scale. *Science of the Total Environment, 752*, 141917.

Coyle, R., Hardiman, G., & Driscoll, K. O. (2020). Microplastics in the marine environment: A review of their sources, distribution processes, uptake and exchange in ecosystems. *Case Studies in Chemical and Environmental Engineering, 2*, 100010.

Curren, E., & Leong, S. C. Y. (2018). Profiles of bacterial assemblages from microplastics of tropical coastal environments. *Science of the Total Environment, 655*, 313–320.

Danopoulos, E., Jenner, L. C., Twiddy, M., & Rotchell, J. M. (2020). Microplastic contamination of seafood intended for human consumption: A systematic review and meta-analysis. *Environmental Health Perspectives, 128*(12), 126002.

de Souza Machado, A. A., Kloas, W., Zarfl, C., Hempel, S., & Rillig, M. C. (2018). Microplastics as an emerging threat to terrestrial ecosystems. *Global Change Biology, 24*(4), 1405–1416.

de Souza Machado, A. A., Lau, C. W., Kloas, W., Bergmann, J., Bachelier, J. B., et al. (2019). Microplastics can change soil properties and affect plant performance. *Environmental Science & Technology, 53*(10), 6044–6052.

Dong, H., Chen, Y., Wang, J., Zhang, Y., Zhang, P., et al. (2021). Interactions of microplastics and antibiotic resistance genes and their effects on the aquaculture environments. *Journal of Hazardous Materials, 403*, 123961.

Duis, K., & Coors, A. (2016). Microplastics in the aquatic and terrestrial environment: Sources (with a specific focus on personal care products), fate and effects. *Environmental Sciences Europe, 28*(1), 1–25.

Dussud, C., Meistertzheim, A. L., Conan, P., Pujo-Pay, M., George, M., et al. (2018). Evidence of niche partitioning among bacteria living on plastics, organic particles and surrounding seawaters. *Environmental Pollution, 236*, 807–816.

Founou, L. L., Founou, R. C., & Essack, S. Y. (2016). Antibiotic resistance in the food chain: A developing country-perspective. *Frontiers in Microbiology, 7*(NOV), 1881.

Gasperi, J., Wright, S. L., Dris, R., Collard, F., Mandin, C., et al. (2018). Microplastics in air: Are we breathing it in? *Current Opinion in Environmental Science & Health, 1*, 1–5.

GESAMP. (2015). *Sources, fate and effects of microplastics in the marine environment: A global assessment*. IMO, FAO, UNESCO-IOC, UNIDO, WMO, IAEA, UN, UNEP, UNDP. ISSN: 1020-4873.

Guo, X., & Wang, J. (2019). The chemical behaviors of microplastics in marine environment: A review. *Marine Pollution Bulletin, 142*, 1–14.

Guo, J. J., Huang, X. P., Xiang, L., Wang, Y. Z., Li, Y. W., et al. (2020). Source, migration and toxicology of microplastics in soil. *Environment International, 137*, 105263.

Hale, R. C., Seeley, M. E., La Guardia, M. J., Mai, L., & Zeng, E. Y. (2020). A global perspective on microplastics. *Journal of Geophysical Research: Oceans, 125*(1), e2018JC014719.

He, L., Li, P., Yu, L. H., Li, L., Zhang, Y., et al. (2018). Protective effects of proanthocyanidins against cadmium-induced testicular injury through the modification of Nrf2-Keap1 signal path in rats. *Environmental Toxicology and Pharmacology, 57*, 1–8.

He, P., Chen, L., Shao, L., Zhang, H., & Lü, F. (2019). Municipal solid waste (MSW) landfill: A source of microplastics? – Evidence of microplastics in landfill leachate. *Water Research, 159*, 38–45.

Helmberger, M. S., Tiemann, L. K., & Grieshop, M. J. (2020). Towards an ecology of soil microplastics. *Functional Ecology, 34*(3), 550–560.

Hirt, N., & Body-Malapel, M. (2020). Immunotoxicity and intestinal effects of nano- and microplastics: A review of the literature. *Particle and Fibre Toxicology, 17*(1), 1–22.

Hu, H., Jin, D., Yang, Y., Zhang, J., Ma, C., & Qiu, Z. (2021). Distinct profile of bacterial community and antibiotic resistance genes on microplastics in Ganjiang River at the watershed level. *Environmental Research, 200*, 111363.

Imran, M., Das, K. R., & Naik, M. M. (2019). Co-selection of multi-antibiotic resistance in bacterial pathogens in metal and microplastic contaminated environments: An emerging health threat. *Chemosphere, 215*, 846–857.

Keswani, A., Oliver, D. M., Gutierrez, T., & Quilliam, R. S. (2016). Microbial hitchhikers on marine plastic debris: Human exposure risks at bathing waters and beach environments. *Marine Environmental Research, 118*, 10–19.

Kesy, K., Oberbeckmann, S., Kreikemeyer, B., & Labrenz, M. (2019). Spatial environmental heterogeneity determines young biofilm assemblages on microplastics in Baltic Sea Mesocosms. *Frontiers in Microbiology, 10*, 1665.

Kirstein, I. V., Kirmizi, S., Wichels, A., Garin-Fernandez, A., Erler, R., et al. (2016). Dangerous hitchhikers? Evidence for potentially pathogenic Vibrio spp. on microplastic particles. *Marine Environmental Research, 120*, 1–8.

Laermanns, H., Lehmann, M., Klee, M., Löder, M. G. J., Gekle, S., & Bogner, C. (2021). Tracing the horizontal transport of microplastics on rough surfaces. *Microplastics and Nanoplastics, 1*(1), 1–12.

Laganà, P., Caruso, G., Corsi, I., Bergami, E., Venuti, V., et al. (2019). Do plastics serve as a possible vector for the spread of antibiotic resistance? First insights from bacteria associated to a polystyrene piece from King George Island (Antarctica). *International Journal of Hygiene and Environmental Health, 222*(1), 89–100.

Lear, G., Kingsbury, J. M., Franchini, S., Gambarini, V., Maday, S. D. M., et al. (2021). Plastics and the microbiome: Impacts and solutions. *Environmental Microbiome, 16*(1), 1–19.

Lehel, J., & Murphy, S. (2021). Microplastics in the food chain: Food safety and environmental aspects. *Reviews of Environmental Contamination and Toxicology, 259*, 1–49.

Lewis, K. (2013). Platforms for antibiotic discovery. *Nature reviews Drug discovery, 12*(5), 371–387.

Li, W. C., Tse, H. F., & Fok, L. (2016). Plastic waste in the marine environment: A review of sources, occurrence and effects. *Science of the Total Environment, 566–567*, 333–349.

Li, J., Cao, J., Zhu, Y. G., Chen, Q. L., Shen, F., et al. (2018). Global survey of antibiotic resistance genes in air. *Environmental Science & Technology, 52*(19), 10975–10984.

Li, C., Busquets, R., & Campos, L. C. (2020). Assessment of microplastics in freshwater systems: A review. *Science of the Total Environment, 707*, 135578.

Li, R., Zhu, L., Yang, K., Li, H., Zhu, Y. G., & Cui, L. (2021). Impact of urbanization on antibiotic resistome in different microplastics: Evidence from a large-scale whole river analysis. *Environmental Science & Technology, 55*(13), 8760–8770.

Lin, D., Yang, G., Dou, P., Qian, S., Zhao, L., et al. (2020). Microplastics negatively affect soil fauna but stimulate microbial activity: insights from a field-based microplastic addition experiment. *Proceedings of the Royal Society B, 287*(1934), 20201268.

Liu, Y., Liu, W., Yang, X., Wang, J., Lin, H., & Yang, Y. (2021). Microplastics are a hotspot for antibiotic resistance genes: Progress and perspective. *Science of the Total Environment, 773*, 145643.

Lu, J., Zhang, Y., Wu, J., & Luo, Y. (2019). Effects of microplastics on distribution of antibiotic resistance genes in recirculating aquaculture system. *Ecotoxicology and Environmental Safety, 184*, 109631.

Lu, J., Wu, J., & Wang, J. (2021). Metagenomic analysis on resistance genes in water and microplastics from a mariculture system. *Frontiers of Environmental Science & Engineering, 16*(1), 1–13.

Marathe, N. P., & Bank, M. S. (2022). The microplastic-antibiotic resistance connection. In M. S. Bank (Ed.), *Microplastic in the environment: Pattern and process* (pp. 311–322). Springer.

McEachran, A. D., Blackwell, B. R., Hanson, J. D., Wooten, K. J., Mayer, G. D., et al. (2015). Antibiotics, bacteria, and antibiotic resistance genes: Aerial transport from cattle feed yards via particulate matter. *Environmental Health Perspectives, 123*(4), 337–343.

Niegowska, M., Sanseverino, I., Navarro, A., & Lettieri, T. (2021). Knowledge gaps in the assessment of antimicrobial resistance in surface waters. *FEMS Microbiology Ecology, 97*(11), 140.

Okelly, B. C., El-Zein, A., Liu, X., Patel, A., Fei, X., et al. (2021). Microplastics in soils: an environmental geotechnics perspective. *Environmental Geotechnics, 8*(8), 586–618. https://doi.org/10.1680/jenge.20.00179

Oliveira, J., Belchior, A., da Silva, V. D., Rotter, A., Petrovski, Ž., et al. (2020). Marine environmental plastic pollution: Mitigation by microorganism degradation and recycling valorization. *Frontiers in Marine Science, 7*, 1007.

Parthasarathy, A., Tyler, A. C., Hoffman, M. J., Savka, M. A., & Hudson, A. O. (2019). Is plastic pollution in aquatic and terrestrial environments a driver for the transmission of pathogens and the evolution of antibiotic resistance? *Environmental Science & Technology, 53*, 1744.

Partridge, S. R., Kwong, S. M., Firth, N., & Jensen, S. O. (2018). Mobile genetic elements associated with antimicrobial resistance. *Clinical Microbiology Reviews, 31*(4), e00088-17.

Pham, D. N., Clark, L., & Li, M. (2021). Microplastics as hubs enriching antibiotic-resistant bacteria and pathogens in municipal activated sludge. *Journal of Hazardous Materials Letters, 2*, 100014.

Prata, J. C. (2018). Airborne microplastics: Consequences to human health? *Environmental Pollution, 234*, 115–126.

Razanajatovo, R. M., Ding, J., Zhang, S., Jiang, H., & Zou, H. (2018). Sorption and desorption of selected pharmaceuticals by polyethylene microplastics. *Marine Pollution Bulletin, 136*, 516–523.

Redhead, S., Nieuwland, J., Esteves, S., Lee, D. H., Kim, D. W., et al. (2020). Fate of antibiotic resistant E. coli and antibiotic resistance genes during full scale conventional and advanced anaerobic digestion of sewage sludge. *PLoS One, 15*(12), e0237283.

Rezania, S., Park, J., Md Din, M. F., Mat Taib, S., Talaiekhozani, A., et al. (2018). Microplastics pollution in different aquatic environments and biota: A review of recent studies. *Marine Pollution Bulletin, 133*, 191–208.

Rillig, M. C., & Bonkowski, M. (2018). Microplastic and soil protists: A call for research. *Environmental Pollution, 241*, 1128–1131.

Robinson, T. P., Bu, D. P., Carrique-Mas, J., Fèvre, E. M., Gilbert, M., et al. (2016). Antibiotic resistance is the quintessential one health issue. *Transactions of the Royal Society of Tropical Medicine and Hygiene, 110*(7), 377–380.

Rodrigues, A., Oliver, D. M., McCarron, A., & Quilliam, R. S. (2019). Colonisation of plastic pellets (nurdles) by E. coli at public bathing beaches. *Marine Pollution Bulletin, 139*, 376–380.

Rogers, K. L., Carreres-Calabuig, J. A., Gorokhova, E., & Posth, N. R. (2020). Micro-by-micro interactions: How microorganisms influence the fate of marine microplastics. *Limnology and Oceanography Letters, 5*(1), 18–36.

Sathicq, M. B., Sabatino, R., Corno, G., & Di Cesare, A. (2021). Are microplastic particles a hotspot for the spread and the persistence of antibiotic resistance in aquatic systems? *Environmental Pollution, 279*, 116896.

Shahul Hamid, F., Bhatti, M. S., Anuar, N., Anuar, N., Mohan, P., & Periathamby, A. (2018). Worldwide distribution and abundance of microplastic: How dire is the situation? *Waste Management & Research, 36*(10), 873–897.

Shi, J., Wu, D., Su, Y., & Xie, B. (2020). (Nano)microplastics promote the propagation of antibiotic resistance genes in landfill leachate. *Environmental Science: Nano, 7*(11), 3536–3546.

Silva, M. M., Maldonado, G. C., Castro, R. O., de Sá, F. J., Cardoso, R. P., et al. (2019). Dispersal of potentially pathogenic bacteria by plastic debris in Guanabara Bay, RJ, Brazil. *Marine Pollution Bulletin, 141*, 561–568.

Sun, X., Chen, B., Xia, B., Li, Q., Zhu, L., et al. (2020). Impact of mariculture-derived microplastics on bacterial biofilm formation and their potential threat to mariculture: A case in situ study on the Sungo Bay, China. *Environmental Pollution, 262*, 114336.

Sun, Y., Cao, N., Duan, C., Wang, Q., Ding, C., & Wang, J. (2021). Selection of antibiotic resistance genes on biodegradable and non-biodegradable microplastics. *Journal of Hazardous Materials, 409*, 124979.

Syranidou, E., & Kalogerakis, N. (2022). Interactions of microplastics, antibiotics and antibiotic resistant genes within WWTPs. *Science of the Total Environment, 804*, 150141.

The Royal Society. (2019). *Microplastics in freshwater and soil*. The Royal Society.

Thompson, R. C., Olson, Y., Mitchell, R. P., Davis, A., Rowland, S. J., et al. (2004). Lost at sea: Where is all the plastic? *Science, 304*(5672), 838.

Vázquez, O. A., & Rahman, M. S. (2021). An ecotoxicological approach to microplastics on terrestrial and aquatic organisms: A systematic review in assessment, monitoring and biological impact. *Environmental Toxicology and Pharmacology, 84*, 103615.

Verla, A. W., Enyoh, C. E., Verla, E. N., & Nwarnorh, K. O. (2019). Microplastic–toxic chemical interaction: A review study on quantified levels, mechanism and implication. *SN Applied Sciences, 1*(11), 1–30.

Wang, J., Qin, X., Guo, J., Jia, W., Wang, Q., et al. (2020). Evidence of selective enrichment of bacterial assemblages and antibiotic resistant genes by microplastics in urban rivers. *Water Research, 183*, 116113.

Wang, C., Yu, J., Lu, Y., Hua, D., Wang, X., & Zou, X. (2021a). Biodegradable microplastics (BMPs): A new cause for concern? *Environmental Science and Pollution Research, 28*(47), 66511–66518.

Wang, H., Hou, L., Liu, Y., Liu, K., Zhang, L., et al. (2021b). Horizontal and vertical gene transfer drive sediment antibiotic resistome in an urban lagoon system. *Journal of Environmental Sciences, 102*, 11–23.

Wang, Y., Yang, Y., Liu, X., Zhao, J., Liu, R., & Xing, B. (2021c). Interaction of microplastics with antibiotics in aquatic environment: Distribution, adsorption, and toxicity. *Environmental Science & Technology, 55*(23), 15579–15595.

Wu, X., Pan, J., Li, M., Li, Y., Bartlam, M., & Wang, Y. (2019). Selective enrichment of bacterial pathogens by microplastic biofilm. *Water Research, 165*, 114979.

Xie, J., Jin, L., Luo, X., Zhao, Z., & Li, X. (2018). Seasonal disparities in airborne bacteria and associated antibiotic resistance genes in PM2.5 between urban and rural sites. *Environmental Science & Technology Letters, 5*(2), 74–79.

Yang, Y., Liu, G., Song, W., Ye, C., Lin, H., et al. (2019). Plastics in the marine environment are reservoirs for antibiotic and metal resistance genes. *Environment International, 123*, 79–86.

Yang, K., Chen, Q. L., Chen, M. L., Li, H. Z., Liao, H., et al. (2020a). Temporal dynamics of antibiotic resistome in the plastisphere during microbial colonization. *Environmental Science & Technology, 54*(18), 11322–11332.

Yang, Y., Liu, W., Zhang, Z., Grossart, H. P., & Gadd, G. M. (2020b). Microplastics provide new microbial niches in aquatic environments. *Applied Microbiology and Biotechnology, 104*(15), 6501–6511.

Yuan, Q., Sun, R., Yu, P., Cheng, Y., Wu, W., et al. (2022). UV-aging of microplastics increases proximal ARG donor-recipient adsorption and leaching of chemicals that synergistically enhance antibiotic resistance propagation. *Journal of Hazardous Materials, 427*, 127895.

Zarei-Baygi, A., & Smith, A. L. (2021). Intracellular versus extracellular antibiotic resistance genes in the environment: Prevalence, horizontal transfer, and mitigation strategies. *Bioresource Technology, 319*, 124181.

Zettler, E. R., Mincer, T. J., & Amaral-Zettler, L. A. (2013). Life in the "plastisphere": Microbial communities on plastic marine debris. *Environmental Science & Technology, 47*(13), 7137–7146.

Zhang, H. (2017). Transport of microplastics in coastal seas. *Estuarine, Coastal and Shelf Science, 199*, 74–86.

Zhang, Y. J., Hu, H. W., Chen, Q. L., Singh, B. K., Yan, H., et al. (2019). Transfer of antibiotic resistance from manure-amended soils to vegetable microbiomes. *Environment International, 130*, 104912.

Zhang, Y., Kang, S., Allen, S., Allen, D., Gao, T., & Sillanpää, M. (2020a). Atmospheric microplastics: A review on current status and perspectives. *Earth-Science Reviews, 203*, 103118.

Zhang, Y., Lu, J., Wu, J., Wang, J., & Luo, Y. (2020b). Potential risks of microplastics combined with superbugs: Enrichment of antibiotic resistant bacteria on the surface of microplastics in mariculture system. *Ecotoxicology and Environmental Safety, 187*, 109852.

Zhang, Q., Fan, D., Pang, X., Zhu, W., Zhao, J., & Xu, J. (2021a). Effects of polyethylene microplastics on the fate of antibiotic resistance genes and microbial communities in anaerobic digestion of dairy wastes. *Journal of Cleaner Production, 292*, 125909.

Zhang, Z., Liu, H., Wen, H., Gao, L., Gong, Y., et al. (2021b). Microplastics deteriorate the removal efficiency of antibiotic resistance genes during aerobic sludge digestion. *Science of The Total Environment, 798*, 149344.

Zhao, T., Lozano, Y. M., & Rillig, M. C. (2021). Microplastics increase soil pH and decrease microbial activities as a function of microplastic shape, polymer type, and exposure time. *Frontiers in Environmental Science, 9*, 235.

Zhu, D., Xiang, Q., Yang, X. R., Ke, X., O'Connor, P., & Zhu, Y. G. (2019). Trophic transfer of antibiotic resistance genes in a soil detritus food chain. *Environmental Science & Technology, 53*(13), 7770–7781.

Zhu, G., Wang, X., Yang, T., Su, J., Qin, Y., et al. (2020). Air pollution could drive global dissemination of antibiotic resistance genes. *The ISME Journal, 15*(1), 270–281.

Zhu, D., Ma, J., Li, G., Rillig, M. C., & Zhu, Y. G. (2021). Soil plastispheres as hotspots of antibiotic resistance genes and potential pathogens. *The ISME Journal, 16*(2), 521–532.

Part II
Trophic Transfer of Micro and Nanoplastics

Chapter 8
Phytoaccumulation of Micro- and Nanoplastics: Root Uptake

S. A. Aransiola, A. A. Ikhumetse, B. R. Babaniyi, O. P. Abioye, O. J. Oyedele, and N. O. Falade

Abstract Accumulation of microplastics and nanoplastics in the environment accounts for some of the ecological and health issues in plants. They move in the environment through ocean current and atmospheric circulation, leading to their abundant accumulation in the major sinks such as soil, rivers, oceans, organisms, and plants. Also, phytotoxicity in plants and its environmental implications are posing challenges for academics and policymakers. Plastic pollution in the agroecosystem is a legitimate ecotoxicological hazard for food web exchanges. Micro- and nanoplastics are produced in agriculture through a variety of agricultural management strategies, including the use of sludge, mulching, sewage, and compost. Their presence has an impact on a variety of soil parameters and plant features. These pollutant materials are increasingly having major consequences for essential soil ecosystem activities like nutrient cycling and microbial activity. Micro- and nanoplastics uptake by plants rely on anatomical and physiological properties of the plant species; also, plastics components, most importantly, those of eco-corona and environmental aging ones affect the surface chemistry and behavior of plants. Accumulation and uptake of negatively charged nanoplastics concentrations are much higher in roots compared to positively charged nanoplastics, these resulted due to their high affinity to bind with radical mucilage and increase in their size through the hetero-aggregation induction by root exudates. As a result of phytotoxicity of micro- and nanoplastics in plants, impeded enzymatic activities, abnormal morphology, oxidative stress, and nutrients absorption interference with lot more issues in plants have been identified. Many plastic materials such as polypropylene, polybrominated diphenyl ethers, and polyethylene, among others, have been detected in plants.

S. A. Aransiola (✉) · B. R. Babaniyi · O. J. Oyedele · N. O. Falade
Bioresources Development Centre, National Biotechnology Development Agency, Ogbomoso, Nigeria

A. A. Ikhumetse · O. P. Abioye
Department of Microbiology, Federal University of Technology, Minna, Nigeria

N. R. Maddela et al. (eds.), *Micro and Nanoplastics in Soil*,
https://doi.org/10.1007/978-3-031-21195-9_8

Keyword Microplastic · Nanoplastic · Plants · Nutrients · Contamination · Environment

8.1 Introduction

Global usage of plastics has resulted in huge amounts of plastic litter in the environment (Lambert & Wagner, 2018). These plastics littered in the environment have fragments or smaller particles (<5 mm), called "microplastic" (Law & Thompson, 2014). Microplastics and nanoplastics size are not the same, this contributed to their surface/volume ratio. Pollution caused by microplastics is a global environmental issue (Plastics Europe, 2015). Results of previous researches provided overwhelming evidence on the direct and indirect impacts of microplastics and nanoplastics on aquatic networks (Lambert & Wagner, 2018). The abundance of microplastic in soils has gotten to a state that needs to be checked (Kim & Rillig, 2021). The absence of limited land spaces and the indiscriminate discharge of plastic wastes have made plastic waste pollution a serious problem in recent years (Auta et al., 2022). Plastic pollution in the environment has gained more attention; it has been revealed that plastic accumulation and persistency in the environment can last several hundreds of years beneath low-light and low-oxygen conditions (Horton et al., 2017). The sources of microplastics in the soil are primarily from pellets and resins used during industrial production of plastics, plastic exfoliators, fragmentation and weathering of larger plastic materials, and improper management and disposal of plastic wastes. Also, the plastic fragmentation and surface ablation due to UV oxidation reactions, radiation, drying cycles, wetting, oxidative reactions, human mechanical activities such as tillage, degradative actions of microbes, and penetration of plant roots increases the occurrence of not only microplastics but also nanoplastics. The size of plastic debris depends on thickness and surface heterogeneity of the plastic materials. The quality of the layer between the surface and the underlying virgin plastic, results in fragmentation by delamination (Pathan et al., 2020). The fate, components and determination of microplastics (MPs) and nanoplastics (NPs) in roots uptake are barely identified. It's evident that about 300 million tons of plastics manufactured yearly ends up in the environment, and the soil acts as a long-term sink for these plastic debris. The abundance of microplastics and nanoplastics in roots' uptake is vehemently examined by physical components of plastics, while effect exerted by chemical structures is negligible. The processes of plastic degradation, called aging, generate micro-and nano-size particles of plastic, which can effectively induce significant changes in their chemical and physical properties with relevant effects on their reactivity. In addition, these processes could lead to the emission of harmful oligomeric and monomeric substances and toxic additives from plastics that are culpable of migration into the food chain through root uptake. These in turn, could possibly be hazardous to human health and can potentially affect the fauna and flora in the environment. With regard to plastics and plastic debris persistence in the soil, bacteria, fungi, and insect resistance to effects posed on them by plastics is increasing daily. Some of these microbes now feeds on plastics wastes.

Soil biota, particularly collembola and earthworms can be both microplastics and nanoplastics transporters by means of soil profile (Zhu et al., 2018a, b). Moreover, bisphenol A and styrene are toxic; this is because, as monomers of polyvinyl chloride (PVC), they are capable of disrupting the endocrine network and are carcinogenic (Bläsing & Amelung, 2018). So many plastic additives and plasticizers are harmful substances, such as phthalates ester and brominated flame retardants (Rist et al., 2018; Pathan et al., 2020).

Contamination of plastics in the soil could trigger a direct and indirect effect on cultivated plants due to available nutrients being contaminated by toxic substances plastics emit and which are taken up by the roots. This can go on to affect chemical, physical, and biological composition of the soil (Abioye et al., 2014). There has been an overwhelming evidence of the direct impacts of microplastics and nanoplastics contamination in plants (Huerta Lwanga et al., 2017). Nanoplastics have higher chemical reactivity and mobility than microplastics and thus have a different colloidal behavior (Hüffer et al., 2018). The colloidal property triggers stable or unstable hetero-aggregation of nanoplastics, which also depends on the ionic strength and pH value of the solution and thus relies on the mineral and organic matter makeup of the soil. Root metabolism of pollutants is a major challenge in the mode of nutrient transmission in plants (Sandermann, 1992). Root uptake of microplastics and nanoplastics relies on the integrity of anatomical and physiological components of the plant in relation to debris and chemicals emitted by plastics, mostly eco-corona and environmental aging surface and behavior (Ng et al., 2018). Recent study by Sun et al. (2020) revealed that root uptake of nanoplastics enormously relies on the charge in their surface. The study further revealed elevated accumulation of negatively charged nanoplastics in root uptake than positively charged nanoplastics, this could be linked to high cleavage interaction with radical mucilage as well as increase in size via hetero-aggregation induction by root's exudates. Meanwhile, the risk of microplastics migration into food chain via vegetables as a result of attraction to the surfaces of leaves and vegetable roots was elucidated in their findings. In addition, Smith et al. (2018) postulated the possibility of microplastics accumulation in plants grown on industrially produced compost. Excessive accumulation of plastic debris in the root can interrupt transport of nutrients through blockage in the pores in the cell wall and reduce the ability of plant genes to resist disease (Jiang et al., 2019). When the root takes up plastic debris, they are migrated to the stem and leaves through the vascular network in line with the transpiration stream, which can be discovered at intercellular level as "string-like" closely packed structure in dispersed forms (Li et al., 2019). Plant cells take up nanoplastics through endocytosis, by ion transport systems, via proteins carrier. More study is required to properly understand the mechanism of nanoplastics' translocation, storage, and toxicity in plants and the defense approaches of plants against nanoplastics (Wang et al., 2013). Also focus should be on direct effect of microplastics and nanoplastics on cultivated plants, with special reference to soil structure, immobilization of nutrients, and microbial population in the soil, root microbiome, and root symbionts. Therefore, this chapter focuses on the kinds of micro and nanoplastics in plants, mechanism of their phytoaccumulation in plant-root-system, and the factors responsible for this accumulation.

8.2 Types of Micro- and Nanoplastics in Plant

(a) *Polyethylene* is by far the most common type of consumer plastic and is utilized in so many day-to-day materials. As a result of its numerous usages, coupled with improper disposal, this material gets into the environments and persists for a long time. As days go by, disintegration processes begin through mechanical, biological, and chemical mechanism resulting in fragments and debris called microplastics and nanoplastics. The fragments' and debris' half-life are long in the soil which enables plant root uptake. Wang et al. (2021) investigated the phytotoxic effects of polyethylene microplastics on the growth of the food crops soybean (*Glycine max*) (Aransiola et al., 2013) and mung bean (Vigna radiata). It was revealed that the presence of polyethylene promotes phytotoxicity in soybean (*Glycine max*) was greater while the effect was prominent in mung bean leading to increase in root length, and the promotion degree was positively correlated with the concentration of polyethylene (Wang et al., 2021).
(b) *Phthalates esters (PAES)* are synthetic compounds which are added to plastics as an additive for producing plastic materials. Phthalates are used in many products, including food packaging materials, cosmetics, toys, and medical equipment. There have been many itches about the possibility of this compound's migration into the water and food chains. Inappropriate storage conditions such as freezing, sunlight, high temperatures, and storage time with disposal of materials after use are proposed as the major paths of migration of these compounds into the content and soil (Yousefi et al., 2019). Uptake and bioaccumulation by plants are important pathways of migration of phthalate esters in the soil (Feng et al., 2017: Ma et al., 2017). Presence of phthalate esters has been discovered in vegetables like lettuce, which was reported to disrupt carotenoids that protect cells from mutagen by preventing mutation in the body cells of the vegetable consumers (Ma et al., 2017). Furthermore, phthalate esters also inhibit plants' Vitamin C (Ascorbic acid) which is responsible for prevention of oxidative stress in cell division and elongation in higher plants (Ma et al., 2018). Vitamin C (Ascorbic acid) contributes hugely to the regulation of plant cell oxidation-reduction balancing as a prominent redox buffer. Vitamin C also regulates transcription of some genes and also function as a co-enzyme. Phthalate esters in plants and hinder the proper functioning of plants wholeness. Some metabolites of these compounds have also been detected in the leaves of the plants, manifesting in mass reduction of leaf, shoot elongation, and bioaccumulation of proline in plants (Li et al., 2016). Ma et al. (2018) concluded that

decrease in leaf biomass is inevitable in as much phthalate esters reduced chlorophyll content of the plant. Hence, decrease in chlorophyll content will result in inefficient photosynthetic process of the plant, leading to the plant's inability to yield organic matter and dry matter bioaccumulation in plant biomass (Ma et al., 2018).

(c) *Polystyrene (PS)* is a stiff, hard, luminously translucent artificial resin formed by the polymerization of styrene (Auta et al., 2022). As a rigid container, PS is mostly used in the food industry as disposable container for plates, bowls, eating utensils, and foamed cups (Zong et al., 2021). Investigation of the adsorption properties of toxic elements onto PS as well as the toxicity and bioavailability of microplastics and toxic elements by hydroponic wheat seedlings were experimented by Zong et al. (2021). The study revealed the presence of polystyrene at 0.5 μm, 100 mg/L, but the reactive oxygen species (ROS), photosynthesis, and effect on wheat seedlings growth had no significant difference. Albeit, polystyrene had the capacity to adsorb cadmium and copper with a predominantly chemisorption. For instance, 100-nm polystyrene nanoplastics block cell connections or cell wall pores, causing oxidative damage and genotoxicity in fava beans (*Vicia faba*) (Jiang et al., 2019).

(d) *Polybrominated diphenyl ethers (PBDEs)* have been employed for decades in numerous applications, such as flame retardants, in a wide variety of products. Some remarkable applications are: high-impact polystyrene (HIPS), electrical insulations, polybutylene terephthalate (PBT), vehicles and aircrafts, upholstered furniture, foams, building insulations, electrical goods, and a variety of technical plastics such as acrylonitrile-butadiene-styrene copolymers (ABS) (Dobslaw et al., 2021). As a result of wide range of molecular weight of PBDEs (328–959 g mol^{-1}), lipophilicity (log KOW = 6–10), and volatility (log KOA = 9–16) (Zhu et al., 2018a, b). BDE plant uptake mechanisms and congener-specific transport (soil-air-plant versus soil–soil moisture-root-plant) are highly varied and rely on vapor pressure, Henry coefficient, air-plant partition coefficient, KOW value, and KOA value as well as meteorological parameters such as kinetics of gaseous deposition and particulate BDEs, rainfall, temperature, wind, long distance transport, precipitation, and plant-specific characteristics like species, carbohydrate content, lipid content, fiber content, non-lipid plant parts, rind consistency, rhizosphere factors, and foliage morphology (Klinčić et al., 2020). Because of high-molecular weight, lipophilicity, bromination, and low mobility BDE-209 as the dominant PBDE in soil is only marginally available for plants at levels of 0.3–0.5% of the initial concentration (Wu et al., 2018). Soil, soil moisture, and root uptake is vital, as tests with living and non-living roots of different plants revealed 3.5–6 times higher BDE-209 levels in their living tissues (Chow et al., 2017). Another analysis of small-scale soil-based BDE gradients within the root plexus showed active BDE-209 uptake by plants (Zhang et al., 2015) and was established through greenhouse experiments by cultivating six different plant species in contaminated and noncontaminated soil in parallel (Jiang et al., 2018). Many studies are based on physicochemical and major components militating plant uptake and biodegradation parameters of PBDE, such as vapor pressure, temperature, wind, and precipitation as well as plant species, lipid content, foliage morphology, ratio of non-lipid plant parts, rind thickness, contents of both sugar and fibers as well as the presence of microbial active rhizosphere which are considered to be highly germane for plant to take up micro and nanoplastics (Zhu et al., 2018a, b).

(e) *Polypropylene* is a thermoplastic used in a wide variety of applications. It is produced through chain growth polymerization from the monomer propylene. Polypropylene belongs to the group of polyolefins and is partially crystalline and non-polar. Polypropylene is generally considered safe for use, but the chemical additives such as bis-phenol A (BPA), lead, and cadmium found in propylene are proven to contribute to issues in plants. However, polypropylene is non-biodegradable, and its increasing accumulation in the environment has been a threat to the ecology foot print. In an open environment, bio-degradation of BPA compounds by extracellular and intracellular fungal enzymes (*peroxidases, laccase, cytochrome P450 mono-oxygenases*) is well documented (Hofmann & Schlosser, 2016). BPA being persistent and unstable in nature can facilitate its leaching and thereby high absorption in the aquatic environments and plants (Vom Saal & Myers, 2008). The presence of BPA (1–729.9 ng/g) on MPs was studied for the first time in samples collected from the open, remote oceans, and urban beaches from America and Europe (Wang et al., 2021). Rehse et al. (2018) showed the effects of BPA present in non-suspended polyamide microplastic particles on freshwater zooplankton (*Daphnia magna*) and its absorption in plastic fragments from remote coasts and open ocean shores (Yu et al., 2020; Esterhuizen & Kim, 2021).

(f) *Polyurethane* is inherently more eco-friendly than most other plastics, it also does not contain any addictives that interfere with endocrine and hormone systems, nor does it contribute to pH changes in soil or water. Polyurethane lasts longer than most thermoplastics in the environment, and there are variety of ways by which polyurethane can be recycled (Plastics Europe, 2019). According to Esterhuizen and Kim (2021) presence of polyurethane was reported in Lotus plant. Micro and nanoplastics are being detected in nearly all ecosystems across the globe in which plants are no exception, therefore, more studies are needed to ascertain and quantify presence of more micro and nanoplastics in plants and other ecology products.

8.3 Phytotoxicity of Micro- and Nanoplastics

The contamination of soil by plastics has a direct and indirect effect on plants due to the root uptake or problems caused by soil chemical on the physical and biological components of the plants. Recently, there has been evidence of direct effects of micro and nanoplastics contamination in plants (Huerta Lwanga et al., 2017). The way in which plants eliminate pollutants coupled with the plants' ability to store and convert recalcitrant contaminants are the prime factor to proper functioning of plants (Sandermann, 1992). Micro and nanoplastics uptake by plants rely on anatomical and physiological properties of the plant species also, plastic components, most importantly, those of eco-corona, and environmental aging affect the surface chemistry and behavior of plants (Ng et al., 2018). According to Li et al. (2019) plastic debris with sizes up to 0.2 μm accumulate in lettuce roots, this triggers the intercellular presence of plastic cluster in "grape-like" form. Sun et al. (2020) recently reported that accumulations of nanoplastics in plants significantly attributed to their surface charge. Furthermore, accumulation and uptake of negatively charged nanoplastics concentrations are much higher in roots compared to positively charged nanoplastics, these could be a result of their high binding-affinity with radical mucilage and increase in their size through the hetero-aggregation induction by root's exudates. Although, these findings also emphasize the possibility of microplastics entering into the human food chain through vegetables as a result of their adhesion to the surfaces of salad and root vegetables (Sun et al., 2020). Microplastics of industrial-based compost could accumulate in plants that are grown on those compost, excessive accumulation of plastic debris in the root is associated with several issues in plants, for instance, the hampering of the nutrient transport system by the blockage of the pores in the cell wall (Smith et al., 2018), more so, the excessive production of reactive oxygen species (ROS) (Jiang et al., 2019) and the reduction of the plant disease resistance by inhibiting downregulation of disease resistance genes are due to phytotoxicity of micro and nanoplastic accumulation in plants (Sun et al., 2020). Uptake and accumulation of plastic debris into roots which later migrates to the stem and leaves through the vascular system following the transpiration path, are detected at the intercellular level as "string-like" cluster and dispersed forms (Li et al., 2019). Notably, intracellular nanoplastics uptake is expected in plant but the microplastics cannot pass through the cell membrane because of their higher size (from 0.1 μm to 5 mm) than nanoplastics (<0.1 μm) (Hrda et al., 2018; Bandmann et al., 2012). Plant cells accumulate nanoplastics through endocytosis and ion transport paths, by proteins carrier, or through aquaporins. Nanoplastics' size is essential for effective plant uptake in respect to nano-polystyrene beads with a diameter of 20–40 nm permeate tobacco cells, regardless of those of 100 nm (Bandmann et al., 2012). Threshold value of 50 nm was projected by Kettler et al. (2014).

Indirect toxicity effects of microplastics and nanoplastics on plants are attributed to their effects on soil structure, nutrient immobilization, contaminant adsorption and diffusion, soil microbial community root-associated microbiome, and root symbionts.

(a) *Oxidative Stress:* oxidative stress causes interference in the metabolic pathways and damage to macromolecules in plants. When the concentrations of nanoplastics exceed optimal levels, plants could be affected adversely, directly or indirectly, and the direct toxic effects caused by high levels of plastic concentration include inhibition of cytoplasmic enzymes and damage to cell structures while the indirect adverse effect is on the replacement of essential nutrients at cation exchange sites of plants. Oxidative stress by micro and nanoplastic could cause lipid peroxidation that accounts for serious cell membrane damage in plants (Asati et al., 2016).

(b) *Impeded Enzymatic Activities:* enzymes are essential for plant metabolism, soil microorganisms responsible for enzymatic functions could be hampered due to plastics residues interference will result to toxic effects culpable of reduced plant growth and plant death in the long run. High level of plastic debris prevents catalase activity in some plants' leaves (Xie et al., 2016).

(c) *Abnormal Morphology:* plants morphology is affected by uptake of micro and nanoplastics. Being ubiquitous, plastics are widely present in the environment. Plastic contamination exert a threat on morphology, growth, and photosynthetic processes of plants. Inhibition of seed germination like *Pinus helipensis* is related to nanoplastics effect, germination of seed inhibition may result from the interaction of nanoplastics with important enzymes for instance, in rice endosperm, contamination may inhibit protease and amylase (Hongwei et al., 2019). Early seedling inhibition in leguminous plants may be linked to plastic contamination, defect in elongation of roots and expansion of some plants' leaves. High level of plastic debris in the soil causes abnormal morphology in some plants, such as cell walls of the endodermis, lignification of cortical parenchyma, and irregular radial thickening in pea roots. Nanoplastics contamination could be responsible for the proliferation of the repair process of vascular plants, chlorosis, oxidative stress, reduction in number of leaves and leaf area, and plants growth retardation (Fonge et al., 2021).

(d) *Physiological Disorders:* phytotoxicity in plants could lead to physiological disorder. Plastic debris persist in soil for many years and are capable of bioaccumulation in plants thereby, linking the food chain. Toxic level of plastic causes injuries and physiological disorders in plants. They bind and link with water pathways and proteins, thereby closing leaf stomata and this leads to impediment of water flow in plants. Decrease in plant height, panicle formation and reduce tiller, low yield, reduction in germination percentage, reduction in flowering and fruit weight, and high rate of bioaccumulation in shoot and root of seedling are caused by excess of plastic fragments, which lead to chlorosis and retardation of photosynthetic activities, thereby impairing plant parts and causing death in the long run. Elevated concentration of plastic in soil causes several physiological alterations and various toxicity such as chlorosis and necrosis in some plants, mostly cereal. Necrotic brown and petioles affecting older leaves, crinkle leaf seen in younger leaf, stem, and petiole tissue are related to plastic toxicity in plants (Kumari & Mishra, 2021).

(e) *Nutrients Absorption Interference:* elevated concentration of micro and nanoplastics accumulation could result in plant injury in terms of chlorosis, growth retardation, browning of root tips, and finally death. Transport and absorption of nitrate from roots to shoots is reduced, and high level of plastic debris inhibit nitrate reductase activity in the plant shoots. The phytotoxicity effect of plastic could result in a decrease in plant nutrient acquisition, inhibition of germination process and reduction of plant biomass, and decreased shoot and root growth.

Excess plastic debris may decrease chlorophyll content, reduce plant nutrient content and antioxidant enzymatic activity, and reduce sugar, amino acid, and protein content of plants. This also causes prevention of Fe concentration, chlorophyll, protein, and translocation of P, S, Mn, Zn, and Cu from roots to the plants (Rillig, 2018; Mahmud et al., 2021).

(f) *Altered Photosynthesis:* photosynthetic processes are hampered by a high level of plastic contamination by interacting with chloroplast ultra-structure. Oxidative stress induced by plastic toxicity triggers inactive photosynthetic pigments, causing perceived growth and low or inefficient seed germination. Stress due to microplastics in plants is one of the major agents affecting photosynthesis in terms of CO_2 fixation, transport of electron, photophosphorylation, and enzyme activities in plants. Stimulation and switching of superoxide dismutase and antioxidant catalase are major ways of contamination detoxification responses in plants, which could be altered by plastics toxicity (Biber et al., 2019; Pichhode & Nikhil, 2015).

(g) *Plasma Membrane Disruption:* plastic toxicity affects the plasma membrane by causing hydration in the water content; notably, plastic nanoparticle is culpable of reacting with the quantity of water in plants, which could result in reduction of ATPase activity of the plasma membrane fractionally in plants, especially wheat and sunflower roots. High concentrations of contaminants decrease seed germination, diminish plant nutrient availability, and impair shoot and root length of plants like garlic, maize, and wheat. High concentrations of nanoplastics could interact with the mitochondrial activity and cause oxidative stress by triggering the generation of ROS. This leads to the disruption of bio-membrane lipids and cellular metabolism in plants. Increased concentration of microplastic could also cause nutrient imbalance, which results in disorder of cell membrane functions. Plastic debris damage the lipid composition and altered H-ATPase activity of plant plasma membrane. High concentration nanoplastics uptakes by roots and transportation to leaves through transpiration stream could stimulate free radical production which impairs cellular structure permanently and destroys membranes, DNA, and proteins in plants (Ahmad et al., 2012; Nizzetto et al., 2016).

Low Yield, plastic toxicity, in summary, reduces plant yields. Decrease in various plants' fruit yields are associated with micro and nanoplastic accumulation and also with reduction of fresh leaf weight, leading to stunted growth, chlorosis, wilting, and reduced seed germination due to alteration in their physiological and biochemical processes (Wagner & Lambert, 2018).

8.4 Mechanisms of Phytoaccumulation of Micro- and Nanoplastics by Plant-Root System

Researchers have recently begun to unravel the mechanisms of microplastics (MPs) uptake and translocation in plants based on preliminary understanding of the impacts of microplastics (MPs) and nanoplastics (NPs) in plants (Azeem et al., 2021; Ullah et al., 2021). MPs are thought not to enter (accumulate) in terrestrial plant tissues directly because their big size or high molecular weight prevents them from doing so (Teuten et al., 2009; Azeem et al., 2021; Ullah et al., 2021), whereas NPs can enter directly through the plant cell walls (Bandmann et al., 2012; Azeem et al., 2021). According to recent research, cell wall pores can be flexible, allowing MPs particles to enter more easily (Li et al., 2020a; Sun et al., 2020; Ullah et al., 2021). Because tiny MPs can penetrate the plant cell wall and membrane barriers or block cell pores, this is theoretically possible (Enyoh et al., 2019; Jiang et al., 2019; Zhu et al., 2019; Ullah et al., 2021). Organic pollutants such as plastics are taken in, translocated, and accumulated differently in different plant species (Ng et al., 2018), depending on anatomical and physiological characteristics (Ullah et al., 2021). MPs and NPs have been found in plants, and plastic particles are particularly absorbed on root hairs (Azeem et al., 2021). However, MPs can permeate seeds, stems, leaves, fruits, and plant cells, depending on their size and type (Dietz & Herth, 2011; Ullah et al., 2021).

Plants can absorb or adsorb MPs and NPs in aggregate form (Mateos-Cárdenas et al., 2019), and transpiration pull plays a key role in plastic particle uptake and translocation (Azeem et al., 2021). Polystyrene (PS) MPs (0.2–1.0 μm) have been found to be taken up and stored in the roots of raw vegetables, as well as translocated from root to shoot tissues (Li et al., 2019; Ullah et al., 2021). On carrot (*Daucus carota* L. var. sativus Hoffm.) plants, Dong et al. (2021) used similar-size polystyrene (PS)-MPs in addition to As (III) and discovered that 0.2-μm PS can penetrate root cells and translocate to leaves. Wheat (*Triticum aestivum*) and lettuce (*Lactuca sativa*) roots also had Polystyrene (PS) MPs beads along the whole lateral root cap and inside the apical meristem (Ullah et al., 2021). Hydroponically grown lettuce plants were cultivated in treated wastewater and sand matrices or in sandy soil. While casparian is not fully mature, plastic particles enter the epidermal tissue of wheat's primary and secondary roots, are stimulated through the pericycle, and are transferred into the xylem (Li et al., 2020a). These particles can migrate to the plant's aerial section via the xylem inside the central cylinder (Li et al., 2020a; Azeem et al., 2021). *Triticum aestivum* (wheat) and *Lactuca sativa* (lettuce) are two crop plants that can absorb 200 nm PS nanobeads and 2.0 mm polymethylmethacrylate microbeads (Li et al., 2020a). These beads were able to penetrate the root stele and subsequently transmit from root to shoot with the help of transpirational pull, according to the authors (Mateos-Cárdenas et al., 2021). They believe that these particles in the root cap mucilage enhance their perception of the cell wall, which is very insecure due to active cell division, and that this allows diffusion through the apical meristem tissue.

Confocal images show that PS luminescence signals were mostly found in the wheat root's vascular system. These signals were apparent in the epidermis and xylem arteries after 2 hours, and they were also visible in wheat cortical tissue after 12 hours. They appear to be restricted to vascular tissues in lettuce (*Lactuca sativa*) plants. Because the casparian band was persistent, PS beads penetrated the cortex through gaps in epidermal cells but did not enter the endo-epidermis. Strong PS luminescence signals were eliminated in cracks where these particles penetrated the cortex and endo-epidermis in the lateral root apex, indicating that small PS bead crack entrances were major sites into the root xylem of lettuce (*Lactuca sativa*) and wheat (*Triticum aestivum*) (Li et al., 2020a). PS beads were seen in the stele on secondary roots after 12 hours and in the epidermis and vascular tissue of wheat after 48 hours of observation using a scanning electron microscope (SEM) (Li et al., 2020a) (Fig. 8.1).

Nanopalstics (NPs) can infiltrate plant cell walls directly, and a research of tobacco (*Nicotiana tabacum*) plant cells found that tobacco plants did not assimilate 100-nm nano polystyrene beads, while 20- to 40-nm beads did (Bandmann et al., 2012). Li et al. (2020a) looked at roots that had been treated with 0.2-μm PS micro-beads with tagged fluorescence and found that the 0.2-μm PS fluorescent micro-beads were stuck in cells outside the root cap mucilage, which is generally visible

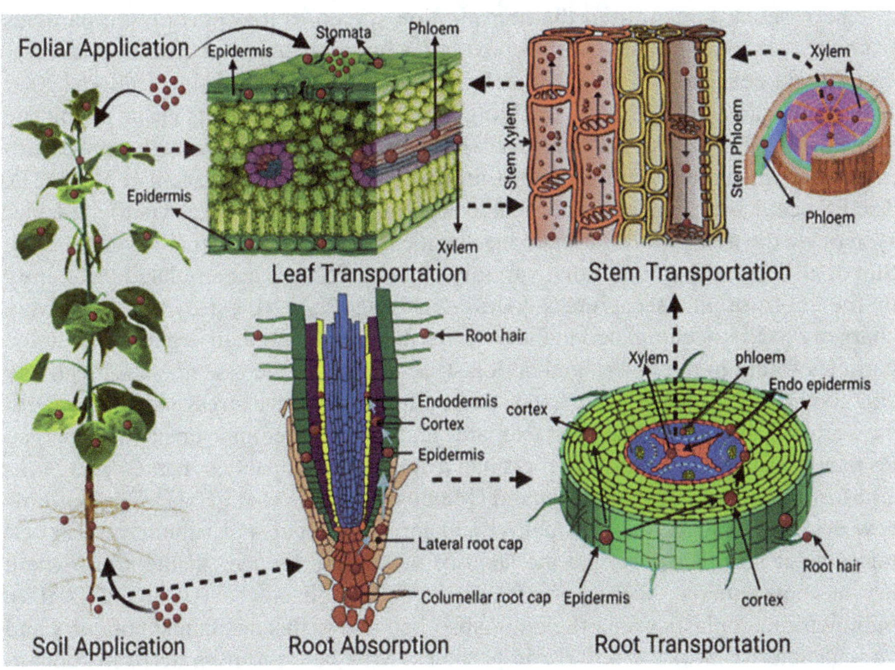

Fig. 8.1 Plastics uptake by plant mechanism through soil via transport pathways and root absorption through root to stem, leaves, and fruits. By foliar application, plastic enters the leaf stomata and moves to other parts of the plant. Curved arrow indicates the availability of plastic to plant and the dashed arrow indicates transportation within the plant. (Source: Azeem et al. (2021))

to the naked eye. Mucilage and exudates, which operate as the first layer of protection in plants and are negatively charged, have been shown to hinder positively charged metal uptake of NPs at the cell wall's outer side (Avellan et al., 2017; Azeem et al., 2021). The 0.2-μm PS luminescence signals were mostly seen in the vascular system and on the cell walls of the cortical tissue of the roots, indicating that the beads traveled through the intercellular channels, the apoplastic transport system (Li et al., 2020a). This pathway follows the apoplast flow of water through cells of juvenile root zones that have not yet produced the casparian strip and band within the radial cell walls of the root's endodermis, which contains the hydrophobic polymer suberin. Thus, root openings developed as a result of aging processes or were damaged by below-ground pathogens/herbivores and mechanical injuries, allowing MPs to enter the roots (Li et al., 2020a; Ullah et al., 2021), whereas the proposed mechanisms for the entry and uptake of engineered nanoparticles are stomatal openings, endocytosis via plasmodesmata (Ng et al., 2018; Ullah et al., 2021). These mechanisms are thought to provide insight into prospective MPs and nanoplastics (NPs) entrance pathways in plants. Rice, *Arabidopsis*, maize (*Zea mays* L.), and soybean [*Glycine max* (L.) Merr.] are some examples of C-based nanoparticles uptake by whole plants (Zhao et al., 2017; Ullah et al., 2021), whereas nano-sized MPs have been found to enter tobacco (*Nicotiana glutinosa* L.) cells via endocytosis (Bandmann et al., 2012; Ullah et al., 2021) (Fig. 8.2).

The vascular system used the transpiration stream to transport plastic particles from the root to the shoot. The presence of PS luminescence signals in the vascular system and cell walls of root cortical tissues was confirmed using confocal laser scanning microscopy (CLSM), implying that PS beads enter plants via an apoplastic pathway through an intercellular space. MPs and NPs can pass through small extracellular channels to reach the water-transporting vasculature (Azeem et al., 2021). The PS beads were carried from roots to stems and finally to leaves via the vascular system via the transpiration stream once inside the central cylinder (Li et al., 2020a). Sun et al. (2020) also found that variably charged PS NPs accumulated differently in the roots of mouseear cress [*Arabidopsis thaliana* (L.) Heynh.]. Positively charged PS-NPs accumulated in the root tips at lower levels than negatively charged particles found in the xylem and apoplast, according to the researchers. Positively charged PS-NPs, on the other hand, caused more oxidative stress since they gathered around the root due to root exudates and growth medium. Positively charged PS nanoplastics accumulated in roots had a deleterious impact on *Arabidopsis thaliana* seedling growth and development (Mateos-Cárdenas et al., 2021). There is now new evidence that PS and polymethylmethacrylate particles of submicrometer and micrometer size can penetrate the stele of wheat and lettuce. At the areas where lateral roots emerge, these particles entered through cracks (Li et al., 2020a). Submicrometer plastics are efficiently absorbed due to this mechanism of entry and the characteristics of the polymeric particles. MPs begin to migrate from roots to shoots as soon as they enter. Overall, roots appear to be important in the internalization of plastics. *Ceratopteris pteridoides* can internalize 100 nm PS nanobeads in their roots, according to Yuan et al. (2019). In addition, 20 nm PS nanobeads were collected in the veins and vessel walls of *Vigna radiata*, a mung bean (Chae & An,

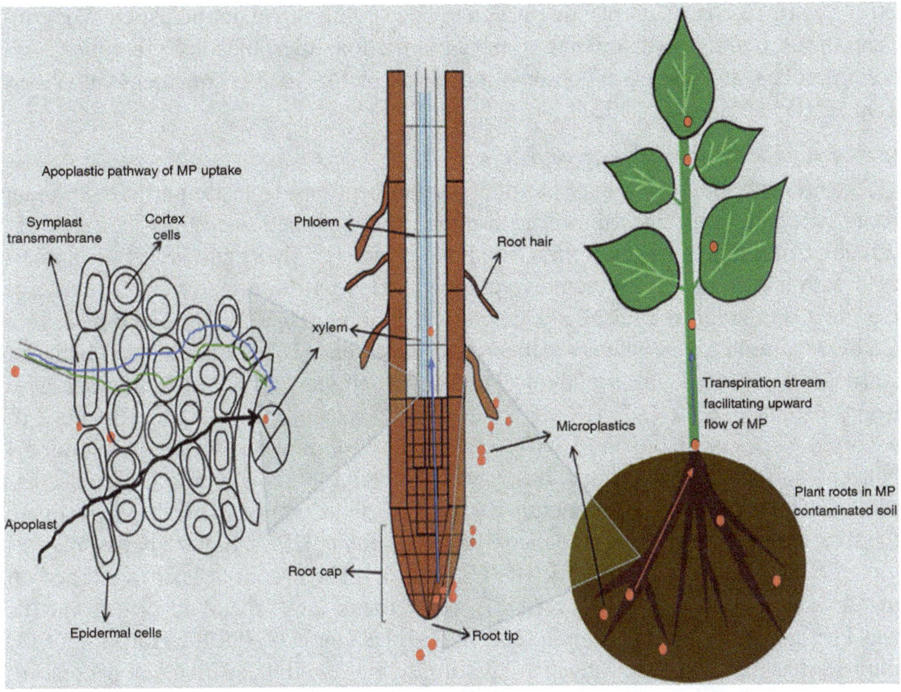

Fig. 8.2 Microplastics uptake by plants. Pathways of apoplastics are shown by which MPs go into the root cells to vascular bundles (xylem); whole plant revealed how the roots interact with MPs-contaminated soil and ultimately uptakes the MPs. (Source: Ullah et al. (2021))

2020). Plants can translocate nanoplastics given to the soil via the roots into the leaves, according to the later study (Chae & An, 2020; Lian et al., 2020a, b; Mateos-Cárdenas et al., 2021). Higher transpiration rates aided the uptake of plastic particles, indicating that transpiration pull was the primary driving mechanism behind their migration (Li et al., 2020a; Ullah et al., 2021).

8.5 Factors Responsible for Micro- and Nanoplastics Phytoaccumulation

The uptake, transport, and accumulation of micro and nanoplastics have a favorable or detrimental impact on plants, and this differs from one species to the next (Ng et al., 2018; Ullah et al., 2021), depending on anatomical and physiological characteristics. Root properties (surface area, density, volume), xylem properties (surface area, volume), growth rate, transpiration, water and lipid fractions, tonoplast potential, plasma membrane potential, and the pH of vacuoles and cytoplasm are all important factors that influence MP and nanoplastic uptake in plants (Ullah et al.,

2021). Surface charge on plastic particles, size of micro and nanoplastics, MP concentrations, plant growth and net primary production, microplastics interaction with soil microbes, and chemical composition are all factors to also consider (Van-Weert et al., 2019; Ullah et al., 2021).

(a) *Surface Charge on Plastic Particles*

Electrostatic attraction may facilitate the adsorption of plastic particles in plant roots, affecting nutrient immobilization and photosynthetic activities (Lian et al., 2020b). Cadmium (Cd) adsorption into Polystyrene (PS) MPs and PS NPs is associated with a decrease in the negative charge carried by these materials. Because of the low Cd content in a PS NPs–Cd solution, Lian et al. (2020a) showed increased Cd bioaccumulation in wheat seedlings. In the presence of sodium dodecyl benzene sulfonate from pentachlorophenol (PCPs), polyethylene (PE) MPs interact with heavy metals like Cr (VI) (Zhang et al., 2020; Azeem et al., 2021). Because of adsorption sites available on PE MPs in increasing competition with sodium dodecyl benzene sulfonate, Cr (VI) adsorption was limited at $pH > 6$, while it increased at $pH < 6$. Three types of heavy metals (Cd^{2+}, Pb^{2+}, and Cu^{2+}) were used to investigate the four varieties of MPs [low-density polyethylene (LDPE), high-density polyethylene (HDPE), polyvinyl chloride (PVC), and PE]. The order of plastic adsorbed on metal was PE > PVC > HDPE > LDPE, while the order of metal adsorbed on MPs was $Pb^{2+} > Cu^{2+} > Cd^{2+}$ (Zou et al., 2020). The efficiency of HDPE adsorption of Cd improved as pH increased; however, as salinity increased, the efficiency decreased. Desorption had a strong preference for adsorbed Cd (Azeem et al., 2021), posing a bigger danger to the biotic environment (Wang et al., 2019).

(b) *Size of Micro- and Nanoplastics*

As particle size has a crucial role in plastic internalization (Qi et al., 2018; Bosker et al., 2019; Jiang et al., 2019; Ullah et al., 2021), microplastics (MPs) size is an important factor for phytoaccumulation. The accumulation of nanoplastics in cellular compartments (vacuoles and cytoplasm) in the root epidermis of the garden onion *Allium cepa* (Mateos-Cárdenas et al., 2021) was explained by the small size of the particles that can cross membranes. However, the cut-off point for cellular absorption is unknown at this time. Plant cells, which are generally larger than animal cells, may be able to take in larger particles. This, however, will be determined by the cellular absorption mechanism. It's also been claimed that aquaporins may let nanoplastics get into rice plant *Oryza sativa* root cells, while long-term exposure could trigger a partial shut-down of these channels as a stress reaction (Mateos-Cárdenas et al., 2021). As a result, another important subject for future research is how much control plants have over plastic uptake in cells.

A few studies have recently demonstrated the impact of MPs on various plant species, including *Lepidium sativum, Triticum aestivum,* and *Vicia faba* (Bosker et al., 2019; Jiang et al., 2019; Qi et al., 2018). Bosker et al. (2019) found that different sizes of MPs reduced the germination rates of *L. sativum* seeds, while Qi et al. (2018) found that low-density polyethylene (LDPE) and biodegradable PMF residues had negative effects on the root and shoot parts of the wheat plant *Triticum aestivum* during vegetative and reproductive growth. For *Allium fistulosum* grown in the presence of different MPs – namely, PEST fibers, PA beads, PE, PP, PS, and PEST terephthalate – significant changes in plant biomass, elemental tissue composition, root attributes, leaf traits, and soil microbial activity were detected (De-Souza-Machado et al., 2019). Another study found a substantial impact on shoot lengths, dry root biomass, dry root/shoot ratio, and chlorophyll a/b ratio after seeding and growing *Lolium perenne* (perennial ryegrass) in soils containing biodegradable poly (lactic acid) (PLA), high-density PE, and MP-clothing fibers. In the presence of MP-clothing fibers or PLA, seed germination success was much lower than in the control soil (Boots et al., 2019).

Taylor et al. (2020) found no uptake of MPs in the internal root structure of Arabidopsis and wheat plant species but rather an accumulation of PS beads near the root surface in both. In contrast, using fluorescent PS markers, Li et al. (2019) established the uptake, dispersion, transportation, and accumulation of 0.2 mm size PS microbeads in an edible plant species (*Lactuca sativa*). The migration of PS microbeads across intercellular spaces of the vascular system, propelled along the transpiration stream, was also demonstrated by microscopic inspection of the shoot system. Similarly, when *Vicia faba* roots were exposed to fluorescent MPs of various sizes, the buildup of these polymers in their roots resulted in reduced growth (Jiang et al., 2019). The latter investigations have defied traditional wisdom – that polymer molecules, being larger than plant cells, should not collect in plants (Li et al., 2020b) – and have opened up a new frontier in the interaction of MPs with plants in terms of their fate and transmission via the food chain. Further research is needed to understand the processes of interactions as well as the level of vulnerability of different plant species to MP contamination.

8.6 Concentrations of MPs

The effects of MPs concentration on phytoaccumulation have been studied by a number of researchers. Microplastic particles collected in the rhizosphere are likely adsorbed to plant below-ground tissues, causing nutrition and water uptake problems, and eventually reaching root eaters (Jiang et al., 2019). To examine soil characteristics and plant biomass, Lozano et al. (2020) looked at numerous types, forms, and concentrations of MPs. All MPs resulted in increased shoot and root mass overall. Some MPs, on the other hand, produce an increase in plant biomass as concentration rises, while others cause a loss. Some MPs have also been shown to impair microbial activity in soil by reducing the diversity and richness of microorganisms (Guo et al., 2020). Using a metabolic approach, Wu et al. (2020) evaluated the impact of low, medium, and high doses of PS MPs on rice plants cultivated in a hydroponic solution. PS treatment resulted in a drop in shoot biomass, a modification in antioxidant activity, and a reduction in leaf metabolites such as sugars, amino acids, organic acids, and a variety of others. Harvest yields decreased as a result of the metabolic system change.

According to Manjate et al. (2020), MPs have no effect on the phytoremediation potential (the use of hyperaccumulator plants to extract toxic heavy metals from the environment) of common reeds [*Phragmites australis* (Cav.) Trin. ex Steud.], which can accumulate metals in their root tissues with concentrations up to 1 mg g^{-1} Cu and 70 g g^{-1} Cd in contaminated media (Tang, 2019). Zhang et al. (2020) investigated the effect of MPs on Cd absorption and desorption in agricultural soil and discovered that greater MPs concentrations resulted in higher Cd desorption, increasing the metal's mobility in the soil and posing potential dangers to crop plants and humans. Metal desorption, on the other hand, is dependent on MPs concentration, soil particle size, and solution pH. Despite several publications on MPs' effects on terrestrial plants, nothing is known about how MPs affect plant growth and development. By increasing reactive oxygen species production and inhibiting the cyclin-dependent cytokinase 2 gene, polystyrene MPs cause lower root length and cytotoxicity (Maity & Pramanick, 2020).

8.7 Plant Growth and Net Primary Production

MP has the ability to influence plant growth through a variety of *methods* (Rillig et al., 2019; Rillig et al., 2021), the majority of which are assumed to be indirect via the effect MP has on soil and soil biota. Changes in soil structure and bulk density, for example, might alter root penetration resistance, water holding capacity, and other indirect effects. As these carbon-rich particles are degraded by the soil microbial population, biodegradable plastics may cause nutrient immobilization. MPs or their impacts on soil physicochemical qualities may also alter essential plant symbionts, such as root-colonizing mycorrhizal fungi. The effects of MP on plant growth have been good in certain situations (Lozano et al., 2020), but there have also been reports of negative effects (Kleunen et al., 2020). Different MPs (with their chemical additions, some of which may be harmful), soils, and plants were used in these trials, which explains the discrepancies, but it's unclear how each of these elements contributed to the reported results. It's also unclear whether the key processes through which MP might affect plant growth are known. MP has distinct impacts on different plant species in a plant community, which could explain the change in grassland plant community composition seen when MP fibers were added to the soil (Lozano & Rillig, 2020).

8.8 Microplastics Interaction with Soil Microbes

Soil bacteria play an important role in the biogeochemical cycling of elements and food production. Understanding how MPs react to soil microorganisms will help researchers better forecast the potential implications of MP pollution. Microorganisms residing in soil–plastic interfaces may find a new home in MPs, resulting in the establishment of distinct microbial communities (Zhou et al., 2021). Soil MPs and NPs alter the bacterial and fungal communities' diversity. Several types of MPs have been found to both encourage and inhibit the bacterial population, and enzymatic activity, such as Bacteroidetes and Actinobacteria, has been found to boost the bacterial community on the surface of PE MPs (Ren et al., 2020). Microbes' metabolic activity is reduced by polyacrylic and polyester fibers (Judy et al., 2019). PVC and PE NPs alter microbial communities from Gram-positive to Gram-negative in wheat soil systems, as well as reduce xylosidase and glucosidase activity by 16–43%. (Zang et al., 2020). Due to some antimicrobials that may be linked to plastic additives, polyvinyl chloride (PVC) promotes Desulfobulbaceae and Desulfobacteraceae while lowering Sedimenticolaceae and Chromatiaceae (Judy et al., 2019). Furthermore, Bacteroidetes, Proteobacteria, and Firmicutes are inhibited by PS microbeads due to a probable interaction with reduced soil nutrients, as found in arsenic-polluted paddy soils (Dong et al., 2020). Soil characteristics and bacterial populations are altered by MPs and NPs. Rhizobia can be affected by changes in the soil matrix (Maity & Pramanick, 2020). MPs may alter bacterial diversity in soil organisms; for example, Zhu et al. (2018a, b) found that MPs may increase bacterial diversity in the collembolan stomach, presumably due to a shift in feeding after MP exposure. Some researchers have claimed that PLA, PCL, PHA, PBAT, and starch-based biopolymers have been approved as fixed C sources to increase the concentration of fungal species such as Fusarium, Aspergillus, and Penicillium (Accinelli et al., 2020).

Because of the difference in surface area, different sizes of plastic have distinct impacts on microorganisms (Brodhagen et al., 2017). Due to filamentous and yeast fungus bioaccumulation in the cell, NPs (<0.1 μm) in particular could breach the cell membrane and have harmful effects (Lei et al., 2018; De-Souza-Machado et al., 2018). NPs' ability to infiltrate and accumulate in soil organic detritus has biological effects on bacteria, although NPs may be less relevant for changing soil characteristics. Furthermore, water-stable aggregates are crucial for microbial activity, and PS fibers reduce water-stable aggregates, thus impacting plant soil health (Machado et al., 2020). Increased MP and NP abundance in soil can change microbial communities by raising the proportion of microbial communities chosen by MPs and NPs. As a result, freshly introduced MPs have an impact on environmental and ecological functions when combined with other natural chemicals and in conjunction with the plastisphere. According to the explanation above, the interaction between plastic and microbes may have a substantial impact on soil-plant interaction, fauna development, and nutrient recycling. Otherwise, it is suggested that the MP's surface serve as a microbiological development hotspot. As a result, understanding MP and NP environmental behaviors is crucial to better understand the reaction and function of soil microorganisms. However, little is known about the mechanism of interaction between the microbial population and MPs, which is a critical research gap that must be bridged in order to adequately analyze the environmental impact of MPs on soil (Azeem et al., 2021).

8.9 Conclusion

Root uptake and accumulation of microplastics and nanoplastics rely on the integrity of anatomical and physiological components of the plant in relation to debris and chemicals emitted by plastics, mostly eco-corona, and environmental aging surface and behavior. Oxidative stress causes interference in the metabolic pathways and damage to macromolecules in plants, when the concentrations of nanoplastics exceed optimal levels, plants are affected adversely, direct toxic effects caused by a high level of plastic concentration include inhibition of cytoplasmic enzymes and damage to cell structures, while the indirect adverse effect is on the replacement of essential nutrients at cation exchange sites of plants. Oxidative stress by micro and nanoplastics could cause lipid peroxidation that accounts for serious cell membrane damage in plants. Also, plastic toxicity affects the plasma membrane by causing hydration in water content, photosynthetic processes are hampered by high level of plastic contamination by interacting with chloroplast ultra-structure. Induced oxidative stress by plastics toxicity triggers inactive photosynthetic pigments causing perceived growth and low or inefficient seed germination. Stress due to microplastics in plants is one of the major agents affecting photosynthesis in terms of CO_2 fixation, transport of electron, photophosphorylation, and enzyme activities in plants. Stimulation and switching of superoxide dismutase and antioxidant catalase are major ways of contamination detoxification responses in plants, which could be altered by plastic toxicity.

References

Abioye, O. P., Mustapha, O. T., & Aransiola, S. A. (2014). Biological treatment of textile effluent using *Candida zeylanoides* and *Saccharomyces cerevisiae* isolated from soil. *Advances in Biology, 670394*, 4. https://doi.org/10.1155/2014/670394

Accinelli, C., Abbas, H. K., Bruno, V., Nissen, L., Vicari, A., Bellaloui, N., Little, N. S., & Shier, W. T. (2020). Persistence in soil of microplastic films from ultra-thin compostable plastic bags and implications on soil *Aspergillus flavus* population. *Waste Management, 113*, 312–318.

Ahmad, I., Akhtar, M. J., Zahir, Z. A., & Jamil, A. (2012). Effect of cadmium on seed germination and seedling growth of four wheats (*Triticum aestivum L.*) cultivars. *Pakistan Journal of Botany, 44*(5), 1569–1574.

Aransiola, S. A., Ijah, U. J. J., & Abioye, O. P. (2013). Phytoremediation of lead polluted Soil by *Glycine max* L. *Applied and Environmental Soil Science, 2013*, 631619. https://doi.org/10.1155/2013/631619

Asati, A., Pichhode, M., & Nikhil, K. (2016). Effect of heavy metals on plants: An overview. *International Journal of Application or Innovation in Engineering & Management, 5*(3), 56–66. https://doi.org/10.13140/RG.2.2.27583.87204

Auta, H. S., Abioye, O. P., Aransiola, S. A., Bala, J. D., Chukwuemeka, V. I., Hassan, A., Aziz, A., & Fauziah, S. H. (2022). Enhanced microbial degradation of PET and PS microplastics under natural conditions in mangrove environment. *Journal of Environmental Management, 304*(2022), 114273. https://doi.org/10.1016/j.jenvman.2021.114273

Avellan, A., Schwab, F., Masion, A., Chaurand, P., Borschneck, D., Vidal, V., Rose, J., Santaella, C., & Levard, C. (2017). Nanoparticle uptake in plants: Gold nanomaterial localized in roots of *Arabidopsis thaliana* by X-ray computed nanotomography and hyperspectral imaging. *Environmental Science and Technology, 51*, 8682–8691.

Azeem, I., Adeel, M., Ahmad, M. A., Shakoor, N., Jiangcuo, G. D., Azeem, K., Ishfaq, M., Shakoor, A., Ayaz, M., Xu, M., & Rui, Y. (2021). Uptake and accumulation of nano/microplastics in plants: A critical review. *Nanomaterials, 11*, 2935. https://doi.org/10.3390/nano11112935

Bandmann, V., Müller, J. D., Köhler, T., & Homann, U. (2012). Uptake of fluorescent nano beads into BY2-cells involves clathrin-dependent and clathrin-independent endocytosis. *FEBS Letters, 586*, 3626–3632.

Biber, N. F. A., Foggo, A., & Thompson, R. C. (2019). Characterizing the deterioration of different plastics in air and seawater. *Marine Pollution Bulletin, 141*, 595–602.

Bläsing, M., & Amelung, W. (2018). Plastics in soil: Analytical methods and possible sources. *Science of the Total Environment, 612*, 422–435.

Boots, B., Russell, C. W., & Green, D. S. (2019). Effects of microplastics in soil ecosystems: Above and below ground. *Environmental Science & Technology, 53*(19), 11496–11506.

Bosker, T., Bouwman, L. J., Brun, N. R., Behrens, P., & Vijver, M. G. (2019). Microplastics accumulate on pores in seed capsule and delay germination and root growth of the terrestrial vascular plant *Lepidium sativum*. *Chemosphere, 226*, 774–781.

Brodhagen, M., Goldberger, J. R., Hayes, D., Inglis, D. A., Marsh, T. L., & Miles, C. (2017). Policy considerations for limiting unintended residual plastic in agricultural soils. *Environmental Science and Policy, 69*, 81–84.

Chae, Y., & An, Y. J. (2020). Nanoplastic ingestion induces behavioral disorders in terrestrial snails: Trophic transfer effects via vascular plants. *Environmental Science: Nano, 7*, 975–983.

Chow, K. L., Man, Y. B., Tam, N. F. Y., Liang, Y., & Wong, M. H. (2017). Removal of decabromodiphenyl ether (BDE-209) using a combined system involving TiO2 photocatalysis and wetland plants. *Journal of hazardous materials, 322*, 263–269. https://doi.org/10.1016/j.jhazmat.2016.05.097

De-Souza-Machado, A. A., Lau, C. W., Till, J., Kloas, W., Lehmann, A., Becker, R., & Rillig, M. C. (2018). Impacts of microplastics on the soil biophysical environment. *Environmental Science and Technology, 52*, 9656–9665.

De-Souza-Machado, A. A., Lau, C. W., & Kloas, W. (2019). Microplastics can change soil properties and affect plant performance. *Environmental Science & Technology, 53*(10), 6044–6052.

Dietz, K. J., & Herth, S. (2011). Plant nanotoxicology. *Trends in Plant Science, 16*, 582–589.

Dobslaw, D., Kuch, B., Woiski, C., & Kiel, M. (2021). *Plant uptake, translocation and metabolism of PBDEs in plants* (pp. 1–21). Intechopen. https://doi.org/10.5772/intechopen.95790

Dong, Y., Gao, M., Liu, X., Qiu, W., & Song, Z. (2020). The mechanism of polystyrene microplastics to affect arsenic volatilization in arsenic-contaminated paddy soils. *Journal of Hazardous Materials, 398*, 122896.

Dong, Y., Gao, M., Qiu, W., & Song, Z. (2021). Uptake of microplastics by carrots in presence of as (III): Combined toxic effects. *Journal of Hazardous Materials, 411*, 125055.

Enyoh, C. E., Verla, A. W., Verla, E. N., Ibe, F. C., & Amaobi, C. E. (2019). Airborne microplastics: A review study on method for analysis, occurrence, movement, and risks. *Environmental Monitoring and Assessment, 191*, 1–17.

Esterhuizen, M., & Kim, Y. J. (2021). Efects of polypropylene, polyvinyl chloride, polyethylene terephthalate, polyurethane, high-density polyethylene, and polystyrene microplastic on Nelumbo nucifera (Lotus) in water and sediment. *Environmental Science and Pollution Research, 29*, 17580. https://doi.org/10.1007/s11356-021-17033-0

Feng, Y. H., Zhang, Y., Zheng, L. P., Ying, R. R., Lin, Y. S., & Wang, G. Q. (2017). Distribution characteristics of phthalic acid esters in soil and vegetable under greenhouse in different areas of Jiangsu Province, China. *Journal of Ecology and Rural Environment, 33*, 308–316.

Fonge, B. A., Larissa, M. T., Egbe, A. M., Afanga, Y. A., Fru, N. G., & Ngole-Jeme, V. M. (2021). An assessment of heavy metal exposure risk associated with consumption of cabbage and carrot grown in a tropical Savannah region. *Sustainable Environment, 7*, 1. https://doi.org/10.1080/27658511.2021.1909860

Guo, J. J., Huang, X. P., Xiang, L., Wang, Y. I. Z. E., Li, Y. W., Li, H., Cai, Q. Y., Mo, C. E. H., & Wong, M. H. (2020). Source, migration and toxicology of microplastics in soil. *Environment International, 137*, 105263.

Hofmann, U., & Schlosser, D. (2016). Biochemical and physicochemical processes contributing to the removal of endocrine-disrupting chemicals and pharmaceuticals by the aquatic ascomycete Phoma sp. UHH 5-1-03. *Applied Microbiology and Biotechnology, 100*, 2381–2399.

Hongwei, L., Xiang, Y., He, D., Li, Y., Zhao, Y., Wang, S., & Pan, X. (2019). Leaching behavior of fluorescent additives from microplastics and the toxicity of leachate to *Chlorellavulgaris*. *Science of the Total Environment, 678*, 1–9.

Horton, A. A., Walton, A., Spurgeon, D. J., Lahive, E., & Svendsen, C. (2017). Microplastics in freshwater and terrestrial environments: Evaluating the current understanding to identify the knowledge gaps and future research priorities. *Science of the Total Environment, 586*, 127–141.

Hrda, K., Pouzar, M., & Knotek, P. (2018). Study of zinc oxide nanoparticles and zinc chloride toxicity to annelid Enchytraeus crypticus in modified agar-based media. *Environmental Science and Pollution Research, 25*, 22702–22709.

Huerta Lwanga, E., Mendoza Vega, J., Ku Quej, V., de los Angeles Chi, J., Sanchez del Cid, L., Chi, C., Escalona Segura, G., Gertsen, H., Salánki, T., & van der Ploeg, M. (2017). Field evidence for transfer of plastic debris along a terrestrial food chain. *Scientific Reports, 7*, 14071.

Hüffer, T., Weniger, A. K., & Hofmann, T. (2018). Sorption of organic compounds by aged polystyrene microplastic particles. *Environmental Pollution, 236*, 218–225.

Jiang, J., Zhao, H., Sun, S., Wang, Y., Liu, S., Xie, Q., & Li, X. (2018). Occurrence and profiles of halogenated phenols, polybrominated diphenyl ethers and hydroxylated polybrominated diphenyl ethers in the effluents of waste water treatment plants around Huang-Bo Sea, North China. *Science of the Total Environment., 622-623*, 1–7. https://doi.org/10.1016/j.scitotenv.2017.11.323

Jiang, X., Chen, H., Liao, Y., Ye, Z., Li, M., & Klobučar, G. (2019). Ecotoxicity and genotoxicity of polystyrene microplastics on higher plant *Vicia faba*. *Environmental Pollution, 250*, 831–838.

Judy, J. D., Williams, M., Gregg, A., Oliver, D., Kumar, A., Kookana, R., & Kirby, J. K. (2019). Microplastics in municipal mixed-waste organic outputs induce minimal short to long-term toxicity in key terrestrial biota. *Environmental Pollution, 252*, 522–531.

Kettler, K., Veltman, K., van de Meent, D., van Wezel, A., & Hendriks, A. J. (2014). Cellular uptake of nanoparticles as determined by particle properties, experimental conditions, and cell type. *Environmental Toxicology and Chemistry, 33*, 481–492.

Kim, S. W., & Rillig, M. C. (2021). Research trends of microplastics in the soil environment: Comprehensive screening of effects. *Soil Ecology Letters, 4*, 1–10. https://doi.org/10.1007/s42832-021-0077-3

Kleunen, M., Van-Brumer, A., Gutbrod, L., & Zhang, Z. (2020). A microplastic used as infill material in artificial sport turfs reduces plant growth. *Plants, People, Planet, 2*, 157–166.

Klinčić, D., Dvoršćak, M., Jagić, K., Mendaš, G., & Herceg, R. S. (2020). Levels and distribution of polybrominated diphenyl ethers in humans and environmental compartments: A comprehensive review of the last five years of research. *Environmental Science and Pollution Research, 27*, 5744–5758. https://doi.org/10.1007/s11356-020-07598-7

Kumari, S., & Mishra, A. (2021). Heavy metal contamination. In M. L. Larramendy & S. Soloneski (Eds.), *Soil contamination – threats and sustainable solutions*. IntechOpen. https://doi.org/10.5772/intechopen.93412

Lambert, S., & Wagner, M. (2018). Microplastics are contaminants of emerging concern in freshwater environments: An overview. In M. Wagner & S. Lambert (Eds.), *Freshwater microplastics. Emerging environmental contaminants?* (pp. 1–23). Springer.

Law, K. L., & Thompson, R. C. (2014). Oceans. Microplastics in the seas. *Science, 345*, 144–145.

Lei, L., Liu, M., Song, Y., Lu, S., Hu, J., Cao, C., & He, D. (2018). Polystyrene (nano) microplastics cause size-dependent neurotoxicity, oxidative damage and other adverse effects in Caenorhabditis elegans. *Environmental Science: Nano, 5*, 2009–2020.

Li, K. K., Ma, D., Wu, J., Chai, C., & Shi, Y. X. (2016). Distribution of phthalate esters in agricultural soil with plastic film mulching in Shandong Peninsula, East China. *Chemosphere, 164*, 314–321. https://doi.org/10.1016/j.chemosphere.2016.08.068

Li, L., Zhou, Q., Yin, N. A., Tu, C., & Luo, Y. (2019). Uptake and accumulation of microplastics in an edible plant. *Chinese Science Bulletin, 64*, 928–934.

Li, L., Luo, Y., Li, R., Zhou, Q., Peijnenburg, W. J. G. M., Yin, N. A., Yang, J., Tu, C., & Zhang, Y. (2020a). Effective uptake of submicrometre plastics by crop plants via a crack-entry mode. *Nature Sustainability, 3*, 929–937.

Li, L., Yang, J., Zhou, Q., Peijnenburg, W. J. G. M., & Luo, Y. (2020b). Uptake of microplastics and their effects on plants. In D. He & Y. Luo (Eds.), *Microplastics in terrestrial environments* (pp. 279–298). Springer.

Lian, J., Wu, J., Xiong, H., Zeb, A., Yang, T., Su, X., Su, L., & Liu, W. (2020a). Impact of polystyrene nanoplastics (PSNPs) on seed germination and seedling growth of wheat (*Triticum aestivum* L.). *Journal of Hazardous Materials, 385*, 121620.

Lian, J., Wu, J., Zeb, A., Zheng, S., Ma, T., Peng, F., Tang, J., & Liu, W. (2020b). Do polystyrene nanoplastics affect the toxicity of cadmium to wheat (*Triticum aestivum* L.)? *Environmental Pollution, 263*, 114498.

Lozano, Y. M., & Rillig, M. C. (2020). Effects of microplastic fibers and drought on plant communities. *Environmental Science and Technology, 54*, 6166–6173.

Lozano, Y. M., Lehnert, T., Linck, L. T., Lehmann, A., & Rillig, M. C. (2020). Microplastic shape, concentration and polymer type affect soil properties and plant biomass. *bioRxiv*. 202007.27.223768. https://doi.org/10.1101/2020.07.27.223768

Ma, T., Zhou, W., Chen, L., Wu, L., Christie, P., & Zhang, H. (2017). Toxicity effects of di-(2-ethylhexyl) phthalate to *Eisenia fetida* at enzyme, cellular and genetic levels. *PLoS One, 12*(3), 0173957. PMID:28319143

Ma, T. T., Zhou, W., Chen, L. K., Wu, L. H., Christie, P., & Liu, W. X. (2018). Toxicity of phthalate esters to lettuce (Lactuca sativa) and the soil microbial community under different soil conditions. *PLoS One, 13*, 0208111.

Machado, A. A. D. S., Horton, A. A., Davis, T., & Maaß, S. (2020). Microplastics and their effects on soil function as a life-supporting system. In *Microplastics in terrestrial environments* (pp. 199–222). Springer.

Mahmud, U., Salam, M. T. B., Khan, A. S., & Rahman, M. M. (2021). Ecological risk of heavy metal in agricultural soil and transfer to rice grains. *Discover Materials, 1*, 10. https://doi.org/10.1007/s43939-021-00010-2

Maity, S., & Pramanick, K. (2020). Perspectives and challenges of micro/nanoplastics-induced toxicity with special reference to phytotoxicity. *Global Change Biology, 26*, 3241–3250.

Manjate, E., Ramos, S., & Almeida, C. M. R. (2020). Potential interferences of microplastics in the phytoremediation of cd and cu by the salt marsh plant Phragmites australis. *Journal of Environmental Chemical Engineering, 8*, 103658. https://doi.org/10.1016/j.jece.2020.103658

Mateos-Cárdenas, A., Scott, D. T., Seitmaganbetova, G., Van-Pelt, F. N. A. M., O'Halloran, J., & Jansen, M. A. K. J. (2019). Polyethylene microplastics adhere to Lemna minor (L.), yet have no effects on plant growth or feeding by Gammarus duebeni (Lillj.). *Science of the Total Environment, 689*, 413–421.

Mateos-Cárdenas, A., Frank, N. A. M., Van-Pelt, F. N. A. M., O'Halloran, J., & Marcel Jansen, A. K. (2021). Adsorption, uptake and toxicity of micro- and nanoplastics: Effects on terrestrial plants and aquatic macrophytes. *Environmental Pollution, 284*, 1–10.

Ng, E. L., Huerta Lwanga, E., Eldridge, S. M., Johnston, P., Hu, H. W., Geissen, V., & Chen, D. (2018). An overview of microplastic and nanoplastic pollution in agroecosystems. *Science of the Total Environment, 627*, 1377–1388.

Nizzetto, L., Bussi, G., Futter, M. N., Butterfield, D., & Whitehead, P. G. (2016). A theoretical assessment of microplastic transport in river catchments and their retention by soils and river sediments. *Environmental Science: Processes and Impacts, 18*, 1050–1059.

Pathan, S. I., Arfaioli, P., Bardelli, T., Ceccherini, M. T., Nannipieri, P., & Pietramellara, G. (2020). Review: Soil pollution from micro- and Nanoplastic debris: A hidden and unknown biohazard. *Sustainability, 12*, 7255. https://doi.org/10.3390/su12187255

Pichhode, M., & Nikhil, K. (2015). Effect of copper dust on photosynthesis pigments concentrations in plants species. *International Journal of Engineering Research and Management (IJERM), 2*, 63–66.

Plastics Europe. (2015). *Plastics – The facts 2014/2015: An analysis of European plastics production, demand and waste data* (pp. 1–34). Plastics Europe.

Plastics Europe. (2019). *Plastics – The facts 2020: An analysis of European plastics production, demand and waste data*. https://www.plasticseurope.org/download_fle/force/4261/181. Accessed 26 December 2021.

Qi, Y., Yang, X., & Pelaez, A. M. (2018). Macro- and micro-plastics in soil–plant system: Effects of plastic mulch film residues on wheat (*Triticum aestivum*) growth. *Science of the Total Environment, 645*, 1048–1056.

Rehse, S., Kloas, W., & Zarfl, C. (2018). Microplastics reduce short-term effects of environmental contaminants. Part I: Effects of bisphenol A on freshwater zooplankton are lower in presence of polyamide particles. *International Journal of Environmental Research and Public Health, 15*, 280.

Ren, X., Tang, J., Liu, X., & Liu, Q. (2020). Effects of microplastics on greenhouse gas emissions and the microbial community in fertilized soil. *Environmental Pollution, 256*, 113347.

Rillig, M. C. (2018). Microplastic disguising as soil carbon storage. *Environmental Science & Technology, 52*, 6079–6080.

Rillig, M. C., Lehmann, A., Machado, A. A. D. S., & Yang, G. (2019). Microplastic effects on plants. *New Phytologist, 223*, 1066–1070.

Rillig, M. C., Leifheit, E., & Lehmann, J. (2021). Microplastic effects on carbon cycling processes in soils. *PLoS Biology, 19*(3), e3001130. https://doi.org/10.1371/journal.pbio.3001130

Rist, S., Carney Almroth, B., Hartmann, N. B., & Karlsson, T. M. (2018). A critical perspective on early communications concerning human health aspects of microplastics. *Science of the Total Environment, 626*, 720–726.

Sandermann, H. (1992). Plant metabolism of xenobiotics. *Trends in Biochemical Sciences, 2*, 82–84.

Smith, M., Love, D. C., Rochman, C. M., & Neff, R. A. (2018). Microplastics in seafood and the implications for human health. *Current Environmental Health Reports, 5*, 375–386.

Sun, X. D., Yuan, X. Z., Jia, Y., Feng, L. J., Zhu, F. P., Dong, S. S., Liu, J., Kong, X., Tian, H., & Duan, J. L. (2020). Differentially charged nanoplastics demonstrate distinct accumulation in Arabidopsis thaliana. *Nature Nanotechnology, 15*, 755–760. Available online: https://www.nature.com/articles/s41565-020-0707-4

Tang, K. H. D. (2019). Phytoremediation of soil contaminated with petroleum hydrocarbons: A review of recent literature. *Global Journal of Civil and Environmental Engineering, 1*, 33–42.

Taylor, S. E., Pearce, C. I., & Sanguinet, K. A. (2020). Polystyrene nano- and microplastic accumulation at Arabidopsis and wheat root cap cells, but no evidence for uptake into roots. *Environmental Science: Nano, 7*, 1942–1953.

Teuten, E. L., Saquing, J. M., Knappe, D., Barlaz, M. A., Jonsson, S., Björn, A., Rowland, S. J., Thompson, R., Galloway, T. S., & Yamashita, R. (2009). Transport and release of chemicals from plastics to the environment and to wildlife. *Philosophical Transactions of the Royal Society B: Biological Sciences, 364*, 2027–2045.

Ullah, R., Tsui, M. T. K., Chen, H., Chow, A., Williams, C., & Ligaba-Osena, A. (2021). Microplastics interaction with terrestrial plants and their impacts on agriculture. *Journal of Environmental Quality, 50*, 1–18.

Van-Weert, S., Redondo-Hasselerharm, P. E., Diepens, N. J., & Koelmans, A. A. (2019). Effects of nanoplastics and microplastics on the growth of sediment-rooted macrophytes. *Science of the Total Environment, 654*, 1040–1047.

Vom Saal, F. S., & Myers, J. P. (2008). BisphenolA and risk of metabolic disorders. *Journal of the American Medical Association, 300*, 1353–1355.

Wagner, M., & Lambert, S. (2018). Freshwater microplastics. Emerging environmental contaminants. In *The handbook of environmental chemistry* (Vol. 58). Springer Open.

Wang, J., Luo, Y., Teng, Y., Ma, W., Christie, P., & Li, Z. (2013). Soil contamination by phthalate esters in Chinese intensive vegetable production systems with different modes of use of plastic film. *Environmental Pollution, 180*, 265–273.

Wang, F., Yang, W., Cheng, P., Zhang, S., Zhang, S., Jiao, W., & Sun, Y. (2019). Adsorption characteristics of cadmium onto microplastics from aqueous solutions. *Chemosphere, 235*, 1073–1080.

Wang, L., Liu, Y., Kaur, M., Yao, Z., Chen, T., & Xu, M. (2021). Phytotoxic effects of polyethylene microplastics on the growth of food crops soybean (Glycine max) and mung bean (vigna radiata). *International Journal of Environmental Research and Public Health, 18*, 10629. https://doi.org/10.3390/ijerph182010629

Wu, X., Wang, W., & Zhu, L. (2018). Enhanced organic contaminants accumulation in crops: Mechanisms, interactions with engineered nanomaterials in soil. *Environmental Pollution, 240*, 51–59. https://doi.org/10.1016/j.envpol.2018.04.072

Wu, X., Liu, Y., Yin, S., Xiao, K., Xiong, Q., Bian, S., Liang, S., Hou, H., Hu, J., & Yang, J. (2020). Metabolomics revealing the response of rice (Oryza sativa L.) exposed to polystyrene microplastics. *Environmental Pollution, 266*, 115159.

Xie, Y., Fan, J., Zhu, W., Amombo, E., Lou, Y., Liang, C., & Jinmin, F. (2016). Effect of heavy metals pollution on soil microbial diversity and bermudagrass genetic variation. *Frontiers in Plant Science, 7*, 755. https://doi.org/10.3389/fpls.2016.00755

Yousefi, Z., Ala, A., Babanezhad, E., & Mohammadpour, R. A. (2019). Evaluation of exposure to phthalate esters through the use of various brands of drinking water bottled in polyethylene terephthalate (PET) containers under different storage conditions. *Environmental Health Engineering and Management Journal, 6*(4), 247–255. https://doi.org/10.15171/EHEM.2019.28

Yu, J., Tian, J.-Y., Xu, R., Zhang, Z.-Y., Yang, G.-P., Wang, X.-D., Lai, J.-G., & Chen, R. (2020). Effects of microplastics exposure on ingestion, fecundity, development, and dimethylsulfide production in Tigriopus japonicus (Harpacticoida, copepod). *Environmental Pollution, 267*, 115429.

Yuan, W., Zhou, Y., Liu, X., & Wang, J. (2019). New perspective on the nanoplastics disrupting the reproduction of an endangered fern in artificial freshwater. *Environmental Science and Technology, 53*, 12715e12724.

Zang, H., Zhou, J., Marshall, M. R., Chadwick, D. R., Wen, Y., & Jones, D. L. (2020). Microplastics in the agroecosystem: Are they an emerging threat to the plant-soil system? *Soil Biology and Biochemistry, 148*, 107926.

Zhang, Y., Luo, X.-J., Mo, L., Wu, J.-P., Mai, B.-X., & Peng, Y.-H. (2015). Bioaccumulation and translocation of polyhalogenated compounds in rice (Oryza sativa L.) planted in paddy soil collected from an electronic waste recycling site, South China. *Chemosphere, 137*, 25–32. https://doi.org/10.1016/j.chemosphere.2015.04.029

Zhang, S., Han, B., Sun, Y., & Wang, F. (2020). Microplastics influence the adsorption and desorption characteristics of Cd in an agricultural soil. *Journal of Hazardous Materials, 388*, 121775. https://doi.org/10.1016/j.jhazmat.2019.121775

Zhao, Q., Ma, C., White, J. C., Dhankher, O. M. P., Zhang, X., Zhang, S., & Xing, B. (2017). Quantitative evaluation of multi-wall carbon nanotube uptake by terrestrial plants. *Carbon, 114*, 661–670.

Zhou, J., Gui, H., Banfield, C. C., Wen, Y., Zang, H., Dippold, M. A., Charlton, A., & Jones, D. L. (2021). The microplastisphere: Biodegradable microplastics addition alters soil microbial community structure and function. *Soil Biology and Biochemistry, 156*, 108211.

Zhu, D., Chen, Q. L., An, X. L., Yang, X. R., Christie, P., Ke, X., Wu, L. H., & Zhu, Y. G. (2018a). Exposure of soil collembolans to microplastics perturbs their gut microbiota and alters their isotopic composition. *Soil Biology and Biochemistry, 116*, 302–310.

Zhu, H., Sun, H., Yao, Y., Gan, Z., Wang, Y., & Kannan, K. (2018b). Legacy and alternative brominated flame retardants in outdoor dust and pine needles in mainland China: Spatial trends, dust plant partitioning and human exposure. *Environmental Pollution*. (Barking, Essex: 1987)*, 243*, 758–765. https://doi.org/10.1016/j.envpol.2018.08.097

Zhu, F., Zhu, C., Wang, C., & Gu, C. (2019). Occurrence and ecological impacts of microplastics in soil systems: A review. *Bulletin of Environmental Contamination and Toxicology, 102*, 741–749.

Zong, X., Zhang, J., Zhu, J., Zhang, L., Jiang, L., Yin, Y., & Guo, H. (2021). Effects of polystyrene microplastic on uptake and toxicity of copper and cadmium in hydroponic wheat seedlings (Triticum aestivum L.). *Ecotoxicology and Environmental Safety, 217*, 112217.

Zou, J., Liu, X., Zhang, D., & Yuan, X. (2020). Adsorption of three bivalent metals by four chemical distinct microplastics. *Chemosphere, 248*, 126064.

Part III
Toxicity of Micro and Nanoplastics

Chapter 9
Toxicity Effects of Micro- and Nanoplastics in Terrestrial Environment

A. Vamshi Krishna Reddy, Golla Shankaraiah, and Palakeerti Srinivas Kumar

Abstract Plastics are synthetic materials composed of organic polymers and chemical additives such as bisphenols, phthalates, and flame retardants, which give them their unique properties. Plastics are used in a wide range of commercial applications due to their low cost, ease of manufacture, flexibility, and hydrophobicity. Every year, the amount of plastic produced rises; nevertheless, reusing, recycling, and repurposing solutions are not adopted quickly enough, especially in poor countries. However, only 21–26% of the plastic waste was adequately recycled and burnt. The rest is discarded in the environment or burnt in open pits, resulting in plastic contamination of the water, air, and soil, among other things.

After entering the environment, interactions between plastic waste and environmental components can break down big pieces of plastic into smaller plastic particles. Micro- or nano-sized plastic particles are produced during the degradation of plastic trash and are referred to as micro- or nanoplastics (MNPs). The diameter of plastic fragments or particles determines whether they are microplastics (MPs) or nanoplastics (NPs), with MPs having a diameter of less than 5 mm and NPs having a diameter of 1 to 100 or 1000 nm. We summarised MNP pollution in the environment in this chapter before evaluating their health impacts based on available MNP research.

MNPs are prevalent in both terrestrial and marine settings all over the world, and they can be ingested and retained by animals at all levels of the food chain. Furthermore, evidence is accumulating that micro- and nanoplastics interact with terrestrial organisms that mediate critical ecosystem services and activities, such as soil-dwelling invertebrates, terrestrial fungi, and plant pollinators. As a result, further study is needed to establish the destiny and environmental implications of micro- and nanoplastics. Because of their widespread distribution, environmental

A. V. K. Reddy (✉) · G. Shankaraiah
Centre for Environment, Institute of Science & Technology, Jawaharlal Nehru technological University Hyderabad, Kukatpally, Hyderabad, Telangana, India

P. S. Kumar
Centre for Biotechnology, Jawaharlal Nehru Technological University Hyderabad, Institute of Science & Technology, Hyderabad, Telangana, India

N. R. Maddela et al. (eds.), *Micro and Nanoplastics in Soil*,
https://doi.org/10.1007/978-3-031-21195-9_9

durability, and diverse interactions with continental biota, micro and nanoplastic pollution may constitute a rising global change threat to terrestrial ecosystems.

Keywords Plastic waste · Microplastics · Nanoplastics · Environment · Toxicity · Health impacts.

9.1 Introduction

Plastics are synthetic materials made primarily of organic polymers and a variety of chemical additives such as bisphenols, phthalates, and flame retardants, which give them their unique properties (Gigault et al., 2018). Plastics are used in a wide range of commercial applications due to their low cost, ease of manufacture, flexibility, and hydrophobicity. Every year, the amount of plastic produced rises; nevertheless, reusing, recycling, and repurposing solutions are not adopted quickly enough, especially in poor countries. Globally, an estimated 6.3 billion tonnes of plastic waste was produced between the 1950s and 2015. If current trends continue, that amount will have increased to 26 billion tonnes by 2050. However, only 21–26% of the plastic waste was adequately recycled and burnt (Wang et al., 2019). The rest is discarded in the environment or burnt in open pits, resulting in plastic contamination of the water, air, and soil, among other things.

After entering the environment, interactions between plastic waste and environmental components can break down big pieces of plastic into smaller plastic particles. Furthermore, tiny plastic particles are routinely manufactured and added to consumer items such as personal care products, which are then discarded, resulting in yet another significant source of plastic pollution in the environment (Lehner et al., 2019). The diameter of plastic fragments or particles determines whether they are microplastics (MPs) or nanoplastics (NPs), with MPs having a diameter of less than 5 mm and NPs having a diameter of 1 to 100 or 1000 nm.

Micro and nanoplastics (MNPs) have been discovered in both marine and terrestrial ecosystems, including oceans, rivers, air, drinking water, sediments, and food, all over the world. According to prior studies, MNP exposure has been connected to reproductive toxicity in oysters, liver toxicity in zebrafish (Lu et al., 2016), and tissue bioaccumulation and potential organ toxicity in mice. These findings indicate that MNP contamination is widespread and that the biological harm caused by MNPs to humans and other living species cannot be ignored. The resulting experimental results, however, are equivocal; the outcomes of various investigations are frequently conflicting; and the underlying mechanisms of detected toxicities are still little understood.

In addition, MPs have recently been identified in human faeces, showing that people are exposed to MNPs via the food chain or food web (Prata et al., 2020). Despite this, little study on the impact of MNPs on human health has been conducted. Furthermore, these microscopic plastic particles can leak plastic additives and/or absorb other ambient chemicals, many of which have been shown to have endocrine disrupting and other hazardous effects. However, how MNPs will alter the toxicity of these additives and adsorbents is still unknown. As a result, the fundamental characteristics, sources, and abundance of MNPs in the environment were studied first in this review before moving on to the health impacts of pure MNPs and their associated adsorbents and additives.

9.2 Global Production of Plastics and Generation of Waste

In today's world, plastics are everywhere. It was initially employed circa 1600 B.C. in prehistoric Mesoamerica when human hands shaped natural rubber and polymerized it into a variety of useful objects (Hosler et al., 1999). In 1839, the invention of polystyrene (PS) and vulcanised rubber transformed the way plastics and plastic items were used and made. Bakelite, the world's first completely synthetic polymer, was developed in Belgium in 1907. Bakelite, on the other hand, had become widely used by 1930, particularly in the fashion, communication, electrical, and car sectors. After that, it took a decade for mass production of plastics to begin, and it has been steadily increasing since. In 2008, global plastic production was predicted to reach 245 million tonnes (Plastics Europe, 2009).

Consumer goods, building materials, automotive, electrical, and agricultural applications account for 22%, 20%, 9%, 6%, and 3% of total plastic consumption in Europe, respectively. Asia was predicted to have the greatest rate of production in 2015 (49%of total world output, with China as the top producer (28%), followed by North America and Europe, each with 19%). In terms of plastic manufacture, the rest of the globe is less significant, but not necessarily in terms of plastic consumption (Worm et al., 2017).

9.3 Current World Production Rate of Plastics

Plastic output is expected to reach 380 million tonnes globally in 2018. Around 6.3 billion tonnes of plastic were produced globally between 1950 and 2018, with 9% and 12% of that being recycled and burned, respectively. Over 5 million tonnes of plastic are consumed in the United Kingdom each year, with only about a quarter being recycled and the rest being dumped in landfills. In terms of weight, researchers anticipate that by 2050, the oceans will contain more plastic than fish (Sutter, 2016). Each year, 500 billion plastic bags are used, with an estimated 13 million tonnes of them ending up in the ocean, killing 100,000 marine species.

9.4 Future Projection of Production of Plastic

Plastic output has more than doubled since 1964. Globally, around 311 million tonnes of plastics were produced in 2014, with output expected to double in the following 20 years and maybe quadruple by 2050 (World Economic Forum, 2016). According to the International Energy Agency's World Energy Outlook 2015, the largest application, plastic packaging (26%of total volume), is expected to grow rapidly, potentially doubling in size within 15 years and quadrupling in size by 2050, to approximately 318 million tonnes annually, which is more than the entire plastic industry today.

9.5 Types of Plastics

Plastics are characterised generally according to their form durability or non-durability, as well as whether they are thermosets or thermoplastics. Thermosets, such as polyurethane, epoxy, and alkyd, are extensively used as insulators, adhesives, and plywood. The thermosetting process uses heat to establish new and irreversible covalent bonds, which makes thermosets robust and difficult to dissolve (Rudyak et al., 2018).

Thermoplastics, on the other hand, have no newly formed chemical bonds and can be recycled and remoulded, making them more commonly used in consumer products than thermosets. Thermoplastics are classified into four types: Polyethylene (PE), Polypropylene (PP), Polystyrene (PS), and Polyvinyl chloride (PVC) are the four main types of plastics (PVC).

1. PE may be found in a variety of low-cost plastic goods, including bags and bottles. PE is classified into two types: HDPE, which is usually found in detergent bottles, milk cans, and moulded plastic cases, and LDPE, which is commonly found in outdoor furniture, siding, floor tiles, shower curtains, and clamshell packaging.
2. Plastic pressure pipe systems, bottle caps, drinking straws, yoghurt containers, appliances, automobile bumpers, fishing lines, and bottle caps are all constructed of PP.
3. PS is used in the manufacture of foam peanuts, food containers, plastic tableware, disposable cups, plates, cutlery, CDs, and cassette boxes.
4. PVC is used for plumbing pipes and gutters, shower curtains, window frames, and flooring.

In addition to the normal plastic classes mentioned above, microplastic fibres (MFs) made of polyester (PES) or polypropylene (PP) are one of the most prevalent forms of MPs found in the environment. MFs are used in clothing, agriculture, industry, and home textiles, as well as various textile goods, semi-finished or auxiliary products used in other industries.

The three most prevalent types of MPs used in scientific investigations are PE, PS, and PVC. The most often utilised plastic materials in consumer items are PE and PS, which have shorter service lifetimes than other types of plastics. PVC is also extensively used for data cable jackets and plastic wire insulation. Metals in cables are recycled at the end of their lives, but PVC-containing plastic parts are frequently thrown away due to the high cost of separation and low recycling value. According to studies, 82% of PVC waste is disposed of in landfills, 15% is burned, and only 3% is recycled. (Suresh et al., 2017).

Because of their high production, short life cycle, and broad environmental discharge, these polymers have been the topic of scientific investigation. Plastics are categorised based on their chemical makeup and the materials used to make them. Table 9.1 shows the many kinds of plastics, as well as their characteristics and their uses.

9.5.1 Polyethylene Terephthalate (PET)

PET is a smooth, transparent, and relatively thin material that stands for polyethylene terephthalate. It is also known as stomach plastics. PET is commonly used in the manufacturing of disposable salad dressing, juice, mouthwash, vegetable oil, cosmetics, soft drinks, margarine, and water bottles because it is anti-inflammatory and completely liquid. PET is also anti-air, meaning it prevents oxygen from entering the body (Proshad et al., 2018). In the production of PET and rubber vulcanization, an inorganic chemical called antimony trioxide is used as a catalyst. When utilising PET plastics, high temperatures must be avoided to avoid the leaching of hazardous compounds such as acetaldehyde, antimony, and phthalates. In humans, antimony has the potential to cause cancer. PET is frequently created for one-time use only.

9.5.2 High-density Polyethylene

The most extensively used plastic in the globe is polyethylene. HDPE (high-density polyethylene) is a heat-resistant plastic made from petroleum. Refrigerators, detergent bottles, toys, milk containers, and a variety of plastic shopping bags are among the places where it may be found. There are no phthalates or BPA in high-density polyethylene. Despite the fact that certain studies have demonstrated that long-term exposure to sunlight can damage plastics, high-density polyethylene containers are generally recognised as appropriate for beverages and food entering the country (Proshad et al., 2018).

Table 9.1 Types of plastics, their features, and common applications

Types of plastics	Symbols	Properties	Common uses	Recycled into
Polyethylene terephthalates (PET)	1 PETE	Clear, tough, solvent resistant, barrier to gas and moisture, softens at 80 °C	Soft drinks, water bottles, containers, salad dressing, biscuit trays, and salad domes	Pillow and sleeping bag filling, clothing, soft drink bottles, carpeting, building insulation
High-density polyethylene (HDPE)	2 HDPE	Hard to semi-flexible, resistant to chemicals and moisture, waxy surface, opaque, softens at 75 °C, easily coloured, processed, and formed	Shopping bags, freezer bags, buckets, shampoo, milk bottles, ice cream containers, juice bottles, chemical and detergent bottles, rigid agricultural pipes, crates	Recycling bins, compost bins, buckets, detergent containers, posts, fencing, pipes, plastic timber
Polyvinyl chloride (PVC) Plasticized polyvinyl chloride PVC-P.	3 V	Strong, tough, softens at 80 °C, can be clear, can be solvent welded Flexible, clear, elastic, can be solvent welded	Cosmetic containers, plumbing pipes and fittings, electrical conduct, blister packs, wall cladding, roof sheeting, bottles Garden hose, shoe soles, cable sheathing, blood bags and tubing	Compost bin, Flooring, film and sheets, cables, speed bumps, packaging, binders, mud flaps and mats, new gumboots and shoes
Low-density polyethylene (LDPE)	4 LDPE	Soft flexible, waxy surface, translucent, softens at 70 °C, scratches easily	Refuse bags, Irrigation tubings, mulch film, cling wrap, garbage bags, squeeze bottles	Bin liners, pallet sheets
Polypropylene (PP)	5 PP	Hard and translucent, soften at 140 °C, translucent, withstands solvents, versatile	Microwave dishes, lunch boxes, packaging tape, garden furniture, kettles, bottles and ice cream tubs, potato chip bags, straws	Pegs, bins, pipes, pallet sheets, oil funnels, car battery cases, trays
Polystyrene (PS) Expanded polystyrene (PS-E)	6 PS	Clear, glassy rigid, opaque, semi-tough, soften at 95 °C, Affected by fat, acids, and solvents, but resistant to alkalis, salt solutions, low water absorption, when not pigmented is clear, is odour and taste free Special types of Polystyrene (PS) are available for special applications	CD cases, plastic cutlery, imitation glassware, low-cost brittle toys, video cases/foamed polystyrene cups, protective packaging, building and food insulation	Recycle bin, Coat hangers, coasters, white, ware components, stationery trays and accessories, picture frames, seed trays, building products
Other – letter below indicate ISO code for plastic type. E.g.-SAN,ABS, PC, Nylon	7 OTHER	Includes all resins and multi-materials (e.g. laminates) properties dependent on plastic or combination of plastics	Automotive and appliance components, computers, electronics, cooler bottles, packaging	Recycle bins, automotive components, plastic timber

Modified from Alabi et al. (2019)

9.5.3 Polyvinyl Chloride (PVC)

Polyvinyl chloride (PVC), a heat-resistant polymer, is used to package fruit juice, cooking oil, and other liquids. PVC is considered extremely hazardous due to the presence of chemical constituents such as heavy metals, dioxins, BPA, and phthalates. PVC is flexible due to the presence of phthalates and is dependent on non-plasticization. Phthalates are poisonous to humans. Because the whole PVC life cycle, including manufacture, use, and disposal, poses major environmental and public health risks, its use has been severely restricted. Due to its low cost and versatility, PVC is still frequently employed in the manufacturing of consumer goods. PVC has been related to chronic bronchitis, birth abnormalities, genetic modifications, cancer, skin disorders, deafness, vision loss, ulcers, liver malfunction, and indigestion (Proshad et al., 2018).

9.5.4 Low-density Polyethylene

Low-density polyethylene is known for its heat resistance, fragility, flexibility, and stiffness. Milk, frozen goods, and juice packaging are common sources. Because the plastic does not contain any harmful components, it may be utilised to make liquids and food enter the body (Proshad et al., 2018).

9.5.5 Polypropylene

Polypropylene, for example, is a tough and semi-transparent plastic. It is more robust and thicker than polyethylene. Medicine, yoghurt, ketchup, and drinks are all packaged in it. Polypropylene polymers do not contain any dangerous compounds, thus polypropylene containers, like polyethylene containers, are deemed safe for use as food and beverage containers by humans.

9.5.6 Polystyrene

Polystyrene, a petroleum-based material, contains benzene, a carcinogen that is dangerous to humans. Polystyrene is a popular polymer used in the manufacture of insulators and packaging. Products containing styrene are harmful to one's health. During long-term exposure, Dowty et al. discovered that a little quantity of styrene can be neurotoxic and produce cytogenetic, carcinogenic, and haematological effects. The International Agency for Research on Cancer (IARC) has categorised styrene as a human carcinogen (Proshad et al., 2018).

9.5.7 *Polycarbonate*

Polycarbonates are used to package consumer goods such as reusable bottles. It contains BPA. When polycarbonated containers are subjected to high temperatures, BPA can leak into the liquid or food they hold. The usage of polycarbonated plastics has decreased drastically as a result of the health concerns connected with BPA, which have been demonstrated in various studies (Proshad et al., 2018).

9.6 Micro and Nanoplastics (MNPs) in the Environment

MNPs, which may be created or synthesised from a variety of polymers, are everywhere. Researchers can better understand their effects on the environment and human health by knowing where they come from and where they go. Plastic particles having a diameter of less than 5 mm are now often referred to as MPs. Nanoparticles (NPs) are tiny polymeric particles of 1 to 100 or 1000 nanometres in diameter (Chae & An, 2017).

(a) *What are Microplastics?*

Microplastics are extremely small plastic particles that have the potential to harm the environment. These materials are not classed as a single category of plastic, however, they can be defined as materials having particles smaller than 5 millimetres. Microplastics can be found in a number of areas. Just a few examples include cosmetics, textiles, and industrial activities. Primary and secondary microplastics are the two forms of microplastics. These two categories are separated by the particle size of the microplastic material before and after it enters the environment. Primary microplastics comprise particles smaller than 5 millimetres before entering the environment, whereas secondary microplastics occur when larger plastic objects penetrate the environment. Both of these microplastic forms may be found in large quantities in the environment, mostly in aquatic and marine environments. Plastic materials deteriorate slowly over time, in general. Microplastics are digested, absorbed into, and deposited in the bodies and tissues of a wide spectrum of organisms as a result. Microplastics may be found in rivers and seas, as well as on the seafloor, in soil, and in human tissues (Madhu, 2020).

(b) *What are Nanoplastics?*

Polymer materials having particle sizes of less than 100 nanometres are known as nanoplastics. As a result of microplastic fragmentation, this material can reach the environment and become an invisible environmental problem in potentially vast numbers.

Because nanoplastics are so tiny, they may pass through cellular membranes and impair cell function, they represent a concern to the environment and human health. Polyethylene nanoplastics have been revealed to be able to integrate into the

Table 9.2 The differences between micro and nanoplastics

Microplastics VS nanoplastics			
	Definition	Particle size	Harm on environment
Microplastics	Microplastic materials are very small pieces of plastic that can pollute the environment	Less than 5 millimetre particle size	Considered as a harmful pollutant in the environment
Nanoplastics	Nanoplastics are polymer materials that contain less than 100 nanometre particle size	Less than 100 nanometre particle size	There is little information on adverse health effects of these materials in organisms, including humans

hydrophobic core of lipid bilayers, according to current study. These chemicals tend to penetrate through the epithelial barrier of fish, accumulating in organs such as the gall bladder, pancreas, and brain. Polystyrene nanoparticles have been found to trigger a stress response in Zebrafish, altering glucose and cortisol levels. However, there is no information on the negative effects of these chemicals on creatures, including humans (Madhu, 2020). The differences between micro- and nanoplastics are seen in Table 9.2.

9.7 Sources and Information of Micro and Nanoplastics (MNPs)

MNPs are categorised as primary or secondary based on how they were created. As main MNPs, processed plastic particles are commonly encountered in personal care products. PE microbeads are used as exfoliants in cosmetics, detergents, toothpastes, scrub face cleansers, and pharmaceutical carriers. Because the main MNPs used in consumer items primarily serve as a physical stimulant and a cleaning agent, they are easily discharged into the environment once they have been used. Furthermore, glitters, which are widely used in cosmetics, crafts, and textiles, are another substantial source of plastic pollution produced by primary MNPs, according to a recent research (Yurtsever, 2019).

The second source of MNPs is plastic debris, which dissolves from big pieces of plastic due to UV radiation, physical wear, and biodegradation in the environment. Once in the environment, plastics are exposed to UV radiation, which catalyses the photo-oxidation of polymers, making them brittle. As a consequence of additional contact with the wind, waves, and other abrasive interactions, the structural integrity of the plastics degrades even more, and MNPs are formed and released from the plastic surface via delimitation. These findings imply that both MPs and NPs can develop and accumulate during the degradation of throwaway plastic garbage (Kole et al., 2017). We described the environmental degradation of plastics as well as the formation of MPs and NPs in Fig. 9.1 based on these data.

9.8 Fate of Microplastics and Nanoplastics in the Environment

Microplastics are generated on land more than 80% of the time, with water accounting for less than 20% of the total. Because microplastics are light, indestructible, and float, they may travel great distances. The bulk of plastics that pollute the ocean comes from land, fishing and other aquaculture operations, and coastal tourism; in fact, about 800 million tonnes of plastics in the sea are considered to have begun on land (Jambeck et al., 2015). Because micro- and nanoplastics are so small, wastewater treatment methods can't filter them out, allowing them to enter rivers and seas, as well as the fresh water supply system.

Furthermore, micro- and nanoplastics are prevalent in soil, and natural erosion will lead to their entry into rivers and seas. According to UN Environment Program estimates, 275 million tonnes of plastic waste were produced in 2010, with an estimated 4.8–12.7 million tonnes entering water systems (Mattsson et al., 2018).

Micro- and nanoplastics are sourced from both main and secondary sources (Fig. 9.1). Exfoliants in cleansers and cosmetics, drug delivery particles in pharmaceuticals, and industrial air blasting are all examples of micro- and nanoplastics that

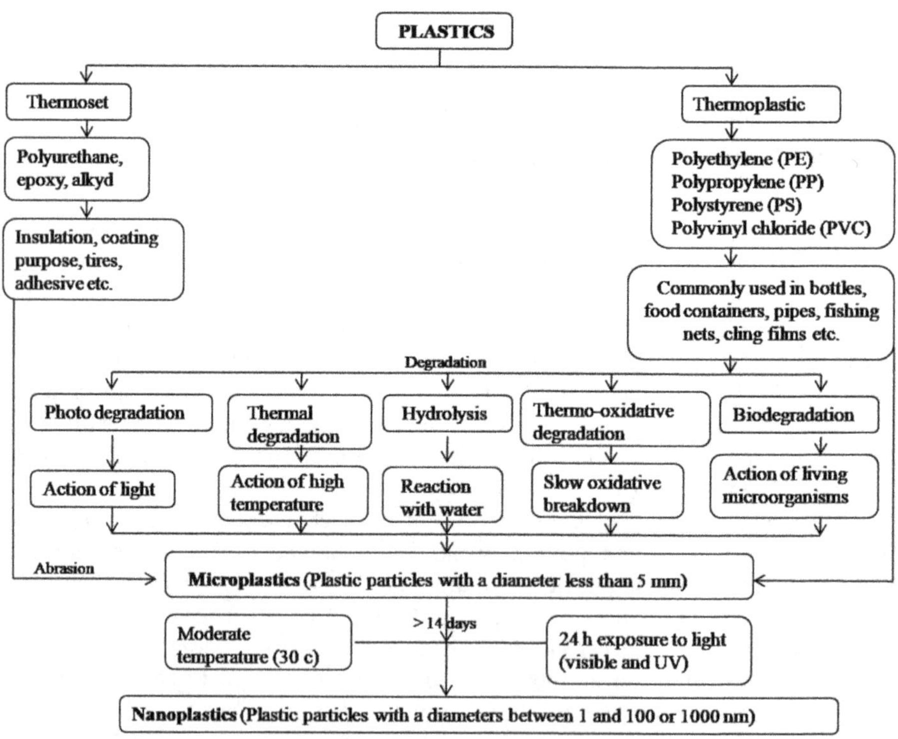

Fig. 9.1 The environmental degradation of plastics as well as the formation of MPs and NPs. (Modified from Baorong Jiang et al. 2020)

have been purposely created for consumer and industrial use. Secondary sources of micro- and nanoplastics include macroplastic products that break down into micron-sized and smaller particles that can be found on land and in water (Karbalaei et al., 2013).

Plastics can breakdown into micro- and nanoplastics through a number of mechanisms that are classed as biodegradation or non-biodegradation (Fig. 9.1). Thermal degradation, physical deterioration, light degradation, thermo-oxidative degradation, and hydrolysis are examples of non-biodegradation processes. Larger polymers break down into smaller bits as a result of physical deterioration, or weathering. Thermal deterioration, also known as heat degradation, is a man-made, commercial process, while thermal degradation, also known as heat degradation, is a man-made commercial process. Hydrolysis and photodegradation, on the other hand, are naturally occurring chemical reactions that use water molecules and UV-visible light to break down chemical bonds in plastics and convert them to monomeric forms. Non-biodegradation methods break down polymeric structures, altering mechanical properties and increasing specific surface area, allowing for better physical-chemical reactions and interactions with microbes. (Lucas et al., 2008).

Bacteria and other microorganisms found in the environment can also help with plastic biodegradation. Extracellular enzymes in these living animals have the ability to break down chemical bonds in plastics (Yuan et al., 2020). Smaller plastic particles with changed molecular structures are produced in this process, eventually leading to nano-sized plastics; a single gramme of macroplastic can generate billions of nanoplastic particles with dramatically increased surface area. These nanoplastics must be abundant in the marine ecology, given the vast volume of plastic that enters the oceans every day.

Furthermore, plastic rubbish fragmentation is thought to occur faster along the coast than in the oceans. One of the most prevalent methods for plastic to degrade is oxidation, which is caused by UV exposure from the sun. The process is accelerated when plastic is more directly exposed to UV radiation and higher temperatures on the shore than in the oceans (Corcoran et al., 2009). Furthermore, the presence of salt accelerates the breakdown of plastic in some coastal locations. Because of the high salt content and naturally occurring bacteria, plastics break down quicker in marine settings than in terrestrial ecosystems. The sources and fate of micro- and nanoplastics in the environment are depicted in Fig. 9.2.

9.9 Classification of MNPs and Their Potential Toxic Effects on Human Health

Several in vitro and in vivo investigations have revealed that micro- and nanoplastics can induce physical stress and damage, as well as apoptosis, necrosis, inflammation, oxidative stress, and immunological responses in humans (Table 9.3).

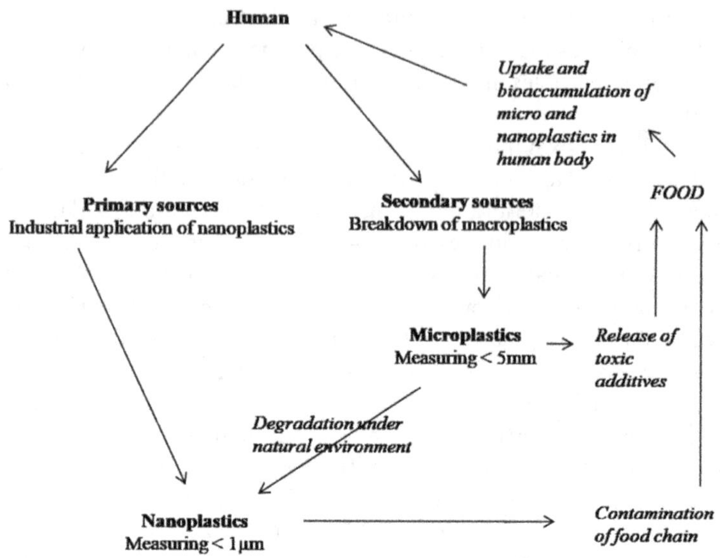

Fig. 9.2 Sources and fate of micro- and nanoplastics in the environment

MNPs are separated into main and secondary MNPs, as illustrated in Fig. 9.3, since they are either directly generated or derived from the fragmentation of larger polymers over time, as previously stated. Primary MNPs are mostly found in plastic pellets and personal care items, including microbeads. Primary MNPs can also be found in artificial grass, paints, washed fabric and wastewater, sewage sludge, plastic running tracks in schools, rubber roads, and vehicle tyre wear (Smyth et al., 2021). Microbeads are small plastic particles used in cosmetics and other personal care products. They are usually made of polyethylene, polypropylene, and polystyrene. They're employed in personal care products, as well as biomedical and health-science investigations, as scouring and exfoliating agents.

MNP microbeads are also used in personal care products and cosmetics as film-forming agents, functionalized polymers, hydrophilic agents, and silicones. Sphericity and particle size homogeneity provide a ball-bearing effect, which results in a smooth texture and spreadability, both of which are desirable aesthetic properties. These MNPs can be elliptical, irregularly frayed, and thread-like, and they can replace natural materials such as pumice stone and activated carbon.

Personal care products with coloured microbeads have a more attractive look. They've been recognised as a source of microplastics because, after being washed down the drain, they move unhindered through sewage treatment facilities, ending up in canals, rivers, streams, and other bodies of water. MNP microbeads are projected to account for 11% (2300 t/a) of the plastic waste dumped into the North Sea (Brzuska et al., 2015). MNPs debris has been recognised as a key source of pre-production resin pellets (granulate), which are principally used in the fabrication of industrial plastics. Cleaning, crushing, melting, sorting, and final moulding procedures all contribute to the production of these plastic pellets (Duncan et al., 2018).

Table 9.3 Summary of potential toxic effects of micro and nanoplastics on human health

Toxic effects	Details	Particle size	Characteristics of plastic particles
Inflammation	Upregulation of IL-8 expression Induced inflammation in human A549 lung cells	202 nm and 535 nm	Polystyrene particles
	Upregulation of IL-6 and IL-8 expression Enhanced inflammation in multiple human malignancies	20 nm, 44 nm, 500 nm, and 1000 nm	Unaltered/ Carboxylated polystyrene nanoparticles
	Altered expression of scavenger receptors M2 cells increased IL-10 production Increased TGFβ1 (M1) and energy metabolism (M2)	120 nm	Carboxylated and amino-modified polystyrene particles
	Increased the secretion of IL-6, IL-1β, and TNFα in murine macrophages	0.3 μm, 10 μm	Unaltered polyethylene particles
	Induced the expression of TNFα, IL-1, and RANKL Resulted in periprosthetic bone resorption	0.2 μm and 10 μm	Polyethylene particles from plastic prosthetic implants
	Induced inflammatory response at the implant area		
	Induced inflammation in the liver Induced adverse effects on neurotransmission	5 μm and 20 μm	Polystyrene microplastics particles
Oxidative stress and apoptosis	Strong interaction and aggregation with mucin Induced apoptosis in all intestinal epithelial cells	60 nm	Amine-modified polystyrene nanoparticles
	Induced ROS generation and ER stress Induced autophagic cell death of mouse macrophages and lung epithelial cells	60 nm	Cationic polystyrene nanoparticles
	Induced apoptosis of several human cell types	20 nm, 40 nm, 50 nm, and 100 nm	Unaltered or functionalized polystyrene
	Reduced cell viability with a reduction of ATP and increase of ROS concentrations	120 nm, 140 nm	Polyvinyl chloride (PVC) and poly (methyl methacrylate) (PMMA)

(continued)

Table 9.3 (continued)

Toxic effects	Details	Particle size	Characteristics of plastic particles
Metabolic homeostasis	Changes in amino acid and bile acid metabolism Induced gut microbiota dysbiosis and intestinal barrier dysfunction	5 μm	Pristine and fluorescent polystyrene microplastics
	Altered ion channel function and ionic homeostasis Activated basolateral K^+ channels Induced Cl^- and HCO^{3-} ion efflux	20 nm	Anionic carboxylated polystyrene nanoparticles
	Blocked vesicle transport and the distribution of cytokinesis-associated proteins	30 nm	Polystyrene nanoparticles
	Disrupted intestinal iron transport and cellular uptake	50 nm and 200 nm	Cationic polystyrene nanoparticles
	Reduction in hepatic ATP levels Impairment of energy metabolism	5 μm and 20 μm	Pristine polystyrene microparticles
	Metabolic disorder associated with gut microbiota dysbiosis and gut barrier dysfunction Increased the risks of metabolic disorder in the offspring	0.5 μm and 5 μm	Microplastics

Modified from Yee et al. (2021)

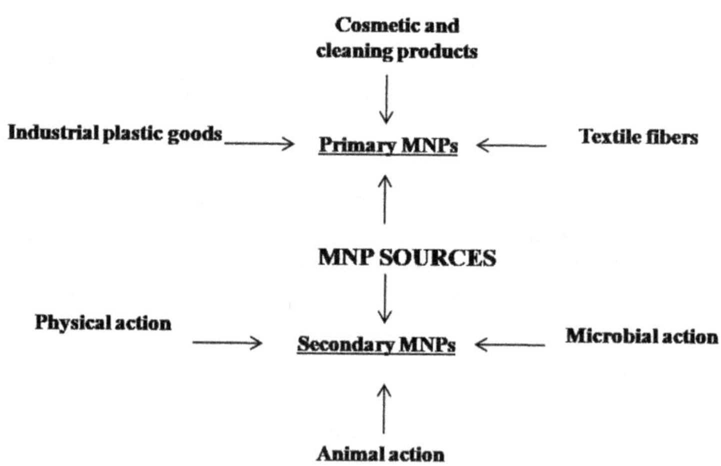

Fig. 9.3 Primary and secondary sources of MNPs

Secondary MNPs are formed when macroplastic polymers are broken by biodegradation, chemical (corrosion, photooxidation, temperature), and mechanical (abrasion erosion, wave action) breakdown processes (Karbalaei et al., 2018). Agriculture films, plastic bags and bottles, fishing equipment, shipping, vehicle tyre wear, and other large-scale plastic wastes have all been recognised as important sources of secondary MNPs. According to current estimations, secondary sources of MNPs account for the majority of MNPs in both aquatic and terrestrial contexts.

Vehicle tyre wear and road marking abrasions are some of the most prominent sources of environmental MNPs, thanks to the increased number of cars on the road (An et al., 2020; Kitahara and Nakata, 2020). Similarly, synthetic textile fibres have been proven to shed significant quantities of MNPs during laundry, which end up in water bodies and/or wastewater treatment facilities. According to De Falco et al. (2019), each kilogramme of washed fabric, around 124 to 308 mg of microplastics, or 640,000–1,500,000 MNPs particles, are released, depending on the kind of textile material.

In the construction business, plastic polymers used in cladding, insulating materials, and pipes emit a considerable amount of MNPs; nonetheless, it is assumed that MNPs are mostly created on construction sites as a consequence of carelessness or wrong storage (Battulga et al., 2019). MNPs are also used in specific applications such as paint removal, cleaning, roughening, and refining surfaces using sandblasting blasting chemicals (Battulga et al., 2019). The main sources of MNPs and their routes of introduction into different ecosystems are summarised in Table 9.1.

9.10 Occurrence and Effects of MNPs

9.10.1 Terrestrial Habitats

MNP pollution contributes significantly to one of the most pervasive and long-term human alterations to the terrestrial environment on the planet. As a result, substantial evidence of MNP pollution's direct and indirect harmful effects on a variety of terrestrial ecosystems has emerged in recent years (Ambrosini et al., 2019). It's important to remember that the vast majority of the plastic wastes that end up in aquatic bodies were made, used, and disposed of in an improper manner on land. As a result, terrestrial ecosystems are seen as vast MNP reservoirs that might expose terrestrial biota to a variety of MNPs, potentially changing geochemistry and causing environmental toxicity (Allouzi et al., 2021).

As a result, regular testing for the presence of MNPs in terrestrial materials, particularly soil samples, has gotten a lot of attention. The total amount and physicochemical characteristics of MNPs from terrestrial environments are shown in Table 9.4. In agricultural soil samples from South-eastern Germany, MNP abundance ranged from 0 to 1.25 particles/kg of dry soil samples, with a mean abundance of 0.34 particles/kg of soil (Piehl et al., 2018). Another research, which used four different soil samples from Shanghai suburbs, discovered a considerably higher amount of MNPs. Floodplain soil (256.7 62.2 particles/kg) had the greatest concentration of MNPs, followed by paddy soil (190 31.2 particles/kg), yellow-brown soil (155 95.2 particles/kg), and agricultural soil (36.6 41.7 particles/kg) (Liu et al., 2019a). MNPs in Swiss floodplain soils were reported to have a mean abundance of 593 particles/kg of soil samples in another investigation (Scheurer & Bigalke, 2018).

Recent studies have also discovered an increase in MNP contamination, which might be due to the continued rise in plastic pollution. For example, agricultural soil samples from China (Li et al., 2019) and Chile have higher MNP concentration, ranging from 80 to 3500 particles/kg. MNPs were identified at even higher quantities (320–12,560 particles/kg of soil) in vegetable fields in Wuhan, China, as well as another test location in the same city (22,000–690,000 particles/kg of soil). In addition to the normal soil samples, MNPs accumulate in wastewater treatment facilities, as demonstrated by their high content in sludge.

In Nanjing, China, for example, MNP particles per kilogramme of sludge ranged from 5553 to 13,460. (Li et al., 2019). Surprisingly, China has conducted the vast bulk of research on MNPs'terrestrial abundance, highlighting a data gap and the need for greater study in other regions of the world. MNPs have unquantifiable deleterious impacts on soil systems; they combine with other potentially hazardous elements and organic contaminants, multiplying their potential and having a substantial influence on the diverse terrestrial biota (Chai et al., 2020).

MNPs have been observed to interact with organic matter in the soil, altering soil physiochemical properties and polluting groundwater, resulting in decreased plant growth and overall productivity. MNPs also have a negative impact on soil fauna, specifically earthworms and nematodes, affecting their growth, reproduction, lifespan, and survival via a variety of toxicity mechanisms such as bioaccumulation, DNA damage, genotoxicity, gut microbiota dysbiosis, histopathological damage, metabolic disorders, neurotoxicity, oxidative stress, and reproductive toxicity. Sources of microplastics and nanoplastics in the environment are shown in Table 9.5.

This will have a severe impact on the normal ecological processes of these species, such as waste breakdown, nutrient cycling, and energy flow, posing significant environmental and health problems. Because of their high surface area-to-volume ratio and hydrophobicity, MNPs may potentially operate as pathogen and organic pollutant transporters on land, as they do in aquatic settings (Atugoda et al., 2021). Microorganisms associated with MNPs are particularly problematic for the environment because they act as a conduit for MNPs to be transmitted from the soil to plants and subsequently to other living beings via the food chain (Chai et al., 2020).

9.10.2 Food Chain

As plastic debris accumulates, the presence of micro- and nanoplastics in the food chain is becoming a health concern. Because of their extensive bioavailability and ubiquity in both aquatic and terrestrial habitats, micro- and nanoplastics are extremely likely to be discovered in a wide range of food products.

According to various studies, micro- and nanoplastics enter the human food chain through a number of routes, including animals ingesting them in their natural habitat, contamination during food preparation operations, and/or leaching from food and drink packaging. Honey, beer, salt, sugar, fish, shrimp, and bivalves have all been discovered to contain microplastic pieces. Using Fourier-transform infrared

Table 9.4 Abundance and physicochemical characteristics of MNPs in terrestrial habitats

MNPs abundance	Sample/ study region	MNPs size	MNPs shape	Method	Composition	References
320–12,560 particles/kg	Vegetable farmland Wuhan, China	<0.2–5 mm	Fibres microbeads	Micro-Raman spectroscopy	PA PP	Chen et al. (2020)
18,000–32,070 particles/kg	Sewage sludge Spain	NA	Fragment (80%) fibres films	μFTIR	PP PVC	van den Berg et al. (2020)
800 particles/kg	Farmland Heilongjiang, China	0.05–5 mm	NA	NA	LDPE	Zhang et al. (2020a)
22,000–690,000 particles/kg	Vegetable farmland Wuhan, China	10–500 μm	Fragment (52%) Bead (14%) Fibres (13.8%)	Stereo microscope	PP PS PA PVC	Zhou et al. (2019)
0–1.25 particles/kg	Farmland Franconia, Germany	1–5 mm	Films (43.75%), fragments (43.75%), fibres	FTIR	PP PS	Piehl et al. (2018)
593 particles/kg	Floodplain soils Switzerland	<500 μm–5 mm	NA	FTIR	PP PS PVC	Scheurer and Bigalke (2018)
930–1100 particles/kg	Agricultural soil Spain	150–250 μm	Fragment (80%) fibres films	μFTIR	PP PVC	van den Berg et al. (2020)
7387–47,047 m^{-2}	Vegetated wetland Washington, USA	<75 μm–5 mm	Fibres fragments	FTIR	PE Synthetic rubber	Helcoski et al. (2020)
80.3–1075.6 particles/kg	Agricultural soil Xinjiang, China	<5 mm	Films	μ-FTIR	PE	Huang et al. (2020)
420–1290 particles/kg	Agriculture soil Nanjing and Wuxi, China	0.02–0.25 mm	Fibres (38.9–65.1%) fragment	Stereo microscope	PP	Li et al. (2019)
5553–13,460 particles/kg	Sludge samples Nanjing, China	0.02–0.25 mm	Fibres(75.8–88.8%)	Stereo microscope	PE PET PAN	Li et al. (2019)
136.6–256.7 particles/kg	Vegetable farmland Shanghai, China	0.03–4.76 mm	Fibres 54 Films 7 fragments 38 granules	μ-FTIR	PP	Liu et al. (2019b)

(continued)

Table 9.4 (continued)

MNPs abundance	Sample/ study region	MNPs size	MNPs shape	Method	Composition	References
1100–3500 particles/kg	Agricultural fields Mellipilla, Chile	<2 mm	Fibres (97%), pellets films	Stereo microscope	Acryli Polyester Nylon LDPE PVC	Corradini et al. (2019)
1430–3410 particles/kg	Agricultural soil Shaanxi, China	<5 mm	Fibres granules	FTIR	PE PP PS PVC HDPE	Ding et al. (2020)
300 $mg{\cdot}kg^{-1}$–67,500 $mg{\cdot}kg^{-1}$	Soil of industrial area Sydney, Australia	20–40 μm	NA	FTIR	PVC PE PS	Fuller and Gautam (2016)

Modified from Amobonye et al. (2021)

spectroscopy, microplastics were discovered in tap, bottled, and spring water samples (FTIR). Microplastic particles less than 5 mm were found in 81%of tap water samples from 159 different locations throughout the world.

According to the findings, microplastic particles were identified in 93% of 259 individual bottles of water from 11 different brands and 27 different batches (Mason et al., 2018). The typical levels of microplastic contamination in food, according to data, are as follows: Seafood = 1.48 particles/g, sugar = 0.44 particles/g, honey = 0.10 particles/g, salt = 0.11 particles/g, alcohol = 32.27 particles/L, bottled water = 94.37 particles/L, tap water = 4.23 particles/L, and air = 9.80 particles/m3. According to these figures, the average person eats 39,000 to 52,000 microplastic particles each year, with age and gender influencing the total quantity. When plastic particle inhalation is taken into account, the total number of particles per year climbs to between 74,000 and 121,000. Furthermore, compared to those who exclusively drink tap water, who will absorb only 4000 extra particles, people who only drink bottled water may ingest an additional 90,000 particles (Cox et al., 2019). These data imply that the majority of microplastics consumed by humans come from the human food chain.

There is no data on the presence of nanoplastics in food since analytical tools are not yet accessible. However, it appears that nanoplastics will reach the food chain as a result of microplastic waste degradation. According to scientific investigations, nanoplastics were formed over time as the polystyrene drinking cup lids disintegrated. Microbial degradation may also occur in seas as a result of the presence of hydrocarbon-degrading microorganisms that have been shown to flourish on plastic garbage, forming a "plastisphere" environment. Because of the huge amount of plastic debris in the seas, microplastics will continue to disintegrate after they reach the water, resulting in the formation of new nanoplastic particles. Many things include commercially produced nanoplastics, which will eventually wind up as plastic trash in the seas and on land, making their way into the food supply chain (Contam E.P.O.C.I.T.F.C, 2016).

Table 9.5 Sources of microplastics and nanoplastics into the environment

Entry point into the environment	MNPs' source	Properties	Application	References
Drifting/surface runoff/loss	Plastic pellets	Granular plastics, commonly with a diameter of 2–5 mm and a regular shape	Raw materials and building blocks for nearly every plastic product	Karkanorachaki et al. (2018), Mendoza et al. (2018)
Drifting/loss	Fishing gears	Polyethylene, polyamide (nylon), and polypropylene monofilaments of between 0.1 and 5 mm	N/A	Dowarah and Devipriya (2019), Xue et al. (2020)
Drifting/surface runoff	Farming films	Microfilms more commonly from polyethylene between 0.03 and 10 mm	N/A	Liu et al. (2018), Zhang and Liu (2018)
Drifting/surface runoff	Construction industry	Fragments of typically polyamide, polyethylene, polyvinylchloride, and polyurethane polymers	N/A	Dehghani et al. (2017), Xu et al. (2020)
N/A	Sewage treatment effluents	Different kinds of MNPs from automobile tire wear, industrial production of plastic, personal care products, chemical laundry products, urban debris, etc. Ranging from 0.1 μm to 5 mm	N/A	Domogalla-Urbansky et al. (2019), Mak et al. (2020)
Drifting/surface runoff	Sports ground (artificial turfs and running tracks)	MNPs from propylene, polyamide 6 (PA6), PE, or polyurethane, styrene-butadiene rubber, thermoplastic elastomer, and green rubber and ethylene propylene diene monomer, which is made of EPDM. Usually between 0.5–2.5 mm	N/A	Wang et al. (2019), van Kleunen et al. (2020)

(continued)

Table 9.5 (continued)

Entry point into the environment	MNPs' source	Properties	Application	References
Surface runoff	Vehicle tire wear	Roundish, kidney-shaped, or elongated particles from styrene-butadiene rubber and natural rubber particles 0.01–350 μm	N/A	Sommer et al. (2018), Järlskog et al. (2020)
Drifting/surface runoff	Municipal debris	Fragments of plastic bags, plastic bottles, and other packaging materials. Of varying particle sizes from 0.1 μm to 5 mm	N/A	Welle and Franz (2018), Sobhani et al. (2020)
Wastewater/sewage sludge	Microbead personal care products	Microbeads varying in colour and ~100–1000 μm	Exfoliating, film-forming, hydrophilic, scrubbing agents, and functionalised polymers in personal care products as well as in biomedical applications	Napper et al. (2015), Nel et al. (2019)
Surface runoff	Paint	Between 0.3 and 5 mm from synthetic polymers mainly alkyds, epoxy resins, poly(acrylate/styrene) and polyurethane	Architectural coatings, marine coatings, automotive coatings, and road-marking paint	An et al. (2020), Gaylarde et al. (2021)
Wastewater/sewage sludge	Textile fabric	100–1000 μm MNPs mainly from acrylic, polyethylene terephthalate, and nylon fabrics	Enhanced appeal and functionality in synthetic fabrics	Carney Almroth et al. (2018), Cai et al. (2020)

Modified from Amobonye et al. (2021)

9.11 Leaching of Toxic Chemicals from Plastics

Plastics are normally made up of chemicals obtained from raw monomers as well as a number of additives to improve their properties. Plastics also absorb chemicals from their surroundings. As a result, these chemicals may leach from the polymer and into the environment. Polycyclic aromatic hydrocarbons (PAHs), for example, have been shown to be absorbed by microplastics and to have a variety of negative impacts on numerous species when ingested. A gradient function can speed up the diffusion of chemical species from a particle's core to its surface, where they can then leak into the surrounding environment. Despite the fact that these chemical species are ephemeral and dissolve swiftly in the human body, these plastic particles serve as a long-term "reservoir" for chemical leaks into tissues and bodily fluids (Sun et al., 2021).

Figure 9.4 depicts the dangerous chemical additions in plastic that have been related to human health, including bisphenol A (BPA), phthalates, triclosan, bisphenone, organotins, and brominated flame retardants (BFR). Although it is unknown if these chemicals truly enter biological tissues, some additives, such as nonylphenol and BPA, have been found to be eaten by marine biota. Leached BPA, a chemical additive widely utilised in the production of polycarbonate (PC) plastics and epoxy resin as the inner layer of food and beverage cans, has been linked to endocrine disorders and a severe impact on human health (Rani et al., 2015).

Importantly, studies have found that BPA leaches from PC into food and drinks and that BPA poisoning causes changes in liver function and insulin resistance, as well as harm to a developing foetus and abnormalities in reproductive and brain functioning. BPA binds to oestrogen receptors and acts as an antagonist, decreasing thyroid hormone-mediated transcription and influencing pancreatic beta-cell activity. Individuals who are exposed to BPA at concentrations ranging from 0.2 to 20 ng/mL are more likely to develop obesity, cardiovascular disease, and a number of other reproductive and developmental issues (Galloway, 2015).

Phthalate esters are used as plasticisers in the manufacturing of PVC polymers and plastisol to promote flexibility and durability. Humans may be harmed by phthalate esters, which can cause abnormal sexual development and birth defects. Furthermore, the US Environmental Protection Agency has designated butyl benzyl

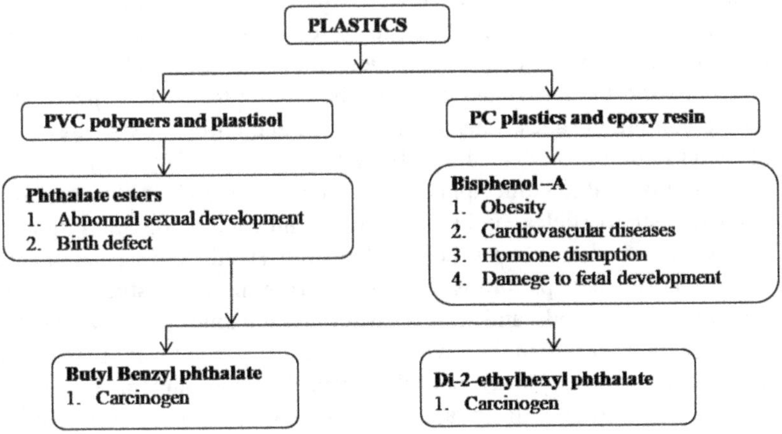

Fig. 9.4 Overview of the toxic effects of chemicals leaching from plastics

phthalate (BBP) as a likely carcinogen and di-2-ethylhexyl phthalate (DEHP) as a possible carcinogen.

9.12 Environmental Behaviour of MNPs and Its Effects on the Ecosystem

9.12.1 Nanoplastics

Currently, little is known about the origins of nanoplastic particles and their environmental behaviour, which encompasses all transit and transformation processes. Lambert et al. (2013), attribute this to the vast diversity of nanoplastic particle sources, physical properties, degradation types and timelines, and modes of movement. Because of the large number of large plastic compartments present in the environment, the amount of nano- and microplastic particles would surely rise.

Nanoplastic particles also undergo environmental transformations such as agglomeration with other particles, causing them to collect in diverse environmental compartments. They can also bind and release toxic substances into the environment, such as flame retardants or plasticizers. However, the contribution of nano- and microplastic-mediated chemicals to overall ambient creature exposure levels is minimal, and these organisms are not at risk (Koelmans et al., 2016).

Nanoplastic particles have the ability to interact with a wide range of environmental organisms. The bulk of research efforts has employed primary polystyrene nanoparticles in laboratory short-term investigations. Nanoplastic particles have been demonstrated to stick to organism surfaces and be absorbed into the gut, posing a risk to their regular function. They do not have significant acute effects, but they do have sublethal repercussions when exposed over long periods of time. In certain animals, nanoplastic particles have distinct effects from microplastic particles (Triebskorn et al., 2019). According to current estimates, nanoplastic levels in the environment are too low to cause problems in natural environments. However, because nanoplastic particle emissions into the environment are expected to skyrocket in the coming decades, long-term research and chronic exposure levels are required for a thorough risk evaluation (Besseling et al., 2018).

As a result, it is vital to reduce plastic emissions in general, and hence indirectly those of nanoplastic particles, and thus the environmental load. Building effective waste management systems across the world, as well as reducing mishandled plastic garbage, are all critical steps in this direction. Furthermore, substituting or forbidding one-way plastic goods and microplastics in consumer items would help to reduce the amount of plastic in the environment and, as a result, nanoplastics 'development. Ways to remove plastic particles from the environment are also being developed, however, these will not apply to nanoplastic particles (Besseling et al., 2018; Talvitie et al., 2017a, 2017b).

9.12.2 Microplastic Effects in Ecosystems

Plastics'chemical makeup, as well as their intrinsic link to human activities, can have a substantial influence on ecosystem functioning. Pathogenic and opportunistic organisms may be enhanced on microplastic surfaces in wastewater treatment plants (Kirstein et al., 2016). Microplastics that escape sewage treatment plants may end up in freshwater systems, where they might disseminate microorganisms (Talvitie et al., 2017a, 2017b). As a result, microplastics discharged from sewage treatment facilities into continental waters interact with a microbiome that is unique from that found on natural particles and possibly harmful (Kirstein et al., 2016). In this way, continental microplastics, like those found in the marine environment, might serve as a vector for disease outbreaks that have lately been detected (Kirstein et al., 2016). The effects of microplastics on terrestrial microbiomes are largely unknown and should be investigated further in the future.

Impacts are also conceivable in particle matter-dominated settings. According to Fuller and Gautam, topsoil near roadways and industrial areas in Sydney (Australia), may contain up to 7% microplastics by weight (Fuller & Gautam, 2016). The leaching of nonvolatile organochlorines from polyvinyl chloride (PVC) and other chlorinated microplastics (found to be 300 times greater than inorganic chloride in the examined region) induced geochemical alterations in soils at this pollution level (Fuller & Gautam, 2016). Microplastic levels of up to 60% in top soil in polluted regions, according to some experts, may be environmentally viable (Huerta Lwanga et al., 2017).

Due to low light and oxygen conditions, microplastics may remain for more than 100 years in soils (Horton et al., 2017). As a result, microplastics may interact with soil fauna by modifying their biophysical environment, posing a risk to their fitness and soil function (Huerta Lwanga et al., 2017). Microplastics have been shown to be transported horizontally and vertically inside soil by springtails and earthworms, for example. Microplastic exposure was connected to structural alterations in earthworm burrows, a measure of soil aggregation and function (Huerta Lwanga et al., 2017). Alterations in the biophysical environment influenced springtail activity, causing changes in their gut microbiomes (Zhu et al., 2018). As a result, even in the absence of clear evidence of ingestion, microplastic-exposed springtails had altered gut microflora, a changed isotopic signature (d15N and d13C), and negative effects on growth and reproduction (Zhu et al., 2018).

Other terrestrial creatures' biophysical environments may also be affected by microplastics. Microplastics may be present in commercially accessible (industrialised and locally manufactured) honey, for example (Liebezeit & Liebezeit, 2013). Microplastics may be ubiquitous in the inflorescences of numerous species, according to an examination into the origins of contamination.

As a result, plant-pollinator interactions are likely to be hampered by microplastics. The particles were actively translocated by the plants to the ovary when 6 lm polyester beads were placed into transmitting tracts of styles of several species' inflorescences (Sanders & Lord, 1989). Pollen tubes and plastic beads with appropriate pollen sizes can therefore flow unidirectionally (and occasionally intercellularly) to the ovules (Sanders & Lord, 1989). Microplastics'potential for detrimental environmental consequences on essential plant and pollinator ecological processes, on the other hand, has yet to be measured.

9.13 Uptake and Bioaccumulation of Microplastics and Nanoplastics in the Human Body

Inhalation, ingestion, and skin contact are the three major ways microplastics and nanoplastics enter the human body. Inhaled airborne microplastics from urban dust include synthetic textiles and rubber tyres, for example. Microplastics will be eaten since, as previously indicated, they are abundant in the food chain and water sources. Microplastics and nanoplastics cannot penetrate the epidermal barrier, but they can enter the body through wounds, sweat glands, and hair follicles. Although all three routes contribute to the overall quantity of microplastics and nanoplastics in the human body, the particles found in seafood and the environment provide the highest risk of absolute exposure. This is owing to the presence of long-term weathering of polymers, leaching of polymer chemical additives, residual monomers, pollution exposure, and pathogenic microbes in these settings.

9.14 Conclusion

Many governments are striving to prevent plastic pollution by decreasing the manufacturing of plastics and plastic items, restricting excessive packaging, trash collecting, and recycling. The tips below may be helpful in the battle against plastic pollution.

Policy Making Realistic regulations must be implemented and enforced to address and minimise persistent environmental contamination caused by plastics. This should include the need for a worldwide convention on plastic pollution, which would oblige plastic manufacturers to list all chemicals in their goods and warn customers about the possible health risks of such materials. Some of the toxic elements contained in plastic items should be classified using policies. There have been successful examples, such as the 1989 reclassification of chlorofluorocarbons (CFCs) as harmful (Montreal Protocol) and the 2004 reclassification of persistent organic pollutants (Stockholm Convention). Over the next seven years, over 200 nations will phase down the manufacture of CFCs and 30 other harmful compounds.

This categorization may also encourage research into innovative and safe alternatives, thus improving our plastic waste management and avoiding the buildup of plastic trash in the environment. It is also vital for the government to adopt and implement restrictions that limit the manufacturing, consumption, use, and final disposal of plastics, regardless of whether they are dangerous. To avoid zero diversion to landfills and indiscriminate disposal to the environment, reduce, reuse, and recycle must be employed at all phases (Comanita et al., 2016).

Plastic waste Management and Recycling Waste management is essential for decreasing the harmful impacts of plastic waste on the environment and public health. In order to prevent global litter and ocean pollution, appropriate plastic trash collection, treatment, and disposal must be improved. Inadequate landfill management will cause hazardous chemicals in plastic trash to leak into the environment, contaminating the soil, air, and groundwater.

If wastewater is effectively handled, microplastics will not reach the environment from landfills. Because the majority of treated wastewaters are released into rivers or seas, a prohibition, such as Annex V of the International Convention for the Prevention of Pollution from Ships (MARPOL) accord, is necessary to restrict the disposal of plastic trash into the sea (Mouat et al., 2010).

Education and Public Awareness The general public must be educated about the possible environmental and public health repercussions of plastic trash contamination. This will aid in the reduction of pollution and the preservation of environmental quality. The chemical components of plastic goods, as well as their health repercussions, must be understood. Educational curriculum at all levels must incorporate solutions for decreasing plastic pollution and waste management systems as information resources.

Bioplastics as Alternative Bioplastics are polymers manufactured from cellulose, which was developed by a British scientist in the 1850s. Weeds, hemp, plant oil, potato starch, cellulose, maize starch, and other biodegradable and non-biodegradable materials may now be used to make bioplastics (Reddy et al., 2013). Sugar-based bioplastics can break down under regular composting conditions. Bioplastics are more ecologically friendly than other forms of plastic since they consume fewer fossil fuels during manufacture.

Bioplastics are commonly utilised in consumer products for disposable items such as cutlery, bowls, pots, crockery, straws, and packaging, despite the fact that they have only been used commercially in a few applications. Bioplastics can theoretically replace petroleum-derived polymers in a variety of applications; however, bioplastics' cost and performance are concerns. Bioplastics may not be advantageous if there are no particular restrictions in place throughout the world to limit the usage of conventional plastics. Since 2011, Italy, for example, has made the use of biodegradable plastic bags for shopping mandatory.

In the manufacturing of bioplastics, wood, cellulose, sugar, and starch are utilised as alternatives for fossil fuels. This has made bioplastic manufacture more ecologically friendly and sustainable than traditional plastic production. Bioplastic manufacture minimises the usage of nonrenewable energy and greenhouse gas emissions (Gironi & Vincenzo, 2011).

We assumed that if manufacturers all over the globe adopted bioplastics, the problem of plastic waste creation, as well as the accompanying environmental and public health consequences, could be solved. Biodegradability with few or no hazardous consequences will go a long way toward maintaining our natural environment, protecting our planet's species, and making the globe a safer place for people.

References

Alabi, O. A., Ologbonjaye, K. I., Awosolu, O., & Alalade, O. E. (2019). Public and environmental health effects of plastic wastes disposal: A review. *Journal of Toxicology and Risk Assessmenmt, 5*(2), 1–13.

Allouzi, M. M. A., Tang, D. Y. Y., Chew, K. W., Rinklebe, J., Bolan, N., Allouzi, S. M. A., et al. (2021). Micro (nano) plastic pollution: The ecological influence on soil-plant system and human health. *Science of the Total Environment, 788*, 147815.

Ambrosini, R., Azzoni, R. S., Pittino, F., Diolaiuti, G., Franzetti, A., & Parolini, M. (2019). First evidence of microplastic contamination in the supraglacial debris of an alpine glacier. *Environmental Pollution, 253*, 297–301.

Amobonye, A., Bhagwat, P., & Raveendran, S. (2021). *Suren Singh and Santhosh Pillai*. Frontiers in Microbiology.

An, L., Liu, Q., Deng, Y., Wu, W., Gao, Y., & Ling, W. (2020). Sources of microplastic in the environment. In D. He & Y. Luo (Eds.), *Microplastics in terrestrial environments: Emerging contaminants and major challenges* (pp. 143–159). Springer.

Atugoda, T., Vithanage, M., Wijesekara, H., Bolan, N., Sarmah, A. K., Bank, M. S, et al. (2021). Interactions between microplastics, pharmaceuticals and personal care products: Implications for vector transport. *Environment International, 149*, 106367.

Baorong Jiang., Kauffman, A. E., Li, L., McFee, W., Cai, B., Weinstein, J., ... & Xiao, S. (2020). Health impacts of environmental contamination of micro-and nanoplastics: A review. *Environmental Health and Preventive Medicine, 25*, 1–15.

Battulga, B., Kawahigashi, M., & Oyuntsetseg, B. (2019). Distribution and composition of plastic debris along the river shore in the Selenga River basin in Mongolia. *Environmental Science and Pollution Research, 26*, 14059–14072.

Besseling, E., et al. (2018). Quantifying ecological risks of aquatic micro- and nanoplastic. *Critical Reviews in Environmental Science and Technology, 49*(1), 32–80.

Brzuska, K., Graaf, J. E. D., Kaumanns, J., Meyberg, M., & Schlatter, H. (2015). Use of microplastic beads in cosmetic products in Europe and their estimated emissions to the North Sea environment. *SOFW Journal, 141*, 40–46.

Cai, Y., Yang, T., Mitrano, D. M., Heuberger, M., Hufenus, R., & Nowack, B. (2020). Systematic study of microplastic fiber release from 12 different polyester textiles during washing. *Environmental Science & Technology, 54*, 4847–4855. https://doi.org/10.1021/acs.est.9b07395

Carney Almroth, B. M., Åström, L., Roslund, S., Petersson, H., Johansson, M., & Persson, N.-K. (2018). Quantifying shedding of synthetic fibers from textiles; a source of microplastics released into the environment. *Environmental Science and Pollution Research, 25*, 1191–1199. https://doi.org/10.1007/s11356-017-0528-7

Chae, Y., & An, Y. J. (2017). Effects of micro- and nanoplastics on aquatic ecosystems: Current research trends and perspectives. *Marine Pollution Bulletin, 124*, 624–632.

Chai, B., Wei, Q., She, Y., Lu, G., Dang, Z., & Yin, H. (2020). Soil microplastic pollution in an e-waste dismantling zone of China. *Waste Management, 118*, 291–301.

Chen, Y., Leng, Y., Liu, X., & Wang, J. (2020). Microplastic pollution in vegetable farmlands of suburb Wuhan, Central China. *Environmental Pollution, 257*, 113449. https://doi.org/10.1016/j.envpol.2019.113449

Comanita, E. D., Hillhor, R. M., Ghinia, C., & Garvilescu, M. (2016). Occurance of plastic waste in the environment: Ecological and health risks. *Environmental Engineering & Management Journal, 15*, 675–685.

Contam, E. P. O. C. I. T. F. C. (2016). Presence of microplastics and nanoplastics in food, with particular focus on seafood. *EFSA Journal, 14*, 14.

Corcoran, P. L., Biesinger, M. C., & Grifi, M. (2009). Plastics and beaches: A degrading relationship. *Marine Pollution Bulletin, 58*, 80–84. https://doi.org/10.1016/j.marpolbul.2008.08.022

Corradini, F., Meza, P., Eguiluz, R., Casado, F., Huerta-Lwanga, E., & Geissen, V. (2019). Evidence of microplastic accumulation in agricultural soils from sewage sludge disposal. *Science of the Total Environment, 671*, 411–420. https://doi.org/10.1016/j.scitotenv.2019.03.368

Cox, K. D., Covernton, G. A., Davies, H. L., Dower, J. F., Juanes, F., & Dudas, S. E. (2019). Human consumption of microplastics. *Environmental Science & Technology, 53*, 7068–7074. Difference Between Microplastics and Nanoplastics | Compare the Difference Between Similar Terms.

De Falco, F., Di Pace, E., Cocca, M., & Avella, M. (2019). The contribution of washing processes of synthetic clothes to microplastic pollution. *Scientific Reports, 9*, 1–11. https://doi.org/10.1038/s41598-019-43023-x

Dehghani, S., Moore, F., & Akhbarizadeh, R. (2017). Microplastic pollution in deposited urban dust, Tehran metropolis, Iran. *Environmental Science and Pollution Research, 24*, 20360–20371. https://doi.org/10.1007/s11356-017-9674-1

Ding, L., Zhang, S., Wang, X., Yang, X., Zhang, C., Qi, Y., et al. (2020). The occurrence and distribution characteristics of microplastics in the agricultural soils of Shaanxi Province, in North-Western China. *Science of the Total Environment, 720*, 137525. https://doi.org/10.1016/j.scitotenv.2020.137525

Domogalla-Urbansky, J., Anger, P. M., Ferling, H., Rager, F., Wiesheu, A. C., Niessner, R., et al. (2019). Raman microspectroscopic identification of microplastic particles in freshwater bivalves (Unio pictorum) exposed to sewage treatment plant effluents under different exposure scenarios. *Environmental Science and Pollution Research, 26*, 2007–2012. https://doi.org/10.1007/s11356-018-3609-3

Dowarah, K., & Devipriya, S. P. (2019). Microplastic prevalence in the beaches of Puducherry, India and its correlation with fishing and tourism/recreational activities. *Marine Pollution Bulletin, 148*, 123–133. https://doi.org/10.1016/j.marpolbul.2019.07.066

Duncan, E. M., Arrowsmith, J., Bain, C., Broderick, A. C., Lee, J., Metcalfe, K., et al. (2018). The true depth of the Mediterranean plastic problem: Extreme microplastic pollution on marine turtle nesting beaches in Cyprus. *Marine Pollution Bulletin, 136*, 334–340.

Fuller, S., & Gautam, A. (2016). A procedure for measuring microplastics using pressurized fluid extraction. *Environmental Science & Technology, 50*, 5774–5780.

Galloway, T. S. (2015). *Marine anthropogenic litter* (pp. 343–366). Springer. Micro- and Nanoplastics and Human Health.

Gaylarde, C. C., Neto, J. A. B., & Da Fonseca, E. M. (2021). Paint fragments as polluting microplastics: A brief review. *Marine Pollution Bulletin, 162*, 111847. https://doi.org/10.1016/j.marpolbul.2020.111847

Gigault, J., ter Halle, A., Baudrimont, M., Pascal, P. Y., Gauffre, F., Phi, T. L., et al. (2018). Current opinion: What is a nanoplastic? *Environmental Pollution, 235*, 1030–1034.

Gironi, F., & Vincenzo, P. (2011). Bioplastics and petroleum based plastics: Strengths and weaknesses. *Energy Sources, Part A: Recovery, Utilization and Environmental Effects, 33*, 1949–1959.

Helcoski, R., Yonkos, L. T., Sanchez, A., & Baldwin, A. H. (2020). Wetland soil microplastics are negatively related to vegetation cover and stem density. *Environmental Pollution, 256*, 113391. https://doi.org/10.1016/j.envpol.2019.113391

Horton, A. A., Svendsen, C., Williams, R. J., Spurgeon, D. J., & Lahive, E. (2017). Large microplastic particles in sediments of tributaries of the River Thames, UK – Abundance, sources and methods for effective quantification. *Marine Pollution Bulletin, 114*, 218–226.

Hosler, D., Burkett, S. L., & Tarkanian, M. J. (1999). Prehistoric polymers: Rubber processing trans R Soc ancient Mesoamerica. *Science, 284*, 1991–1998.

Huang, Y., Liu, Q., Jia, W., Yan, C., & Wang, J. (2020). Agricultural plastic mulching as a source of microplastics in the terrestrial environment. *Environmental Pollution, 260*, 114096. https://doi.org/10.1016/j.envpol.2020.114096

Huerta Lwanga, E., Gertsen, H., Gooren, H., Peters, P., Salanki, T., van der Ploeg, M., et al. (2017). Incorporation of microplastics from litter into burrows of Lumbricusterrestris. *Environmental Pollution, 220*, 523–531.

Jambeck, J. R., Geyer, R., Wilcox, C., Siegler, T. R., Perryman, M., Andrady, A., Narayan, R., & Law, K. L. (2015). Plastic waste inputs from land into the ocean. *Science, 347*, 768–771. https://doi.org/10.1126/science.1260352

Järlskog, I., Strömvall, A.-M., Magnusson, K., Gustafsson, M., Polukarova, M., Galfi, H., et al. (2020). Occurrence of tire and bitumen wear microplastics on urban streets and in sweep-sand and washwater. *Science of the Total Environment, 729*, 138950. https://doi.org/10.1016/j.scitotenv.2020.138950

Karbalaei, S., Golieskardi, A., Watt, D. U., Boiret, M., Hanachi, P., Walker, T. R., et al. (2020). Analysis and inorganic composition of microplastics in commercial Malaysian fish meals. *Marine Pollution Bulletin, 150*, 110687.

Karkanorachaki, K., Kiparissis, S., Kalogerakis, G. C., Yiantzi, E., Psillakis, E., & Kalogerakis, N. (2018). Plastic pellets, meso-and microplastics on the coastline of northern Crete: Distribution and organic pollution. *Marine Pollution Bulletin, 133*, 578–589. https://doi.org/10.1016/j.marpolbul.2018.06.011

Kirstein, I. V., Kirmizi, S., Wichels, A., Garin-Fernandez, A., Erler, R., Loder, M., & Gerdts, G. (2016). Dangerous hitchhikers? Evidence for potentially pathogenic Vibrio spp. on microplastic particles. *Marine Environmental Research, 120*, 1–8.

Kitahara, K.-I., & Nakata, H. (2020). Plastic additives as tracers of microplastic sources in Japanese road dusts. *Science of the Total Environment, 736*, 139694. https://doi.org/10.1016/j.scitotenv.2020.139694

Koelmans, A. A., et al. (2016). *Environmental Science & Technology, 50*(7), 3315–3326.

Kole, P. J., Lohr, A. J., Van Belleghem, F., & Ragas, A. M. J. (2017). Wear and tear of tyres: A stealthy source of microplastics in the environment. *International Journal of Environmental Research and Public Health, 14*, 1265.

Lambert, S., et al. (2013). *Science of the Total Environment, 447*, 225–234.

Lehner, R., Weder, C., Petri-Fink, A., & Rothen-Rutishauser, B. (2019). Emergence of nanoplastic in the environment and possible impact on human health. *Environmental Science & Technology, 53*, 1748–1765.

Li, Q., Wu, J., Zhao, X., Gu, X., & Ji, R. (2019). Separation and identification of microplastics from soil and sewage sludge. *Environmental Pollution, 254*, 113076. https://doi.org/10.1016/j.envpol.2019.113076

Liebezeit, G., & Liebezeit, E. (2013). Non-pollen particulates in honey and sugar. *Food Additives and Contaminants Part A-Chemistry Analysis Control Exposure & Risk Assessment, 30*, 2136–2140.

Liu, M., Lu, S., Song, Y., Lei, L., Hu, J., Lv, W., et al. (2018). Microplastic and mesoplastic pollution in farmland soils in suburbs of Shanghai, China. *Environmental Pollution, 242*, 855–862. https://doi.org/10.1016/j.envpol.2018.07.051

Liu, C., Li, J., Zhang, Y., Wang, L., Deng, J., Gao, Y., et al. (2019a). Widespread distribution of PET and PC microplastics in dust in urban China and their estimated human exposure. *Environment International, 128*, 116–124.

Liu, M., Song, Y., Lu, S., Qiu, R., Hu, J., Li, X., et al. (2019b). A method for extracting soil microplastics through circulation of sodium bromide solutions. *Science of the Total Environment, 691*, 341–347. https://doi.org/10.1016/j.scitotenv.2019.07.144

Lu, Y., Zhang, Y., Deng, Y., Jiang, W., Zhao, Y., Geng, J., et al. (2016). Uptake and accumulation of polystyrene microplastics in zebrafish (Danio rerio) and toxic effects in liver. *Environmental Science & Technology, 50*, 4054–4060.

Lucas, N., Bienaime, C., Belloy, C., Queneudec, M., Silvestre, F., & Nava-Saucedo, J.-E. (2008). Polymer biodegradation: Mechanisms and estimation techniques—A review. *Chemosphere, 73*, 429–442. https://doi.org/10.1016/j.chemosphere.2008.06.064

Madhu. (2020). Difference between microplastics and nanoplastics, A review artcle.

Mak, C. W., Tsang, Y. Y., Leung, M. M. L., Fang, J. K.-H., & Chan, K. M. (2020). Microplastics from effluents of sewage treatment works and stormwater discharging into the Victoria Harbor, Hong Kong. *Marine Pollution Bulletin, 157*, 111181. https://doi.org/10.1016/j.marpolbul.2020.111181

Mason, S. A., Welch, V. G., & Neratko, J. (2018). Synthetic polymer contamination in bottled water. *Frontiers in Chemistry, 6*, 407. https://doi.org/10.3389/fchem.2018.00407

Mattsson, K., Jocic, S., Doverbratt, I., & Hansson, L.-A. (2018). *Microplastic contamination in aquatic environments* (pp. 379–399). Elsevier. Nanoplastics in the Aquatic Environment.

Mendoza, L. M. R., Karapanagioti, H., & Álvarez, N. R. (2018). Micro (nanoplastics) in the marine environment: Current knowledge and gaps. *Current Opinion in Environmental Science & Health, 1*, 47–51. https://doi.org/10.1016/j.coesh.2017.11.004

Mouat, J., Lozano, R. L., & Bateson, H. (2010). *Economic impacts of marine litter*. KIMO International conference 105.

Napper, I. E., Bakir, A., Rowland, S. J., & Thompson, R. C. (2015). Characterisation, quantity and sorptive properties of microplastics extracted from cosmetics. *Marine Pollution Bulletin, 99*, 178–185. https://doi.org/10.1016/j.marpolbul.2015.07.029

Nel, H. A., Dalu, T., Wasserman, R. J., & Hean, J. W. (2019). Colour and size influences plastic microbead underestimation, regardless of sediment grain size. *Science of the Total Environment, 655*, 567–570. https://doi.org/10.1016/j.scitotenv.2018.11.261

Piehl, S., Leibner, A., Löder, M. G., Dris, R., Bogner, C., & Laforsch, C. (2018). Identification and quantification of macro-and microplastics on an agricultural farmland. *Scientific Reports, 8*, 1–9.

Plastics Europe. (2009). The compelling facts about plastics. An analysis of European plastics production, demand and recovery for 2008. Plastics Europe, European plastics converters, European association of plastics recycling and recovery organizations and European Plastics Recyclers.

Prata, J. C., da Costa, J. P., Lopes, I., Duarte, A. C., & Rocha-Santos, T. (2020). Environmental exposure to microplastics: An overview on possible human health effects. *Science of the Total Environment, 702*, 134455.

Proshad, R., Islam, M. S., Kormoker, T., Haque, M. A., Mahfuzur Rahman, M. D., et al. (2018). Toxic effects of plastic on human health and environment: A consequences of health risk assessment in Bangladesh. *International Journal of Health, 6*, 1–15.

Rani, M., Shim, W. J., Han, G. M., Jang, M., Al-Odaini, N. A., Song, Y. K., & Hong, S. H. (2015). Qualitative analysis of additives in plastic marine debris and its new products. *Archives of Environmental Contamination and Toxicology, 69*, 352–366.

Reddy, R. L., Reddy, S. V., & Gupta, G. A. (2013). Study of bio-plastics as green and sustainable alternative to plastics. *International Journal of Emerging Technology & Advanced Engineering, 3*, 82–89.

Rudyak, V. Y., Efimova, E. A., Guseva, D. V., & Chertovich, A. V. (2018). Thermoset polymer matrix structure and properties: Coarse-grained simulations. *Polymers (Basel)., 11*, 36.

Sanders, L. C., & Lord, E. M. (1989). Directed movement of latex-particles in the gynoecia of 3 species of flowering plants. *Science, 243*, 1606–1608.

Scheurer, M., & Bigalke, M. (2018). Microplastics in Swiss floodplain soils. *Environmental Science & Technology, 52*, 3591–3598.

Smyth, K., Drake, J., Li, Y., Rochman, C., Van Seters, T., & Passeport, E. (2021). Bioretention cells remove microplastics from urban stormwater. *Water Research, 191*, 116785.

Sobhani, Z., Lei, Y., Tang, Y., Wu, L., Zhang, X., Naidu, R., et al. (2020). Microplastics generated when opening plastic packaging. *Scientific Reports, 10*, 1–7. https://doi.org/10.1038/s41598-020-61146-4

Sommer, F., Dietze, V., Baum, A., Sauer, J., Gilge, S., Maschowski, C., et al. (2018). Tire abrasion as a major source of microplastics in the environment. *Aerosol and Air Quality Research, 18*, 2014–2028. https://doi.org/10.4209/aaqr.2018.03.0099

Sun, K., Song, Y., He, F., Jing, M., Tang, J., & Liu, R. (2021). A review of human and animals exposure to polycyclic aromatic hydrocarbons: Health risk and adverse effects, photo-induced toxicity and regulating effect of microplastics. *Science of the Total Environment, 773*, 145403.

Suresh, S. S., Mohanty, S., & Nayak, S. K. (2017). Composition analysis and characterization of waste polyvinyl chloride (PVC) recovered from data cables. *Waste Management, 60*, 100–111.

Sutter, J. D. (2016). *How top stop the sixth mass extinction*. CNN.

Talvitie, J., et al. (2017a). *Water Research, 123*, 401–407.

Talvitie, J., Mikola, A., Setala, O., Heinonen, M., & Koistinen, A. (2017b). How well is microlitter purified from wastewater? A detailed study on the stepwise removal of microlitter in a tertiary level wastewater treatment plant. *Water Research, 109*, 164–172.

Triebskorn, R., et al. (2019). *TrAC-Trends in Analytical Chemistry, 110*, 375–392.

Van Den Berg, P., Huerta-Lwanga, E., Corradini, F., & Geissen, V. (2020). Sewage sludge application as a vehicle for microplastics in eastern Spanish agricultural soils. *Environmental Pollution, 261*,114198. https://doi.org/10.1016/j.envpol.2020.114198

Van Kleunen, M., Brumer, A., Gutbrod, L., & Zhang, Z. (2020). A micro plastic used as infill material in artificial sport turfs reduces plant growth. *Plants, People, Planet, 2*, 157–166. https://doi.org/10.1002/ppp3.10071

Wang, T., Li, B., Zou, X., Wang, Y., Li, Y., Xu, Y., et al. (2019). Emission of primary microplastics in mainland China: Invisible but not negligible. *Water Research, 162*, 214–224.

Welle, F., & Franz, R. (2018). Microplastic in bottled natural mineral water– Literature review and considerations on exposure and risk assessment. *Food Additives & Contaminants. Part A, Chemistry, Analysis, Control, Exposure & Risk Assessment, 35*, 2482–2492. https://doi.org/10.1080/19440049.2018.1543957

World Economic Forum. (2016). *The new plastic economy, rethinking the future of plastics* (pp. 1–36).

Worm, B., Lotze, H. K., Jubinville, I., Wilcox, C., & Jambeck, J. (2017). Plastic as a persistant marine pollutant. *The Annual Review of Environment and Resources, 42*, 1–26.

Xu, C., Zhang, B., Gu, C., Shen, C., Yin, S., Aamir, M., et al. (2020). Are we underestimating the sources of microplastic pollution in terrestrial environment? *Journal of Hazardous Materials, 400*, 123228. https://doi.org/10.1016/j.jhazmat.2020.123228

Xue, B., Zhang, L., Li, R., Wang, Y., Guo, J., Yu, K., et al. (2020). Underestimated microplastic pollution derived from fishery activities and "hidden" in deep sediment. *Environmental Science & Technology, 54*, 2210–2217. https://doi.org/10.1021/acs.est.9b04850

Yee, M. S.-L., Hii, L.-W., Looi, C. K., Lim, W.-M., Wong, S.-F., Kok, Y.-Y., Wong, B.-K. T. . C.-Y., & Leong, C.-O. (2021). Impact of microplastics and nanoplastics on human health. *Nano Materials MDPI, 11*(2), 496.

Yuan, J., Ma, J., Sun, Y., Zhou, T., Zhao, Y., & Yu, F. (2020). Microbial degradation and other environmental aspects of microplastics/plastics. *Science of the Total Environment, 715*, 136968. https://doi.org/10.1016/j.scitotenv.2020.136968

Yurtsever, M. (2019). Tiny, shiny, and colorful microplastics: Are regular glitters a significant source of microplastics? *Marine Pollution Bulletin, 146*, 678–682.

Zhang, G. S., & Liu, Y. F. (2018). The distribution of microplastics in soil aggregate fractions in southwestern China. *Science of the Total Environment, 642*, 12–20. https://doi.org/10.1016/j.scitotenv.2018.06.004

Zhang, D., Ng, E. L., Hu, W., Wang, H., Galaviz, P., Yang, H., et al. (2020a). Plastic pollution in croplands threatens long-term food security. *Global Change Biology, 26*, 3356–3367. https://doi.org/10.1111/gcb.15043

Zhou, Y., Liu, X., & Wang, J. (2019). Characterization of microplastics and the association of heavy metals with microplastics in suburban soil of Central China. Science of the Total Environment, 694, 133798. https://doi.org/10.1016/j.scitotenv.2019.133798

Zhu, D., Chen, Q.-L., An, X.-L., Yang, X. R., Christie, P., Ke, X., et al. (2018). Exposure of soil collembolans to microplastics perturbs their gut microbiota and alters their isotopic composition. *Soil Biology and Biochemistry, 116*, 302–310.

Chapter 10
Ecological Impacts and Toxicity of Micro- and Nanoplastics in Agroecosystem

S. A. Aransiola, M. O. Victor-Ekwebelem, A. E. Ajiboye, S. S. Leh-Togi Zobeashia, U. J. J. Ijah, and O. J. Oyedele

Abstract Micro- and nanoplastics are fragments of small plastics that are of sizes 1–5000 microns and <1 μm and consist of carbon and hydrogen atoms chained together by polymer. Micro- and nanoplastics are environmental pollutants, and their degradation depends on the properties of plastics, soil type, environmental condition, and microbial community. Their presence in the agricultural system is an emerging concern, which is basically attributed to the ability of the plastics to penetrate the soil and contaminate the soil plants, and microflora and fauna which thereby affect the food chain and security. Micro- and nanoplastics pollution in agrosystems originates from human activities (agricultural practices and anthropogenic sources) and natural sources (atmospheric inputs and flooding). Micro- and nanoplastics contamination of soil plants alters the chemical, physical, and biological properties of the soil ecosystem due to increased adsorption capacity when in combination with another organic contaminant. In agricultural ecosystems, micro- and nanoplastics affect soil microbial activity, microbial biomass, functional diversity, and the cycling process of plant nutrient elements in the soil, which have an indirect effect on plant seed germination and growth. When ingested or in association with the soil biota, micro- and nanoplastics can influence the agro-functionality through effects on soil root-associated microbiome and root symbionts, soil

S. A. Aransiola (✉) · O. J. Oyedele
Bioresources Development Centre, National Biotechnology Development Agency, Ogbomoso, Nigeria

M. O. Victor-Ekwebelem
Department of Microbiology, Alex Ekwueme Federal University, Ndufu-Alike, Abakaliki, Nigeria

A. E. Ajiboye
Minerals and Metals Initiative Department of Chemical Engineering, University of Cape Town, Cape Town, South Africa

S. S. L.-T. Zobeashia
National Biotechnology Development Agency, Abuja, Nigeria

U. J. J. Ijah
Department of Microbiology, Federal University of Technology, Minna, Nigeria

N. R. Maddela et al. (eds.), *Micro and Nanoplastics in Soil*,
https://doi.org/10.1007/978-3-031-21195-9_10

structure, nutrient immobilization, contaminant adsorption, and diffusion which can directly impact the fertility of the agricultural soil, plant qualities, and its yield. Microplastics excessive accumulation can directly result in toxic risk effects, including the interruption of the nutrient transport system by the obstruction of the pores in the cell wall, alter the community diversity, activity of the soil biota, and inhibition of nitrification. Microplastics and nanoplastics contribute to a major distribution of toxic and harmful compounds to soil plants, soil fauna, and photosynthetic organisms.

Keywords Environment · Contamination · Microplastics · Nanoplastics · Agriculture · Soil

10.1 Introduction

It is undoubtedly true that the need for hygienic products and equipment for people's daily lives has led to an astronomical increase in the demand for plastics. It is also evident that the problem will only grow as almost 400 million tons of plastics are produced annually, with a mass projected to be more than double by 2050 (Lim, 2021; Auta et al., 2022). Though plastics are discarded within 3 years of their production, above one-third of the plastics is used in disposables (Paul et al., 2020). Indiscriminate disposal of these materials recently become an issue of concern globally because of their potential environmental hazard due to their resistance to degradation and long-term persistence in the environment. Doubtfully, if all plastic production were magically stopped from now on, the existing plastics; that is, a sizeable number of debris that has already accumulated in landfills and the ecosystem would continue degrading into tiny fragments that are impossible to collect or clean up, constantly raising micro and nanoplastic levels. This global problem affects probably all ecosystems as well as the complete food chain (Abioye et al., 2015a).

Officially, there is no published definition for micro and nanoplastics but they are generally considered to vary in size from 0.1 to 5000 mm and 1–100 nm respectively (EFSA, 2016; Hardy et al., 2018). Microplastics are primary and secondary by classification according to their source into the environment. The key source of primary microplastics is the raw materials used in the manufacture of plastic items, poor handling, accidental loss, run-off from processing facilities, and residues from the production process while secondary microplastics comes from fragmentation of larger plastic particles when exposed to the physical, chemical, and biological processes (Gouin et al., 2015). The first part of the environment at the receiving end of micro and nanoplastics is soil. Farming remains an important activity on soil, as food is the main sustenance of human beings. However, farmland may be particularly vulnerable to accumulation of micro and nanoplastics (Nizzetto et al., 2016) because agricultural plastics remain valuable items in farming, particularly in sustainable agriculture (United States Government, 2018). Applications include mulch films, high tunnel coverings, drip tape, row covers, silage films, packaging use for seedlings, fertilizers, etc. (Scarascia-Mugnozza et al., 2011; Steinmetz et al., 2016). In practice, these agricultural plastic materials employed are mostly polyethylene and non-biodegradable. After some time, the plastics become brittle because of weather-related effects and form small fragments that disperse in the soil that house plants and living organisms. More so, pesticides can adsorb plastic fragments which could be used in plasticizers or production by plastics manufacturing companies that may be released during breakdown of plastics, resulting in soil contamination (Bouwmeester et al., 2015).

However, toxicological effects of micro and nanoplastics on humans and animals have become a great concern to researchers globally because of their interconnection with the food chain in relation to the environment (Verma et al., 2016). On the terrestrial animals, recent reports suggest that microplastics in soil affect soil geochemistry and microorganisms (de Souza Machado et al., 2017). Earthworms and collembolans (hexapods) exposed to MPs underwent increased mortality and reduced growth and reproductive rates (Huerta et al., 2017; Zhu et al., 2018), and this will deprive the soil of its fertility and retard the plant growth. In all studies, terrestrial micro and nanoplastics have received less attention and their occurrence in soil is at higher levels than in marine systems, by at least a factor of four (Nizzetto et al., 2016; Horton et al., 2017; Alimi et al., 2018). In this regard, since everyone eats foods and inhales sand and dust, and it's not clear if an extra diet of plastic specks will harm us, it has become imperative to reveal the findings on the threat the micro and nanoplastics would have on plant-based food. This article also provides information into sources of micro and nanoplastics in soil, the potential effect on soil microflora and fauna, soil properties and toxicity, and evaluating the plant performance in a soil containing micro and nanoplastics.

10.2 Sources of Micro and Nanoplastics in Soil

Soil is a critical component of nearly every ecosystem but is often taken for granted. It plays a significant role in sustaining life on earth. More importantly, most of the foods that humans consume, except for what is harvested from marine environments, are grown in the earth's soils. The soil consists of chemical, physical, and biological environment leading to material transformation, possibly rendering initially harmful materials less dangerous and immobilizing others as a result of the interactions between these added materials and the organic and inorganic soil constituents (Nortcliff, 2012). However, numerous human activities result in different forms of soil pollution when materials are indiscriminately disposed on the soil.

10.2.1 *Micro- and Nanoplastics in Soil*

Globally, the pollutants of major concern in soil are micro and nanoplastics. In recent years, most of the reports in the scientific and popular press have focused upon the accumulation and fate of micro and nanoplastics in marine environments, particularly oceans whereas micro and nanoplastics are usually transported from land to other parts of ecosystems. Our major interest is on micro and nanoplastics in soil, and this chapter addressed the sources of micro and nanoplastics in soil. The problem of microplastic pollution in the soil is extremely serious. Horton et al. (2017) summarized the sources and hazardous maturing of micro and nanoplastics in the soil environment in recent years. However, one of the most serious risks is that microplastics may be ingested by humans and other organisms via the food chain. It is important to note that micro and nanoplastics are easily transported from their sources into soil environment and get transformed via the soil chemistry and impact negatively (Fig. 10.1).

Around the world, various sources of micro and nanoplastics in the soil have been identified to include agricultural production activities; that is, the use of agricultural films, and the addition of organic fertilizers, the industrial production activities, urban construction, daily life, atmospheric subsidence, automobile tire wear, among others.

(a) *Micro- and Nanoplastics from Industrial Activities*

About 9.7 billion people would share the world by 2050 (United Nation, 2019) with food supplies needed globally projected to increase by 50% (Guillard et al., 2018). As a result of this geometric increase in population, there will definitely be

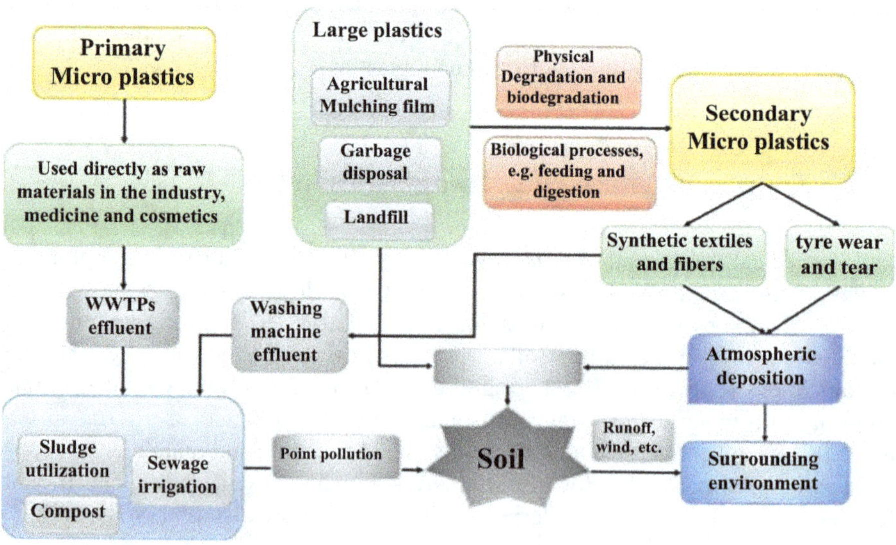

Fig. 10.1 Schematic diagram of the sources of microplastic in soil ecosystem (Yu et al., 2022)

an increase in food demand, which could drastically lead to an increase in food packaging material usage (Ncube et al., 2021).

Long ago, polymers have been beneficial to man and plastics appeared as the most important polymer helping human to survive (Al-Salem et al., 2009; Rahimi & García, 2017; Payne et al., 2019; Horodytska et al., 2019; Papadopoulou et al., 2019; European Bioplastics, 2021; Auta et al., 2022). These waste plastics are discarded indiscreetly, leading to soil contamination (Aarnio & Hämäläinen, 2008; Aransiola et al., 2013, 2021). Plastics used in packaging of materials often served a purpose but many are discarded and become post-consumer waste (Tencati et al., 2016; Ragaert et al., 2017). Discarded plastics find their way into incinerating plants, landfills, recycling plants, or the environment (Geyer et al., 2017; Abioye et al., 2015b). However, during recycling of plastics by mechanical operations, micro and nanoplastics could escape by contaminant separation, cutting/shredding, milling, floating, drying, washing, extrusion, quenching, and agglutination into the soil (Kumar et al., 2016).

(b) *Micro- and Nanoplastics from Agricultural Activities*

Farmlands have been identified to be vulnerable to accumulation of micro and nanoplastics (Nizzetto et al., 2016). Because most agricultural activities nowadays involve valuable uses of plastics, particularly in sustainable agriculture (United States Government, 2018). Agricultural film manufactured from polyethylene and polyvinyl chloride is commonly employed in agriculture. Applications include mulch films, high tunnel coverings, drip tape, row covers, silage films, packaging for seeds, seedlings, or fertilizers, among others (Scarascia-Mugnozza et al., 2011). Most plastics used for this purpose of production are non-biodegradable. The plastics become brittle due to sunlight and other weather-related effects and form small fragments that disperse in the environment due to flowing water and wind (Benedict, 2018). Often, plastic fragments become incorporated into the soil due to incomplete retrieval of the mulch film when it is being removed or recovered prior to disposal. Fragments of polyethylene are frequent in the soil in high concentrations of up to 60—300 kg/ha, which could rise to 500 kg/ha as reported in China (Bloomberg, 2017; Tremblay, 2018; Bouwmeester et al., 2015). The long-term fate of plastic fragments in soil is unknown. Recent reports predict that plastic fragments may reside in soils for over 100 years due to the near absence of oxygen and ultraviolet radiation from the sun (de Souza Machado et al., 2017).

Another aspect of agriculture that introduced micro and nanoplastics into soil is irrigation of farmland with wastewater and sewage sludges. Wastewater serves as a medium that transfers a large part of micro and nanoplastics materials from the sources; soil, industrial environment, roads to surface water bodies, and domestic environment (Carr et al., 2016; Ziajahromi et al., 2017; Mahon et al., 2017; Sun et al., 2019). Comparatively, more than 90% of microplastics found in wastewater are accumulated in sewage sludge, which in turn is used for land applications: the annual amount of microplastics entering the soil in this way is greater than that entering the oceans (Zhang et al., 2020a, b; Hurley & Nizzetto, 2018). Microplastic sources in domestic sewage are detergents and personal care products. About 20 million hectares of arable land worldwide are reported to be irrigated with untreated or partially treated sewage, and an estimated 10% of the world's population depends on food grown with contaminated wastewater (Abioye et al., 2021).

(c) *Micro- and Nanoplastics from Other Sources*

Runoff from roads or urban areas that is not captured by sewer systems can contaminate surrounding soils. Moreover, atmospheric transport has the potential to move plastics in the smallest size classes over long distances and likely contributes to a proportion of micro and nanoplastics in soils. Atmospheric deposition has been demonstrated in urban environments (Dris et al., 2016) and the transport of particles from landfill sites to soils has also been discussed (Rillig, 2012; Rocha & Duarte, 2015). More so, overbank deposition likely enriches alluvial soils with micro and nanoplastic particles. It has been shown that fluvial sediments comprise of high concentrations of microplastics (Castañeda et al., 2014; Leslie et al., 2017) which gathered during flooding (Veerasingam et al., 2016). This leads to accumulation of plastics in the soils. This likely represents a significant, albeit localized, source of microplastics.

10.3 Effects of Micro- and Nanoplastics to Soil Microflora and Fauna

Microflora plays a major role in biogeochemical transformation in the soil ecosystem. The activities carried out by soil microflora helps in the availability of nutrients to soil biota and also affect the physical and chemical properties of the soil ecosphere (Rillig et al., 2017b; Huerta Lwanga et al., 2018; Li et al., 2020). The microflora is affected when the soil environment is contaminated with plastics such as macro and nano.

Microplastics act as a vector for transport of harmful substances and microbes in soil. The movement of microplastics will affect the soil microflora as microbes attach to the plastics, colonize the surface area of the plastics, and interact with the pollutant. The harmful interaction of the plastic-microbial association will affect the ecological functions of the microflora, retard the growth of some organisms and alter the microbial community composition and density (Judy et al., 2019; Chai et al., 2020; Atugoda et al., 2021). In addition to microbial dispersal and DNA transfer in biofilm formation on microplastics, microbial attachment to microplastics can act as a vehicle of transport of plastics to plants (Hoellein et al., 2019; Chai et al., 2020). Soil contaminated with nanoplastics affect the metabolic activities and function of the microflora when the plastics (nano) enters the lipid membrane of the microflora (Rossi et al., 2014), which can be prevented by the microbes through protection mechanisms such as secretion of extracellular molecules that degrades the plastics contaminant or through changes of cell membrane structure (Henriques & Lov, 2007). In addition, nanoplastics can induce redundancy and resilience in the functional properties of the microorganism in the soil flora, which can impact the ecological activities including nutrient cycling, decomposition of organic matters, energy flow, and biofilm formation of the organism (Tang et al., 2018; Wang et al., 2020).

Microplastics can affect soil microflora via changes in the soil structure. Changes in soil structure can have direct effects on soil parameters, which can result in a shift in microbial community composition, abundance, and distribution. Microplastics contamination of the soil can change the soil porosity through oxygen flow and can also alter the soil profile (soil pore space), leading to loss of inherent soil microbes and an alteration in microbial structure (Machado et al., 2018; Judy et al., 2019).

Micro and nanoplastics are high in carbon content and contribute to carbon sources in the soil which impact the microbial biomass and also result in microbial immobilization (Rillig, 2018). The carbon in the plastics is relatively inert, which is due to slow decomposition of the plastics, especially microplastics. When degraded, the C:N ratio increases, this will lead to increase in microbial activities (Qi et al., 2018). As reported, increase in abundances and activity of Ascomycota fungi in the presence of readily degradable microplastics (polylactic acid).

Micro- and nanoplastics impact the symbiotic relationship between plant and microorganism in the soil. The plant growth, reproduction, and cycling of nutrients depend heavily on the interaction of soil biota and the root of a plant, especially on root colonizing microbes, which include mycorrhizal fungi-fixers and pathogens (Wagg et al., 2014; Powell & Rillig, 2018). The change caused by micro and nanoplastics in soil structure, affects the community diversity of the soil, rate of decomposition, and also the community abundance and distribution of root symbionts (Vallespir Lowery & Ursell, 2019). For instance, nanoplastics contamination affect the soil-borne stage root symbionts of arbuscular mycorrhizal fungi via toxic effects and functional activities of mycorrhizal. (Feng et al., 2013). Macro- and nanoplastics association in the rhizosphere affect root exudate quality and quantity by altering the length of the root, the weight, and oxidative responses to stress, cell wall pores disruption, and cell-to-cell relationship used for transport of nutrients (Jiang et al., 2019). Plastics also impact the ability of plants to uptake some soil microbiome and promote the expression of genes, including those required for chemotaxis and biofilm formation (Jing et al., 2014).

Soil microfauna plays an important role in decomposition of organic compounds, nutrient cycling, and food sources for lower trophic levels and are major drivers of chemical and biological processes in the soil. Micro and nanoplastics pollution of the soil microfauna can impede the growth rate, reproduction, lifespan, and survival of the fauna biota through ingestion, bioaccumulation, oxidative stress, DNA damage, neurotoxicity, genotoxicity, reproductive toxicity, histopathological damage, gut microbiota dysbiosis, and metabolic disorders. Micro and nanoplastics interact with other soil contaminants to produce combined toxicity to soil fauna; their presence reduces the abundance of microfauna such as soil microarthropod, nematodes, and protists. Higher concentration of plastics and continuous exposure have a greater negative impact on the soil fauna composition (Zhu et al., 2018). The exposure of Collembola and Nematodes to increased concentrations of microplastics results in high mortality and decreased growth and rate of reproduction (Zhu et al.,

2018). Nematode's consumption of plastics (microplastics) results in oxidative and intestinal damage that leads to a reduction in the level of calcium and an increase in the oxidative stress gene gst-4 in nematodes (Liu et al., 2018; Zhu et al., 2018). Earthworms are vehicles of movement for plastics especially, microplastics. Their burrowing activities transport plastics from the soil surface to the in-depth layers, promoting distribution and pollution which stunts their growth and development resulting from obstruction and irritation of the digestive tract, limiting nutrient absorption. A study reported by Cao et al. (2017) indicated that the growth of earthworms was significantly inhibited at concentrations of 1% and 2% and posed a toxic effect to them. A study carried out by Rodriguez-Seijo et al. (2017) reported an increase in lipids, polysaccharide and protein content, histopathological damage and immune system of earthworms at 10% concentration of polyethylene. In micro-arthropods, microplastics can prevent migration by filling up soil pores, while protists can easily absorb plastic fragments (nanoplastics) and colonize their surface and increase their abundance in the soil. But for microplastic, their uptake rate depends on the type, age, nutritional status, and the microplastics concentration (Rillig & Bonkowski, 2018; Lin et al., 2020).

10.4 Soil Properties and Micro- and Nanoplastics Toxicity

Nanoplastics are the smaller nanoscale fraction of plastics (defined as particles with a diameter below 100 nm) and are most likely to be incidentally produced from the fragmentation of larger plastic debris. The fragmentation of plastic debris down to the nanoscale may be caused by mechanical wear, heat, UV degradation, and, in some cases, biological factors (Ekvall, 2019; Hernandez, 2019; Lambert and Wagner (2016); Dawson, 2018). Microplastics (MPs), as defined by Frias and Nash (2018), are "synthetic solid particles or polymeric matrices, with regular or irregular shape and with size ranging from 1 μm to 5 mm, of either primary or secondary manufacturing origin, which are insoluble in water." Nanoplastics have traditionally been treated as a size-dependent extension of microplastics, but their size-dependent properties distinguish them from microplastics in terms of transport properties, interactions with light and natural colloids, analytical challenges, bioavailability, potential toxicity, and additive leaching times. In contrast to engineered nanomaterials (ENMs), which can include polymer formulations, accidentally produced nanoplastics in the environment are essentially debris from the environmental fragmentation of larger plastic objects (Gigault et al., 2021).

MPs are common contaminants that are causing increasing concern in aquatic and terrestrial ecosystems (Zhang et al., 2021a, b). MPs can harm organisms if they are released into the environment (Teuten et al., 2009). Depending on the properties of the microplastic, microplastics accumulation in soil could have an impact on the characteristics of the soil (Liu et al., 2017; Yi et al., 2020; Lozano et al., 2021a). The shape of microplastics may influence how it interacts with soil particles; for example, once fused into the aggregate soil, fibers have the ability to undermine the structure of the soil due to their linear shape (de Souza Machado et al., 2018; Rillig et al., 2017a; Lehmann et al., 2020; Rillig & Lehmann, 2020). Furthermore, microplastics' chemical properties, such as molecular chain arrangement and functional group, may affect their ability to absorb other chemicals, like antibiotics and toxic elements (Fred-Ahmadu et al., 2020), with potential consequences for the properties of soil and the activities of microorganisms (Pathan et al., 2020). For example, polyethylene (PE) had a high sorption capacity for phenanthrene (Wang & Wang, 2018), which could inhibit the activities of microorganisms in soil when combined with its nitrogen heterocyclic analogs (Zhao et al., 2021). Similarly, PVC, PP, and PE could have dissimilar capacities of chemical sorption according to research (Teuten et al., 2009; Wang et al., 2018). PE, for example, had a higher hydrophobic sorption capacity for organic compounds like pesticides and solvents than PE, PVC, or PET (Teuten et al., 2009; Fred-Ahmadu et al., 2020), while PS had a higher sorption capacity for Polycyclic Aromatic Hydrocarbons than PVC, PET, PP, or PE (Teuten et al., 2009; Fred-Ahmadu et al., 2020). PVC, on the other hand, could absorb more Cu than PS. As a result, the polymer type of microplastics may influence their effects on soil enzymatic activities. A Similarly, different polymer types (e.g., PE, PP, and PVC) may have different chemical sorption capacities, according to research (Teuten et al., 2009; Wang et al., 2018). PE, for example, had a higher sorption capacity for hydrophobic organic compounds like pesticides and solvents than PET, PVC, PE, or PP (Teuten et al., 2009; Fred-Ahmadu et al., 2020), while PS had a higher sorption capacity for Polycyclic Aromatic Hydrocarbons than PET, PVC, PE, or PP (Teuten et al., 2009; Fred-Ahmadu et al., 2020). PVC, on the other hand, could absorb more Cu than PS. As a result, the polymer type of microplastics may influence their effects on soil enzymatic activities. Soil properties: little is known about microplastics' effects on soil pH, a key soil parameter that could impact a range of microbial processes (Zhao et al., 2021).

Microplastics could alter the soil microbial communities (Huang et al., 2019; Fei et al., 2020), suggesting potential effects on soil respiration (Lozano et al., 2021a, b), affecting enzymatic activities. Microplastics have been shown to affect nutrient and/or substrate availability, most likely due to microplastic absorption or competition for physicochemical niches with microorganisms (Lozano et al., 2021b). The shape of the microplastics and the type of polymer it is made of may also play a role. According to the polymer type, PE and polyvinyl chloride (PVC) microplastics can enhance enzymes like urease and acid phosphatase (Huang et al., 2019; Fei et al., 2020), whereas PP, PES, and PVC can inhibit or enhance soil fluorescein diacetate hydrolase activity depending on the polymer type (Liu et al., 2017; Fei et al., 2020). Likewise, enzymes such as β-D-glucosidase and cellobiosidase (involved in cellulose degradation), N-acetyl-β-glucosaminidase (involved in chitin degradation), and phosphatase, which are related to C, N, P-cycling, could be negatively affected by microplastics (Lozano et al., 2021b).

10.5 Micro- and Nanoplastics Toxicity and Plant Performance

MPs pose a risk to human health because they are harmful to soil flora, which could affect plant growth and development. Sludge composts may act as a vehicle of MPs into soils and then enter soil biota, which in turn can influence the spread of MPs in the environment (Zhang et al., 2020a, b). Meanwhile, MPs can change the structure and properties of soil and the performance of plants. The effects of MPs on the physicochemical properties of soil adversely affect the root properties, growth, and nutrient absorption of plants (de Souza Machado et al., 2018).

Numerous studies validated that MPs delayed the germination of seeds, reduced plant growth, and induced the ecotoxicity and genetic toxicity of plants (Jiang et al., 2019), depending on the amounts of MPs present in the soil (Wang et al., 2020). Plants are the initial source of energy and organic matter in all ecosystems. MPs in the soil are migrated and accumulated in plants, and then transported into humans through the food chain, ultimately posing risks to the ecological environment and human health.

In general, toxicity mechanisms of MNPs hinge on the polymer size, surface characteristics, and type of the polymer. Plausible toxicity mechanisms mainly include membrane disruption, extracellular polymeric substance disruption, reactive oxygen species generation, DNA damage, cell pore blockage, lysosome destabilization, and mitochondrial depolarization. Positively charged nanoplastics accumulated in the root tips at lower levels than negatively charged sulfonic-acid-modified nanoplastics, but they induced a higher accumulation of reactive oxygen species and inhibited plant growth and seedling development. Negatively charged nanoplastics, on the other hand, were found frequently in the apoplast and xylem, implying that nanoplastics can accumulate in plants based on their surface charge (Sun et al., 2020).

10.6 Conclusion

In agricultural ecosystems, micro and nanoplastics affect soil microbial activity, microbial biomass, functional diversity, and the cycling process of plant nutrient elements in the soil which have an indirect effect on plant seed germination and growth. When ingested or in association with the soil biota, micro and nanoplastics can influence the agro-functionality through effects on soil root-associated microbiome and root symbionts, soil structure, nutrient immobilization, contaminant adsorption, and diffusion which can directly impact the fertility of the agricultural soil, quality of crops, and its yields.

References

Aarnio, T., & Hämäläinen, A. (2008). Challenges in packaging waste management in the fast-food industry. *Resources, Conservation and Recycling, 52*, 612–621. [CrossRef].

Abioye, O. P., Iroegbu, V. T., & Aransiola, S. A. (2015a). Biodegradation of methyl red by *Staphylococcus aureus* isolated from waste dump site. *Journal of Environmental Science and Technology, 8*, 131–138. https://doi.org/10.3923/jest.2015.131.138. https://scialert.net/fulltext mobile/?doi=jest.2015.131.138

Abioye, O. P., Adefisan, A. E., Aransiola, S. A., & Damisa, D. (2015b). Biosorption of chromium by *Bacillus subtilis* and *Pseudomonas aeruginosa* isolated from waste dump site. *Expert Opinion on Environmental Biology, 4*, 1. https://doi.org/10.4172/2325-9655.1000112. https://www.scitechnol.com/biosorption-of-chromium-by-bacillus-subtilis-and-pseudomonas-aeruginosa-isolated-from-waste-dump-site CNqL.php?article_id=2619

Abioye, O. P., Ijah, U. J. J., Aransiola, S. A., Auta, S. H., & Ojeba, M. I. (2021). Bioremediation of toxic pesticides in soil using microbial products. In R. Prasad, S. C. Nayak, R. N. Kharwar, & N. K. Dubey (Eds.), *Mycoremediation and environmental sustainability* (Fungal Biology). Springer. https://doi.org/10.1007/978-3-030-54422-5_1

Alimi, O. S., Farner, B. J., Hernandez, L. M., & Tufenkji, N. (2018). Microplastics and nanoplastics in aquatic environments. Aggregation, deposition, and enhanced contaminant transport. *Environmental Science & Technology, 52*, 1704–1724.

Al-Salem, S. M., Lettieri, P., & Baeyens, J. (2009). Recycling and recovery routes of plastic solid waste (PSW): A review. *Waste Management, 29*, 2625–2643. [CrossRef].

Aransiola, S. A., Ijah, U. J. J., & Abioye, O. P. (2013). Phytoremediation of lead polluted Soil by *Glycine max L. Applied and Environmental Soil Science, 2013*, 631619. https://doi.org/10.1155/2013/631619. https://www.hindawi.com/journals/aess/2013/631619/abs/

Aransiola, S. A., Ijah, U. J. J., Abioye, O. P., & Victor-Ekwebelem, M. O. (2021). ANAMMOX in wastewater treatment. In N. R. Maddela, L. C. García Cruzatty, & S. Chakraborty (Eds.), *Advances in the domain of environmental biotechnology* (Environmental and microbial biotechnology). Springer. https://doi.org/10.1007/978-981-15-8999-7_15

Atugoda, T., Vithanage, M., Wijesekara, H., Bolan, N., Sarmah, A. K., Bank M, You, S., & Ok, Y. S. (2021). Interactions between microplastics, pharmaceuticals and personal care products: Implications for vector transport. *Environment International, 149*, 106367. https://doi.org/10.1016/j.envint.2020.106367

Auta, S. H., Abioye, O. P., Aransiola, S. A., Bala, J. D., Chukwuemeka, V. I., Hassan, A., Aziz, A., & Fauziah, S. H. (2022). Enhanced microbial degradation of PET and PS microplastics under natural conditions in mangrove environment. A critical review. *Journal of Environmental Management, 304*(2022), 114273. https://doi.org/10.1016/j.jenvman.2021.114273

Benedict, C. (2018). *Biodegradable mulch breakdown in soil: Role of microbiology* [Online]. Available: https://www.youtube.com/embed/-EqrF2y9lh0

Bloomberg News. (2017). Plastic film covering most of China's farmland pollutes soil [Online]. *Bloomberg News*. Available: https://www.bloomberg.com/news/articles/2017-09-05/plastic film covering 12 of China farmland contaminates soil. Accessed 2 August 2018.

Bouwmeester, H., Hollman, P. C. H., & Peters, R. J. B. (2015). Potential health impact of environmentally released micro- and Nanoplastics in the human food production chain: Experiences from Nanotoxicology. *Environmental Science & Technology, 49*, 8932–8947.

Cao, D. D., Wang, X., Luo, X. X., Liu, G. C., & Zheng, H. (2017). Effects of polystyrene microplastics on the fitness of earthworms in an agricultural soil. *IOP Conference Series: Earth and Environmental Science, 61*, 012148.

Carr, S. A., Liu, J., & Tesoro, A. G. (2016). Transport, and fate of microplastic particles in wastewater treatment plants. *Water Research, 91*, 174–182. [CrossRef].

Castañeda, R. A., Avlijas, S., Simard, M. A., & Ricciardi, A. (2014). Microplastic pollution in St. Lawrence River sediments. *Canadian Journal of Fisheries and Aquatic Sciences, 71*, 1767–1771.

Chai, B., Wei, Q., She, Y., Lu, G., Dang, Z., & Yin, H. (2020). Soil microplastic pollution in an e-waste dismantling zone of China. *Waste Management, 118*, 291–301. https://doi.org/10.1016/j.wasman.2020.08.048

Dawson, A. L. (2018). Turning microplastics into nanoplastics through digestive fragmentation by Antarctic krill. *Nature Communications, 9*, 1001.

de Souza Machado, A. A., Kloas, W., Zarfl, C., Hempel, S., & Rillig, M. C. (2017). Microplastics as an emerging threat to terrestrial ecosystems. *Global Change Biology, 24*, 1405–1416.

de Souza Machado, A. A., Lau, C. W., Till, J., Kloas, W., Lehmann, A., Becker, R., et al. (2018). Impacts of microplastics on the soil biophysical environment. *Environmental Science & Technology, 52*(17), 9656–9665. https://doi.org/10.1021/acs.est.8b02212. [CrossRef].

Dris, R., Gasperi, J., Saad, M., Mirande, C., & Tassin, B. (2016). Synthetic fibers in atmospheric fallout: A source of microplastics in the environment? *Marine Pollution Bulletin, 104*, 290–293.

EFSA. (2016). Presence of microplastics and nano-plastics in food, with particular focus on seafood. *EFSA Journal, 14*(6), e04501.

Ekvall, M. T. (2019). Nanoplastics formed during the mechanical breakdown of daily-use polystyrene products. *Nanoscale Advances, 1*, 1055–1061.

European Bioplastics. (2021). Bioplastics facts and figures. *Recycling, 6*(12), 20–25. Available online: https://docs.european-bioplastics.org/publications/EUBP Facts and figures

Fei, Y., Huang, S., Zhang, H., Tong, Y., Wen, D., Xia, X., et al. (2020). Response of soil enzyme activities and bacterial communities to the accumulation of microplastics in an acid cropped soil. *Science of The Total Environment, 707*, 135634. https://doi.org/10.1016/j.scitotenv.2019.135634

Feng, Y., Cui, X., He, S., Dong, G., Chen, M., Wang, J., & Lin, X. (2013). The role of metal nanoparticles in influencing arbuscular mycorrhizal fungi effects on plant growth. *Environmental Science & Technology, 47*, 9496–9504.

Fred-Ahmadu, O. H., Bhagwat, G., Oluyoye, I., Benson, N. U., Ayejuyo, O. O., & Palanisami, T. (2020). Interaction of chemical contaminants with microplastics: Principles and perspectives. *Science of The Total Environment, 706*, 135978. https://doi.org/10.1016/j.scitotenv.2019.135978

Frias, J., & Nash, R. (2018). Microplastics: Finding a consensus on the definition. *Marine Pollution Bulletin, 138*, 145–147.

Geyer, R., Jambeck, J. R., & Law, K. L. (2017). Production, use, and fate of all plastics ever made. *Science Advances, 3.7*(2017), e1700782.

Gigault, J., El Hadri, H., Nguyen, B., Grassl, B., Rowenczyk, L., Tufenkji, N., Siyuan, F., & Wiesner, M. (2021). Nanoplastics are neither microplastics nor engineered nanoparticles. *Nature Nanotechnology, 16*, 501–507.

Gouin, T., Avalos, J., Brunning, I., Brzuska, K., de Graaf, J., Kaumanns, J., Koning, T., Meyberg, M., Rettinger, K., Schlatter, H., Thomas, J., van Welie, R., & Wolf, T. (2015). Use of microplastic beads in cosmetic products in Europe and their estimated emissions to the North Sea environment. *International Journal for Applied Science., 141*(4), 40–46.

Guillard, V., Gaucel, S., Fornaciari, C., Angellier-Coussy, H., Buche, P., & Gontard, N. (2018). The next generation of sustainable food packaging to preserve our environment in a circular economy context. *Frontiers in Nutrition, 5*, 121. [CrossRef] [PubMed].

Hardy, A., Benford, D., Halldorsson, T., Jeger, M. J., Knutsen, H. K., More, S., Naegeli, H., Noteborn, H., Ockleford, C., Ricci, A., Rychen, G., Schlatter, J. R., Silano, V., Solecki, R., Turck, D., Younes, M., Chaudhry, Q., Cubadda, F., Gott, D., Oomen, A., Weigel, S., Karamitrou, M., Schoonjans, R., & Mortensen, A. (2018). Guidance on risk assessment of the application of nanoscience and nanotechnologies in the food and feed chain: Part 1, human and animal health. *EFSA Journal, 16*(7), e05327. https://doi.org/10.2903/j.efsa.2018.5327

Henriques, I. D. S., & Lov, N. G. (2007). The role of extracellular polymeric substances in the toxicity response of activated sludge bacteria to chemical toxins. *Water Research, 41*, 4177–4185.

Hernandez, L. M. (2019). Plastic teabags release billions of microparticles and nanoparticles into tea. *Environmental Science & Technology, 53*, 12300–12310.

Hoellein, T. J., Shogren, A. J., Tank, J. L., Risteca, P., & Kelly, J. J. (2019). Microplastic deposition velocity in streams follows patterns for naturally occurring allochthonous particles. *Scientific Reports, 9*, 3740.

Horodytska, O., Cabanes, A., & Fullana, A. (2019). Plastic waste management: Current status and weaknesses. In F. Stock, G. Reifferscheid, N. Brennholt, & E. Kostianaia (Eds.), *The handbook of environmental chemistry* (pp. 1–18). Springer.

Horton, A. A., Walton, A., Spurgeon, D. J., Lahive, E., & Svendsen, C. (2017). Microplastics in freshwater and terrestrial environments: Evaluating the current understanding to identify the knowledge gaps and future research priorities. *Science of the Total Environment, 586*, 127–141.

Huang, Y., Zhao, Y., Wang, J., Zhang, M., Jia, W., & Qin, X. (2019). LDPE microplastic films Alter microbial community composition and enzymatic activities in soil. *Environmental Pollution, 254*(PtA), 112983. https://doi.org/10.1016/j.envpol.2019.112983

Huerta Lwanga, E., Thapa, B., Yang, X., Gertsen, H., Salánki, T., Geissen, V., & Garbeva, P. (2018). Decay of low-density polyethylene by bacteria extracted from earthworm's guts: A potential for soil restoration. *Science of the Total Environment, 624*, 753–757.

Huerta, L. E., Gertsen, H., Gooren, H., Peters, P., Salánki, T., van der Ploeg, M., Besseling, E., Koelmans, A. A., & Geissen, V. (2017). Incorporation of microplastics from litter into burrows of *Lumbricus terrestris*. *Environmental Pollution, 220*, 523–531.

Hurley, R. R., & Nizzetto, L. (2018). Fate and occurrence of micro(nano)plastics in soils: Knowledge gaps and possible risks. *Current Opinion in Environmental Science & Health, 1*, 6–11. https://doi.org/10.1016/j.coesh.2017.10.006

Jiang, X., Chen, H., Liao, Y., Ye, Z., Li, M., & Klobuˇcar, G. (2019). Ecotoxicity and genotoxicity of polystyrene microplastics on higher plant Vicia faba. *Environmental Pollution, 250*, 831–838.

Jing, H., Mezgebe, B., Aly Hassan, A., Sahle-Demessie, E., Sorial, G. A., & Bennett-Stamper, C. (2014). Experimental and modeling studies of sorption of ceria nanoparticle on microbial biofilms. *Bioresource Technology, 161*, 109–117.

Judy, J. D., Williams, M., Gregg, A., Oliver, D., Kumar, A., Kookana, R., & Kirby, J. K. (2019). Microplastics in municipal mixed-waste organic outputs induce minimal short to long-term toxicity in key terrestrial biota. *Environmental Pollution, 252*, 522–531.

Kumar, G. M., Irshad, A., Raghunath, B., & Rajarajan, G. (2016). Waste management in food packaging industry. In *Integrated waste management in India* (pp. 265–277). Springer.

Lambert, S., & Wagner, M. (2016). Formation of microscopic particles during the degradation of diferent polymers. *Chemosphere, 161*, 510–517.

Lehmann, A., Leifheit, E. F., Gerdawischke, M., & Rillig, M. C. (2020). Microplastics have shape- and polymer-dependent effects on soil processes. *BioRxiv, 12*, 339–344. https://doi.org/10.1101/2020.06.02.130054

Leslie, H. A., Brandsma, S. H., van Velzen, M. J. M., & Vethaak, A. D. (2017). Microplastics enroute: Field measurements in the Dutch river delta and Amsterdam canals, wastewater treatment plants, North Sea sediments and biota. *Environment International, 101*, 133–142.

Li, M., Yu, H., Wang, Y., Li, J., Ma, G., & Wei, X. (2020). QSPR models for predicting the adsorption capacity for microplastics of polyethylene, polypropylene and polystyrene. *Scientific Reports, 10*, 14597.

Lim, X. (2021). Microplastics are everywhere, but are they harmful? *Nature, 593*, 22–25.

Lin, D., Yang, G., Dou, P., Qian, S., Zhao, L., Yang, Y., & Fanin, N. (2020). Data from: Microplastics negatively affect soil fauna but stimulate microbial activity: Insights from a field-based microplastic addition experiment. *Dryad Digital Repository*. https://doi.org/10.5061/dryad.1zcrjdfpw

Liu, H., Yang, X., Liu, G., Liang, C., Xue, S., Chen, H., et al. (2017). Response of soil dissolved organic matter to microplastic addition in Chinese loess soil. *Chemosphere, 185*, 907–917. https://doi.org/10.1016/j.chemosphere.2017.07.064

Liu, M., Lu, S., Song, Y., Lei, L., Hu, J., Lv, W., Zhou, W., Cao, C., Shi, H., Yang, X., & He, D. (2018). Microplastic and mesoplastic pollution in farmland soils in suburbs of Shanghai, China. *Environmental Pollution, 242*, 855–862.

Lozano, Y. M., Lehnert, T., Linck, L. T., Lehmann, A., & Rillig, M. C. (2021a). Microplastic shape, polymer type, and concentration affect soil properties and plant biomass. *Frontiers in Plant Science, 12*, 616645. https://doi.org/10.3389/fpls.2021.616645

Lozano, Y. M., Aguilar-Trigueros, C. A., Onandia, G., Maaß, S., Zhao, T., & Rillig, M. C. (2021b). Effects of microplastics and drought on soil ecosystem functions and multifunctionality. *Journal of Applied Ecology, 58*, 988–996. https://doi.org/10.1111/1365-2664.13839

Machado, A. A. S., Kloas, W., Zarfl, C., Hempel, S., & Rillig, M. C. (2018). Microplastics as an emerging threat to terrestrial ecosystems. *Global Change Biology, 24*, 1405–1401.

Mahon, A. M., O'Connell, B., Healy, M. G., O'Connor, I., Officer, R., Nash, R., & Morrison, L. (2017). Microplastics in sewage sludge: Effects of treatment. *Environmental Science & Technology, 51*, 810–818. [CrossRef].

Ncube, L. K., Ude, A. U., Ogunmuyiwa, E. N., Zulkifli, R., & Beas, I. N. (2021). An overview of plastic waste generation and Management in Food Packaging Industries. *Recycling, 6*, 12. https://doi.org/10.3390/recycling6010012

Nizzetto, L., Futter, M., & Langaas, S. (2016). Are agricultural soils dumps for microplastics of urban origin? *Environmental Science & Technology, 50*, 10777–10779.

Nortcliff, S. (2012). Definition, function, and utilization of soil. In *Ullmann's encyclopaedia of industrial chemistry* (Vol. 33(1)). Wiley-VCH Verlag GmbH & Co. KGaA. https://doi.org/10.1002/14356007.b07_613.pub3

Papadopoulou, E. L., Paul, U. C., Tran, T. N., Suarato, G., Ceseracciu, L., Marras, S., d'Arcy, R., & Athanassiou, A. (2019). Sustainable active food packaging from poly(lactic acid) and cocoa bean shells. *Acs Applied Materials & Interfaces, 11*, 31317–31327. [CrossRef] [PubMed].

Pathan, S. I., Arfaioli, P., Bardelli, T., Ceccherini, M. T., Nannipieri, P., & Pietramellara, G. (2020). Soil pollution from micro- and Nanoplastic debris: A hidden and unknown biohazard. *Sustainability, 12*, 7255. https://doi.org/10.3390/su12187255

Paul, M. B., Stock, V., Cara-Carmona, J., Lisicki, E., Shopova, S., Fessard, V., Braeuning, A., Sieg, H., & Bashert, L. (2020). Micro- and nanoplastics current state of knowledge with the focus on oral uptake and toxicity. *Nanoscale Advances, 2*(10), 4350–4367. https://doi.org/10.1039/d0na00539h

Payne, J., McKeown, P., & Jones, M. D. (2019). A circular economy approach to plastic waste. *Polymer Degradation and Stability, 165*, 170–181. [CrossRef].

Powell, J. R., & Rillig, M. C. (2018). Biodiversity of arbuscular mycorrhizal fungi and ecosystem function. *New Phytologist, 220*, 1059–1075.

Qi, Y., Yang, X., Pelaez, A. M., Huerta Lwanga, E., Beriot, N., Gertsen, H., Garbeva, P., & Geissen, V. (2018). Macro- and micro- plastics in soil-plant system: Effects of plastic mulch film residues on wheat (Triticum aestivum) growth. *Science of the Total Environment, 645*, 1048–1056.

Ragaert, K., Delva, L., & Van Geem, K. (2017). Mechanical and chemical recycling of solid plastic waste. *Waste Management, 69*, 24–58. [CrossRef] [PubMed].

Rahimi, A., & García, J. M. (2017). Chemical recycling of waste plastics for new materials production. *Nature Reviews Chemistry, 1*, 1–11. [CrossRef].

Rillig, M. C. (2012). Microplastic in terrestrial ecosystems and the soil. *Environmental Science & Technology, 46*, 6453–6454.

Rillig, M. C. (2018). Microplastic disguising as soil carbon storage. *Environmental Science & Technology, 52*, 6079–6080.

Rillig, M. C., & Bonkowski, M. (2018). Microplastic and soil protists: A call for research. *Environmental Pollution, 241*, 1128–1131.

Rillig, M. C., & Lehmann, A. (2020). Microplastic in terrestrial ecosystems. *Science, 368*(6498), 1430–1431. https://doi.org/10.1126/science.abb5979

Rillig, M. C., Ziersch, L., & Hempel, S. (2017a). Microplastic transport in soil by earthworms. *Scientific Reports, 7*, 1362.

Rillig, M. C., Muller, L. A. H., & Lehmann, A. (2017b). Soil aggregates as massively concurrent evolutionary incubators. *The ISME Journal, 11*, 1943–1948.

Rocha, S. T., & Duarte, A. C. (2015). A critical overview of the analytical approaches to the occurrence, the fate and behaviour of microplastics in the environment. *TrAC Trends in Analytical Chemistry, 65*, 47–53.

Rodriguez-Seijo, A., Lourenço, J., Rocha-Santos, T. A. P., da Costa, J., Duarte, A. C., Vala, H., & Pereira, R. (2017). Histopathological and review molecular effects of microplastics in Eisenia andrei Bouché. *Environmental Pollution, 220*, 495–503.

Rossi, G., Barnoud, J., & Monticelli, L. (2014). Polystyrene nanoparticles perturb lipid membranes. *Journal of Physical Chemistry Letters, 5*, 241–246.

Scarascia-Mugnozza, G., Sica, C., & Russo, G. (2011). Plastic materials in European agriculture: Actual use and perspectives. *Journal of Agricultural Engineering, 42*, 15–28.

Steinmetz, Z., Wollmann, C., Schaefer, M., Buchmann, C., David, J., Tröger, J., Muñoz, K., Frör, O., & Schaumann, G. E. (2016). Plastic mulching in agriculture. Trading short-term agronomic benefits for long-term soil degradation? *Science of the Total Environment, 550*, 690–705.

Sun, J., Dai, X., Wang, Q., van Loosdrecht, M. C. M., & Ni, B. J. (2019). Microplastics in wastewater treatment plants: Detection, occurrence, and removal. *Water Research, 152*, 21–37. [CrossRef] [PubMed].

Sun, X. D., Yuan, X. Z., Jia, Y., et al. (2020). Differentially charged nanoplastics demonstrate distinct accumulation in *Arabidopsis thaliana*. *Nature Nanotechnology, 15*, 755–760. https://doi.org/10.1038/s41565-020-0707-4

Tang, J., Wu, Y., Esquivel-Elizondo, S., Sørensen, S. J., & Rittmann, B. E. (2018). How microbial aggregates protect against nanoparticle toxicity. *Trends in Biotechnology, 36*, 1171–1182.

Tencati, A., Pogutz, S., Moda, B., Brambilla, M., & Cacia, C. (2016). Prevention policies addressing packaging and packaging waste: Some emerging trends. *Waste Management, 56*, 35–45. [CrossRef] [PubMed].

Teuten, E. L., Saquing, J. M., Knappe, D. R., Barlaz, M. A., Jonsson, S., Bjorn, A., Rowland, S. J., Thompson, R. C., Galloway, T. S., Yamashita, R., Ochi, D., Watanuki, Y., Moore, C., Viet, P. H., Tana, T. S., Prudente, M., Boonyatumanond, R., Zakaria, M. P., Akkhavong, K., Ogata, Y., Hirai, H., Iwasa, S., Mizukawa, K., Hagino, Y., Imamura, A., Saha, M., & Takada, H. (2009). Transport and release of chemicals from plastics to the environment and to wildlife. *Philosophical Transactions of the Royal Society B: Biological Sciences, 364*, 2027–2045.

Tremblay, J. F. (2018). Agriculture trying new films on Chinese lands Conventional plastic coverings are a boon to China's agriculture but are taking an environmental toll. *Chemical & Engineering News, 96*, 18–19.

United Nations Department of Economic and Social Affairs, Population Division, UNDESAPD. (2019). *World population prospects, highlights*.

United States Government. (2018). *Agriculture Section 3103* [Online]. United State Code (2011) Title 7. Available: https://www.gpo.gov/fdsys/pkg/USCODE-2011-title7/html/USCODE-2011-title7-chap64.htm

Vallespir Lowery, N., & Ursell, T. (2019). Structured environments fundamentally alter dynamics and stability of ecological communities. *Proceedings of the National Academy of Sciences USA, 116*, 379–388.

Veerasingam, S., Mugilarasan, M., Venkatachalapathy, R., & Vethamony, P. (2016). Influence of 2015 flood on the distribution and occurrence of microplastic pellets along the Chennai coast, India. *Marine Pollution Bulletin, 109*, 196–204.

Verma, R., Vinoda, K. S., Papireddy, M., & Gowda, A. N. S. (2016). Toxic pollutants from plastic waste: A review. *Procedia Environmental Sciences, 35*, 701–708. https://doi.org/10.1016/j.proenv.2016.07.069

Wagg, C., Bender, S. F., Widmer, F., & van der Heijden, M. G. A. (2014). Soil biodiversity and soil community composition determine ecosystem multifunctionality. *Proceedings of the National Academy of Sciences USA, 111*, 5266–5270.

Wang, W., & Wang, J. (2018). Different partition of polycyclic aromatic hydrocarbon on environmental particulates in freshwater: Microplastics in comparison to natural sediment. *Ecotoxicology and Environmental Safety, 147*, 648–655. https://doi.org/10.1016/j.ecoenv.2017.09.029

Wang, F., Wang, F., & Zeng, E. Y. (2018). Sorption of toxic chemicals on microplastics. In E. Zeng (Ed.), *Microplastic contamination in aquatic environments* (1st ed., pp. 225–247). Elsevier. https://doi.org/10.1016/B978-0-12-813747-5.00007-22018

Wang, W., Ge, J., Yu, X., & Li, H. (2020). Environmental fate and impacts of microplastics in soil ecosystems: Progress and perspective. *Science of the Total Environment, 708*, 134841. https://doi.org/10.1016/j.scitotenv.2019.134841

Yi, M., Zhou, S., Zhang, L., & Ding, S. (2020). The effects of three different microplastics on enzyme activities and microbial communities in soil. *Water Environment Research, 93*(1), 24–32. https://doi.org/10.1002/wer.1327

Yu, J. R., Adingo, S., Liu, X. L., Li, X. D., Sun, J., & Zhang, X. N. (2022). Micro plastics in soil ecosystem – A review of sources, fate, and ecological impact. *Plant, Soil and Environment, 68*, 1–17.

Zhang, B., Yang, X., Chen, L., Chao, J., Teng, J., & Wang, Q. (2020a). Microplastics in soils: A review of possible sources, analytical methods and ecological impacts. *Journal of Chemical Technology & Biotechnology, 95*, 2052–2068. [CrossRef].

Zhang, L., Xie, Y., Liu, J., Zhong, S., Qian, Y., & Gao, P. (2020b). An overlooked entry pathway of microplastics into agricultural soils from application of sludge-based fertilizers. *Environmental Science & Technology, 54*(7), 4248–4255.

Zhang, L., Liu, J., Xie, Y., Zhong, S., & Gao, P. (2021a). Occurrence and removal of microplastics from wastewater treatment plants in a typical tourist city in China. *Journal of Cleaner Production, 291*, 125968.

Zhang, L., Xie, Y., Zhong, S., Liu, J., Qin, Y., & Gao, P. (2021b). Microplastics in freshwater and wild fishes from Lijiang River in Guangxi, Southwest China. *Science of the Total Environment, 755*(Pt 1), 142428.

Zhao, T., Lozano, Y. M., & Rillig, M. C. (2021). Microplastics increase soil pH and decrease microbial activities as a function of microplastic shape, polymer type, and exposure time. *Frontiers in Environmental Science, 9*, 675803. https://doi.org/10.3389/fenvs.2021.675803

Zhu, D., Chen, Q. L., An, X. L., Yang, X. R., Christie, P., Ke, X., Wu, L. H., & Zhu, Y. G. (2018). Exposure of soil collembolans to microplastics perturbs their gut microbiota and alters their isotopic composition. *Soil Biology and Biochemistry, 116*, 302–310.

Ziajahromi, S., Neale, P. A., Rintoul, L., & Leusch, F. D. L. (2017). Wastewater treatment plants as a pathway for microplastics: Development of a new approach to sample wastewater-based microplastics. *Water Research, 112*, 93–99. [CrossRef].

Chapter 11
Micro- and Nanoplastics on Plant Functionalities

A. Srividya, Adityah Ganesh, and D. Rathnaprabha

Abstract Plastic pollution has become a major environmental concern of the globe. The increasing plastic pollution has captured the attention of many researchers. Disposal of plastic on land and in water, in due course, leads to the production of various plastic particles like micro (1–5 mm) and nano (1–100 nm) plastics by diverse physico-chemical processes. These micro- and nanoplastics are ubiquitous and have become the major contributors of pollution to aquatic and terrestrial ecosystems. Plants, which are considered as the main producers of the terrestrial and aquatic ecosystems, are vulnerable to the plastic pollutants. Terrestrial and aquatic plants are exposed to different types of plastics, leading to altered physiological and metabolic functions. This chapter has highlighted the impact of micro- and nanoplastics on plants during seed germination and growth. The altered responses of plants are because of the imbalances in soil microbial community, anti-oxidative enzymes and photosynthetic and metabolic activities. Various characteristics of autotrophic macrophytes on exposure to plastic pollutants are also being focussed on. Many of these studies have shown the inhibitory effects of these particles, which are dependent on their size, shape, charge and concentration used. Based on the size, they get adsorbed or internalised by the plant, which reduces its growth and photosynthetic activity primarily by inducing oxidative stress. The accumulated particles block the pores on the seed and root surface thereby affecting seed germination and also nutrient uptake by roots. This chapter covers the major research topics that investigated the effects of micro- and nanoplastics on various parameters of plant growth and functions. The effects of micro- and nanoplastics on plant functionalities are being discussed.

Keywords Microplastics · Nanoplastics · Functionalities · Polyethylene · High-density polyethylene · Low-density polyethylene · Polystyrene · Oxidative stress

---

A. Srividya · A. Ganesh · D. Rathnaprabha (✉)
Department of Biotechnology, National Institute of Technology, Warangal, Telangana, India
e-mail: rathnaprabhad@nitw.ac.in

N. R. Maddela et al. (eds.), *Micro and Nanoplastics in Soil*,
https://doi.org/10.1007/978-3-031-21195-9_11

11.1 Introduction

We live in a world surrounded by plastic. Usage of plastic has become convenient as it is economical and reliable (Boucher & Friot, 2017). The usage of plastic has increased by 25 folds over last 40 years (Sutherland et al., 2014). Because of its extensive and uncontrollable use, plastic pollution has become a global issue. Plastic production was estimated to be more than 359 million tons as of 2019 (Europe, 2017). The estimated plastic production would reach 12,000 million tons by 2050 (Geyer et al., 2017). In spite of recycle and reuse, 32% of the plastic waste still remains in the natural habitat (Geyer et al., 2017; de Souza Machado et al., 2018).

Plastic contamination is omnipresent, on land and in water, namely, oceans, lakes, estuaries and shores (Alomar et al., 2016; Browne et al., 2011; Naidoo et al., 2015). Plastic gets degraded in course of time by physical and biological processes. Plastic particles that are synthesised primarily for certain purposes and those that originated from degradative process will eventually lead to marine water with run-off, which is the main source of plastic in marine water (Andrady, 2011). So, most of the focus was on marine plastic pollution but terrestrial environment is also a major recipient of plastic pollutants (Khalid et al., 2020; Zhou et al., 2021). Plastic pollution generates particles of various sizes, ranging from micro to nano size in course of time.

Soil is the source of many plastic particles in terrestrial ecosystem, which hints us of the safety of agricultural crops and food. Soil polluted with various plastic particles like micro and nanoplastics will affect plants negatively or positively (de Souza Machado et al., 2018). Micro and nanoplastic pollution is widespread. Microplastics (MPs) are found in various terrestrial systems like agricultural fields and industrial areas (Piehl et al., 2018; Fuller & Gautam, 2016). Microplastics are becoming a threat to terrestrial ecosystem (de souza Machado et al., 2018). They alter the soil structure and plant performance (de souza Machado et al., 2019). They change the physical and chemical properties of the soil and affect microbial activities and plant performance (Xu et al., 2020). These effects are dependent on their shape and size rather than on their chemical composition (Rillig & Lehmann, 2020). Khalid et al. (2020), have described various direct and indirect ways of microplastic effects on terrestrial plants. Microplastics show direct effects on plant growth and functionalities; and indirectly affect plants by altering soil properties and soil-microbe interactions. Zhou et al. (2021) have focussed on the effects of microplastics on soil properties and microbial communities. Microplastics are found to alter soil properties like aggregation (Rillig et al., 2017) and soil pH (Boots et al., 2019). Because of the altered properties, plant performance would be affected.

Various kinds of microplastics have shown variable effects on plant growth, as noticed in *Triticum aestivum, Allium fistulosum, Lactuca sativa and Phaseolus vulgaris,* under different conditions (de Souza Machado et al., 2019; Jiang et al., 2019; Meng et al., 2021; Qi et al., 2018).

Nanoparticles are inherently difficult to detect and analyse as the minimum size limits of most of the commonly used sampling and analysis techniques fall in the range of 10–100 μm (Li et al., 2020c, 2021a; vanWeert et al., 2019). This creates the additional challenge of detecting and tracing nanoplastics in human food items and animal feeds. The nanoplastics should therefore be evaluated for presence in food production chain, mode of entry of the plastic into the production chain, and effect of the particles on production process and potential risks that may develop from it (Yin et al., 2021; Mateos-Cárdenas et al., 2021). Various effects of MPs and NPs on agro-ecosystems were reported by Ng et al. (2018). As plants are primary producers and main living constituents of the environment, much attention is needed to understand the effects of micro plastics (MPs) and nanoplastics (NPs) on them.

11.2 Types of Plastics

There are various classes of plastic depending on their physicochemical properties, namely, polyethylene terephthalate (PET), high-density polyethylene (HDPE), low-density polyethylene (LDPE), Polystyrene (PS), polypropylene (PP), polyvinyl chloride (PVC) and others (Fig. 11.1). These are produced to a great extent and commonly found in the environment. Plastic span over different shapes and sizes, Microplastics (MPs) are those plastic particles ranging from 1 to 5 mm (Frias & Nash, 2019) and particles ranging from 1 to 100 nm that exhibit colloidal behaviour are Nanoplastics (NP) (Gigault et al., 2018). Because of their extensive usage, plastic particles are widespread across the world. They have many effects on living beings, including human beings.

Different types of MPs have been detected in the environment (Mintenig et al., 2019; Sharma & Chatterjee, 2017) these plastic particles cause serious environmental problems. Primary microplastics are those that are synthesised for some purposes like fabrication, as cosmetics and in the form of microbeads. Secondary microplastics emerge from the fragmentation of larger plastics by physical, chemical and biological processes (Duis & Coors, 2016; De Falco et al., 2019). These microplastics are of various shapes, like beads, fibres, films and fragments, which are carriers of various toxic substances (Wang et al., 2018). Depending on their shape, they show various effects on plants, directly or indirectly (Rillig et al., 2019).

Fig. 11.1 Types of plastics

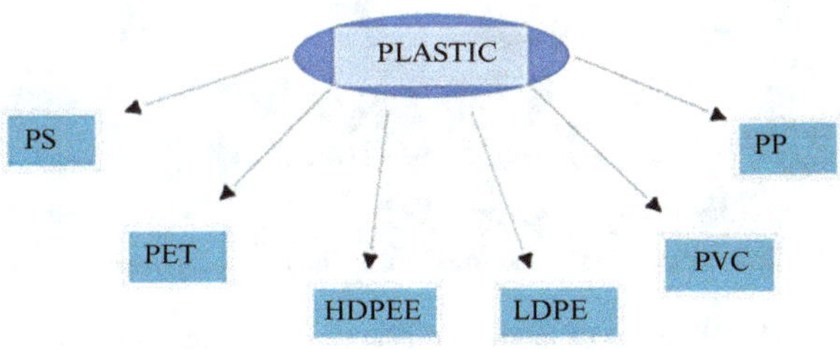

11.2.1 Sources of MPs and NPs

Microplastics/Nanoplastics enter into plants from the soil, which acts as the medium of transfer. An ample focus was drawn to microplastics pollution because of the reports that stated much of microplastics that were released were from personal care products (Duis & Coors, 2016). Soil polluted with microplastics will affect plant performance (de Souza Machado et al., 2019). Main sources of microplastics in soil arise from municipal sludge, irrigation water and plastic mulching. Plastic used in agriculture (agroplastic), is considered to be an important source of plastic in terrestrial ecosystem, whose usage is expected to increase 69% from 2012 to 2019 (Sintim & Flury, 2017). Wastewater treatment plants remove waste from water and prevent them from entering water sources. The remaining sludge is used as fertiliser in agricultural fields, which is the main source of plastic particles in agricultural fields (Alimi et al., 2018; Hurley & Nizzetto, 2018). Corradini et al. (2019) have emphasised on the role of municipal sludge in contaminating agricultural soil with plastic particles. While wastewater treatment methods can remove up to 90% of the micro and nanoplastics that exist in the wastewater, this amount is not satisfactory and the use of this treated water in agriculture for irrigation has effectively introduced micro and nanoplastic from the water into the agricultural ecosystem (Fig. 11.2).

Increasing microplastics are an alarm to indicate the threat imposed on environment. Leaching of plasticisers and polymers from municipal waste disposal sites into ground water impose threat to terrestrial ecosystem (Teuten et al., 2009). Huang et al. (2020) have described plastic mulching as the source of microplastics in the soil. They are also commonly found in the aquatic ecosystem like oceans, lakes and

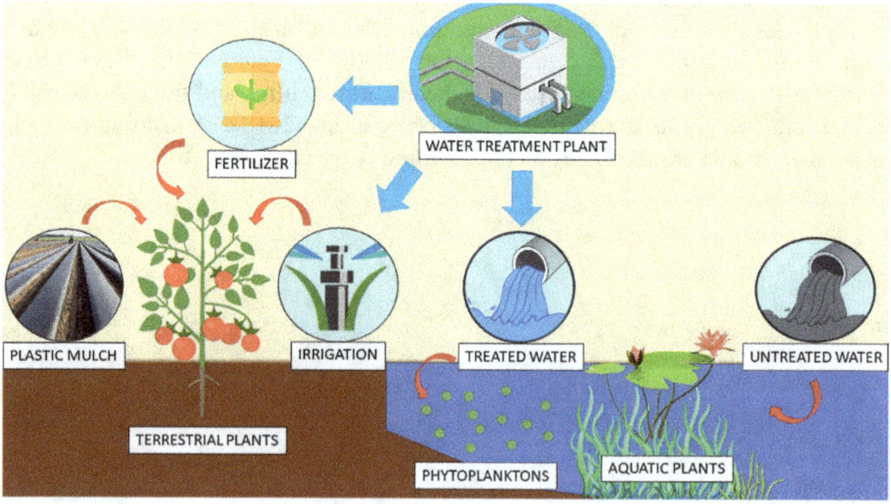

Fig. 11.2 Sources of micro and nanoplastics to terrestrial and aquatic plants

estuaries, which have been accumulating over last four decades (Thompson et al., 2004). Runoff and weathering are the main sources of microplastics in marine waters (Andrady, 2011).

There are various sources, through which NPs enter the terrestrial environment. Nanoplastic sources can be broadly classified as point sources and non-point sources. Point sources include sources from which free nanoparticles are directly dispersed into the surrounding environment. This includes direct release of nanoplastics into the environment as a result of untreated release and undesired leakage or spill out during the production and transport of nanoplastics; and products that use micro and nanoplastics (Yano et al., 2021). The advancing trend in the consumer industry (Yoshino et al., 2012) to lean towards nanoparticles has resulted in the production of nano particulate polymers which can add to the pool of nanoplastic in the environment thus acting as non-point sources of nanoplastic pollution into the environment.

11.2.2 Accumulation of MPs and NPs in Plants

MPs and NPs can adsorb on the surface of the plants and enter into plant system. Some of these particles are usually trapped on the root surface and enter root cells (Li et al., 2019). Various fluorescently labelled tags were used to track their transport, NPs and micro-sized MPs were found to travel from root to leaf (Li et al., 2020a, b). They travel to the upper parts of the plant by transpirational pull (Li et al., 2020a). Depending on their size and shape, these MPs and NPs can affect plants. They may have a positive or negative effect on plants depending on the species (Rillig et al., 2019). Vascular plant tissues have been observed to adsorb nanoplastics under specific conditions. Studies using scanning electron microscopy (SEM) and laser scanning confocal microscopy (LSCM) have observed the location specificity and internalisation process of nanoplastics in plants (Khalid et al., 2020; Lian et al., 2020; Yan et al., 2021a). These studies have shown that absorption of nanoplastics depends on the plant species and characteristic properties of the plastic.

Positively and negatively charged NPs accumulated differently in *A. thaliana* (Sun et al., 2020). PS beads have been found to accumulate on the root surface but are not absorbed in floating plants like *Spirodela polyrhiza* (Mateos-Cárdenas et al., 2019). MPs were detected on seagrass and macroalgae by Seng et al. (2020), these macrophytes act as glue for various plastic particles and serve as temporary sink for MPs (Sfriso et al., 2021), which hold MPs in aquatic environment. Fibres are more predominant in MPs, retained by filamentous algae (Esiukova et al., 2021). Internalisation of PS microbeads was studied using fluorescently labelled particles. These MPs (80 and 1 μm) are being absorbed by roots and translocated to the aerial parts through the vascular system in rice. This study highlights the accumulation of MPs in crop plants and their probable transfer in food chain (Liu et al., 2022; Li et al., 2020c). Physical deformations or damages in the plant roots can provide pathways for the entry of plant nanoparticles into the plant. A study conducted by Li et al. (2020a) revealed that cracks in the lateral roots of lettuce (*Lactuca sativa*) and wheat (*Triticum aestivum*) provided a pathway for the entry of nanoparticles of PMMA (polymethylmethacrylate) and PS (polystyrene). Another pathway that can lead to nanoplastic internalisation is the endocytosis through a liquid phase. Zhou et al. (2021) showed that aquaporins can assist in the nanoplastic intake.

Airborne MPs are gaining attention for being a threat to human health (Cox et al., 2019). Terrestrial plants act as a sink for various particulate matter, including MPs (Rindy et al., 2019). Depending on the leaf trait (leaf area and hairiness), particulate matter is being deposited on the urban trees, which are usually considered to reduce the pollutants in the air (Chiam et al., 2019). So, terrestrial plants act as an important sink for MPs and influence their accumulation in plants and soil (Bi et al., 2020). Airborne transport and accumulation of NPs have also been demonstrated in *Zea mays* L. by Sun et al. (2021). They exposed leaf of *Zea mays* L. seedlings to both positively and negatively charged PS particles. Positively charged PS particles aggregation was more notable and travelled from leaf to the roots through vascular bundle, which induced inhibitory effect on photosynthesis. MPs in combination with other chemical compounds have more adverse effects on plants (Prata et al., 2018). MPs/NPs in the soil serve as reservoirs of many pollutants. Cyanotoxins (CTX) released by bacteria would accumulate on MPs/NPs and serve as co-contaminants. Maity et al. (2021) have presumed the importance of these co-contaminants as phytotoxic. MPs, especially the abundant polystyrene (PS) was found to be accumulated in plants using fluorescent dyes by confocal laser scanning microscope (Li et al., 2020b). These MPs and NPs accumulate in plants by various ways and pose health risks to humans. A study on risk assessment of MPs on edible fruits and vegetables was carried out by Conti et al. (2020). Apple and carrot are the most contaminated fruit and vegetable respectively. There are various studies that concentrated on effects of MPs and NPs on plant performance, like seed germination, photosynthesis, growth and others. The present chapter explains the effect of these MPs/NPs on plant functionalities.

11.3 Effects of MPs and NPs on Plant Functionalities

MPs/NPs affect plants negatively, by reducing growth and photosynthetic activity in terrestrial and aquatic primary producers. Majority of MPs/NPs induce oxidative stress, thereby increasing anti-oxidative enzyme levels. Various effects of MPs/NPs include inhibition of seed germination, reduced or inhibited photosynthetic activity, reduction of root and shoot lengths, plant biomass and others (Fig. 11.3). The direct and indirect effects of MPs/NPs on plants based on the published literature are discussed below:

(i) *Effects of MPs and NPs on Seed Germination:*

MPs and NPs have shown to affect seed germination in various plant species. These effects are dependent on their size and concentration (De Silva et al., 2021)

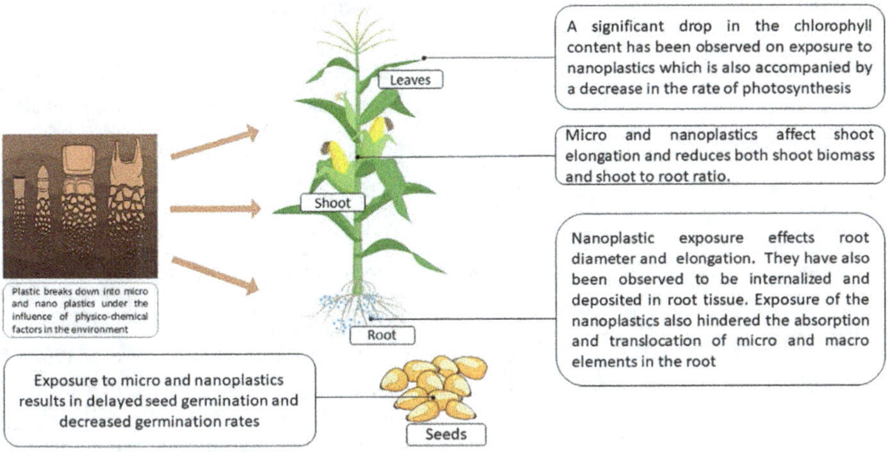

Fig. 11.3 Effects of micro and nanoplastics on various plant functionalities based on the organ they accumulated

as shown in Table 11.1. Soil is the main sink for these particles. MPs (PLA, polylactic acid) accumulated in soil and affected germination in *L. perenne* (Boots et al., 2019). Many of the studies report that MPs adsorb on the plant surface, whereas NPs are being absorbed into the plants. The accumulated particles on the seed surface reduce the germination rate (De Silva et al., 2021). Recent studies have highlighted that seed germination and growth of plants are dependent on the age of decaying plastic. Based on the age of microplastic, its toxicity also varied on the germination, growth and photosynthesis in *L. Sativum* (Pflugmacher et al., 2021). Bosker et al. (2019) studied the effects of differently sized MPs (50, 500 and 4800 nm) of different concentrations on seed germination of *L. sativum*, using 72 h germination assay. Germination was reduced due to the blockage of the seed capsule by accumulated MPs. The impact of PS-MPs (2 μm and 80 nm) on germination of ornamental plants was studied by Guo et al. (2022). It was found that these MPs could inhibit germination in these species depending on their concentration (Table 11.1). Zhang et al. (2021b) studied the effect of PS-MPs (200 nm) on the seed germination of rice. These particles showed no significant effect on seed germination but reduced the activity of anti-oxidative enzymes. Reactive oxygen species (ROS) levels increased in roots and transcriptome analysis revealed that PS-MPs increased the activity of anti-oxidative enzymes in roots.

The composition of MPs also affects the toxicity. Various MPs like PP, PE and PVC were analysed on different growth parameters like seed germination, plant height, biomass production and oxidative stress levels in *Lepidium sativum,* by Pignattelli et al. (2020). The evaluated toxicity was found to be due to PVC on *L. Sativum.* These particles alone or in combination with other toxic elements will affect the plant performance, as witnessed in *L. sativum.* PET alone or with acid rain was found to negatively affect seed germination, plant growth and increased chlb production (Pignattelli et al., 2021). Ethylene vinyl acetate (EVA), LLDPE (Linear

Table 11.1 Effects of MPs/NPs on seed germination of various plant species

Plant	MP/NP type	Size	Effect on seed germination	Reference
L. perenne	Poly lactic acid HDPE	Mean diameter: HDPE: 102.6 μm PLA: 65.6 μm	Germination reduced reduction in shoot length HDPE: decreased soil pH	Boots et al. (2019)
L. sativum	MPs	50, 500 and 4800 nm	Decreased germination rate	Bosker et al. (2019)
L. sativum	PP, PE, PVC	Not mentioned	Oxidative burst induced decreased germination	Pignattelli et al. (2020)
L. sativum	PET, acid rain	60–3000 μm	Reduced germination induced oxidative burst	Pignattelli et al. (2021)
T. aestivum	EVA, LLDPE, PMMA	Not mentioned	Inhibited seed germination	Lian et al. (2019)
Italian lettuce	PSNPs and micro PS	Not mentioned	Germination inhibited	Gong et al. (2021)
Rice	PS-MPs	200 nm	No significant effect increase in length and ROS in roots	Zhang et al. (2021b)
Trifolium ripens, O. Violaceus, Impatiens balsamina	PS-MP	2 μm and 80 nm	Inhibited germination rates	Guo et al. (2022)

MP Microplastic, *PSMP* Polystyrene microplastic, *PSNP* Polystyrene nanoparticle, *PP* Polypropylene, *PE* Polyethylene, *PVC* Polyvinyl chloride, *PET* Poly ethylene terephthalate, *PSNP* Polystyrene nanoplastic, *EVA* Ethylene vinyl acetate, *LLDPE* Linear low density poly ethylene, *HDPE* High density polyethylene, *PMMA* Poly methyl methacrylate, *ROS* Reactive oxygen species

low-density polyethylene) and poly methyl methacrylate (PMMA) have shown to inhibit seed germination in *Triticum aestivum*, at lower concentrations (Lian et al., 2019). MPs derived from fertilisers have exhibited differential effects on wheat growth and soil properties (Lian et al., 2021b).

A systemic study done on the effect of polystyrene nanoplastics (PSNPs) on crop plant *Triticum aestivum* L. (Lian et al., 2020) and on *Allium cepa* (Giorgetti et al., 2020), has shown that PSNPs exhibited no visible effect on seed germination. But there was an increase in root length and decrease in shoot-to-root biomass ratio in *T.aestivum*. PSNPs' effect on seed germination is species dependent. Among the four food crops (Italian lettuce, radish, wheat and corn) exposed to PSNPs and microspheres (micro PS), seed germination in Italian lettuce was inhibited at a very early stage of development. It was also observed that these toxic effects are because of the oxidative stress imposed on plants (Gong et al., 2021). A pot experiment conducted by Li et al. (2021a) on effects of LDPE and bio mulch on germination and growth of *Glycine max* has shown that bio-mulch has negatively affected germination viability and LDPE affected plant height.

(ii) *Effects of Accumulated MPs and NPs on Plant Growth:*

MPs and NPs, when accumulated in plants, affect their growth directly in various ways, like reduction in biomass and by induction of oxidative stress. These plastic pollutants are ubiquitous; they are toxic alone or in combination with other organic or metallic pollutants. They even affect plant performance indirectly by altering soil properties like soil pH or soil microbial community (Rillig et al., 2019). Foliar application of PSNPs have shown to reduce the growth and induce oxidative stress in lettuce. These NPs got absorbed at the leaf and transported to the root part. They reduced nutritional quality, biomass and leaf size of lettuce (Lian et al., 2021a).

(a) *Direct Effects of MPs and NPs on Plant Growth:*

Much of the scientific attention is being driven by these pollutants and their effects on agricultural crops, which were studied using hydroponic cultures in certain crop plants. Accumulation of MPs (PE microbeads) in the rhizosphere and reduction in water and nutrient uptake were observed in hydroponic maize. Upward transport of PE beads to shoot was not observed (Urbina et al., 2020). In hydroponic wheat seedlings, there was no effect of PSMPs on photosynthesis and oxidative stress. Instead, PSMPs reduced the toxicity of cadmium and copper on wheat seedlings (Zong et al., 2021). Rice (*Oryza sativa*) seedlings were exposed to 1 μm and 100 nm PSMPs, to study their toxic effects in hydroponic conditions. 1 μm PS were more toxic than 100 nm particles in rice seedlings (Wu et al., 2021). Differently sized particles would affect plants differently. MPs of 1 μm size accumulated in the intercellular space of carrot roots. But larger MPs would enter roots and leaves in the presence of arsenic. They reduced the quality of carrots and eventually led to health risks (Dong et al., 2021b).

Molecular and physiological effects of differently sized MPs were studied in maize seedlings by Pehlivan and Gedik (2021). They analysed various combinations of MPs differing in their sizes, for their toxicity. They induced xenobiotic stress in the maize. It was observed that the bigger the particle size, the quicker they restore to the normal condition. Their research insists on the toxicity of pollutants on agricultural crops. Likewise, the toxicity of MPs on the leguminous plant soybean was also investigated by Xu et al. (2021). PSMPs decreased the availability of organic pollutants like phenanthrene in soybean seedlings. It was shown that PSMPs along with phenanthrene are harmful to higher plants. Ren et al. (2021) studied the effect of PS beads on the growth of Chinese cabbage. They found that these plastic particles have soil-microbial community that would affect the plant growth.

Nanoplastics can get internalised in plants and affect them adversely. Exposure of corn to PSNPs has altered the plant's metabolic rates because of the elevated antioxidative enzymes. But photosynthetic activity was not altered (Zhang et al., 2021a). The effect of PSNPs was also studied in cucumber by exposing it to different sized particles. It was found that they first accumulated at root but later transported to fruit, flower and leaves. They increased the protein content of fruit and also root activity (Li et al., 2021b). Differently charged nanoplastics accumulated differentially in *Arabidopsis thaliana.* Positively charged particles, though accumulated slowly, could reduce the growth of the plant. Negatively charged particles were found in apoplast and xylem (Sun et al., 2020). Giorgetti et al. (2020) studied the interaction of PSNPs and *Allium cepa*. Various concentrations of PSNPs (0.01, 0.1 and 1 mg/L) were used during germination. All concentrations were shown to inhibit root length. Higher concentrations were found to induce oxidative stress. The internalisation and movement of particles was visualised using TEM. Internalised nanoplastic particles in food crops raise concerns for food safety, as they are transferred in trophic levels. Foliar application of positively and negatively charged PSNPs on *Zeamays* L. was carried out to observe the absorption tendency of air transported NPs carrying different charges. They got internalised and transported to the root leading to photosynthesis inhibition by PS-NH_2 (Sun et al., 2021).

In higher plants like *Vicia faba,* the root tips were exposed to 5 and 100 nm PSMPs of various concentrations. It was observed that 100-nm particles accumulated in roots and blocked intercellular connections. There was a decrease in the growth with 100-nm particles, which have also proven to be genotoxic and imposed oxidative damage (Jiang et al., 2019). In lettuce, MP (PS) under di-n-butyl phthalate (DBP) stress conditions induced toxicity by blocking root pores. PS in lettuce decreased biomass by inducing oxidative stress and damaging leaves and root. (Gao et al., 2021a). Similar research was carried to study the effect of PS and DBP on green and purple lettuce. Toxicity induced was dependent on the variety; purple variety was more sensitive to PS. The toxicity induced was by reducing growth of root, soluble protein and sugar in leaves (Gao et al., 2021b). PE MPs effect on cadmium (Cd) bioavailability was also investigated in lettuce by Wang et al. (2021). They increased the availability of Cd and accumulation in plants. This study suggests that MPs increase the toxicity of heavy metals in contaminated soils. Toxicological effects of MPs on farmland were researched on crops like lettuce. Physiological responses of lettuce in presence of PVC microplastics were studied by applying different sizes of PVC (100 nm–18 μm and 18–150 μm). Various contents were analysed on the root activity, which had no effect. But, root length, diameter and surface were increased. Photosynthetic activity was altered with 100 nm–18 μm particles and 18–150 μm sized particles were correlated to root morphology (Li et al., 2020b).

(b) *Indirect Effects of MPs on Plants:*

Microplastics show altered effects on various plants indirectly by changing soil structure, properties or soil microbial interaction. Boots et al. (2019) have reported the change in soil structure in the presence of HDPE and PLA, which results in altered soil ecosystem. Different MPs show different effects on soil property and plant biomass, based on their concentration (Lozano et al., 2021). These altered structure and physicochemical properties of soil will affect the plant functionalities. de souza Machado et al. (2019) conducted an experiment to study the effects of six different MPs on *Allium fistulosum.* Depending on the type and size, they altered the soil properties, which affected the plant's performance. MPs get integrated into the soil when they reach the soil surface (Rillig et al., 2017). They contain a lot of organic carbon, so they add to the organic part of the soil, which will affect the carbon cycle in the soil and the plant's performance (Rillig et al., 2021). Hydroponic cultures of maize were employed to study the effect of MPs. Isotopic analysis was used to study the effects of MPs and it was noticed that 30% of the carbon in the rhizosphere was derived from microplastic (Urbina et al., 2020). Biodegradable plastic is being employed in the place of plastic mulching, which releases various compounds on degradation. An in vitro toxicity test was conducted to study the effects of biodegradable plastic on crop plants. Among the compounds released from bioplastic, adipic acid was shown to inhibit growth of tomato and lettuce (Martin-Closas et al., 2014). The influence of biodegradable plastics on soil microbial communities and agro-ecosystem was reported by Bandopadhyay et al. (2018).

(iii) *Effects of MPs on Plant Community:*

The effects of microplastics are mainly dealt with in an individual or in plant species. But the effects of microplastics vary with plant species within a community. MPs have been found to show allelopathic effects, which affect plant community structure. In allelopathy, they promote the growth of same species and inhibit the same of neighbour. Lozano and Rillig (2020) have demonstrated this in *Hieracium* and its neighbour *Festuca.* They affect the community structure and plant productivity (Rillig et al., 2019), which is witnessed in root length of *Plantago lanceolata* (forb) and *A. fistulosum.* They have shown opposite responses to microplastic exposure (de Souza machado et al., 2019; Van Kleunen et al., 2020). So, this indicates that microplastics affect varyingly on different plant species within a community. Yu et al. (2021b), highlighted the importance of research on the effects of microplastics on plant biomass and community structure. Lozano and Rillig (2020) have shown the effect of microfibres on productivity and structure of plant community. They witnessed the dominance of an allelopathic species in the presence of microplastics and reduced biomass of the other species in a community.

(iv) *Effects of MPs and NPs on Growth of Aquatic Photosynthetic Organisms:*

The aquatic system receives much of plastic particles by runoff. They adversely affect many aquatic plants in various ways. The toxicity exhibited by these plastic particles is dependent on their properties like charge, type and size (Lagarde et al., 2016). Charge dependent toxicity was witnessed in microalgae, *Dunaliella tertiolecta.* Positively charged nano-PS particles have inhibited the growth when compared to negatively charged PS particles (Bergami et al., 2017). Schiavo et al. (2021) found the toxic effects of PS, PP and PE leachates on growth and DNA of microalgae. As the macrophytes exposed to various concentrations of MPs and NPs, NPs were found to be more effective in reducing shoot to root length compared to Mps (vanWeert et al., 2019). In a comparative study of the effects of MPs and NPs, PSNPs were found to impose a more inhibitory effect on the growth of *Chlamydomonas reinhardtii.* The treated algal species was shown to have increased levels of reactive oxygen species and malonaldehyde. Internalised NPs were visualised in vacuoles. They mainly imposed oxidative stress in the algae (Yan et al., 2021b). Yang et al. (2020) proved the toxic effects of MPs (PS, PE, PA) on growth due to oxidative stress in *Chlorella pyrenoidosa.* Similar effect of PSMPs was observed on the growth of *Chlorella pyrenoidosa*; there was reduction in the growth due to oxidative stress and physical damage. Later, it could regain its growth in late logarithmic phase (Mao et al., 2018). In the floating plant *Salvinia cucullata,* the combined effect of MPs and herbicide (glyphosate) has reduced the growth by activating antioxidative enzymes. PSMP (1 μm) could reduce relative growth and root morphology. These findings indicate the toxic effects of contaminants in aquatic ecosystems (Yu et al., 2021a). A growth inhibition rate of 39.7% was also observed in marine microalgae *Skeletonema costatum*, on 96 h exposure to MPs. The toxicity was dependent on particle size and concentration of MPs on interaction with microalgae (Zhang et al., 2017). The toxicity of MP in combination with Cu nanoparticles and triclosan was also studied in *Skeletonema costatum* (Zhu et al., 2019; Zhu et al., 2020), shown in Table 11.2. *Chlamydomonas rheinhardtii* was also affected by an inhibition rate of 45.8%, on exposure to PSMPs (Li et al., 2020d). Kalčíková et al. (2017) studied the effect of PE microbeads on freshwater duckweed, *Lemna minor.* It was observed that these particles reduced the root length, but the photosynthetic activity was not affected (Table 11.2).

(v) *Effects of MPs and NPs on Photosynthetic Activity of Plants:*

Photosynthetic organisms are present in aquatic and terrestrial systems. They are the primary producers and sinks for CO_2. MPs and NPs have shown to affect photosynthesis by altering chlorophyll content in photosynthetic organisms (Table 11.3). Photosynthetic activity was reduced in cucumber plant when it was exposed to various sizes of PSNPs (100, 300, 500 and 700 nm). Among them 100 nm particles reduced chlorophyll-a and chlorophyll-b content, along with fluorescence in leaves. Exposure to 300 nm particles reduced biomass, while 700 nm particles induced oxidative stress in leaves (Li et al., 2020a). An antagonistic response was observed in wheat seedlings on exposure to PSNPs. The photosynthetic activity was increased at 0.1 mg/L concentration and shoot-to-root ratio was decreased at 0.01 mg/L concentration (Lian et al., 2020).

MPs act as carriers of many pollutants and exert their effects on crop plants. Effects of MPs in combination of other pollutants were being investigated in some of the researches. Growth and photosynthetic activity were reduced in lettuce, on exposure to MP (PE) and di-n-butyl phthalate (DBP). Exogenous MP has further increased the effect of DBP on photosynthesis (Gao et al., 2019).The integrated effect of PS and di-butyl phthalate (DBP) on photosynthesis of red lettuce were studied Dong et al. (2021a). It was noted that PS particles reduced the uptake of DBP; jointly, they could reduce photosynthetic activity and also the quality of red lettuce. This explains the possible risk of microplastics on vegetable crops.

(a) *Effects of MPs and NPs on Photosynthesis of Aquatic Photosynthetic Organisms:*

Microalgae are primary producers in the aquatic ecosystem; study of plastic pollutants on these organisms will give us an insight into how MPs/NPs would affect photosynthetic process (Table 11.3). Small size and positive surface charge on MPs will have more adverse effects on microalgae (Prata et al., 2019). Small-sized PSMPs (0.05 μm) were proven to have adverse effects at certain concentrations, on

Table 11.2 Effects of MPs/NPs on growth of terrestrial and aquatic photosynthetic organisms

Plant	MP/NP	Size	Effect	Reference
Lettuce	PSNP	Not mentioned	Reduction in biomass, quality and leaf size imposed oxidative stress	Lian et al. (2021b)
Cucumber	PSNPs	100 nm	Decrease in chla, chlb in leaves	Li et al. (2020b)
		300 nm	Decrease in biomass	
		500 nm	Altered enzymatic activity	
		700 nm	Increase in anti-oxidative enzymes; induction of oxidative stress	
Cucumber	PSNPs	100, 300, 500,700 nm	Increase in soluble protein in fruits and increase in root activity, MDA and proline content by 300 nm particles	Li et al. (2021b)
Maize	PE microbeads	Not mentioned	Plastic bioaccumulation in rhizosphere water and nutrient uptake reduced	Urbina et al. (2020)
Maize	PP, PET, PVC, PS, PE (PP + PET + PVC + PS + PE)	75–150 μm 150–212 μm	Decreased photosynthetic pigments oxidative stress	Pehlivan and Gedik (2021)
Maize	HDPE, PLA	Not mentioned	High dose of PLA decreased biomass	Yang et al. (2021)
Carrot	MP	1 μm	Accumulated in roots	Dong et al. (2021a)
		Larger MPs and arsenic	Reduced quality of carrot	
Soybean	PSMP + Phe	1, 10, 100 μm	Oxidative stress decreased activity of root	Xu et al. (2021)
Viciafaba	PSMPs	5, 100 nm	Accumulated in roots and blocked intercellular connections	Jiang et al. (2019)
Lettuce	PSMP + DBP	SPS 100–1000 nm LPS >1000 nm	Blocked root pores leaf and root damage oxidative stress	Gao et al. (2021b)
Wheat	PS + degradable mulching film	5 and 70 nm	Decreased base diameter and microbial community increased plant height	Ren et al. (2021)

(continued)

Table 11.2 (continued)

Plant	MP/NP	Size	Effect	Reference
Corn	PSNPs	100, 300, 500 nm	Metabolic rate altered anti-oxidative enzymes increased	Zhang et al. (2021a)
Oryza sativa	PSMP	1, 100 nm	Reduced growth oxidative stress	Wu et al. (2021)
Allium cepa	PSNP	50 nm	Reduction in root length at concentrations (0.1 and 1 mg/L) oxidative stress at higher concentrations cytotoxic and genotoxic	Giorgetti et al. (2020)
Allium fistulosum	PE fibres, PA beads, PP, PS, Poly ester terephthalate (PET) and PE	PA: 15–20 μm PE fibres: length 5000 μm; diameter 8 μm PEHD: 643 μm PET: 376 μm PP: 816 μm PS: 754 μm	PA: increase in soil nitrogen content; increase in total biomass increase in onion bulb biomass PE fibres: altered soil biophysical properties, increase in root biomass and decrease in diameter PEHD: decrease in soil bulk density	de Souza Machado et al. (2019)
Arabidopsis thaliana	NP	–	Positively charged reduced the growth internalisation of negatively charged particles	Sun et al. (2020)
Zea mays	PSNP PS-COOH PS-NH_2	–	PS-NH_2 inhibited photosynthesis	Sun et al. (2021)
Myriophyllumspicatum and Elodea sp	PSNPs	50–190 nm	Reduced shoot to root biomass	vanWeert et al. (2019)
	PSMPs	20–500 μm	Increase in shoot length in *M. spicatum*	
Dunaliella tertiolecta	PS, PP and PE	5 mm ± 0.3	Growth inhibited, induced oxidative stress DNA damaged	Schiavo et al. (2021)
Chlorella pyrenoidosa	MP (PS, PE, PA)	PE1000,13 μm and PE150 μm PA1000,13 μm and PA150 μm PS 150 μm	Growth inhibition oxidative stress	Yang et al. (2020)
Lemna minor	PE microbeads	30–600 μm	Root length affected viability of root cells decreased with sharp particles	Kalčíková et al. (2017)

(continued)

Table 11.2 (continued)

Plant	MP/NP	Size	Effect	Reference
Dunaliella tertiolecta	PSNP PS-COOH PS-NH2	40 and 50 nm	PS-NH2 (aggregates of <200 nm) inhibited algal growth	Bergami et al. (2017)
Salvinia cucullata	PSMP	1 μm	PSMP reduced relative growth and root morphology	Yu et al. (2021a)
Skeletonema costatum	mPVC	1 μm	Growth was inhibited at high concentrations photosynthesis was reduced	Zhang et al. (2017)
Skeletonema costatum	MP and TCS	PE, PS, PVC: 74 μm PVC 800: 1 μm	TCS is more inhibitory than other MPS PVC + PVC800 + TCS: toxicity reduced	Zhu et al. (2019)
Skeletonema costatum	mPVC, Cu nano	–	Growth inhibition cell membrane damage mPVC reduced the toxicity of Cu nanoparticles	Zhu et al. (2020)

PSMP Polystyrene microplastic, *PSNP* Polystyrene nanoparticle, *PP* Polypropylene, *PE* Polyethylene, *PVC* Polyvinyl chloride, *PET* Poly ethylene terephthalate, *PEHD* Poly ethylene high density, *PSNP* Polystyrene nanoplastic, *PLA* Poly lactic acid, *chla* chlorophyll a, *TCS* Triclosan, *Phe* Phenanthrene

microalgae by inhibiting growth, but photosynthesis was not affected (Sjollema et al., 2016). MPs have proven to be toxic to the freshwater algae by negatively affecting the photosynthesis, which was dependent on the size, in a study dealing with the interaction of MPs and microalgae *Skeletonema costatum* (Zhang et al., 2017). Toxicity of PVC was greater than PP at certain concentrations by affecting chlorophyll-a content in *Chlorella pyrenoidosa* and *Microcystis (M.) flosaquae.* These results indicate the risk of MPs on growth of algae (Wu et al., 2019). Similar effects of PSMPs were observed in *Chlamydomonas rheinhardtii,* in all different concentrations tested (Li et al., 2020d). It could recover from the toxic effect of MPs thereafter. The effects of MPs on the entire growth cycle of *Chlorella pyrenoidosa* was studied by Mao et al. (2018). It was observed that the photosynthetic activity was decreased initially but later, after logarithmic phase, an increase in the photosynthetic activity was observed. MPs/NPs do not necessarily reduce photosynthetic activity in all, some plants like *Spirodela polyrhiza* are not affected by plastic (Dovidat et al., 2020). PE MPs adhered to duckweed species, *Lemma minor* have imposed no effect on photosynthesis and growth of the weed (Mateos-Cárdenas et al., 2019). Reduced chla and growth were noticed in case of *Chlorella vulgaris* on exposure to PSMP and metals (Cu, Zn and Mn). Combined effect showed more inhibition on growth and chla content (Tunali et al., 2020). Long-term exposure of differently sized PSMP and PSNP on *Chlorella vulgaris* were proven to reduce cell viability, chla and also induced stress. PS particles of varying sizes 20, 50 and

Table 11.3 Effect of MPs/NPs on photosynthesis

Plant	MP/NP	Size	Effect	Reference
Cucumber	PSNPs	100, 300, 500, 700 nm	Reduction of chla and chlb Oxidative stress	Li et al. (2020a, b)
Lettuce	MP PE and DBP	Not mentioned	Reduced photosynthesis	Gao et al. (2019)
Microalgae	PSMPs	0.05, 0.5, 6 μm	Photosynthesis not effected Growth affected at high concentrations of uncharged particles	Sjollema et al. (2016)
Skeletonema costatum	mPVC	1 μm	Photosynthesis reduced Growth inhibited	Zhang et al. (2017)
Chlorella pyrenoidosa Microcystis (M.) flosaquae	PVC, PP	Not mentioned	Reduction in chla	Wu et al. (2019)
Chlorella pyrenoidosa	NP	80 nm	Growth inhibited Photosynthesis reduced Amino acyl tRNA synthetase blocked	Yang et al. (2021)
Chlamydomonas rheinhardtii	PSMP	Not mentioned	Reduction in chla	Li et al. (2020d)
Mycrocystis aeruginosa	PSNP (nPS-NH2) + glyphosate	200 nm	Combined effect decreased chla content but exhibited antagonistic effects PSNP decreased in chla content at 10 and 20 mg/ L Growth inhibition was observed in single agent treatment	Zhang et al. (2018)
Prochlorococcus	HDPE and PVC	–	Phosynthetic activity declined Growth retarded Oxygen production reduced	Tetu et al. (2019)

(continued)

Table 11.3 (continued)

Plant	MP/NP	Size	Effect	Reference
Chlorella vulgaris	PSMP, Cu Zn and Mn	0.5 μm	Reduced growth and chla content Joint effect of PSMP and metal is more toxic	Tunali et al. (2020)
Chlorella vulgaris	PSNP PSMP	20, 50, 500 nm	20 and 50 nm particles reduced chla concentration and cell viability Oxidative stress is induced by 20 nm	Hazeem et al. (2020)
Chlorella vulgaris	PS-NH_2	90, 200, 300 nm	Decrease in chla and cell biomass	Khoshnamvand et al. (2021)

PSMP Polystyrene microplastic, *PSNP* Polystyrene nanoplastic, *PE* Polyethylene, *PVC* Polyvinyl chloride, *PP* Poly propylene, *DBP* Di-butyl pthalate, *chla* Chlorophyll a, *chlb* Chlorophyll b

500 nm were used along with PS-COOH. Smaller-sized particles declined the amount of chla and cell viability. 20 nm particles induced stress and small-sized PSNPs were responsible for cell wall damage, which was visualised by SEM and TEM (Hazeem et al., 2020). A similar study was conducted by Khoshnamvand et al. (2021) using PSNP (PS-NH2) with diameter ranging 90, 200 and 300 nm. Diameters of 90 and 200 nm decreased chla and algal biomass. These studies convey the need to focus on the adverse effects of various plastic particles on the photosynthesis of phytoplankton, which is the primary producer and major source of oxygen in aquatic ecosystems.

Yang et al. (2021) studied the interaction between nanoplastics (80 nm) and *Chlorella pyrenoidosa.* They found the inhibitory effect of nanoplastics was greater than microplastics. Nanoplastics inhibited the growth by blocking gene expression of aminoacyl t-RNA synthetase enzyme at low concentrations. At high concentrations, photosynthesis was affected. The combined effect of PSNP (nPS-NH2) and glyphosate was studied in blue-green algae, *Mycrocystis aeruginosa.* PSNP at 5 mg/L was not as toxic as glyphosate at the same concentration. But after 96 h of exposure, chla content decreased, indicating a decrease in photosynthetic activity. The integrated effect on growth indicated an antagonistic effect of PS-NH2 and glyphosate (Zhang et al., 2018). The most widespread photosynthetic organism, *Prochlorococcus,* was used to study the effect of plastic leachates, HDPE and PVC. A rapid decline in the photosynthetic activity was observed at various concentrations used. Growth and oxygen production were also hindered (Tetu et al., 2019). In another research conducted by Sarker et al. (2020) on the impact of weathering on toxicity of these leachates, it was observed that toxicity declined gradually on weathering. But leachates produced even after weathering had shown to affect the growth and photosynthesis negatively. The toxicity of weathered and unweathered leachates varied on two strains of *prochlorococcus* tested. Zinc, which is the common additive in many plastics, was found to leach after 112 days of its entry into the environment. The impact of varying concentrations of zinc on growth of *Prochlorococuus* and *Synechococcus* was also focussed on (Sarker et al., 2021).

11.4 Conclusion

Plants, both terrestrial and aquatic, are the primary producers and act as sinks for these plastic particles, which in turn are severely affected by them. The outcomes of the exposure of Micro and nanoplastics on photosynthetic autotrophs are focussed on in the present chapter. Most of the research claimed the toxic effects of MPs/NPs on plants. Their toxicity is dependent on the type, size and concentration being employed. MPs/NPs exhibit direct effects by altering plant growth or indirectly by altering soil properties or soil microbial interaction, which will impair plant performance. The positive impact of these particles on plant growth is also witnessed in certain cases. Phytoplankton, which is an important primary producer and source of oxygen, is also negatively affected by the MPs/NPs that get accumulated in aquatic ecosystems in various ways. The effects of weathering on toxicity of plastic leachates on phytoplankton are also being investigated. MPs/NPs act as carriers of many toxic compounds that exhibit harmful effects on plants. The researches focussed on the effects of micro and nanoplastics on agricultural crops recommend focusing on safety of food and human health, as the internalised particles might get transferred at trophic levels. So, there is a need to encourage and focus our research on the impact of plastic pollution on plant functionalities.

References

Alimi, O. S., FarnerBudarz, J., Hernandez, L. M., & Tufenkji, N. (2018). Microplastics and nanoplastics in aquatic environments: Aggregation, deposition, and enhanced contaminant transport. *Environmental Science & Technology, 52*(4), 1704–1724.

Alomar, C., Estarellas, F., & Deudero, S. (2016). Microplastics in the Mediterranean Sea: Deposition in coastal shallow sediments, spatial variation and preferential grain size. *Marine Environmental Research, 115*, 1–10.

Andrady, A. L. (2011). Microplastics in the marine environment. *Marine Pollution Bulletin, 62*(8), 1596–1605.

Bandopadhyay, S., Martin-Closas, L., Pelacho, A. M., & DeBruyn, J. M. (2018). Biodegradable plastic mulch films: Impacts on soil microbial communities and ecosystem functions. *Frontiers in Microbiology, 9*, 819.

Bergami, E., Pugnalini, S., Vannuccini, M. L., Manfra, L., Faleri, C., Savorelli, F., Dawson, K. A., & Corsi, I. (2017). Long-term toxicity of surface-charged polystyrene nanoplastics to marine planktonic species Dunaliellatertiolecta and Artemiafranciscana. *Aquatic Toxicology, 189*, 159–169.

Bi, M., He, Q., & Chen, Y. (2020). *What roles are terrestrial plants playing in global microplastic cycling?* (Vol. 54, pp. 5325–5327).

Boots, B., Russell, C. W., & Green, D. S. (2019). Effects of microplastics in soil ecosystems: Above and below ground. *Environmental Science & Technology, 53*(19), 11496–11506.

Bosker, T., Bouwman, L. J., Brun, N. R., Behrens, P., & Vijver, M. G. (2019). Microplastics accumulate on pores in seed capsule and delay germination and root growth of the terrestrial vascular plant Lepidiumsativum. *Chemosphere, 226*, 774–781.

Boucher, J., & Friot, D. (2017). *Primary microplastics in the oceans: A global evaluation of sources* (Vol. 43). IUCN.

Browne, M. A., Crump, P., Niven, S. J., Teuten, E., Tonkin, A., Galloway, T., & Thompson, R. (2011). Accumulation of microplastic on shorelines woldwide: Sources and sinks. *Environmental Science & Technology, 45*(21), 9175–9179.

Chiam, Z., Song, X. P., Lai, H. R., & Tan, H. T. W. (2019). Particulate matter mitigation via plants: Understanding complex relationships with leaf traits. *Science of the Total Environment, 688*, 398–408.

Conti, G. O., Ferrante, M., Banni, M., Favara, C., Nicolosi, I., Cristaldi, A., Fiore, M., & Zuccarello, P. (2020). Micro-and nano-plastics in edible fruit and vegetables. The first diet risks assessment for the general population. *Environmental Research, 187*, 109677.

Corradini, F., Meza, P., Eguiluz, R., Casado, F., Huerta-Lwanga, E., & Geissen, V. (2019). Evidence of microplastic accumulation in agricultural soils from sewage sludge disposal. *Science of the Total Environment, 671*, 411–420.

Cox, K. D., Covernton, G. A., Davies, H. L., Dower, J. F., Juanes, F., & Dudas, S. E. (2019). Human consumption of microplastics. *Environmental Science & Technology, 53*(12), 7068–7074.

De Falco, F., Di Pace, E., Cocca, M., & Avella, M. (2019). The contribution of washing processes of synthetic clothes to microplastic pollution. *Scientific Reports, 9*(1), 1–11.

De Silva, Y. S. K., Rajagopalan, U. M., & Kadono, H. (2021). Microplastics on the growth of plants and seed germination in aquatic and terrestrial ecosystems. *Global Journal of Environmental Science and Management, 7*, 347–368.

de Souza Machado, A. A., Kloas, W., Zarfl, C., Hempel, S., & Rillig, M. C. (2018). Microplastics as an emerging threat to terrestrial ecosystems. *Global Change Biology, 24*(4), 1405–1416.

de Souza Machado, A. A., Lau, C. W., Kloas, W., Bergmann, J., Bachelier, J. B., Faltin, E., Becker, R., Görlich, A. S., & Rillig, M. C. (2019). Microplastics can change soil properties and affect plant performance. *Environmental Science & Technology, 53*(10), 6044–6052.

Dong, Y., Gao, M., Qiu, W., & Song, Z. (2021a). Uptake of microplastics by carrots in presence of As (III): Combined toxic effects. *Journal of Hazardous Materials, 411*, 125055.

Dong, Y., Song, Z., Liu, Y., & Gao, M. (2021b). Polystyrene particles combined with di-butyl phthalate cause significant decrease in photosynthesis and red lettuce quality. *Environmental Pollution, 278*, 116871.

Dovidat, L. C., Brinkmann, B. W., Vijver, M. G., & Bosker, T. (2020). Plastic particles adsorb to the roots of freshwater vascular plant Spirodelapolyrhiza but do not impair growth. *Limnology and Oceanography Letters, 5*(1), 37–45.

Duis, K., & Coors, A. (2016). Microplastics in the aquatic and terrestrial environment: Sources (with a specific focus on personal care products), fate and effects. *Environmental Sciences Europe, 28*(1), 1–25.

Esiukova, E. E., Lobchuk, O. I., Volodina, A. A., & Chubarenko, I. P. (2021). Marine macrophytes retain microplastics. *Marine Pollution Bulletin, 171*, 112738.

Europe, P. (2017). *Plastics-the facts 2017 an analysis of European plastics production, demand and waste data*. Available on the website: http://www.plasticseurope.org

Frias, J. P. G. L., & Nash, R. (2019). Microplastics: Finding a consensus on the definition. *Marine Pollution Bulletin, 138*, 145–147.

Fuller, S., & Gautam, A. (2016). A procedure for measuring microplastics using pressurized fluid extraction. *Environmental Science & Technology, 50*(11), 5774–5780.

Gao, M., Liu, Y., & Song, Z. (2019). Effects of polyethylene microplastic on the phytotoxicity of di-n-butyl phthalate in lettuce (Lactucasativa L. var. ramosaHort). *Chemosphere, 237*, 124482.

Gao, M., Liu, Y., Dong, Y., & Song, Z. (2021a). Effect of polyethylene particles on dibutyl phthalate toxicity in lettuce (Lactucasativa L.). *Journal of Hazardous Materials, 401*, 123422.

Gao, M., Xu, Y., Liu, Y., Wang, S., Wang, C., Dong, Y., & Song, Z. (2021b). Effect of polystyrene on di-butyl phthalate (DBP) bioavailability and DBP-induced phytotoxicity in lettuce. *Environmental Pollution, 268*, 115870.

Geyer, R., Jambeck, J. R., & Law, K. L. (2017). Production, use, and fate of all plastics ever made. *Science Advances, 3*(7), e1700782.

Gigault, J., Ter Halle, A., Baudrimont, M., Pascal, P. Y., Gauffre, F., Phi, T. L., El Hadri, H., Grassl, B., & Reynaud, S. (2018). Current opinion: What is a nanoplastic? *Environmental Pollution, 235*, 1030–1034.

Giorgetti, L., Spanò, C., Muccifora, S., Bottega, S., Barbieri, F., Bellani, L., & Castiglione, M. R. (2020). Exploring the interaction between polystyrene nanoplastics and Allium cepa during germination: Internalization in root cells, induction of toxicity and oxidative stress. *Plant Physiology and Biochemistry, 149*, 170–177.

Gong, W., Zhang, W., Jiang, M., Li, S., Liang, G., Bu, Q., Xu, L., Zhu, H., & Lu, A. (2021). Species-dependent response of food crops to polystyrene nanoplastics and microplastics. *Science of the Total Environment, 796*, 148750.

Guo, M., Zhao, F., Tian, L., Ni, K., Lu, Y., & Borah, P. (2022). Effects of polystyrene microplastics on the seed germination of herbaceous ornamental plants. *Science of the Total Environment, 809*, 151100.

Hazeem, L. J., Yesilay, G., Bououdina, M., Perna, S., Cetin, D., Suludere, Z., Barras, A., & Boukherroub, R. (2020). Investigation of the toxic effects of different polystyrene micro-and nanoplastics on microalgae Chlorella vulgaris by analysis of cell viability, pigment content, oxidative stress and ultrastructural changes. *Marine Pollution Bulletin, 156*, 111278.

Huang, Y., Liu, Q., Jia, W., Yan, C., & Wang, J. (2020). Agricultural plastic mulching as a source of microplastics in the terrestrial environment. *Environmental Pollution, 260*, 114096.

Hurley, R. R., & Nizzetto, L. (2018). Fate and occurrence of micro (nano) plastics in soils: Knowledge gaps and possible risks. *Current Opinion in Environmental Science & Health, 1*, 6–11.

Jiang, X., Chen, H., Liao, Y., Ye, Z., Li, M., & Klobučar, G. (2019). Ecotoxicity and genotoxicity of polystyrene microplastics on higher plant Viciafaba. *Environmental Pollution, 250*, 831–838.

Kalčíková, G., Gotvajn, A. Ž., Kladnik, A., & Jemec, A. (2017). Impact of polyethylene microbeads on the floating freshwater plant duckweed Lemna minor. *Environmental Pollution, 230*, 1108–1115.

Khalid, N., Aqeel, M., & Noman, A. (2020). Microplastics could be a threat to plants in terrestrial systems directly or indirectly. *Environmental Pollution, 267*, 115653.

Khoshnamvand, M., Hanachi, P., Ashtiani, S., & Walker, T. R. (2021). Toxic effects of polystyrene nanoplastics on microalgae Chlorella vulgaris: Changes in biomass, photosynthetic pigments and morphology. *Chemosphere, 280*, 130725.

Lagarde, F., Olivier, O., Zanella, M., Daniel, P., Hiard, S., & Caruso, A. (2016). Microplastic interactions with freshwater microalgae: Hetero-aggregation and changes in plastic density appear strongly dependent on polymer type. *Environmental Pollution, 215*, 331–339.

Li, L., Zhou, Q., Yin, N., Tu, C., & Luo, Y. (2019). Uptake and accumulation of microplastics in an edible plant. *Chinese Science Bulletin, 64*(9), 928–934.

Li, Z., Li, Q., Li, R., Zhao, Y., Geng, J., & Wang, G. (2020a). Physiological responses of lettuce (Lactucasativa L.) to microplastic pollution. *Environmental Science and Pollution Research, 27*, 30306–30314.

Li, Z., Li, R., Li, Q., Zhou, J., & Wang, G. (2020b). Physiological response of cucumber (Cucumissativus L.) leaves to polystyrene nanoplastics pollution. *Chemosphere, 255*, 127041.

Li, L., Luo, Y., Li, R., Zhou, Q., Peijnenburg, W. J., Yin, N., Yang, J., Tu, C., & Zhang, Y. (2020c). Effective uptake of submicrometre plastics by crop plants via a crack-entry mode. *Nature sustainability, 3*(11), 929–937.

Li, S., Wang, P., Zhang, C., Zhou, X., Yin, Z., Hu, T., Hu, D., Liu, C., & Zhu, L. (2020d). Influence of polystyrene microplastics on the growth, photosynthetic efficiency and aggregation of freshwater microalgae Chlamydomonasreinhardtii. *Science of the Total Environment, 714*, 136767.

Li, B., Huang, S., Wang, H., Liu, M., Xue, S., Tang, D., Cheng, W., Fan, T., & Yang, X. (2021a). Effects of plastic particles on germination and growth of soybean (Glycine max): A pot experiment under field condition. *Environmental Pollution, 272*, 116418.

Li, Z., Li, Q., Li, R., Zhou, J., & Wang, G. (2021b). The distribution and impact of polystyrene nanoplastics on cucumber plants. *Environmental Science and Pollution Research, 28*(13), 16042–16053.

Lian, J., Shen, M., & Liu, W. (2019). Effects of microplastics on wheat seed germination and seedling growth. *Journal of Agro-Environment Science, 38*(4), 737–745.

Lian, J., Wu, J., Xiong, H., Zeb, A., Yang, T., Su, X., Su, L., & Liu, W. (2020). Impact of polystyrene nanoplastics (PSNPs) on seed germination and seedling growth of wheat (Triticum aestivum L.). *Journal of Hazardous Materials, 385*, 121620.

Lian, J., Liu, W., Meng, L., Wu, J., Chao, L., Zeb, A., & Sun, Y. (2021a). Foliar-applied polystyrene nanoplastics (PSNPs) reduce the growth and nutritional quality of lettuce (Lactucasativa L.). *Environmental Pollution, 280*, 116978.

Lian, J., Liu, W., Meng, L., Wu, J., Zeb, A., Cheng, L., Lian, Y., & Sun, H. (2021b). Effects of microplastics derived from polymer-coated fertilizer on maize growth, rhizosphere, and soil properties. *Journal of Cleaner Production, 318*, 128571.

Liu, Y., Guo, R., Zhang, S., Sun, Y., & Wang, F. (2022). Uptake and translocation of nano/microplastics by rice seedlings: Evidence from a hydroponic experiment. *Journal of Hazardous Materials, 421*, 126700.

Lozano, Y. M., & Rillig, M. C. (2020). Effects of microplasticfibers and drought on plant communities. *Environmental Science & Technology, 54*(10), 6166–6173.

Lozano, Y. M., Lehnert, T., Linck, L. T., Lehmann, A., & Rillig, M. C. (2021). Microplastic shape, polymer type, and concentration affect soil properties and plant biomass. *Frontiers in Plant Science, 12*, 169.

Maity, S., Guchhait, R., Chatterjee, A., & Pramanick, K. (2021). Co-occurrence of co-contaminants: Cyanotoxins and microplastics, in soil system and their health impacts on plant–A comprehensive review. *Science of the Total Environment, 794*, 148752.

Mao, Y., Ai, H., Chen, Y., Zhang, Z., Zeng, P., Kang, L., Li, W., Gu, W., He, Q., & Li, H. (2018). Phytoplankton response to polystyrene microplastics: Perspective from an entire growth period. *Chemosphere, 208*, 59–68.

Martin-Closas, L., Botet, R., & Pelacho, A. M. (2014). An in vitro crop plant ecotoxicity test for agricultural bioplastic constituents. *Polymer Degradation and Stability, 108*, 250–256.

Mateos-Cárdenas, A., Scott, D. T., Seitmaganbetova, G., van Pelt Frank, N. A. M., & AK, J. M. (2019). Polyethylene microplastics adhere to Lemna minor (L.), yet have no effects on plant growth or feeding by Gammarusduebeni (Lillj.). *Science of the Total Environment, 689*, 413–421.

Mateos-Cárdenas, A., van Pelt, F. N., O'Halloran, J., & Jansen, M. A. (2021). Adsorption, uptake and toxicity of micro-and nanoplastics: Effects on terrestrial plants and aquatic macrophytes. *Environmental Pollution, 284*, 117183.

Meng, F., Yang, X., Riksen, M., Xu, M., & Geissen, V. (2021). Response of common bean (Phaseolus vulgaris L.) growth to soil contaminated with microplastics. *Science of the Total Environment, 755*, 142516.

Mintenig, S. M., Löder, M. G. J., Primpke, S., & Gerdts, G. (2019). Low numbers of microplastics detected in drinking water from ground water sources. *Science of the Total Environment, 648*, 631–635.

Naidoo, T., Glassom, D., & Smit, A. J. (2015). Plastic pollution in five urban estuaries of KwaZulu-Natal, South Africa. *Marine pollution bulletin, 101*(1), 473–480.

Ng, E. L., Lwanga, E. H., Eldridge, S. M., Johnston, P., Hu, H. W., Geissen, V., & Chen, D. (2018). An overview of microplastic and nanoplastic pollution in agroecosystems. *Science of the Total Environment, 627*, 1377–1388.

Pehlivan, N., & Gedik, K. (2021). Particle size-dependent biomolecular footprints of interactive microplastics in maize. *Environmental Pollution, 277*, 116772.

Pflugmacher, S., Tallinen, S., Kim, Y. J., Kim, S., & Esterhuizen, M. (2021). Ageing affects microplastic toxicity over time: Effects of aged polycarbonate on germination, growth, and oxidative stress of Lepidium sativum. *Science of the Total Environment, 790*, 148166.

Piehl, S., Leibner, A., Löder, M. G., Dris, R., Bogner, C., & Laforsch, C. (2018). Identification and quantification of macro-and microplastics on an agricultural farmland. *Scientific Reports, 8*(1), 1–9.

Pignattelli, S., Broccoli, A., & Renzi, M. (2020). Physiological responses of garden cress (L. sativum) to different types of microplastics. *Science of The Total Environment, 727*, 138609.

Pignattelli, S., Broccoli, A., Piccardo, M., Terlizzi, A., & Renzi, M. (2021). Effects of polyethylene terephthalate (PET) microplastics and acid rain on physiology and growth of Lepidium sativum. *Environmental Pollution, 282*, 116997.

Prata, J. C., Lavorante, B. R., Maria da Conceição, B. S. M., & Guilhermino, L. (2018). Influence of microplastics on the toxicity of the pharmaceuticals procainamide and doxycycline on the marine microalgae Tetraselmis chuii. *Aquatic Toxicology, 197*, 143–152.

Prata, J. C., da Costa, J. P., Lopes, I., Duarte, A. C., & Rocha-Santos, T. (2019). Effects of microplastics on microalgae populations: A critical review. *Science of the Total Environment, 665*, 400–405.

Qi, Y., Yang, X., Pelaez, A. M., Lwanga, E. H., Beriot, N., Gertsen, H., Garbeva, P., & Geissen, V. (2018). Macro-and micro-plastics in soil-plant system: Effects of plastic mulch film residues on wheat (Triticum aestivum) growth. *Science of the Total Environment, 645*, 1048–1056.

Ren, X., Tang, J., Wang, L., & Liu, Q. (2021). Microplastics in soil-plant system: Effects of nano/microplastics on plant photosynthesis, rhizosphere microbes and soil properties in soil with different residues. *Plant and Soil, 462*(1), 561–576.

Rillig, M. C., & Lehmann, A. (2020). Microplastic in terrestrial ecosystems. *Science, 368*(6498), 1430–1431.

Rillig, M. C., Ingraffia, R., & de Souza Machado, A. A. (2017). Microplastic incorporation into soil in agroecosystems. *Frontiers in Plant Science, 8*, 1805.

Rillig, M. C., Lehmann, A., de Souza Machado, A. A., & Yang, G. (2019). Microplastic effects on plants. *New Phytologist, 223*(3), 1066–1070.

Rillig, M. C., Leifheit, E., & Lehmann, J. (2021). Microplastic effects on carbon cycling processes in soils. *PLoS Biology, 19*(3), e3001130.

Rindy, J. E., Ponette-González, A. G., Barrett, T. E., Sheesley, R. J., & Weathers, K. C. (2019). Urban trees are sinks for soot: Elemental carbon accumulation by two widespread oak species. *Environmental Science & Technology, 53*(17), 10092–10101.

Sarker, I., Moore, L. R., Paulsen, I. T., & Tetu, S. G. (2020). Assessing the toxicity of leachates from weathered plastics on photosynthetic marine bacteria Prochlorococcus. *Frontiers in Marine Science, 7*, 777.

Sarker, I., Moore, L. R., & Tetu, S. G. (2021). Investigating zinc toxicity responses in marine Prochlorococcus and Synechococcus. *Microbiology, 167*(6), 001064.

Schiavo, S., Oliviero, M., Chiavarini, S., Dumontet, S., & Manzo, S. (2021). Polyethylene, Polystyrene, and Polypropylene leachate impact upon marine microalgae Dunaliella tertiolecta. *Journal of Toxicology and Environmental Health, Part A, 84*(6), 249–260.

Seng, N., Lai, S., Fong, J., Saleh, M. F., Cheng, C., Cheok, Z. Y., & Todd, P. A. (2020). Early evidence of microplastics on seagrass and macroalgae. *Marine and Freshwater Research, 71*(8), 922–928.

Sfriso, A. A., Tomio, Y., Juhmani, A. S., Sfriso, A., Munari, C., & Mistri, M. (2021). Macrophytes: A temporary sink for microplastics in transitional water systems. *Water, 13*(21), 3032.

Sharma, S., & Chatterjee, S. (2017). Microplastic pollution, a threat to marine ecosystem and human health: A short review. *Environmental Science and Pollution Research, 24*(27), 21530–21547.

Sintim, H. Y., & Flury, M. (2017). Is biodegradable plastic mulch the solution to agriculture's plastic problem? *Environmental Science and Technology, 51*, 1068–1069.

Sjollema, S. B., Redondo-Hasselerharm, P., Leslie, H. A., Kraak, M. H., & Vethaak, A. D. (2016). Do plastic particles affect microalgal photosynthesis and growth? *Aquatic Toxicology, 170*, 259–261.

Sun, X. D., Yuan, X. Z., Jia, Y., Feng, L. J., Zhu, F. P., Dong, S. S., Liu, J., Kong, X., Tian, H., Duan, J. L., & Ding, Z. (2020). Differentially charged nanoplastics demonstrate distinct accumulation in Arabidopsis thaliana. *Nature Nanotechnology, 15*(9), 755–760.

Sun, H., Lei, C., Xu, J., & Li, R. (2021). Foliar uptake and leaf-to-root translocation of nanoplastics with different coating charge in maize plants. *Journal of Hazardous Materials, 416*, 125854.

Sutherland, W. J., Aveling, R., Brooks, T. M., Clout, M., Dicks, L. V., Fellman, L., Fleishman, E., Gibbons, D. W., Keim, B., Lickorish, F., & Monk, K. A. (2014). A horizon scan of global conservation issues for 2014. *Trends in Ecology & Evolution, 29*(1), 15–22.

Tetu, S. G., Sarker, I., Schrameyer, V., Pickford, R., Elbourne, L. D., Moore, L. R., & Paulsen, I. T. (2019). Plastic leachates impair growth and oxygen production in Prochlorococcus, the ocean's most abundant photosynthetic bacteria. *Communications biology, 2*(1), 1–9.

Teuten, E. L., Saquing, J. M., Knappe, D. R., Barlaz, M. A., Jonsson, S., Björn, A., Rowland, S. J., Thompson, R. C., Galloway, T. S., Yamashita, R., & Ochi, D. (2009). Transport and release of chemicals from plastics to the environment and to wildlife. *Philosophical Transactions of the Royal Society B: Biological Sciences, 364*(1526), 2027–2045.

Thompson, R. C., Olsen, Y., Mitchell, R. P., Davis, A., Rowland, S. J., John, A. W., McGonigle, D., & Russell, A. E. (2004). Lost at sea: Where is all the plastic? *Science, 304*(5672), 838–838.

Tunali, M., Uzoefuna, E. N., Tunali, M. M., & Yenigun, O. (2020). Effect of microplastics and microplastic-metal combinations on growth and chlorophyll a concentration of Chlorella vulgaris. *Science of the Total Environment, 743*, 140479.

Urbina, M. A., Correa, F., Aburto, F., & Ferrio, J. P. (2020). Adsorption of polyethylene microbeads and physiological effects on hydroponic maize. *Science of the Total Environment, 741*, 140216.

van Kleunen, M., Brumer, A., Gutbrod, L., & Zhang, Z. (2020). A microplastic used as infill material in artificial sport turfs reduces plant growth. *Plants, People, Planet, 2*(2), 157–166.

vanWeert, S., Redondo-Hasselerharm, P. E., Diepens, N. J., & Koelmans, A. A. (2019). Effects of nanoplastics and microplastics on the growth of sediment-rooted macrophytes. *Science of the Total Environment, 654*, 1040–1047.

Wang, F., Wong, C. S., Chen, D., Lu, X., Wang, F., & Zeng, E. Y. (2018). Interaction of toxic chemicals with microplastics: A critical review. *Water Research, 139*, 208–219.

Wang, F., Wang, X., & Song, N. (2021). Polyethylene microplastics increase cadmium uptake in lettuce (Lactucasativa L.) by altering the soil microenvironment. *Science of the Total Environment, 784*, 147133.

Wu, Y., Guo, P., Zhang, X., Zhang, Y., Xie, S., & Deng, J. (2019). Effect of microplastics exposure on the photosynthesis system of freshwater algae. *Journal of Hazardous Materials, 374*, 219–227.

Wu, J., Liu, W., Zeb, A., Lian, J., Sun, Y., & Sun, H. (2021). Polystyrene microplastic interaction with Oryzasativa: Toxicity and metabolic mechanism. *Environmental Science: Nano, 8*(12), 3699–3710.

Xu, B., Liu, F., Cryder, Z., Huang, D., Lu, Z., He, Y., Wang, H., Lu, Z., Brookes, P. C., Tang, C., & Gan, J. (2020). Microplastics in the soil environment: Occurrence, risks, interactions and fate–a review. *Critical Reviews in Environmental Science and Technology, 50*(21), 2175–2222.

Xu, G., Liu, Y., & Yu, Y. (2021). Effects of polystyrene microplastics on uptake and toxicity of phenanthrene in soybean. *Science of the Total Environment, 783*, 147016.

Yan, N., Tang, B. Z., & Wang, W. X. (2021a). Cell cycle control of nanoplastics internalization in phytoplankton. *ACS Nano, 15*, 12237–12248.

Yan, Z., Xu, L., Zhang, W., Yang, G., Zhao, Z., Wang, Y., & Li, X. (2021b). Comparative toxic effects of microplastics and nanoplastics on Chlamydomonasreinhardtii: Growth inhibition, oxidative stress, and cell morphology. *Journal of Water Process Engineering, 43*, 102291.

Yang, W., Gao, X., Wu, Y., Wan, L., Tan, L., Yuan, S., Ding, H., & Zhang, W. (2020). The combined toxicity influence of microplastics and nonylphenol on microalgae Chlorella pyrenoidosa. *Ecotoxicology and Environmental Safety, 195*, 110484.

Yang, W., Gao, P., Li, H., Huang, J., Zhang, Y., Ding, H., & Zhang, W. (2021). Mechanism of the inhibition and detoxification effects of the interaction between nanoplastics and microalgae Chlorella pyrenoidosa. *Science of the Total Environment, 783*, 146919.

Yano, K. A., Geronimo, F. K., Reyes, N. J., & Kim, L. H. (2021). Characterization and comparison of microplastic occurrence in point and non-point pollution sources. *Science of the Total Environment, 797*, 148939.

Yin, L., Wen, X., Huang, D., Du, C., Deng, R., Zhou, Z., Tao, J., Li, R., Zhou, W., Wang, Z., & Chen, H. (2021). Interactions between microplastics/nanoplastics and vascular plants. *Environmental Pollution, 290*, 117999.

Yoshino, M., Osawa, H., & Yamanaka, A. (2012). Effects of process conditions on nano-dot array formation by thermal dewetting. *Journal of Manufacturing Processes, 14*(4), 478–486.

Yu, H., Peng, J., Cao, X., Wang, Y., Zhang, Z., Xu, Y., & Qi, W. (2021a). Effects of microplastics and glyphosate on growth rate, morphological plasticity, photosynthesis, and oxidative stress in the aquatic species Salviniacucullata. *Environmental Pollution, 279*, 116900.

Yu, Z. F., Song, S., Xu, X. L., Ma, Q., & Lu, Y. (2021b). Sources, migration, accumulation and influence of microplastics in terrestrial plant communities. *Environmental and Experimental Botany, 192*, 104635.

Zhang, C., Chen, X., Wang, J., & Tan, L. (2017). Toxic effects of microplastic on marine microalgae Skeletonema costatum: Interactions between microplastic and algae. *Environmental Pollution, 220*, 1282–1288.

Zhang, Q., Qu, Q., Lu, T., Ke, M., Zhu, Y., Zhang, M., Zhang, Z., Du, B., Pan, X., Sun, L., & Qian, H. (2018). The combined toxicity effect of nanoplastics and glyphosate on Microcystis aeruginosa growth. *Environmental Pollution, 243*, 1106–1112.

Zhang, Y., Yang, X., Luo, Z. X., Lai, J. L., Li, C., & Luo, X. G. (2021a). Effects of polystyrene nanoplastics (PSNPs) on the physiology and molecular metabolism of corn (Zea mays L.) seedlings. *Science of the Total Environment, 806*, 150895.

Zhang, Q., Zhao, M., Meng, F., Xiao, Y., Dai, W., & Luan, Y. (2021b). Effect of polystyrene microplastics on Rice seed germination and antioxidant enzyme activity. *Toxics, 9*(8), 179.

Zhou, J., Wen, Y., Marshall, M. R., Zhao, J., Gui, H., Yang, Y., Zeng, Z., Jones, D. L., & Zang, H. (2021). Microplastics as an emerging threat to plant and soil health in agroecosystems. *Science of the Total Environment, 787*, 147444.

Zhu, Z. L., Wang, S. C., Zhao, F. F., Wang, S. G., Liu, F. F., & Liu, G. Z. (2019). Joint toxicity of microplastics with triclosan to marine microalgae Skeletonema costatum. *Environmental Pollution, 246*, 509–517.

Zhu, X., Zhao, W., Chen, X., Zhao, T., Tan, L., & Wang, J. (2020). Growth inhibition of the microalgae Skeletonema costatum under copper nanoparticles with microplastic exposure. *Marine Environmental Research, 158*, 105005.

Zong, X., Zhang, J., Zhu, J., Zhang, L., Jiang, L., Yin, Y., & Guo, H. (2021). Effects of polystyrene microplastic on uptake and toxicity of copper and cadmium in hydroponic wheat seedlings (Triticumaestivum L.). *Ecotoxicology and Environmental Safety, 217*, 112217.

Chapter 12
Cellular and Animal Toxicities of Micro- and Nanoplastics

Manikantha D, C. D. S. L. N. Tulasi, and Kalyani Chepuri

Abstract Plastic-based items are abundantly found on the globe because of their immense utility in daily lives. The poor biodegradability of plastics, particularly micro- and nanoplastics, has recently sparked environmental concerns around the world. These anthropogenic pollutants are either generated, particularly in the tiny size range, for diverse commercial applications or result from the environmental fragmentation of macropolymers. Micro- and nanoplastics are now found in large quantities in the oceans, freshwater bodies, and on land, as well as in food. Micro- and nanoplastics' biological effects on aquatic creatures are extensively known, but their effects on human systems have not been thoroughly examined. The potential pathways of exposure to micro- and nanoplastics, the biological consequences of these particles in human cells, factors influencing toxicity, and the likely mechanisms of cytotoxicity are all discussed in this chapter. In general, cellular toxicity appears to be induced by oxidative stress, membrane damage, immunological response, and genotoxicity in micro/nanoplastics due to their tiny size, positive charge, high dose, and inclusion of hazardous chemicals or contaminants. A thorough understanding of these chemicals' cellular destiny and toxicity may aid in extrapolating dangers to mammals.

Keywords Microplastics · Environmental concerns · Biological effects · Mechanism of toxicity

12.1 Introduction

In our daily lives, plastic and plastic-based goods are extremely useful. From 15 million tonnes in 1964 to 359 million tonnes in 2018, global plastic manufacturing surged by around 24 times (Vidal, 2020; World Economic Forum, 2016). Plastic is now mostly used in packaging (26% of total Vidal, 2020; World Economic Forum,

M. D · C. D. S. L. N. Tulasi · K. Chepuri (✉)
Centre for Biotechnology, Jawaharlal Nehru Technological University Hyderabad, Institute of Science & Technology, Hyderabad, Telangana, India

N. R. Maddela et al. (eds.), *Micro and Nanoplastics in Soil*,
https://doi.org/10.1007/978-3-031-21195-9_12

2016 production), but it is also used in other industries such as electronics, construction, transportation, healthcare, and agriculture. The poor biodegradability of these polymers, on the other hand, poses a significant environmental risk. Every year, an estimated eight million tonnes of plastic waste enters the oceans, with 269,000 tonnes of plastic floating on the surface (National Oceanic and Atmospheric Administration, 2019). Plastics can break down into microplastics (0.1 m–5 mm diameter) or nanoplastics (0.1 m diameter) (NJDEP-Science Advisory Board, 2015), over time due to microbial degradation, extended ultraviolet radiation exposure, or physical wear. Microplastics (MPs)/Nanoplastics (NPs) are also produced for use in air blasting technologies, cleansers, cosmetics, medicine delivery formulations, paints, and toothpaste, adding to the MPs/NPs pool already present in the environment (Cole et al., 2011; Sharma & Chatterjee, 2017). Clothing, cigarette filters, automobile tires, and fishing equipment are also potential sources (Toussaint et al., 2019) (Fig. 12.1).

Plastics in the environment interact with terrestrial and marine biota, causing considerable worry about severe ecological consequences. Plastics consumed by

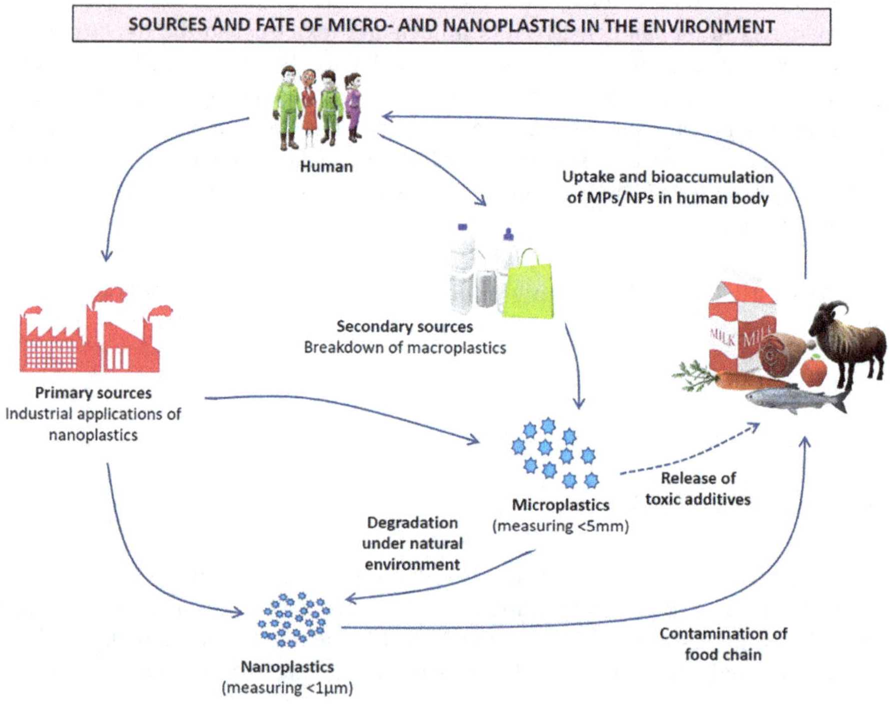

Fig. 12.1 Sources and fate of micro and nanoplastics in the environment. Consumers and industries produce MPs/NPs from primary and secondary sources. Degradation of macroplastic materials that dissolve into micron-sized particles into nanoplastics. MPs/NPs found in both the aquatic and terrestrial environment, eventually entering the food chain and water supplies, resulting in their uptake and bioaccumulation in the human body

organisms can bioaccumulate and make their way up the food chain to humans via trophic transfer (Carbery et al., 2018; Chae et al., 2018; Nelms et al., 2018). In addition, individuals can be exposed to plastic through the eating of plastic-contaminated food or the inhalation of plastic-polluted air (Prata et al., 2020). Seafood (fish, shrimp, mussels), home products (sea salt, honey, sugar, plastic tea bags), tap water, bottled water, beer, construction sites, factories, and agriculture have all been found to contain micronized plastics (Barbosa et al., 2020; de Souza Machado et al., 2018; EFSA Panel on Contaminants in the Food Chain, 2016; Hesler et al., 2019; Karami et al., 2017; Mason et al., 2018). The body's excretory system is estimated to discharge >90% of micro-and nanoplastics in feces once consumed (Smith et al., 2018). The human stool has been found to contain 50–500 m sized PP and PET microplastics (20 particles/10 g stool) (Schwabl et al., 2019).

Due to ethical considerations, no meta-analysis clinical trial has been/can be undertaken to evaluate health risks in humans except from risk assessment data extracted from in vivo experiments (Yang et al., 2019a, b). As a result, the health effects of MPs/NPs on humans are unknown. It is unknown whether MPs/NPs can be absorbed and bioaccumulated by humans through ingestion/inhalation or other modes of exposure. The pharmacokinetics, pharmacodynamics, and variables that determine the pharmacological response to MPs/NPs are still unknown. In the absence of clinical evidence, in vitro investigations in human or other mammalian cells may be able to shed light on these issues. This chapter looks at recent research that looked at the biological effects of MPs/NPs in mammalian cells using various exposure methods. The need of addressing MPs/NPs toxicity about particle size, dose, charge, exposure period, additives/leachates, and/or other co-contaminants has been emphasized.

The cellular pathways that contribute to toxicity after MPs/NPs internalization are also discussed, as well as recommendations for future research. The information offered in this research will aid in a better understanding of the potential implications of human plastic exposure. Because in vitro research using weathered particles is essentially missing, the focus is mostly on studies utilizing MPs/NPs purposefully generated in the micro/nano-size range. Particle preparation techniques were notably noted in the review for experiments involving particles generated from technologies that mimicked the environmental degradation of bigger polymers.

12.2 Polymer Types of MPs/NPs

A variety of pathways could lead to breaking down plastics into macroplastics (>25 mm), mesoplastics (5–25 mm), microplastics (5 mm), and nanoplastics (0.1 μm). Microplastics (MPs) are plastic particles with a diameter of less than 5 mm that can be found in the environment in sizes ranging from a few microns to a few millimeters, and even nano-sized particles, which often have an unevenly mixed state (Boyle & Örmeci, 2020).

12.2.1 Primary Type

Primary (initially and consciously created for industrial and domestic uses within a microscopic size) and secondary (originally and purposefully manufactured for industrial and domestic applications within a microscopic size) origins are both present in microplastics (resulting from the continuous breakdown of large plastic debris). Primary microplastics (microbeads) are widely used in cosmetics formulations such as makeup, sunscreen, nail polish, hair coloring, eye shadow, shower gels, and personal care products containing scrubs and toothpaste, facial cleansers, and air-blasting.

12.2.2 Secondary Type

Secondary microplastics are formed by the breakdown and degradation of large plastic debris into small fragments when exposed to high solar UV radiation and mechanical abrasion as a result of a combination of physical (mechanical), chemical (photolytic), and biological processes and can be transported directly into marine environments from coastlines, rivers, and sewage pipes.

There are numerous different varieties of polymers, each of which can be classified as either natural or synthetic (Koelmans et al., 2015). PET (polyethylene terephthalate), HDPE (high-density polyethylene), PVC (polyvinyl chloride), LDPE (low-density polyethylene), PP (polypropylene), PS (polystyrene), and PU (polyurethane) are examples of synthetic polymers (PUR). Another important factor that affects the floating and sinking of MPs/NPs, as well as the removal rate, is chemical composition. There are currently around 30 different MPs/NPs types available. More than 30 different types of MPs/NPs polymers have indeed been discovered so far (Sun et al., 2019).

12.3 Detection of MPs/NPs

Understanding the behavior and bioavailability of microplastics requires precise knowledge of physical and chemical properties (i.e., form, size, polymer compositions, and functional groups) (Fu et al., 2020). Separation, identification, quantification, and characterization of plastics in terms of physicochemical attributes are all part of the detection process. MPs/NPs can be characterized in a variety of ways, including microscopic, chromatographic, and sophisticated spectroscopic techniques (Mintenig et al., 2018; Shim et al., 2017).

12.3.1 Separation

The initial and most important stage in the separation of MPs/NPs is usually accomplished using sieves with various mesh sizes. For MPs/NPs separation, these can be employed alone or in a sequence (Hollman et al., 2013). Filters with a fine mesh are commonly used to separate small MPs of size 5 μm (Löder & Gerdts, 2015). Furthermore, chromatographic techniques, both active and passive separation, are used to separate the majority of MPs/NPs with a size range of 1 μm that is of plastic origin (Mintenig et al., 2018). Field-Flow Fractionation (FFF) is used in active separation, while Hydrodynamic Chromatography (HDC) is used in passive separation (Mendoza & Jones, 2015). Both methods, when combined with sophisticated techniques such as GC-MS, size-exclusion chromatography, and plasma mass spectroscopy, have been demonstrated to be useful in quantifying and characterizing MPs/NPs of various chemical forms of PS, PE, and PACR with sizes ranging from 50 to 9900 nm and 90 to 106 m (Gigault et al., 2017; Correia & Loeschner, 2018; Philippe et al., 2014; Pirok et al., 2017). In the described study, MPs/NPs were extracted from tap water, surface water, and fish samples.

12.3.2 Visualization

A second phase in the identifying process is visualization. Large MPs are often recognized with a standard microscope, and their shape, color, and light transmittance can later be used to separate them from a combination of non-plastics (Hidalgo-Ruz et al., 2012). The hue of plastic litter may reflect its state of degradation and could be used as a proxy for environmental exposure duration (Marti et al., 2020). Polypropylene fibers, for example, were discovered to be typically hazy or red, whereas milky white color forms PS, and yellow and brown color generates PE, PP, PVC, PS, and PET (Eriksen et al., 2013; Brandon et al., 2016; Vianello et al., 2013). Though large microplastics with distinguishable colors or morphologies can be visually sorted and identified, particles without distinguishable color or form are difficult to sort with the naked eye. To identify confusing plastic-like particles, electron microscopy with magnified pictures is required (Song et al., 2015). SEM-EDS is a technique that combines scanning electron microscopy with energy-dispersive X-ray spectroscopy to characterize the shape of amazingly small materials and estimate their chemical constituents (Goldstein et al., 2017). SEM could produce high-resolution topographical images of objects, allowing microplastics to be distinguished from those other plastic-like particles more easily (Cooper & Corcoran, 2010). By identifying the characteristic X-rays released from the elements well within the specimen by the electron beam, EDS offers elemental information about the samples, allowing for certain characteristics of micro-plastic recognition in sample composites. The identification of several MPs/NPs (PP, PS, PE, PA) in seawater, shallow waterways, and beaches with sizes ranging from 1 to

5 mm was aided by these forms of microscope visualization (Fries et al., 2013). Fluorescence Microscopy has also proven to be a promising method for identifying plastic particles in seawater and studying their effects on marine assemblages and settling rates. For the first time, Qiu et al. (2015) used FCM to examine the presence and prevalence of microplastics (PET, PE, HDPE, and PS) with a size range smaller than 5 mm in China (Qiu et al., 2015). The tendency of zooplankton to consume microplastics was studied using fluorescence microscopy. This study found that marine microplastic debris (especially PS) with sizes ranging from 7.3 to 30.6 μm) can have a deleterious influence on zooplankton physiology and overall health (Cole et al., 2013). Similarly, fluorescence microscopy was employed in another work to discover the existence of fluorescent microplastic beads in copepods (PS with sizes ranging from 0.05 to 6 μm). The findings revealed that micro or nanoscale PS beads may reduce the survival rate and fertility of marine copepods. The colorful plastic fibers in the sample were observed using fluorescence microscopy, and this study successfully quantified the prevalence of MPs of uncertain origin with sizes ranging from 0.5 to 1 mm in Swedish west coast waters (Sweden, 2007). Fluorescence microscopy was utilized to confirm the integration of microspheres (Fluoresbrite carboxylate) with sizes ranging from 3.6 to 11 μm into planktons and to assist in microsphere quantification (Okubo et al., 2018).

12.3.3 Characterization

Physiological features of MPs/NPs can be examined at the third level of characterization. The hydrodynamic size as well as the surface charge of particles has been studied extensively using modern technologies such as Dynamic light scattering (DLS) (Bhattacharjee, 2016). In addition, the DLS approach can be integrated into other systems to provide quick and easy identification of microplastic deterioration. Gigault et al. used a photo-reactor in conjunction with DLS to study the photocatalytic degradation of ocean microplastic particles under various conditions without the need for sample or manipulation (15). In addition to hydrodynamic size measurement, nanoparticles tracking analysis (NTA) is an improved approach for measuring concentrations of poly-dispersed substances. With monochrome photography, NTA illuminates free diffusing particles with strong laser light to trace their Brownian motion (Gigault & Budzinski, 2016). The other sophisticated approach for testing the presence of organic substances or carbon in surface and ground waters is fluorescence spectrophotometry. The use of fluorescence spectroscopy to evaluate toxicity and explore the detrimental impacts of microplastics on microorganisms in soil and water, such as suppression of enzyme activity and energy metabolism, has been demonstrated (Henderson et al., 2009). Chen et al. (2018) mapped and described the microstructure of MPs (PS) with diameters of 20 mm, 6 mm, 500 nm, and 80 nm to analyze MPs' environmental behavior. Similarly, detection of accumulated microplastics, particularly (PS) of size 5 μm in the gills, liver, and gut of crabs, detection of cadmium, lead, and bromine in MPs of uncertain origin with sizes 5–10 mm from beach waters, and measurement of NPs concentration (PS) of sizes 45 μm and 50 nm in zebrafish larvae and nematodes (Yu et al., 2018; Massos & Turner, 2017; Chen et al., 2017, 2018; Zhao et al., 2017).

Finally, advanced technology like Fourier transform infrared (FTIR) spectroscopy, can be used to analyze the chemical/functional group composition of MPs/NPs. It collects chemical data by sensing the modes of vibration of analyte at various infrared frequencies throughout a broad-spectrum range (Stuart, 2005). From sand samples collected in Sishili Bay, North Yellow Sea, FTIR was used to detect eight polymer kinds of MPs (Rayon, PE, PP, PA, PET, PS, PMMA, and PU) with diameters ranging from 34.97 to 4983.73 μm. These findings show that river and sewage discharge, as well as maritime activities, were the main sources of MP pollution (Zhang et al., 2019). Yang et al. (2019a, b) used FTIR to demonstrate the presence of MPs (PET, PS, PP) of varied sizes in municipal wastewater from China's biggest water reclamation plant (Yang et al., 2019a, b). FTIR was also used to determine the existence of airborne MPs (PET, PES, PAN, RY, EVA, PAA, EP, ALK) in China's atmosphere, as well as the polymer kinds of MPs (PET, PP, PS, Nylon) deposits in the Pacific Ocean (Liu et al., 2018a, b; Peng et al., 2020). Raman spectroscopy is another prominent biochemical characterization mapping technique that uses the Raman effect to extract the vibrational modes and identify analytes of samples by using the frequency response of inelastically scattered light from the samples (Araujo et al., 2018). Micro-Raman spectroscopy can detect MPs with a resolution of up to 10 μm, whereas Raman spectroscopy can detect MPs greater than roughly 1 μm (Imhof et al., 2016). Several other methods, such as matrix-assisted laser desorption/ionization time-of-flight mass spectrometry (MALDI-TOF-MS) (Karas & Krüger, 2003), high-performance liquid chromatography (HPLC) (Siddiqui et al., 2017), and atomic force microscopy and its combinations with IR and Raman as a unique scanning probe technique, open up new opportunities in microplastics and nanoplastics characterization. AFM may be used to detect and quantify a variety of material physical and mechanical properties like elasticity, surface electric properties, and chemical properties (Akhatova et al., 2022). Several studies have also shown the presence of different micro and nanoplastics present in various food stuffs (Table 12.1).

12.4 Exposure of MPs/NPs to the Biological System

MPs/NPs can be ingested, inhaled, absorbed through the skin, or administered intravenously (NJDEP-Science Advisory Board, 2015). When particles are swallowed, they first come into contact with the gut mucosa, followed by the epithelia, which together provide a formidable barrier to xenobiotic uptake. Several investigations have revealed, however, that micro/nanoparticles can pass through the intestinal barrier and enter the bloodstream (Jenkins et al., 1994; Reineke et al., 2013; Walczak et al., 2015). Airborne MPs/NPs would come into direct contact with the mucus layer, periciliary layer, ciliated cells, non-ciliated secretory cells, and basal cells of

Table 12.1 Studies assessing the micro- and nanoplastics present in food stuffs

S. No.	Food stuffs	Polymer	Particle size	References
1.	Fish	PE, PP, PA, PS, PET, PVC, PAN	>20 μm	Su et al. (2019)
2.	Clam, oyster, mud snails	PE, PET, PA	10–5000 nm in diameter	Naji et al. (2018)
3.	Mussel	PE CP, PET, PA, PS, PP	150–6000 μm	Nalbone et al. (2021)
4.	Crab, prawn	PE, PP, PET	–	Akhbarizadeh et al. (2019)
5.	Oyster	PET, PA, PE, PS, PP, PPS	4–2100 μm	Teng et al. (2019)
6.	Poultry meat (packed)	Epoxy resin, rayon, PET, PS, PMPS	130–250 μm	Kedzierski et al. (2020)
7.	Apple	PET, poly (ether-urethrane) PET, PVCA, PVC, PES, PEVA, PVK	1.99 μm	Conti et al. (2020)
8.	Pear		2.10 μm	
9.	Broccoli		1.51 μm	
10.	Potato		20 μm	
11.	Lettuce		20 μm	
12.	Sea salt	PP, PET, PS, PP	40–170 μm	Kosuth et al. (2018)
13.	Tap water	MP fragments	50–500 μm	Mintenig et al. (2019)
14.	Bottled water	MP fragments, fibers	–	Schymanski et al. (2018)

Modified from Llorca and Farré (2021)

CP Cellophane, *HDPE* high-density polyethylene, *PA* polyamide, *PAA* polyacrylic acid, *PAN* polyacrylonitrile, *PE* polyethylene, *PES* poly {p-phenylene ether sulfone}, *PET* polyethylene terephthalate, *PEVA* polyethylene-vinyl-acetate, *PMPS* polymethyl pentene, *PP* polypropylene, *MP* microplastics, *PPS* polyphenylene sulphide, *PS* polystyrene, *PVC* polyvinyl chloride, *PVCA* vinyl chloride/vinyl avetate copolymer, *PVK* poly {N-vinyl carbazole}

the respiratory tract (Ganesan et al., 2013; Gasperi et al., 2018). When inhaled repeatedly, plastic fibers can infiltrate lung tissues, causing inflammation and subsequent genotoxicity (Gasperi et al., 2018). In the lungs of textile workers, granulomatous lesions harboring foreign substances (perhaps polyester, nylon, or acrylic dust) were discovered (Pimentel et al., 1975). Because the stratum corneum impedes the passage of molecules larger than 500 Da (1 nm) across the skin layers, transdermal absorption of MPs/NPs via intact skin is unlikely (Bos & Meinardi, 2000). MPs/NPs can also enter the bloodstream through plastic-based intravenous catheters, syringes, and other medication delivery methods (Stapleton, 2019). Because they are "inert and biocompatible," polystyrene micro and nanoparticles are frequently utilized as vectors for medication delivery or to study bio-interactions (Loos et al., 2014a; Poon et al., 2016). Plastic absorption by cells, as well as the release of plastic additives or surface-adsorbed pollutants, can have a deleterious impact on cell function (Bouwmeester et al., 2015).

12.5 Factors That Influence Their Cytotoxicity

Plastic toxicity varies depending on the polymer type. Based on the hazard classification of monomers, polyurethane, polyacrylonitriles, PVC, epoxy resins, and styrene-based copolymers have been classified as the most dangerous (category 1A or 1B mutagen or carcinogen) (Lithner et al., 2011). Several factors can influence particle cytotoxicity within a plastic class.

12.5.1 Size of MPs/NPs and Dosage

Small particles are often internalized to a higher extent by cells than large particles (Florence et al., 1995). Smaller particles can be taken up via endocytic or passive absorption, but bigger particles require phagocytosis by specialized cells (Alberts et al., 2002). Internalization of 44 nm PS particles was an ATP-independent passive process in two primary mammalian cell lines – bovine oviductal epithelial cells and human colon fibroblasts (Fiorentino et al., 2015). In most cases, particle size and toxicity have an inverse relationship. Particles smaller than 10 nanometers are thought to act as a gaseous substance that can easily infiltrate tissues and cause broad damage (Bahadar et al., 2016). When evaluated using 100 L of 1 mg/mL particle solution in monomac-6 human monocytic cells, 64 nm PS particles induced a considerable rise in intracellular Ca^{2+} levels compared to bigger PS particles (202 and 535 nm) (Brown et al., 2001). The 64 nm particles produced higher IL-8 expression in human lung cancer A549 cells after 2 hours than the 202 and 535 nm particles (Brown et al., 2001).

Smaller particles can be more easily absorbed through the colon or lungs, affecting particle cellular fate and biodistribution. Particles larger than 150 nm can have local effects in the stomach, whereas smaller particles can cause toxicity in a variety of secondary organs and tissues (EFSA Panel on Contaminants in the Food Chain, 2016; Rubio et al., 2020). For particle intake and transport, however, a Goldilocks zone for size may exist. For example, 40 nm PS particles were found to have a greater absorption in 1321N1 human astrocytoma and A549 cells than 20 and 100 nm PS particles (Varela et al., 2012). The transfer of 200 nm PS particles through the cells in a microfluidic model of the blood-brain barrier (BBB) utilizing hCMEC/D3 cells was higher than that of 100 and 500 nm particles (Nowak et al., 2019).

High doses and long-term exposure can cause increased cellular absorption and toxicities. For example, at 20 and 50 g/mL doses, 20 nm plain PS particles did not cause toxicity in THP-1 monocytes in 24 hours, but at 200 g/mL, cellular viability was reduced to 12% (Mrakovcic et al., 2014). Cell number reduced considerably after 16 days of incubation with 50 g/mL of 20 nm PS particles, exhibiting dosage- and time-dependent toxicity. Similar findings were made with human umbilical vein EAhy 926 cells, where 20 nm plain PS particles displayed dose-dependent cytotoxicity in 24 hours, with an IC_{50} value of 120 g/mL (Mrakovcic et al., 2013). Cell number dropped by 50% after a 28-day incubation period with a 20 g/mL dosage of these particles.

12.5.2 Charge

Surface charge influences particle absorption, translocation, and toxicity. Positively charged PS particles transported 20 and 100–120 nm aminated or carboxylated PS particles 20–40 times faster than negatively charged PS particles across rat alveolar epithelial cell monolayers (Yacobi et al., 2010). Furthermore, cationic particles exhibit higher cytotoxicity in non-phagocytic cells than their anionic counterparts, owing to plasma membrane damage (Frohlich, 2012). Aminated PS particles (110 nm) decreased THP-1 cell proliferation, although carboxylated PS particles of similar size did not affect cell division (Loos et al., 2014a). The viability of THP-1 cells was similarly dramatically reduced after 72 hours of incubation with 10–100 g/mL aminated particles but not with carboxylated PS particles (Loos et al., 2014b).

12.5.3 Additives

Stabilizers, plasticizers, lubricants, dyes, and flame retardants are among the plastic additives/leachates that make up an average of 4% of microplastic content and can be harmful (Bouwmeester et al., 2015; Campanale et al., 2020; EFSA Panel on Contaminants in the Food Chain, 2016). Bisphenol A, phthalates, and brominated flame-retardants, all of which alter endocrine function, are of special concern (Campanale et al., 2020; De Toni et al., 2017; Legler & Brouwer, 2003; Rubin, 2011). At temperatures above 60 °C, commercially available PET water bottles leached Sb into the water; temperatures that could be reached if bottles were left inside cars and garages during the summer (Westerhoff et al., 2008). Surfactants can lyse cell membranes or affect the structure and function of cell surface receptors, glycoproteins, proteoglycans, signaling molecules, extracellular matrix components, and lipid rafts, to mention a few (Yong et al., 2020). However, the presence and release of additives do not always constitute a health risk, as toxicity is determined by the plastic composition and the velocity of leachate migration, or the amount and solubility of leachate in the surrounding environment. In fatty foods and when stored at high temperatures or for long periods, there is a higher migration of additives from plastics (Hahladakis et al., 2018).

12.5.4 Adsorbed Pollutants

MPs/NPs can absorb additional pollutants such as persistent organic pollutants (POP), heavy metals, and pathogens due to their small size and high surface to volume ratio (de Souza Machado et al., 2018; Yu et al., 2019). Persistent organic pollutants (polychlorinated biphenyls, polycyclic aromatic hydrocarbons, DDT), heavy metals (Cd, Cr, Cu, Zn, Sb, Al, Br, Hg, As, Sn, Ti, Co, Ba, Mn), and microorganisms (pathogenic vibrio spp.) can all be vectored by plastics (Brennecke et al., 2016; Campanale et al., 2020; Kirstein et al., 2016; Prinz & Korez, 2020; Rodrigues et al., 2019; Velzeboer et al., 2014). Pyrene and BDE-47 were carried by polystyrene nanoparticles (100 nm) in saturated soil (Liu et al., 2018a).

12.6 Toxicity Caused to Human Cells (In Vitro)/Potential Effects on Organ System

It's crucial to look at MPs/NPs uptake and biological consequences in cells that are either immediately exposed to them or come into contact with them after systemic absorption. A list compiled of recent MPs/NPs toxicity studies was completed in mammalian cells from 2001 to 2020. MPs/NPs can be internalized by gastrointestinal, airway, immune, and other miscellaneous cell types and induce various cellular responses, the nature and extent of which may be governed by MPs/NPs size, dose, charge, exposure time, and the presence of additives/leachates/co-contaminants, as described in this section (Fig. 12.2).

12.6.1 Immune Cells

Immune cells serve as the body's gatekeepers, assisting in the clearance of infections and xenobiotics. As a result, they are likely to interact with MPs/NPs found in food, water, and air, as well as those absorbed into the systemic circulation via various routes of exposure. Smaller particles can enter immune cells via clathrin/caveolae-mediated endocytosis, clathrin/caveolae-independent internalization, micropinocytosis, and phagocytosis, while microparticles can be ingested by immune cells via phagocytosis or micropinocytosis (Firdessa et al., 2014).

MPs/NPs can be quickly internalized by immune cells such as monocytes, macrophages, fibroblasts, and mast cells (Heinlaan et al., 2020). At dosages less than 100 μg/mL, nano- and sub-micron-sized particles did not appear to cause cytotoxicity in THP-1 macrophages. In most cases, the positive charge and tiny size exacerbated toxicity (Hwang et al., 2019). The viability of cells was harmed by particle leachates and additions. However, PS particles were used in the bulk of this research to explore biological effects in immune cells. To fully comprehend the influence of

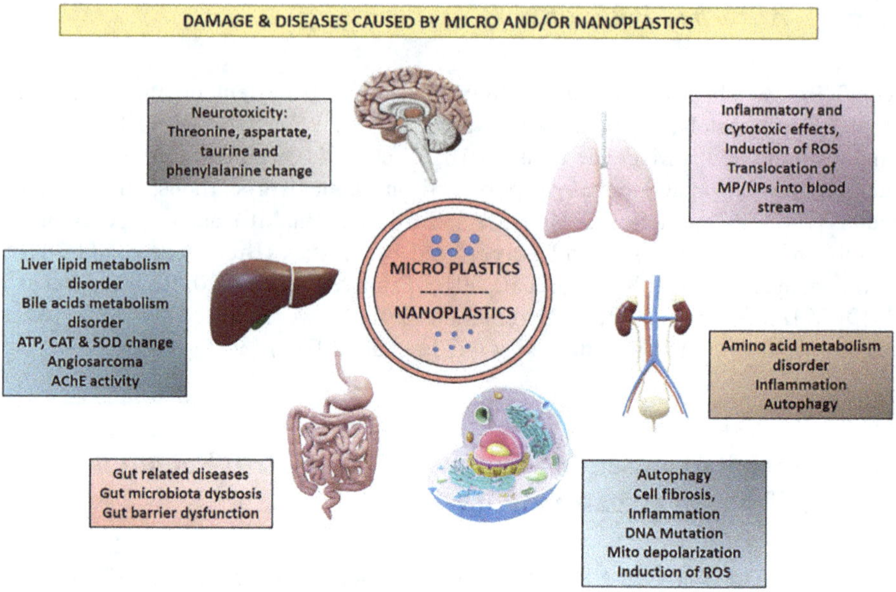

Fig. 12.2 Damage and diseases caused by micro- and/or nanoplastics

diverse plastic types on immune cells, more research employing different plastic polymers is required.

12.6.2 *Gastrointestinal Cells*

MPs/NPs toxicity in gastrointestinal cells has been intensively researched because ingestion of MPs/NPs contaminated food or water is the primary source of exposure to these particles. Enterocytes, mucus-producing goblet cells, and microfold or M cells make up the majority of the intestinal epithelia. Caco-2 cells, a commonly used in vitro model for enterocytes, were found to ingest 100-nm carboxylated PS particles largely by micropinocytosis-mediated uptake at the apical surface, followed by mostly storage or limited exocytosis at the basolateral membrane end (Reinholz et al., 2018). Regular epithelial cell shedding every 4–5 days could remove polystyrene particles retained in the cells (Reinholz et al., 2018). Diffusion across the cell membrane followed by basal exocytosis was a secondary, although limited, mechanism of uptake and excretion.

In vitro investigations of gastric cells show that nanoplastics can be taken up by stomach cells and that smaller particles are potentially more harmful than bigger particles. To better understand MPs/NPs toxicity in the stomach, more research is needed in other gastric cells such as SNU-1, SNU-5, and KATO III, utilizing particles of various sizes, charges, dosages, and exposure times (Liao & Yang, 2020). Numerous researches have looked into the toxicity of MPs/NPs in the intestine, however, no consensus has been reached on size/charge-dependent transport or toxicity (Baos et al., 2012; Chen et al., 2020; Collado-Gonzalez et al., 2019). To reconcile the contradictory data, a factorial study design assessing the effects of various plastic polymers, charges, sizes, doses, and exposure times in intestinal cells is required.

12.6.3 Airway Cells

Clothes, dried sludge, agriculture, tires, manufacturing operations, and sea salt aerosols are all reported sources of airborne microplastics (Wright & Kelly, 2017). Microplastics were found in air fallout in Paris, with a higher quantity in densely populated areas than in less densely populated areas (Gasperi et al., 2018; Wright & Kelly, 2017). When breathed in, plastic particles may pass through mucociliary clearance systems in the respiratory tract, especially if the particle size is more than 1 nm, or (ii) pass through the pleura and be absorbed by lung epithelial cells. As seen among workers in nylon flock, polyester, polyolefin, and polyamide fiber plants, inhaled plastic dust can cause respiratory distress such as irritation of the respiratory tract, dyspnea, decreased lung capacity, coughing, increased phlegm production, interstitial fibrosis, and granulomatous lesions (Wright & Kelly, 2017). Human lung tissues have also been discovered to contain plastic fibers (Pauly et al., 1998). Fibers made of polypropylene, polyethylene, and polycarbonate can last up to 6 months in extracellular lung fluid (Gasperi et al., 2018). The size, type, concentration, and duration of exposure to these particulate materials all influence the health risks they pose (occupational vs occasional). To assess if the airborne MPs/NPs constitute a health risk, researchers must first determine how they interact with respiratory cells (uptake, transport, cytotoxic potential, and metabolic effect).

MPs/NPs can be absorbed by several immortalized and primary airway epithelial (bronchial, alveolar) cells, according to in vitro investigations. Positively charged beads, a greater dose, and the inclusion of additives are all linked to increased cytotoxicity. The influence of nano- and sub-micron-sized PS particles on respiratory cells has been the focus of research. To model chronic exposure to varied MPs/NPs, however, the toxicity of different plastic polymers and at longer time points (72, 96, or more) are required. Furthermore, investigations involving airway cells at the air-lung interface are required for a more realistic understanding of the interaction between MPs/NPs exposure and lung damage.

12.6.4 Mammalian Cells

The toxicity of MPs/NPs has been investigated in a variety of mammalian cells, including blood, cerebral, endothelial, epithelial, hepatic, kidney, melanoma, ovarian, and placental cells, due to the possibility of systemic exposure or absorption.

T98G glioma cells and HeLa epithelial cells were used to test the cytotoxicity of PE microspheres (0.1, 0.6, and 3–16) and PS particles (40–250 nm and 10) (Schirinzi et al., 2017). In 24 hours, the particles (0.05–10 mg/L) did not affect cellular viability in either cell line. In Madin Darby canine kidney cell II monolayers, charge-dependent quicker trafficking of PS particles was reported, with amidine-functionalized 20 and 120 nm PS beads translocating 500 times faster than 20 and 100 nm carboxyl-functionalized beads (Fazlollahi et al., 2011). In ovarian cancer cells SK-OV-3 and NIH-OVCAR3, 50 nm amine-functionalized PS particles were quickly taken up, accumulated in lysosomes, and caused cytotoxicity within 4–8 hours, whereas 30 nm carboxyl functionalized beads did not accumulate in lysosomes and were not cytotoxic even after 24 hours of treatment (Ekkapongpisit et al., 2012). The hemolytic potential of 100 nm PS particles isolated from commercial face washes was compared to 100 nm virgin PS particles in erythrocytes (Gopinath et al., 2019). Overnight incubation with isolated particles at 5 g/mL resulted in 40% hemolysis, but only 22% with virgin particles. The inclusion of additives or other toxic polymers on the surface of isolated particles was blamed for the greater toxicity of isolated PS compared to virgin PS.

These findings show that a variety of non-phagocytic cells can internalize MPs/NPs, with size and charge being the most important factors. When compared to large, negative, or non-functionalized particles, smaller, positively charged particles are more likely to be taken in and cause cellular damage. HeLa and T98 glioblastoma cells, on the other hand, demonstrate no size-dependent toxicity. Small (200 nm) particles can be internalized by red blood cells, whereas large (1000 nm) particles cannot. The toxic effects shown by different micro- and nanoplastics are mentioned below (Table 12.2).

12.6.5 Animals

Micro and nanoplastics have long been recognized as common contaminants in the environment. Their existence has been established in water bodies (fresh & marine), terrestrial systems, as well as the air we breathe (Rillig & Lehmann, 2020). MPs/NPs are taken up by animals, dispersed in their bodies, and deposited in several tissues, from which they were later transported widely through food chains, according to a growing body of evidence (Zhang et al., 2020) As a result, bioaccumulated micro and nanoplastics could endanger human health and the ecosystems (de Souza Machado et al., 2018) (Fig. 12.3).

MPs/NPs have an impact on the growth, development, and reproduction of organisms and can even cause mortality in individuals. These tend to generate oxidative stress reactions in organisms, disrupt pigment formation or enzyme activities, create endocrine and metabolic abnormalities, and cause various degrees of genotoxicity, cytotoxicity, and neurotoxicity due to their small size. Simultaneously, as the food chain spreads, these consequences are aggregated and magnified step by

Table 12.2 Micro- and nanoplastics and their potential toxic effects on human health

S.No.	Characteristics of plastic particles	Particle size	Toxic effects	References
1.	Polystyrene particles	202, 535 nm	Inflammation in A549 cells	Brown et al. (2001)
2.	Carboxylated polystyrene particles	20, 44, 500, 1000 nm	Upregulation of IL-6 & 8 expression	Forte et al. (2016) Prietl et al. (2014)
3.	Carboxylated and amino modified polystyrene particles	120 nm	Altered expression of scavenger receptors, increased TGFβ1 and energy metabolism	Fuchs et al. (2016)
4.	Unaltered polyethylene particles	0.3,10 μm	Increased secretion of IL-6, IL-1β, and TNFα in murine macrophages	Green et al. (1998)
5.	Polyethylene particles from plastic prosthetic implants	0.2,10 μm	Periprosthetic bone resorption	Nich and Goodman (2014)
6.	Polystyrene microplastics particles	5, 20 μm	Adverse effects on neurotransmission, inflammation in the liver	Deng et al. (2017)
7.	Amine modified polystyrene nanoparticles	60 nm	Apoptosis induction in all intestinal epithelial cells	Inkielewicz-Stepniak et al. (2018)
8.	Unaltered/ functionalized polystyrene	20, 40, 50, 100 nm	Apoptosis induction in several human cells	Liu et al. (2018a, b) Paget et al. (2015)
9.	PVC (poly vinyl chloride) PMMA (poly methyl methacrylate)	120, 140 nm	Reduced cell viability with a reduction of ATP and increase of ROS concentrations.	Mahadevan and Valiyaveettil (2021)
10.	Cationic polystyrene nanoparticles	50, 60, 200 nm	ROS generation, ER stress, autophagic cell death of mouse macrophages and lung epithelial cells, disrupted intestinal iron transport and cellular uptake	Xia et al. (2008) Chiu et al. (2015) Mahler et al. (2012)
11.	Pristine and fluorescent polystyrene microplastics	5 μm	Altered amino acid and bile acid metabolism, gut microbiota dysbiosis, intestinal barrier dysfunction	Luo et al. (2019) Jin et al. (2019)
12.	Anionic carboxylated polystyrene nanoparticles	20 nm	Altered ion channel function and ionic homeostasis	McCarthy et al. (2011)

(continued)

Table 12.2 (continued)

S.No.	Characteristics of plastic particles	Particle size	Toxic effects	References
13.	Polystyrene nanoparticles	30 nm	Blocked vesicle transport and the distribution of cytokinesis-associated proteins	Xia et al. (2016)
14.	Pristine polystyrene microparticles	5, 20 μm	Hepatic ATP level reduction and impairment of energy metabolism	Lu et al. (2018)

Modified from Yee et al. (2021)

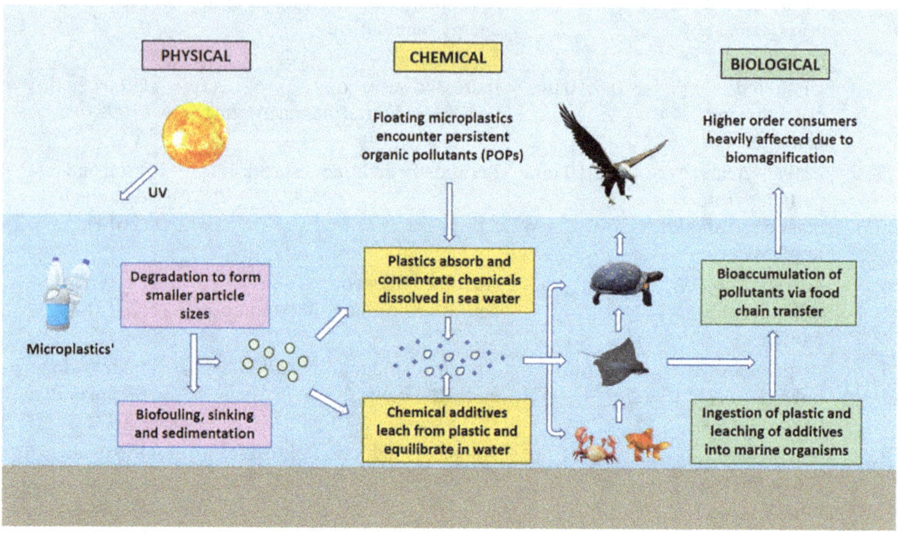

Fig. 12.3 The route of exposure of micro- and nanoplastics into the food chain. (Modified from Braden Wilkinson, 2019)

step from people to populations to communities and finally to ecosystems, thereby worsening the fragile natural system (Ma et al., 2020).

Smaller microplastics can significantly affect algal growth, fertility, and even disturb photosynthesis (Chen et al., 2020). It was observed that HDPE (High-Density Polyethylene) plastic beads of 10–45 μm reduced the filtering rate of *Dreissena bugensis* (quagga mussel) (Pedersen et al., 2020). Many studies have shown that MPs/NPs can pile up in the intestines of earthworms and that the growth activity was significantly lowered and the rate of mortality increased at 28, 45, and 60% w/w microplastics (Gaylor et al., 2013; Wang et al., 2020; Huerta Lwanga et al., 2016). Earthworms' immune systems and pathological responses were also reduced by microplastics (Rodriguez-Seijo et al., 2017). In soil contaminated with microplastics, the mobility of springtails (*Lobella sokamensis*) was reduced (Kim & An, 2019). As a result, microplastics may obstruct the movement of terrestrial animals by blocking gaps.

MPs/NPs can alter the early stages of development in aquatic animals. The embryonic stage is critical for aquatic animal development, and embryonic chorion serves as an effective barrier against exogenous contaminants. It was reported that zebrafish (*Danio rerio*) embryonic chorions had an effective barrier property against micro and nanoparticles. Though embryonic chorions can effectively prevent micro- and nanoplastics, they can however have an impact on aquatic species' early development. MPs/NPs attaching to embryonic chorions may lead to an internal hypoxic microenvironment within embryos, as well as a delay in hatching. When embryos are exposed to polystyrene particles, specifically nanopolystyrene particles, the pathways of biosynthesis of unsaturated fatty acids, linoleic acid metabolism, alanine, and also glutamate, and aspartate metabolism, are significantly altered (Duan et al., 2020).

The absence of tools for characterization and measurement of these particles in complicated biological matrices has impeded research into micro and nanoplastic buildup in animal bodies. In microplastic research, ocular inspection (Sobhani et al., 2020), Fourier-transform infrared spectroscopy (González-Pleiter et al., 2019), and Raman spectroscopy (Gillibert et al., 2019) are being employed. However, these methods are often limited to particles with a diameter of 5 mm to 20 μm, with only a few studies focused on the sub-20 μm portion (Cole et al., 2013). Furthermore, organic debris on the micro and nanoplastics surfaces makes spectroscopic identification and quantification difficult. Pyrolysis combined with gas chromatography-mass spectrometry (Py-GC/MS) is a promising method for quantifying nanoplastics (Fischer & Scholz-Böttcher, 2017; Zhou et al., 2018). Even though this technique may be used to reliably identify and quantify micro and nanoplastics without regard to particle size, extracting and determining micro and nanoplastics from complicated biological matrices is a simple task (Mitrano et al., 2019).

12.7 Cellular and Molecular Interactions Caused by MPs and NPs

MPs/NPs toxicity is thought to be caused by membrane damage, oxidative stress, immunological response, and genotoxicity. Among these, MP/NP's cytotoxicity has been attributed mostly to membrane damage and oxidative stress (Bhattacharjee et al., 2014). Cationic particles, for example, have been known to damage the plasma membrane (Feng et al., 2019; Frohlich, 2012). Polyethylene nanoparticles were discovered to enter the plasma membrane bilayer's hydrophobic milieu and cause structural alterations (Holloczki & Gehrke, 2020). Endocytosed particles can permeabilize the endosomal-lysosomal membrane, allowing them to interact with intracellular organelles (Wang et al., 2018a, b; Yong et al., 2020). ROS are produced during the polymerization and processing of plastic particles, and when they come into contact with the biological environment, they cause cellular stress (Rubio et al., 2020). While big particles can cause inflammation in the gastrointestinal and respiratory tracts, smaller particles can pass through the gut/lung barrier, causing intracellular oxidative stress and cytotoxicity in the organs where they collect (Rubio et al., 2020). Direct or indirect DNA damage caused by particle or ROS translocation into the nucleus, as well as disruption to the DNA replication/repair machinery, can all contribute to particle genotoxicity (Rubio et al., 2020) (Fig. 12.4).

MPs/NPs can disrupt nuclear membranes, generate oxidative stress, release damage-associated molecular patterns, and activate inflammatory and apoptotic/necrotic pathways in mammalian cells (Hwang et al., 2020; Yong et al., 2020). Hepatocytes from 3-month-old mice have been treated with 50 nm PS particles for 24 hours, causing an increase in ROS (superoxide dismutase and malondialdehyde concentration) and DNA damage (Zheng et al., 2019). PS beads caused superoxide radical anion ($O_2^{\cdot -}$) production in human hepatocyte-derived cancer Huh-7 cells,

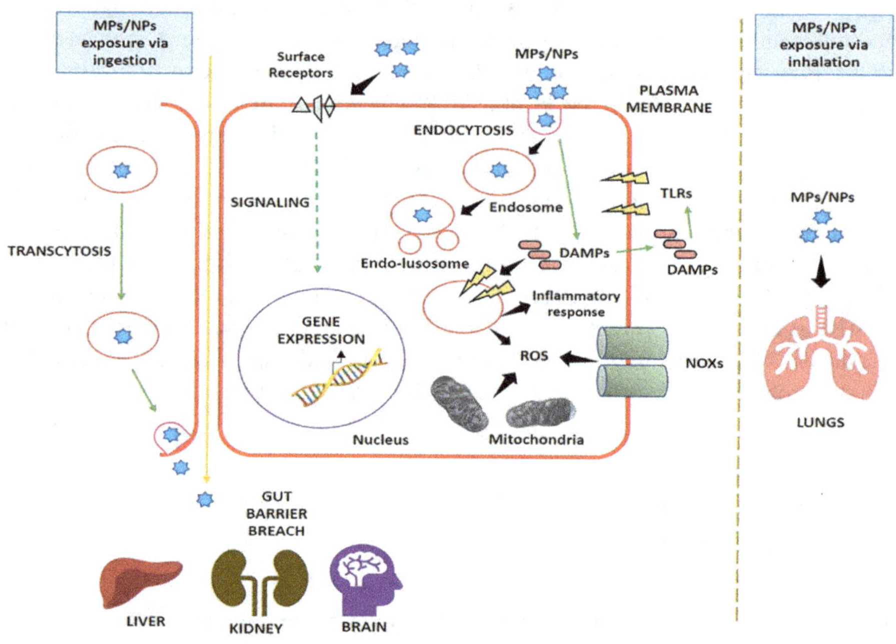

Fig. 12.4 A schematic illustration depicting possible cellular processes of MP/NP toxicity. Ingestion and inhalation are two ways to absorb MPs/NPs. These have the potential to disrupt the plasma membrane and compromise the gut barrier (left). These could also disrupt cell surface receptor signaling and change gene expression in the nucleus. Endocytosed MPs/NPs have the potential to disrupt the endocytic process and impair endosomal membranes. Endogenous and secreted damage-associated molecular patterns (DAMP) triggering the innate immunity-mediating toll-like receptors (TLRs) could activate the cellular innate immune system as a result of the aforesaid stresses. Stress may cause the NADP oxidases to produce reactive oxygen species (ROS) (NOXs). Mitochondrial dysfunction, whether caused by MPs/NPs from endosomes or as a result of stress, could result in an increase in ROS due to a decrease in the efficiency of electron transport chain (ETC) operations. If the gut–vascular barrier is breached, MPs/NPs gain access to the circulation, or transcytosis may occur, allowing them to reach other organs. The lung is more likely to have direct contact with airborne MPs/NPs (right). (Modified from Yong et al., 2020)

according to another study (Liu et al., 2018b). The toxicity of positively charged PS beads in RAW 264.7 and BEAS-2B cells, on the other hand, was found to be attributable to autophagy via the Akt/mTOR and AMPK pathways (Chiu et al., 2015). The majority of research points to an oxidative stress-mediated cellular response to MPs/NPs when taken together. Many of the aforementioned toxicity processes, however, are closely interrelated, and induction of one process might trigger a cascade of toxicological responses.

12.8 Regulatory Policies/International, National, and Regional Instruments

"Microplastics" (MPs) are a hot topic in the media and one of the most fiercely debated environmental issues among the general population. As a result, the public wants policymakers to address and handle the issue as quickly as possible (Sharma et al., 2021). In reality, policymakers are becoming more conscious of the problem. Some of the most powerful and influential international and intergovernmental organizations are debating the global effect of environmental plastics (e.g., G7, World Bank, United Nations, World Economic Forum, etc.) (Brennholt et al., 2018). Aside from that, the (micro)plastic/nanoplastics issue has been addressed in a few international and national rules and policy instruments. Because the majority of environmental MPs are caused by improper disposal and fragmented plastic litter, MP management is directly linked to a variety of policy areas (Coffin et al., 2021). Furthermore, regulatory responsibilities can shift throughout a single plastic product's life cycle and include plastic production process layout, trade and consumer behavior, recycling, and waste management (Deme et al., 2022). It can also be called land-based policies, as well as sewage management and water protection, also called water-based policy (Freeman et al., 2020). As a result, plastics regulation is already addressed in several directives, recommendations, agreements, and other documents addressing the use of plastic products, beginning with restrictions on plastic monomer compositions and the addition of additives. Policy and regulatory instruments are now being established all over the globe at international/ regional/ and national levels to handle the problem of (micro) plastics in the environment (Mitrano & Wohlleben, 2020). National policy instruments are limited to a single country, whereas regional policy instruments address specific issues within a geographical region, such as Europe. International accords and regional treaties, for example, are incorporated into national legislation. For the first time in January 2018, Europe enacted ESPCE (European Strategy for Plastics in a Circular Economy), which altered the way the European Union designs, manufactures, uses, and recycles plastic items (Campanale et al., 2020). Many instruments, such as the G7 Summit (2014), the G20 Summit (2017), and the United Nations Environment Assembly (UNEA) I, II, III in 2014, 2016, and 2017, have been vital in preventing marine litter from land-based sources and in reducing marine plastic litter and microplastics, as well as combating their spread. National instruments in countries such as the United States, the United Kingdom, Sweden, Canada, Italy, New Zealand, South Korea, Australia, Taiwan Province, and China primarily impose laws prohibiting the use of microbeads in rinse-off cosmetics, microbeads in toiletries, plastic cotton buds, microbead scrub particles in cosmetics, and the sale of microbead-containing products (Xu et al., 2021; Deng et al., 2020). Better plastic product design, increased waste plastic recycling rates, etc., and high-quality recycles will all assist to improve the demand for recycled plastics, protecting the environment, reducing marine debris, greenhouse gas emissions, and our dependence on external fossil fuels (Gago et al., 2020).

The Ministry of Ecology and Environment issued "Opinions on Further Strengthening the Control of Plastic Pollution" in January 2020, outlining three phases of action. Prohibiting and regulating the manufacture, sale, and use of certain plastic products, for example, shopping bags that are ultra-thin with 0.025 mm thickness. Simultaneously, we should encourage the use of non-plastic goods (such as paper bags and biodegradable shopping bags) and standardize plastic trash recycling and disposal. In addition, Japan, South Korea, Thailand, and other countries have also implemented harsher rules in recent years. The European Commission adopted new guidelines in May of the same year for ten typical throwaway plastics goods and fishing gear using plastics to minimize or limit the environmental effect of individual plastic products. In addition, Plastics Europe proposed the "Plastic 2030" voluntary commitment, which included "Zero Plastics to Landfill," "Zero Pellet Loss," and other initiatives aimed at preventing plastic leakage into the surroundings and trying to improve the resource productivity of plastic items and concentricity of plastic packaging. Some regions have seen positive results and increased public awareness of environmental preservation after China enacted the "Plastics Restriction Order" in 2008, however, this is far from enough (Wang et al., 2020). These policies, in contrast to previous rules and regulations, focus on the entire life cycle of plastic products, including the overall process and each link of manufacturing, circulation, use, reprocessing (such as mechanical and chemical recycling and energy recovery), and disposal, making it easier to establish a long-term mechanism for controlling plastic pollution. Preventing the flow of low-end plastic items from developed to developing countries, from places with high supervision and utilization competency to those with inadequate supervision and use competence, is critical. This will not only result in a circular economy but will also protect our planet and help us accomplish our sustainable development goals.

12.9 Conclusion/Future Directions

Although their cytotoxicity is mostly determined by their size, surface functionalization, dose, exposure period, and presence of co-contaminants, studies have revealed that micro/nanoplastics are not 'inert' materials. These pollutants have been found in our food, drinking water, and air, implying that we are constantly exposed to them. As has been found with occupational PVC dust exposure, chronic exposure to plastics can lead to bioaccumulation and subsequent biological consequences. Organic contaminants and heavy metals, for example, are plastic additives or compounds that hitchhike on plastic surfaces, posing additional toxicity concerns that require further exploration. Studies on the cytotoxicity of MPs/NPs in mammalian cells revealed that MPs/NPs with small size and positive charge, when delivered at high doses for long periods and including surfactants or other adsorbed contaminants, have higher toxicity. These conclusions, however, may not apply to all cells and particle kinds. Furthermore, the influence of certain physicochemical characteristics of particles in particular cell types has yet to be thoroughly understood. A multi-end-point toxicological investigation utilizing MPs/NPs of varied types and physicochemical features at environmentally realistic concentrations in human cells is required for a complete knowledge of the health impact of MPs/NPs.

Various physicochemical parameters of MPs/NPs have different toxicological implications in different cell types, and even within the same cell type, there are discrepancies. Several of these discrepancies can be attributed to changes in experimental settings, particle type/synthesis/source/extraction method, end-points studied, and the difficulties of regulating other parameters during a study. A multi-factorial study design that uses specific cell types, uniform experimental circumstances, and end-points could help to resolve the inconsistencies in cellular responses.

In many cell types, size has been demonstrated to govern cellular uptake and viability among the other particle properties. Gastric cells, hepatocytes, immune cells, RBCs, squamous carcinoma, melanoma, and umbilical cells, for example, have higher bioreactivity with small particles, but there is no clear consensus with airway, intestinal, or ovarian cancer cells. This could be because various cell types have an optimal size range for enthusiastic internalization. Positive charges on MPs/NPs increased toxicity in the airway, immune, ovarian cancer, MDCK-II kidney cells, adrenal medulla, mammary epithelial cells, and HEPA-1 hepatocytes at particular concentrations.

Positively charged particles, in general, can interact electrostatically with the negatively charged phospholipid cell membrane, resulting in greater internalization than negatively or neutrally charged particles (Foroozandeh & Aziz, 2018). High binding of positively charged particles to the plasma membrane, on the other hand, might raise surface tension and cause membrane portion or deformation (Li & Malmstadt, 2013). High doses boosted cellular responses in the airway, adrenal medulla, RBCs, immunological, intestinal, HEPA-1, and EAhy926 umbilical cells, among other characteristics. The cytotoxicity of BEAS-2B airway cells, on the other hand, increased with a longer incubation period. Internalization of particles can be increased by a high dose and a long exposure duration, resulting in greater toxicity. It should be emphasized that spherical polystyrene has been used as the model MPs/NPs in many of these cytotoxicity experiments, owing to the commercial availability of polystyrene micro/nanobeads. It's also critical to assess the toxicity of various plastic polymers, particularly category 1 carcinogens and mutagens. Toxicity of fibrous or other shaped polymers at sizes that more closely resemble the majority of the MPs/NPs population in the natural environment is also required.

Furthermore, several investigations have been conducted at extremely high particle doses, which may or may not be environmentally relevant.

The toxicity of these particles is not conclusive due to a mismatch between the concentration, size, shape, and type of microplastics examined in the laboratory and those found in nature (Burns & Boxall, 2018). The lack of standardized analytical techniques for detecting and quantifying plastics in various matrices, as well as the establishment of plastic contamination control protocols during the analysis of collected samples, are important hurdles to determining MPs/NPs toxicity (Barbosa et al., 2020). Plastics with a diameter of fewer than 20 μm are particularly difficult to identify and separate (Hale et al., 2020). The development of analytical procedures for MPs/NPs identification and quantitation would pave the way for the use of environmentally realistic dosages, sizes, forms, and kinds to better understand the health effects of these particles.

References

Akhatova, F., Ishmukhametov, I., Fakhrullina, G., & Fakhrullin, R. (2022). Nanomechanical atomic force microscopy to probe cellular microplastics uptake and distribution. *International Journal of Molecular Sciences, 23*(2), 806.

Akhbarizadeh, R., Moore, F., & Keshavarzi, B. (2019). Investigating microplastics bioaccumulation and biomagnification in seafood from the Persian Gulf: A threat to human health? *Food Additives & Contaminants: Part A, 36*(11), 1696–1708.

Alberts, B., Johnson, A., & Lewis, J. (2002). Transport into the cell from the plasma membrane: Endocytosis. In *Molecular biology of the cell*. Garland Science.

Araujo, C. F., Nolasco, M. M., Ribeiro, A. M., & Ribeiro-Claro, P. J. (2018). Identification of microplastics using Raman spectroscopy: Latest developments and future prospects. *Water Research, 142*, 426–440.

Bahadar, H., Maqbool, F., Niaz, K., & Abdollahi, M. (2016). Toxicity of nanoparticles and an overview of current experimental models. *Iranian Biomedical Journal., 20*, 1–11. Pii-IBJ A-10-1853-1.

Baos, S. C., Phillips, D. B., Wildling, L., McMaster, T. J., & Berry, M. (2012). Distribution of sialic acids on mucins and gels: A defense mechanism. *Biophysical Journal, 102*, 176–184. https://doi.org/10.1016/j.bpj.2011.08.058

Barbosa, F., Adeyemi, J. A., Bocato, M. Z., Comas, A., & Campiglia, A. (2020). A critical viewpoint on current issues, limitations, and future research needs on micro- and nanoplastic studies: From the detection to the toxicological assessment. *Environmental Research, 182*, 109089. https://doi.org/10.1016/j.envres.2019.109089

Bhattacharjee, S. (2016). DLS and zeta potential–what they are and what they are not? *Journal of Controlled Release, 235*, 337–351.

Bhattacharjee, S., Ershov, D., Islam, M. A., Kämpfer, A. M., Maslowska, K. A., van der Gucht, J., Alink, G. M., Marcelis, A. T., Zuilhof, H., & Rietjens, I. M. (2014). Role of membrane disturbance and oxidative stress in the mode of action underlying the toxicity of differently charged polystyrene nanoparticles. *RSC Advances, 4*, 19321–19330. https://doi.org/10.1039/C3RA46869K

Bos, J. D., & Meinardi, M. M. (2000). The 500 Dalton rule for the skin penetration of chemical compounds and drugs. *Experimental Dermatology, 9*, 165–169.

Bouwmeester, H., Hollman, P. C., & Peters, R. J. (2015). Potential health impact of environmentally released micro- and Nanoplastics in the human food production chain: experiences from nanotoxicology. *Environmental Science & Technology, 49*, 8932–8947. https://doi.org/10.1021/acs.est.5b01090

Boyle, K., & Örmeci, B. (2020). Microplastics and nanoplastics in the freshwater and terrestrial environment: A review. *Water, 12*(9), 2633.

Brandon, J., Goldstein, M., & Ohman, M. D. (2016). Long-term aging and degradation of microplastic particles: Comparing in situ oceanic and experimental weathering patterns. *Marine Pollution Bulletin, 110*(1), 299–308.

Brennecke, D., Duarte, B., Paiva, F., Caçador, I., & Canning-Clode, J. (2016). Microplastics as vector for heavy metal contamination from the marine environment. *Estuarine, Coastal and Shelf Science, 178*, 189–195. https://doi.org/10.1016/j.ecss.2015.12.003

Brennholt, N., Heß, M., & Reifferscheid, G. (2018). Freshwater microplastics: Challenges for regulation and management. In *Freshwater microplastics* (pp. 239–272). Springer.

Brown, D. M., Wilson, M. R., MacNee, W., Stone, V., & Donaldson, K. (2001). Size-dependent proinflammatory effects of ultrafine polystyrene particles: A role for surface area and oxidative stress in the enhanced activity of ultrafines. *Toxicology and Applied Pharmacology, 175*, 191–199. https://doi.org/10.1006/taap.2001.9240

Burns, E. E., & Boxall, A. B. A. (2018). Microplastics in the aquatic environment: Evidence for or against adverse impacts and major knowledge gaps. *Environmental Toxicology and Chemistry, 37*, 2776–2796. https://doi.org/10.1002/etc.4268

Campanale, C., Massarelli, C., Savino, I., Locaputo, V., & Uricchio, V. F. (2020). A detailed review study on potential effects of microplastics and additives of concern on human health. *International Journal of Environmental Research and Public Health, 17*(4), 1212. https://doi.org/10.3390/ijerph17041212. E1212 [pii].

Carbery, M., O'Connor, W., & Palanisami, T. (2018). Trophic transfer of microplastics and mixed contaminants in the marine food web and implications for human health. *Environment International, 115*, 400–409.

Chae, Y., Kim, D., Kim, S. W., & An, Y. J. (2018). Trophic transfer and individual impact of nano-sized polystyrene in a four-species freshwater food chain. *Scientific Reports., 8*, 284. https://doi.org/10.1038/s41598,017-18849-y

Chen, Q., Gundlach, M., Yang, S., Jiang, J., Velki, M., Yin, D., & Hollert, H. (2017). Quantitative investigation of the mechanisms of microplastics and nanoplastics toward zebrafish larvae locomotor activity. *Science of the Total Environment, 584*, 1022–1031.

Chen, W., Ouyang, Z. Y., Qian, C., & Yu, H. Q. (2018). Induced structural changes of humic acid by exposure of polystyrene microplastics: A spectroscopic insight. *Environmental Pollution, 233*, 1–7.

Chen, Y., Ling, Y., Li, X., Hu, J., Cao, C., & He, D. (2020). Size-dependent cellular internalization and effects of polystyrene microplastics in microalgae P. helgolandica var. tsingtaoensis and S. quadricauda. *Journal of Hazardous Materials, 399*, 123092.

Chiu, H. W., Xia, T., Lee, Y. H., Chen, C. W., Tsai, J. C., & Wang, Y. J. (2015). Cationic polystyrene nanospheres induce autophagic cell death through the induction of endoplasmic reticulum stress. *Nanoscale., 7*, 736–746. https://doi.org/10.1039/c4nr05509h

Coffin, S., Wyer, H., & Leapman, J. C. (2021). Addressing the environmental and health impacts of microplastics requires open collaboration between diverse sectors. *PLoS Biology, 19*(3), e3000932.

Cole, M., Lindeque, P., Halsband, C., & Galloway, T. S. (2011). Microplastics as contaminants in the marine environment: A review. *Marine Pollution Bulletin, 62*, 2588–2597. https://doi.org/10.1016/j.marpolbul.2011.09.025

Cole, M., Lindeque, P., Fileman, E., Halsband, C., Goodhead, R., Moger, J., & Galloway, T. S. (2013). Microplastic ingestion by zooplankton. *Environmental Science & Technology, 47*(12), 6646–6655.

Collado-Gonzalez, M., Gonzalez Espinosa, Y., & Goycoolea, F. M. (2019). Interaction between chitosan and mucin: Fundamentals and applications. *Biomimetics (Basel), 4*. https://doi.org/10.3390/biomimetics4020032. E32 [pii].

Conti, G. O., Ferrante, M., Banni, M., Favara, C., Nicolosi, I., Cristaldi, A., et al. (2020). Micro- and nano-plastics in edible fruit and vegetables. The first diet risks assessment for the general population. *Environmental Research, 187*, 109677.

Cooper, D. A., & Corcoran, P. L. (2010). Effects of mechanical and chemical processes on the degradation of plastic beach debris on the island of Kauai, Hawaii. *Marine Pollution Bulletin, 60*(5), 650–654.

Correia, M., & Loeschner, K. (2018). Detection of nanoplastics in food by asymmetric flow field-flow fractionation coupled to multi-angle light scattering: Possibilities, challenges and analytical limitations. *Analytical and Bioanalytical Chemistry, 410*(22), 5603–5615.

De Souza Machado, A. A., Kloas, W., Zarfl, C., Hempel, S., & Rillig, M. C. (2018). Microplastics as an emerging threat to terrestrial ecosystems. *Global Change Biology, 24*, 1405–1416. https://doi.org/10.1111/gcb.14020

De Toni, L., Tisato, F., Seraglia, R., Roverso, M., Gandin, V., Marzano, C., Padrini, R., & Foresta, C. (2017). Phthalates and heavy metals as endocrine disruptors in food: A study on pre-packed coffee products. *Toxicology Reports, 4*, 234–239. https://doi.org/10.1016/j.toxrep.2017.05.004

Deme, G. G., Ewusi-Mensah, D., Olagbaju, O. A., Okeke, E. S., Okoye, C. O., Odii, E. C., et al. (2022). Macro problems from microplastics: Toward a sustainable policy framework for managing microplastic waste in Africa. *Science of the Total Environment, 804*, 150170.

Deng, Y., Zhang, Y., Lemos, B., & Ren, H. (2017). Tissue accumulation of microplastics in mice and biomarker responses suggest widespread health risks of exposure. *Scientific Reports, 7*(1), 1–10.

Deng, Y., Zhang, J., Zhang, C., Ding, Z., Hao, C., & An, L. (2020). Countermeasures on plastic and microplastic garbage management. In *Microplastics in terrestrial environments: Emerging contaminants and major challenges* (pp. 447–469). Springer.

Duan, Z., Duan, X., Zhao, S., Wang, X., Wang, J., Liu, Y., et al. (2020). Barrier function of zebrafish embryonic chorions against microplastics and nanoplastics and its impact on embryo development. *Journal of Hazardous Materials, 395*, 122621.

EFSA Panel on Contaminants in the Food Chain. (2016). Presence of microplastics and nanoplastics in food, with particular focus on seafood. *EFSA Journal, 14*, 1–30. https://doi.org/10.2903/j.efsa.2016.4501

Ekkapongpisit, M., Giovia, A., Follo, C., Caputo, G., & Isidoro, C. (2012). Biocompatibility, endocytosis, and intracellular trafficking of mesoporous silica and polystyrene nanoparticles in ovarian cancer cells: Effects of size and surface charge groups. *International Journal of Nanomedicine, 7*, 4147–4158. https://doi.org/10.2147/IJN.S33803

Eriksen, M., Mason, S., Wilson, S., Box, C., Zellers, A., Edwards, W., et al. (2013). Microplastic pollution in the surface waters of the Laurentian Great Lakes. *Marine Pollution Bulletin, 77*(1–2), 177–182.

Fazlollahi, F., Angelow, S., Yacobi, N. R., Marchelletta, R., Yu, A. S. L., Hamm-Alvarez, S. F., Borok, Z., Kim, K., & Crandall, E. D. (2011). Polystyrene nanoparticle trafficking across MDCK-II. *Nanomedicine: Nanotechnology, Biology and Medicine, 7*, 588–594. https://doi.org/10.1016/j.nano.2011.01.008

Feng, L., Li, J., Xu, E. G., Sun, X., Zhu, F., Ding, Z., Tian, H., Dong, S., Xiaa, P., & Yuan, X. (2019). Short-term exposure to positively charged polystyrene nanoparticles causes oxidative stress and membrane destruction in cyanobacteria. *Environmental Science: Nano, 6*, 3072–3079. https://doi.org/10.1039/C9EN00807A

Fiorentino, I., Gualtieri, R., Barbato, V., Mollo, V., Braun, S., Angrisani, A., Turano, M., Furia, M., Netti, P. A., Guarnieri, D., Fusco, S., & Talevi, R. (2015). Energy independent uptake and release of polystyrene nanoparticles in primary mammalian cell cultures. *Experimental Cell Research, 330*, 240–247. https://doi.org/10.1016/j.yexcr.2014.09.017

Firdessa, R., Oelschlaeger, T. A., & Moll, H. (2014). Identification of multiple cellular uptake pathways of polystyrene nanoparticles and factors affecting the uptake: Relevance for drug delivery systems. *European Journal of Cell Biology, 93*, 323–337. https://doi.org/10.1016/j.ejcb.2014.08.001

Fischer, M., & Scholz-Böttcher, B. M. (2017). Simultaneous trace identification and quantification of common types of microplastics in environmental samples by pyrolysis-gas chromatography–mass spectrometry. *Environmental Science & Technology, 51*(9), 5052–5060.

Florence, A. T., Hillery, A. M., Hussain, N., & Jani, P. U. (1995). Factors affecting the oral uptake and translocation of polystyrene nanoparticles: Histological and analytical evidence. *Journal of drug targeting, 3*, 65–70. https://doi.org/10.3109/10611869509015936

Foroozandeh, P., & Aziz, A. A. (2018). Insight into cellular uptake and intracellular trafficking of nanoparticles. *Nanoscale Research Letters, 13*, 339. https://doi.org/10.1186/s11671-0182728-6

Forte, M., Iachetta, G., Tussellino, M., Carotenuto, R., Prisco, M., De Falco, M., et al. (2016). Polystyrene nanoparticles internalization in human gastric adenocarcinoma cells. *Toxicology In Vitro, 31*, 126–136.

Freeman, S., Booth, A. M., Sabbah, I., Tiller, R., Dierking, J., Klun, K., et al. (2020). Between source and sea: The role of wastewater treatment in reducing marine microplastics. *Journal of Environmental Management, 266*, 110642.

Fries, E., Dekiff, J. H., Willmeyer, J., Nuelle, M. T., Ebert, M., & Remy, D. (2013). Identification of polymer types and additives in marine microplastic particles using pyrolysis-GC/MS and scanning electron microscopy. *Environmental Science: Processes & Impacts, 15*(10), 1949–1956.

Frohlich, E. (2012). The role of surface charge in cellular uptake and cytotoxicity of medical nanoparticles. *International Journal of Nanomedicine, 7*, 5577–5591. https://doi.org/10.2147/IJN.S36111

Fu, W., Min, J., Jiang, W., Li, Y., & Zhang, W. (2020). Separation, characterization and identification of microplastics and nanoplastics in the environment. *Science of the Total Environment, 721*, 137561.

Fuchs, A. K., Syrovets, T., Haas, K. A., Loos, C., Musyanovych, A., Mailänder, V., et al. (2016). Carboxyl-and amino-functionalized polystyrene nanoparticles differentially affect the polarization profile of M1 and M2 macrophage subsets. *Biomaterials, 85*, 78–87.

Gago, J., Booth, A. M., Tiller, R., Maes, T., & Larreta, J. (2020). Microplastics pollution and regulation. In *Handbook of microplastics in the environment* (pp. 1–27). Springer.

Ganesan, S., Comstock, A. T., & Sajjan, U. S. (2013). Barrier function of airway tract epithelium. *Tissue Barriers, 1*, e24997. https://doi.org/10.4161/tisb.24997

Gasperi, J., Wright, S. L., Dris, R., Collard, F., Mandin, C., Guerrouache, M., Langlois, V., Kelly, F. J., & Tassin, B. (2018). Microplastics in air: Are we breathing it in? *Current Opinion in Environmental Science & Health, 1*, 1–5. https://doi.org/10.1016/j.coesh.2017.10.002

Gaylor, M. O., Harvey, E., & Hale, R. C. (2013). Polybrominated diphenyl ether (PBDE) accumulation by earthworms (Eisenia fetida) exposed to biosolids-, polyurethane foam microparticle-, and penta-BDE-amended soils. *Environmental Science & Technology, 47*(23), 13831–13839.

Gigault, J., & Budzinski, H. (2016). Selection of an appropriate aqueous nano-fullerene (nC60) preparation protocol for studying its environmental fate and behavior. *TrAC Trends in Analytical Chemistry, 80*, 1–11.

Gigault, J., El Hadri, H., Reynaud, S., Deniau, E., & Grassl, B. (2017). Asymmetrical flow field flow fractionation methods to characterize submicron particles: Application to carbon-based aggregates and nanoplastics. *Analytical and Bioanalytical Chemistry, 409*(29), 6761–6769.

Gillibert, R., Balakrishnan, G., Deshoules, Q., Tardivel, M., Magazzù, A., Donato, M. G., et al. (2019). Raman tweezers for small microplastics and nanoplastics identification in seawater. *Environmental Science & Technology, 53*(15), 9003–9013.

Goldstein, J. I., Newbury, D. E., Michael, J. R., Ritchie, N. W., Scott, J. H. J., & Joy, D. C. (2017). *Scanning electron microscopy and X-ray microanalysis*. Springer.

González-Pleiter, M., Tamayo-Belda, M., Pulido-Reyes, G., Amariei, G., Leganés, F., Rosal, R., & Fernández-Piñas, F. (2019). Secondary nanoplastics released from a biodegradable microplastic severely impact freshwater environments. *Environmental Science: Nano, 6*(5), 1382–1392.

Gopinath, P. M., Saranya, V., Vijayakumar, S., Mythili Meera, M., Ruprekha, S., Kunal, R., Pranay, A., Thomas, J., Mukherjee, A., & Chandrasekaran, N. (2019). Assessment on interactive prospectives of nanoplastics with plasma proteins and the toxicological impacts of virgin, coronated and environmentally released-nanoplastics. *Scientific Reports, 9*, 8860, 019-45139-6. https://doi.org/10.1038/s41598-019-45139-6

Green, T. R., Fisher, J., Stone, M., Wroblewski, B. M., & Ingham, E. (1998). Polyethylene particles of a 'critical size'are necessary for the induction of cytokines by macrophages in vitro. *Biomaterials, 19*(24), 2297–2302.

Hahladakis, J. N., Velis, C. A., Weber, R., Iacovidou, E., & Purnell, P. (2018). An overview of chemical additives present in plastics: Migration, release, fate and environmental impact during their use, disposal and recycling. *Journal of Hazardous Materials., 344*, 179–199. https://doi.org/10.1016/j.jhazmat.2017.10.014

Hale, R. C., Seeley, M. E., La Guardia, M. J., Mai, L., & Zeng, E. Y. (2020). A global perspective on microplastics. *Journal of Geophysical Research: Oceans, 125*, 1–40. https://doi.org/10.1029/2018JC014719

Heinlaan, M., Kasemets, K., Aruoja, V., Blinova, I., Bondarenko, O., Lukjanova, A., Khosrovyan, A., Kurvet, I., Pullerits, M., Sihtmae, M., Vasiliev, G., Vija, H., & Kahru, A. (2020). Hazard evaluation of polystyrene nanoplastic with nine bioassays did not show particle-specific acute toxicity. *Science of the Total Environment, 707*, 136073. S0048 9697(19)36069-3 [pii].

Henderson, R. K., Baker, A., Murphy, K. R., Hambly, A., Stuetz, R. M., & Khan, S. J. (2009). Fluorescence as a potential monitoring tool for recycled water systems: A review. *Water Research, 43*(4), 863–881.

Hesler, M., Aengenheister, L., Ellinger, B., Drexel, R., Straskraba, S., Jost, C., Wagner, S., Meier, F., von Briesen, H., Buchel, C., Wick, P., Buerki-Thurnherr, T., & Kohl, Y. (2019). Multi- endpoint toxicological assessment of polystyrene nano- and microparticles in different biological models in vitro. *Toxicology In Vitro, 61*, 104610. S0887-2333(19)30260-7 [pii].

Hidalgo-Ruz, V., Gutow, L., Thompson, R. C., & Thiel, M. (2012). Microplastics in the marine environment: A review of the methods used for identification and quantification. *Environmental Science & Technology, 46*(6), 3060–3075.

Hollman, P. C., Bouwmeester, H., & Peters, R. J. B. (2013). *Microplastics in aquatic food chain: Sources, measurement, occurrence and potential health risks* (No. 2013.003). Rikilt-Institute of Food Safety.

Holloczki, O., & Gehrke, S. (2020). Can nanoplastics alter cell membranes? *ChemPhysChem, 21*, 9–12. https://doi.org/10.1002/cphc.201900481

Huerta Lwanga, E., Gertsen, H., Gooren, H., Peters, P., Salánki, T., Van Der Ploeg, M., et al. (2016). Microplastics in the terrestrial ecosystem: Implications for Lumbricus terrestris (Oligochaeta, Lumbricidae). *Environmental Science & Technology, 50*(5), 2685–2691.

Hwang, J., Choi, D., Han, S., Choi, J., & Hong, J. (2019). An assessment of the toxicity of polypropylene microplastics in human derived cells. *Science of the Total Environment, 684*, 657–669. https://doi.org/10.1016/j.scitotenv.2019.05.071

Hwang, J., Choi, D., Han, S., Jung, S. Y., Choi, J., & Hong, J. (2020). Potential toxicity of polystyrene micro plastic particles. *Scientific Reports, 10*, 7391. https://doi.org/10.1038/s41598-02064464-9

Imhof, H. K., Laforsch, C., Wiesheu, A. C., Schmid, J., Anger, P. M., Niessner, R., & Ivleva, N. P. (2016). Pigments and plastic in limnetic ecosystems: A qualitative and quantitative study on microparticles of different size classes. *Water Research, 98*, 64–74.

Inkielewicz-Stepniak, I., Tajber, L., Behan, G., Zhang, H., Radomski, M. W., Medina, C., & Santos-Martinez, M. J. (2018). The role of mucin in the toxicological impact of polystyrene nanoparticles. *Materials, 11*(5), 724.

Jenkins, P. G., Howard, K. A., Blackhall, N. W., Thomas, N. W., Davis, S. S., & O'Hagan, D. T. (1994). The quantitation of the absorption of microparticles into the intestinal lymph of Wistar rats. *International Journal of Pharmaceutics, 102*, 261–266. https://doi.org/10.1016/0378-5173(94)90064-7

Jin, Y., Lu, L., Tu, W., Luo, T., & Fu, Z. (2019). Impacts of polystyrene microplastic on the gut barrier, microbiota and metabolism of mice. *Science of the Total Environment, 649*, 308–317.

Karami, A., Golieskardi, A., Keong Choo, C., Larat, V., Galloway, T. S., & Salamatinia, B. (2017). The presence of microplastics in commercial salts from different countries. *Scientific Reports, 7*, 46173. https://doi.org/10.1038/srep46173

Karas, M., & Krüger, R. (2003). Ion formation in MALDI: The cluster ionization mechanism. *Chemical Reviews, 103*(2), 427–440.

Kedzierski, M., Lechat, B., Sire, O., Le Maguer, G., Le Tilly, V., & Bruzaud, S. (2020). Microplastic contamination of packaged meat: Occurrence and associated risks. *Food Packaging and Shelf Life, 24*, 100489.

Kim, S. W., & An, Y. J. (2019). Soil microplastics inhibit the movement of springtail species. *Environment International, 126*, 699–706.

Kirstein, I. V., Kirmizi, S., Wichels, A., Garin-Fernandez, A., Erler, R., Loder, M., & Gerdts, G. (2016). Dangerous hitchhikers? Evidence for potentially pathogenic Vibrio spp. on microplastic particles. *Marine Environmental Research, 120*, 1–8. https://doi.org/10.1016/j.marenvres.2016.07.004

Koelmans, A. A., Besseling, E., & Shim, W. J. (2015). Nanoplastics in the aquatic environment. Critical review. *Marine Anthropogenic Litter*, 325–340.

Kosuth, M., Mason, S. A., & Wattenberg, E. V. (2018). Anthropogenic contamination of tap water, beer, and sea salt. *PLoS One, 13*(4), e0194970.

Legler, J., & Brouwer, A. (2003). Are brominated flame retardants endocrine disruptors? *Environment International, 29*, 879–885. S0160-4120(03)00104-1 [pii].

Li, S., & Malmstadt, N. (2013). Deformation and poration of lipid bilayer membranes by cationic nanoparticles. *Soft Matter, 9*, 4969–4976. https://doi.org/10.1039/C3SM27578G

Liao, Y., & Yang, J. (2020). Microplastic serves as a potential vector for Cr in an in-vitro human digestive model. *Science of the Total Environment, 703*, 134805. https://doi.org/10.1016/j.scitotenv.2019.134805

Lithner, D., Larsson, Å., & Dave, G. (2011). Environmental and health hazard ranking and assessment of plastic polymers based on chemical composition. *Science of the Total Environment, 409*, 3309–3324. https://doi.org/10.1016/j.scitotenv.2011.04.038

Liu, X., Tian, X., Xu, X., & Lu, J. (2018a). Design of a phosphinate-based bioluminescent probe for superoxide radical anion imaging in living cells. *Luminescence, 33*(6), 1101–1106.

Liu, J., Ma, Y., Zhu, D., Xia, T., Qi, Y., Yao, Y., Guo, X., Ji, R., & Chen, W. (2018b). Polystyrene nanoplastics-enhanced contaminant transport: Role of irreversible adsorption in glassy polymeric domain. *Environmental Science & Technology, 52*, 2677–2685. https://doi.org/10.1021/acs.est.7b05211

Llorca, M., & Farré, M. (2021). Current insights into potential effects of micro-nanoplastics on human health by in-vitro tests. *Frontiers in Toxicology, 39*.

Löder, M. G., & Gerdts, G. (2015). Methodology used for the detection and identification of microplastics—A critical appraisal. *Marine Anthropogenic Litter*, 201–227.

Loos, C., Syrovets, T., Musyanovych, A., Mailander, V., Landfester, K., Nienhaus, G. U., & Simmet, T. (2014a). Functionalized polystyrene nanoparticles as a platform for studying bio-nano interactions. *Beilstein Journal of Nanotechnology, 5*, 2403–2412. https://doi.org/10.3762/bjnano.5.250

Loos, C., Syrovets, T., Musyanovych, A., Mailänder, V., Landfester, K., & Simmet, T. (2014b). Amino-functionalized nanoparticles as inhibitors of mTOR and inducers of cell cycle arrest in leukemia cells. *Biomaterials, 35*, 1944–1953. https://doi.org/10.1016/j.biomaterials.2013.11.056

Lu, L., Wan, Z., Luo, T., Fu, Z., & Jin, Y. (2018). Polystyrene microplastics induce gut microbiota dysbiosis and hepatic lipid metabolism disorder in mice. *Science of the Total Environment, 631*, 449–458.

Luo, T., Zhang, Y., Wang, C., Wang, X., Zhou, J., Shen, M., et al. (2019). Maternal exposure to different sizes of polystyrene microplastics during gestation causes metabolic disorders in their offspring. *Environmental Pollution, 255*, 113122.

Ma, H., Pu, S., Liu, S., Bai, Y., Mandal, S., & Xing, B. (2020). Microplastics in aquatic environments: Toxicity to trigger ecological consequences. *Environmental Pollution, 261*, 114089.

Mahadevan, G., & Valiyaveettil, S. (2021). Understanding the interactions of poly (methyl methacrylate) and poly (vinyl chloride) nanoparticles with BHK-21 cell line. *Scientific Reports, 11*(1), 1–15.

Mahler, G. J., Esch, M. B., Tako, E., Southard, T. L., Archer, S. D., Glahn, R. P., & Shuler, M. L. (2012). Oral exposure to polystyrene nanoparticles affects iron absorption. *Nature Nanotechnology, 7*(4), 264–271.

Marti, E., Martin, C., Galli, M., Echevarría, F., Duarte, C. M., & Cózar, A. (2020). The colors of the ocean plastics. *Environmental Science & Technology, 54*(11), 6594–6601.

Mason, S. A., Welch, V. G., & Neratko, J. (2018). Synthetic polymer contamination in bottled water. *Frontiers in Chemistry., 6*, 407. https://doi.org/10.3389/fchem.2018.00407

Massos, A., & Turner, A. (2017). Cadmium, lead and bromine in beached microplastics. *Environmental Pollution, 227*, 139–145.

McCarthy, J., Gong, X., Nahirney, D., Duszyk, M., & Radomski, M. W. (2011). Polystyrene nanoparticles activate ion transport in human airway epithelial cells. *International Journal of Nanomedicine, 6*, 1343.

Mendoza, L. M. R., & Jones, P. R. (2015). Characterisation of microplastics and toxic chemicals extracted from microplastic samples from the North Pacific Gyre. *Environmental Chemistry, 12*(5), 611–617.

Mintenig, S. M., Bäuerlein, P. S., Koelmans, A. A., Dekker, S. C., & Van Wezel, A. P. (2018). Closing the gap between small and smaller: Towards a framework to analyse nano-and microplastics in aqueous environmental samples. *Environmental Science: Nano, 5*(7), 1640–1649.

Mintenig, S. M., Löder, M. G. J., Primpke, S., & Gerdts, G. (2019). Low numbers of microplastics detected in drinking water from ground water sources. *Science of the Total Environment, 648*, 631–635.

Mitrano, D. M., & Wohlleben, W. (2020). Microplastic regulation should be more precise to incentivize both innovation and environmental safety. *Nature Communications, 11*(1), 1–12.

Mitrano, D. M., Beltzung, A., Frehland, S., Schmiedgruber, M., Cingolani, A., & Schmidt, F. (2019). Synthesis of metal-doped nanoplastics and their utility to investigate fate and behaviour in complex environmental systems. *Nature Nanotechnology, 14*(4), 362–368.

Mrakovcic, M., Absenger, M., Riedl, R., Smole, C., Roblegg, E., Fröhlich, L. F., & Fröhlich, E. (2013). Assessment of long-term effects of nanoparticles in a microcarrier cell culture system. *Plos One, 8*, e56791. https://doi.org/10.1371/journal.pone.0056791

Mrakovcic, M., Meindl, C., Roblegg, E., & Fröhlich, E. (2014). Reaction of monocytes to polystyrene and silica nanoparticles in short-term and long-term exposures. *Toxicology Research, 3*, 86–97. https://doi.org/10.1039/C3TX50112D

Naji, A., Nuri, M., & Vethaak, A. D. (2018). Microplastics contamination in molluscs from the northern part of the Persian Gulf. *Environmental Pollution, 235*, 113–120.

Nalbone, L., Cincotta, F., Giarratana, F., Ziino, G., & Panebianco, A. (2021). Microplastics in fresh and processed mussels sampled from fish shops and large retail chains in Italy. *Food Control, 125*, 108003.

National Oceanic and Atmospheric Administration. (2019). *A guide to plastic in the ocean*. https://oceanservice.noaa.gov/hazards/marinedebris/plastics-in-the-ocean.html. Accessed 9 Mar 2020.

Nelms, S. E., Galloway, T. S., Godley, B. J., Jarvis, D. S., & Lindeque, P. K. (2018). Investigating microplastic trophic transfer in marine top predators. *Environmental Pollution, 238*, 999–1007. https://doi.org/10.1016/j.envpol.2018.02.016

Nich, C., & Goodman, S. B. (2014). Role of macrophages in the biological reaction to wear debris from joint replacements. *Journal of Long-Term Effects of Medical Implants, 24*(4), 259.

NJDEP-Science Advisory Board. (2015). *Human health impacts of microplastics and nano plastics*.

Nowak, M., Brown, T. D., Graham, A., Helgeson, M. E., & Mitragotri, S. (2019). Size, shape, and flexibility influence nanoparticle transport across brain endothelium under flow. *Bioengineering & Translational Medicine*, 1–11. https://doi.org/10.1002/btm2.10153

Okubo, N., Takahashi, S., & Nakano, Y. (2018). Microplastics disturb the anthozoan-algae symbiotic relationship. *Marine Pollution Bulletin, 135*, 83–89.

Paget, V., Dekali, S., Kortulewski, T., Grall, R., Gamez, C., Blazy, K., et al. (2015). Specific uptake and genotoxicity induced by polystyrene nanobeads with distinct surface chemistry on human lung epithelial cells and macrophages. *PLoS One, 10*(4), e0123297.

Pauly, J. L., Stegmeier, S. J., Allaart, H. A., Cheney, R. T., Zhang, P. J., Mayer, A. G., & Streck, R. J. (1998). Inhaled cellulosic and plastic fibers found in human lung tissue. *Cancer Epidemiology, Biomarkers & Prevention, 7*, 419–428.

Pedersen, A. F., Gopalakrishnan, K., Boegehold, A. G., Peraino, N. J., Westrick, J. A., & Kashian, D. R. (2020). Microplastic ingestion by quagga mussels, Dreissena bugensis, and its effects on physiological processes. *Environmental Pollution, 260*, 113964.

Peng, G., Bellerby, R., Zhang, F., Sun, X., & Li, D. (2020). The ocean's ultimate trashcan: Hadal trenches as major depositories for plastic pollution. *Water Research, 168*, 115121.

Philippe, A., Gangloff, M., Rakcheev, D., & Schaumann, G. E. (2014). Evaluation of hydrodynamic chromatography coupled with inductively coupled plasma mass spectrometry detector for analysis of colloids in environmental media–effects of colloid composition, coating and shape. *Analytical Methods, 6*(21), 8722–8728.

Pimentel, J. C., Avila, R., & Lourenço, A. G. (1975). Respiratory disease caused by synthetic fibers: a new occupational disease. *Thorax, 30*, 204–219. https://doi.org/10.1136/thx.30.2.204

Pirok, B. W., Abdulhussain, N., Aalbers, T., Wouters, B., Peters, R. A., & Schoenmakers, P. J. (2017). Nanoparticle analysis by online comprehensive two-dimensional liquid chromatography combining hydrodynamic chromatography and size-exclusion chromatography with intermediate sample transformation. *Analytical Chemistry, 89*(17), 9167–9174.

Poon, C. K., Tang, O., Chen, X. M., Pham, B. T., Gody, G., Pollock, C. A., Hawkett, B. S., & Perrier, S. (2016). Preparation of inert polystyrene latex particles as MicroRNA delivery vectors by surfactant-free RAFT emulsion polymerization. *Biomacromolecules, 17*, 965–973. https://doi.org/10.1021/acs.biomac.5b01633

Prata, J. C., da Costa, J. P., Lopes, I., Duarte, A. C., & Rocha-Santos, T. (2020). Environmental exposure to microplastics: An overview on possible human health effects. *Science of the Total Environment, 702*, 134455. https://doi.org/10.1016/j.scitotenv.2019.134455

Prietl, B., Meindl, C., Roblegg, E., Pieber, T. R., Lanzer, G., & Fröhlich, E. (2014). Nano-sized and micro-sized polystyrene particles affect phagocyte function. *Cell Biology and Toxicology, 30*(1), 1–16.

Prinz, N., & Korez, Š. (2020). Understanding how microplastics affect marine biota on the cellular level is important for assessing ecosystem function: A review. In S. Jungblut, V. Liebich, & M. Bode-Dalby (Eds.), *YOUMARES 9 – The oceans: Our research, our future* (pp. 101–120). Springer. https://doi.org/10.1007/978-3-030-20389-4_6

Qiu, Q., Peng, J., Yu, X., Chen, F., Wang, J., & Dong, F. (2015). Occurrence of microplastics in the coastal marine environment: First observation on sediment of China. *Marine Pollution Bulletin, 98*(1–2), 274–280.

Reineke, J. J., Cho, D. Y., Dingle, Y. T., Morello, A. P., 3rd, Jacob, J., Thanos, C. G., & Mathiowitz, E. (2013). Unique insights into the intestinal absorption, transit, and subsequent biodistribution of polymer-derived microspheres. *Proceedings of the National Academy of Sciences, 110*, 13803–13808. https://doi.org/10.1073/pnas.1305882110

Reinholz, J., Diesler, C., Schöttler, S., Kokkinopoulou, M., Ritz, S., Landfester, K., & Mailänder, V. (2018). Protein machineries defining pathways of nanocarrier exocytosis and transcytosis. *Acta Biomaterialia, 71*, 432–443. https://doi.org/10.1016/j.actbio.2018.03.006

Rillig, M. C., & Lehmann, A. (2020). Microplastic in terrestrial ecosystems. *Science, 368*(6498), 1430–1431.

Rodrigues, J. P., Duarte, A. C., Santos-Echeandía, J., & Rocha-Santos, T. (2019). Significance of interactions between microplastics and POPs in the marine environment: A critical overview. *TrAC Trends in Analytical Chemistry, 111*, 252–260. https://doi.org/10.1016/j.trac.2018.11.038

Rodriguez-Seijo, A., Lourenço, J., Rocha-Santos, T. A. P., Da Costa, J., Duarte, A. C., Vala, H., & Pereira, R. (2017). Histopathological and molecular effects of microplastics in Eisenia andrei Bouché. *Environmental Pollution, 220*, 495–503.

Rubin, B. S. (2011). Bisphenol A: An endocrine disruptor with widespread exposure and multiple effects. *The Journal of Steroid Biochemistry and Molecular Biology, 127*, 27–34. https://doi.org/10.1016/j.jsbmb.2011.05.002

Rubio, L., Marcos, R., & Hernandez, A. (2020). Potential adverse health effects of ingested micro- and nanoplastics on humans. Lessons learned from in vivo and in vitro mammalian models. *Journal of Toxicology and Environmental Health: Part B, Critical Reviews, 23*, 51–68. https://doi.org/10.1080/10937404.2019.1700598

Schirinzi, G. F., Perez-Pomeda, I., Sanchis, J., Rossini, C., Farre, M., & Barcelo, D. (2017). Cytotoxic effects of commonly used nanomaterials and microplastics on cerebral and epithelial human cells. *Environmental Research, 159*, 579–587. S0013-9351(17)31077-0 [pii].

Schwabl, P., Koppel, S., Konigshofer, P., Bucsics, T., Trauner, M., Reiberger, T., & Liebmann, B. (2019). Detection of various microplastics in human stool: A prospective case series. *Annals of Internal Medicine*. https://doi.org/10.7326/M19-0618

Schymanski, D., Goldbeck, C., Humpf, H. U., & Fürst, P. (2018). Analysis of microplastics in water by micro-Raman spectroscopy: Release of plastic particles from different packaging into mineral water. *Water Research, 129*, 154–162.

Sharma, S., & Chatterjee, S. (2017). Microplastic pollution, a threat to marine ecosystem and human health: A short review. *Environmental Science and Pollution Research International, 24*, 21530–21547. https://doi.org/10.1007/s11356-017-9910-8

Sharma, S., Sharma, V., & Chatterjee, S. (2021). Microplastics in the Mediterranean Sea: Sources, pollution intensity, sea health, and regulatory policies. *Frontiers in Marine Science, 8*, 634934.

Shim, W. J., Hong, S. H., & Eo, S. E. (2017). Identification methods in microplastic analysis: A review. *Analytical Methods, 9*(9), 1384–1391.

Siddiqui, M. R., AlOthman, Z. A., & Rahman, N. (2017). Analytical techniques in pharmaceutical analysis: A review. *Arabian Journal of Chemistry, 10*, S1409–S1421.

Smith, M., Love, D. C., Rochman, C. M., & Neff, R. A. (2018). Microplastics in seafood and the implications for human health. *Current Environmental Health Reports, 5*, 375–386. https://doi.org/10.1007/s40572-018-0206-z

Sobhani, Z., Zhang, X., Gibson, C., Naidu, R., Megharaj, M., & Fang, C. (2020). Identification and visualisation of microplastics/nanoplastics by Raman imaging (i): Down to 100 nm. *Water Research, 174*, 115658.

Song, Y. K., Hong, S. H., Jang, M., Han, G. M., Rani, M., Lee, J., & Shim, W. J. (2015). A comparison of microscopic and spectroscopic identification methods for analysis of microplastics in environmental samples. *Marine Pollution Bulletin, 93*(1–2), 202–209.

Stapleton, P. A. (2019). Toxicological considerations of nano-sized plastics. *AIMS Environmental Science, 6*, 367–378. https://doi.org/10.3934/environsci.2019.5.367

Stuart, B. (2005). *Infrared spectroscopy. Kirk-Othmer encyclopedia of chemical technology*. Wiley.

Su, L., Deng, H., Li, B., Chen, Q., Pettigrove, V., Wu, C., & Shi, H. (2019). The occurrence of microplastic in specific organs in commercially caught fishes from coast and estuary area of east China. *Journal of Hazardous Materials, 365*, 716–724.

Sun, J., Dai, X., Wang, Q., van Loosdrecht, M. C., & Ni, B. J. (2019). Microplastics in wastewater treatment plants: Detection, occurrence and removal. *Water Research, 152*, 21–37.

Sweden, K. I. M. O. (2007). Small plastic particles in Coastal Swedish waters. *N-Research Report*.

Teng, J., Wang, Q., Ran, W., Wu, D., Liu, Y., Sun, S., et al. (2019). Microplastic in cultured oysters from different coastal areas of China. *Science of the Total Environment, 653*, 1282–1292.

Toussaint, B., Raffael, B., Angers-Loustau, A., Gilliland, D., Kestens, V., Petrillo, M., Rio-Echevarria, I. M., & Van den Eede, G. (2019). Review of micro- and nanoplastic contamination in the food chain. *Food Additives & Contaminants: Part A Chemistry, Analysis, Control, Exposure & Risk Assessment., 36*, 639–673. https://doi.org/10.1080/19440049.2019.1583381

Varela, J. A., Bexiga, M. G., Aberg, C., Simpson, J. C., & Dawson, K. A. (2012). Quantifying size- dependent interactions between fluorescently labeled polystyrene nanoparticles and mammalian cells. *Journal of Nanobiotechnology, 10*, 39, 3155-10-39. https://doi.org/10.1186/1477-3155-10-39

Velzeboer, I., Kwadijk, C. J., & Koelmans, A. A. (2014). Strong sorption of PCBs to nanoplastics, microplastics, carbon nanotubes, and fullerenes. *Environmental Science & Technology, 48*, 4869–4876. https://doi.org/10.1021/es405721v

Vianello, A., Boldrin, A., Guerriero, P., Moschino, V., Rella, R., Sturaro, A., & Da Ros, L. (2013). Microplastic particles in sediments of Lagoon of Venice, Italy: First observations on occurrence, spatial patterns and identification. *Estuarine, Coastal and Shelf Science, 130*, 54–61.

Vidal, J. (2020). *The plastic polluters won 2019 – and we're running out of time to stop them.* https://www.theguardian.com/environment/2020/jan/02/year-plastic-pollution-clean-beaches-seas. Accessed 9 Mar 2020.

Walczak, A. P., Hendriksen, P. J., Woutersen, R. A., van der Zande, M., Undas, A. K., Helsdingen, R., van den Berg, H. H., Rietjens, I. M., & Bouwmeester, H. (2015). Bioavailability and biodistribution of differently charged polystyrene nanoparticles upon oral exposure in rats. *Journal of Nanoparticle Research, 17*, 231015-3029-y. Epub 2015 May 22. https://doi.org/10.1007/s11051-015-3029-y

Wang, F., Salvati, A., & Boya, P. (2018a). Lysosome-dependent cell death and deregulated autophagy induced by amine-modified polystyrene nanoparticles. *Open Biology, 8*. https://doi.org/10.1098/rsob.170271. 170271 [pii].

Wang, J., Zheng, L., & Li, J. (2018b). A critical review on the sources and instruments of marine microplastics and prospects on the relevant management in China. *Waste Management & Research, 36*(10), 898–911.

Wang, J., Qin, X., Guo, J., Jia, W., Wang, Q., Zhang, M., & Huang, Y. (2020). Evidence of selective enrichment of bacterial assemblages and antibiotic resistant genes by microplastics in urban rivers. *Water Research, 183*, 116113.

Westerhoff, P., Prapaipong, P., Shock, E., & Hillaireau, A. (2008). Antimony leaching from polyethylene terephthalate (PET) plastic used for bottled drinking water. *Water Research, 42*, 551–556. https://doi.org/10.1016/j.watres.2007.07.048

Wilkinson, B. (2019, July 12). *Conservation captured microplastics- are they in your food?* Two OceansAquarium. https://www.aquarium.co.za/blog/entry/microplastics-are-they-in-your-food

World Economic Forum. (2016). *The new plastics economy: Rethinking the future of plastics*. http://www3.weforum.org/docs/WEF_The_New_Plastics_Economy.pdf. Accessed 25 Mar 2020.

Wright, S. L., & Kelly, F. J. (2017). Plastic and human health: A micro issue? *Environmental Science & Technology, 51*, 6634–6647. https://doi.org/10.1021/acs.est.7b00423

Xia, T., Kovochich, M., Liong, M., Zink, J. I., & Nel, A. E. (2008). Cationic polystyrene nanosphere toxicity depends on cell-specific endocytic and mitochondrial injury pathways. *ACS Nano, 2*(1), 85–96.

Xia, L., Gu, W., Zhang, M., Chang, Y. N., Chen, K., Bai, X., et al. (2016). Endocytosed nanoparticles hold endosomes and stimulate binucleated cells formation. *Particle and Fibre Toxicology, 13*(1), 1–12.

Xu, Y., Chan, F. K. S., He, J., Johnson, M., Gibbins, C., Kay, P., et al. (2021). A critical review of microplastic pollution in urban freshwater environments and legislative progress in China: Recommendations and insights. *Critical Reviews in Environmental Science and Technology, 51*(22), 2637–2680.

Yacobi, N. R., Malmstadt, N., Fazlollahi, F., DeMaio, L., Marchelletta, R., Hamm-Alvarez, S. F., Borok, Z., Kim, K. J., & Crandall, E. D. (2010). Mechanisms of alveolar epithelial translocation of a defined population of nanoparticles. *American Journal of Respiratory Cell and Molecular Biology, 42*, 604–614. https://doi.org/10.1165/rcmb.2009-0138OC

Yang, Y. F., Chen, C. Y., Lu, T. H., & Liao, C. M. (2019a). Toxicity-based toxicokinetic/toxicodynamic assessment for bioaccumulation of polystyrene microplastics in mice. *Journal of Hazardous Materials, 366*, 703–713. S0304-3894(18)31188-9 [pii].

Yang, L., Li, K., Cui, S., Kang, Y., An, L., & Lei, K. (2019b). Removal of microplastics in municipal sewage from China's largest water reclamation plant. *Water Research, 155*, 175–181.

Yee, M. S. L., Hii, L. W., Looi, C. K., Lim, W. M., Wong, S. F., Kok, Y. Y., et al. (2021). Impact of microplastics and nanoplastics on human health. *Nanomaterials, 11*(2), 496.

Yong, C. Q. Y., Valiyaveetill, S., & Tang, B. L. (2020). Toxicity of microplastics and nanoplastics in mammalian systems. *International Journal of Environmental Research and Public Health, 17*. https://doi.org/10.3390/ijerph17051509. E1509 [pii].

Yu, P., Liu, Z., Wu, D., Chen, M., Lv, W., & Zhao, Y. (2018). Accumulation of polystyrene microplastics in juvenile Eriocheir sinensis and oxidative stress effects in the liver. *Aquatic Toxicology, 200*, 28–36.

Yu, F., Yang, C., Zhu, Z., Bai, X., & Ma, J. (2019). Adsorption behavior of organic pollutants and metals on micro/nanoplastics in the aquatic environment. *Science of the Total Environment, 694*, 133643. https://doi.org/10.1016/j.scitotenv.2019.133643

Zhang, B., Wu, D., Yang, X., Teng, J., Liu, Y., Zhang, C., et al. (2019). Microplastic pollution in the surface sediments collected from Sishili Bay, North Yellow Sea, China. *Marine Pollution Bulletin, 141*, 9–15.

Zhang, C., Wang, S., Pan, Z., Sun, D., Xie, S., Zhou, A., et al. (2020). Occurrence and distribution of microplastics in commercial fishes from estuarine areas of Guangdong, South China. *Chemosphere, 260*, 127656.

Zhao, L., Qu, M., Wong, G., & Wang, D. (2017). Transgenerational toxicity of nanopolystyrene particles in the range of μg L− 1 in the nematode Caenorhabditis elegans. *Environmental Science: Nano, 4*(12), 2356–2366.

Zheng, T., Yuan, D., & Liu, C. (2019). Molecular toxicity of nanoplastics involving in oxidative stress and desoxyribonucleic acid damage. *Journal of Molecular Recognition, 32*, e2804. https://doi.org/10.1002/jmr.2804

Zhou, X. X., Hao, L. T., Wang, H. Y. Z., Li, Y. J., & Liu, J. F. (2018). Cloud-point extraction combined with thermal degradation for nanoplastic analysis using pyrolysis gas chromatography–mass spectrometry. *Analytical Chemistry, 91*(3), 1785–1790.

Part IV
Bioremediation of Micro and Nano Plastics-Polluted Soil

Chapter 13
Restoration of Micro-/Nano plastics: Contaminated Soil by Phytoremediation

Yeddanapalli Prathima, Donthamalla Venkatanarasaiah, Dandu Akhil, and Rami Reddy Pallerla

Abstract This chapter gives an overview of an environmental remediation technology, including its principles, applicability, utilization, and advantages/disadvantages. This report is reported for information view only. Information used in this was collected from periodicals, through web searches, and in some cases, personal communications with the involved party. Microplastics are small pieces of plastic found in the environment and are harmful to animals. I wondered if microplastics could affect plants as well. The objective of this project was to determine if plants could absorb microplastics through their roots. Phytoremediation uses plants to clean up contaminated soil and underground water, taking advantage of plants' natural ability to take up constituents of their soil and water environments. Major advantages of this as compared to traditional methods of remediation, which has the possibility of generating less secondary wastes, minimal associated environmental disturbances and possibility of contaminant entrance into the food chain through consumption of plants by animals.

Keywords Phytoremediation · Nano-/microplastics · Methods · Applicability

13.1 Introduction

Phytoremediation is a special application of bioremediation. It is *a natural biological process of degradation of xenobiotic and recalcitrant compounds responsible for environmental pollution* (Cunningham et al., 1996). The word phyto stands for "plant" hence the remediation mediated by plant system. Environmental restoration is a phenomenon required to keep the ecosystem intact or enhance the rejuvenation of impaired environmental media; soil, water, and air (Wikipedia). Various methods

Y. Prathima
St Anna's High School, Guntur, Andhra Pradesh, India

D. Venkatanarasaiah · D. Akhil · R. R. Pallerla (✉)
Indian Immunologicals Limited, Hyderabad, Telangana, India

N. R. Maddela et al. (eds.), *Micro and Nanoplastics in Soil*,
https://doi.org/10.1007/978-3-031-21195-9_13

of remediation exist, yet restoring the environment to the proximal or original state appears elusive to most methods. Interestingly, phytoremediation, which is a biological process, does not only restore the environment in a greener way but can also adopt diverse mechanisms such as phytoextraction, phytodegradation, rhizodegradation, phytostabilization, and phytovolatilization to achieve the desired outcome (Tanee & Akonye, 2009; Tessier et al., 1979; Cunningham & Ow, 1996). The chapter also unlined the merits and a few demerits of this principle, while the identification of sustainable plants and the mitigation of time constraints were the future directions mentioned for the projection of phytoremediation as the ideal approach for the restoration of the environment EPA (2000). Phytoremediation is proposed as a cost-effective plant-based approach of environmental remediation that takes advantage of the ability of plants to concentrate elements and compounds from the environment and to detoxify various compounds. The concentrating effect results from the ability of certain plants, called hyperaccumulators, to bioaccumulate chemicals (Raskin & Ensley, 2000). The remediation effect is quite different. Toxic heavy metals cannot be degraded, but organic pollutants can be and are generally the major targets for phytoremediation. Several field trials confirmed the feasibility of using plants for environmental cleanup (Li et al, 2014).

Microplastics: Microplastics are small pieces of plastic, less than 5 mm (0.2 inch) in length, that occur in the environment as a consequence of plastic pollution. Microplastics are present in a variety of products, from *cosmetics to synthetic clothing to plastic bags and bottles* (Wikipedia).

13.2 Methodology

The mechanisms and efficiency of phytoremediation depend on the type of contaminant, bioavailability, and soil properties (UNEP) (Fig. 13.1). There are many ways in which plants clean up or remediate contaminated sites. The uptake of contaminants in plants occurs primarily through the root system, in which the initial mechanisms for preventing toxicity are found. The root system provides a surface area that absorbs and accumulates water and nutrients essential for growth, along with other unwanted contaminants. This has identified mechanisms by which plants can affect contaminant mass in soil, sediments, and water (USEPA, 2000). Although overlap or similarities can be observed between some of these mechanisms, and the nomenclature varies, this report makes reference to phytoremediation methods, of these mechanisms will have an effect on the volume, mobility, or toxicity of contaminants, as the application of phytoremediation is intended to do (Figs. 13.2 and 13.3).

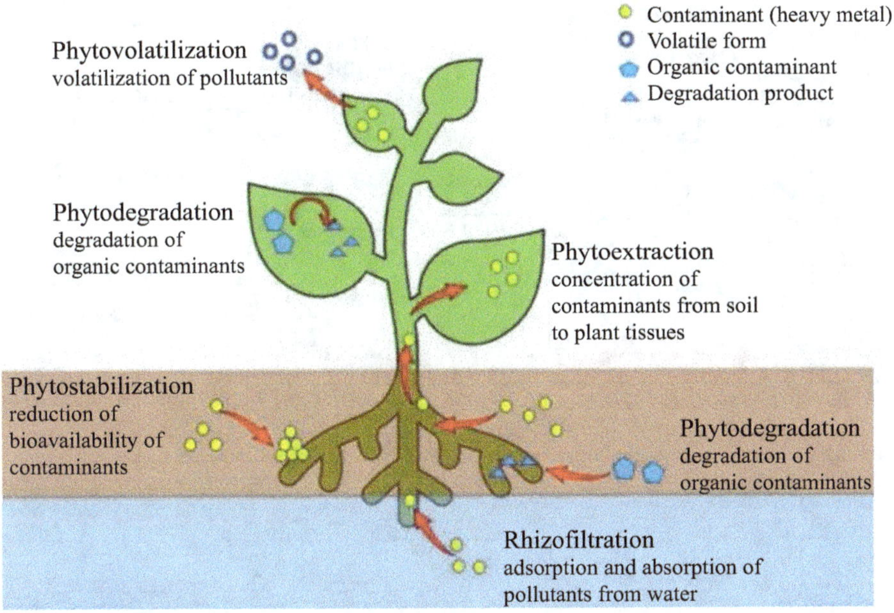

Fig. 13.1 Schematic representation of phytoremediation. (Yadav et al., 2011)

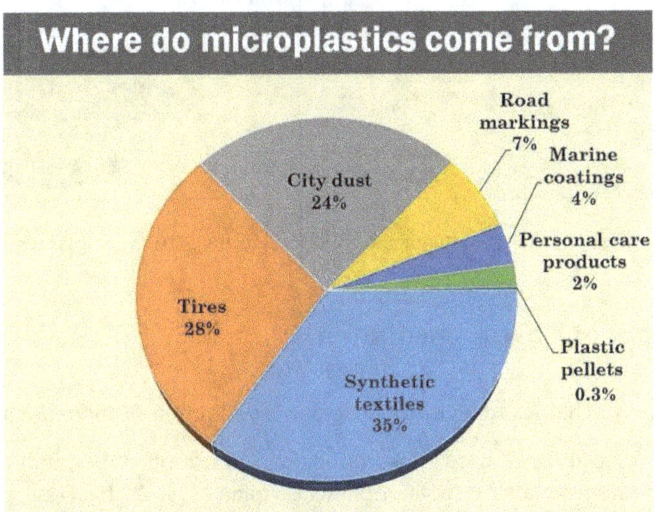

Fig. 13.2 Potential sources of microplastics to contaminate environments. (Source: International Union for Conservation of Nature, 2017)

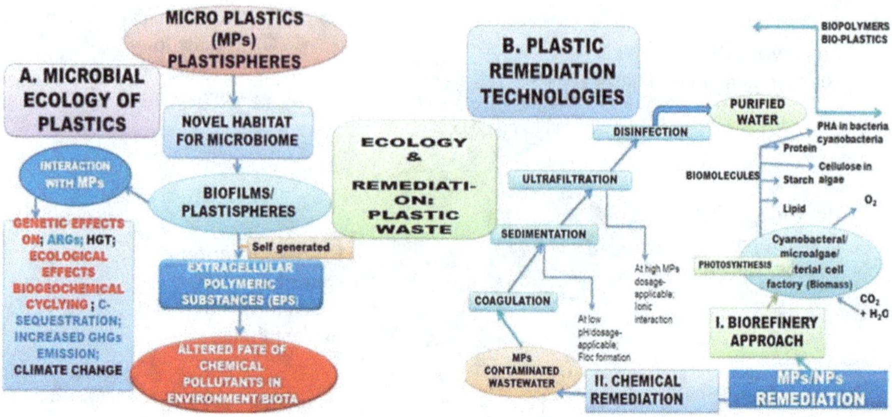

Fig. 13.3 Sources of microplastics in soil. WWTPs indicates wastewater treatment plants. (Andrady, 2011)

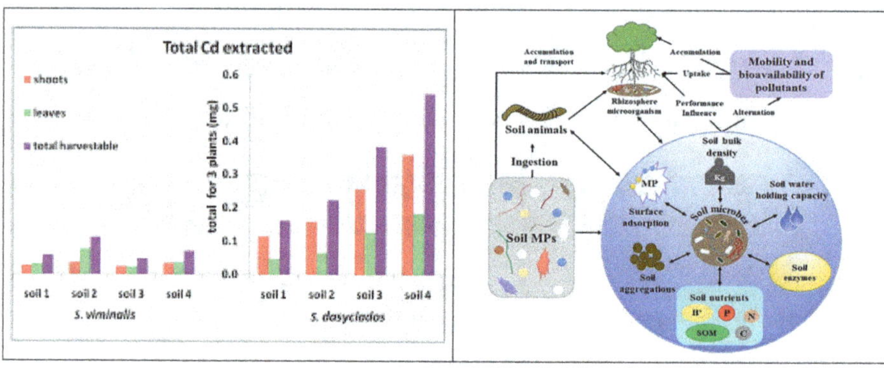

Fig. 13.4 (**a**) Heavy metals: contaminant soils (Wikipedia) (**b**) Microplastics in soil (Gou et al., 2020)

13.3 Types of Phytoremediation

There are many schemes related to this, the most common of them are as follows:

Rhizofiltration: primarily used, absorption, concentration, precipitation of heavy metals and non-metals by roots of plants (Wuana et al., 2010) (Fig. 13.4).

Phytoextraction: contaminants in harvestable plant tissues of roots and surface shoots are extracted and accumulated.

Phytotransformation: The incorporation of simple molecules into plant tissues after degradation of organic compounds..

Phytostimulation: also known as plant-assisted bioremediation. Stimulation of microbial degradation by the release of enzymes into the rhizosphere (Zhuang et al., 2005).

Phytostabilization: by reducing the mobility and preventing their migration to groundwater it involves in absorption, precipitation of contaminants, principally of metals by plants.

13.4 Harvesting/Disposal of Plant Material

Plants shoots are harvested and roots are removed when plants accumulate waste material, and methods used for disposal are based on the toxicity of end products of in-plant organic chemicals and storage locations in plant tissues (Boisson et al., 1999).

If the harmful contaminants are degraded from organic to simpler molecules there is no need for disposal. The significant application takes place only in roots, then those specific things have to be removed. The most common method used is *controlled incineration,* which results in ashes with toxic metals. Other methods of plant tissue treatment currently under investigation include sun, heat, air drying; composting; pressing and compacting; leaching (Li et al, 2014; Miretzky & Fernandez-Cirelli, 2008).

13.5 Soil Remediation Methods

Phytoextraction it involves removal of metals, radionuclides, certain organic compounds like hydrocarbons by direct absorption in plant tissues (Wani et al., 2012). It involves planting of one or more species that are major accumulators of contaminants of concern. Water and fertilizers may be required as primary field testing to ensure successful plant growth. After complete growth and development plant tissue is removed and a new crop is planted Brennan and Shelley (1999).

Characteristics of plants which are able to perform the method are rapid growth rate; high biomass production; ability to tolerate high accumulation of metals in harvestable tissues.

Phytostabilization the use of certain metal-tolerant plant species to absorb and precipitate toxic organic molecules by reducing their bioavailability and reduces the effect on humans. It is used to reestablish a vegetative cover at sites where natural vegetation is lacking due to high metal concentrations in surface soils and materials. Metal-tolerant species can be used to restore vegetation to the sites, thereby decreasing the potential migration of contamination to water bodies (Freitas et al., 2014).

Characteristics of plants suitable for phytostabilization at a particular site: tolerance to high levels of contaminants; high production of root biomass able to immobilize these contaminants through uptake, precipitation, or reduction; retention of applicable contaminants in roots, as opposed to transfer to shoots, to avoid special handling and disposal of shoots (Tokalıoğlu et al., 2010).

13.6 Applicability

Phytoremediation is a plant-based approach which involves the use of plants to extract and remove elemental pollutants or lower their bioavailability in soil (Tokalıoğlu et al., 1979). Or, in other words, phytoremediation is the direct use of living green plants for in situ removal, degradation, or containment of contaminants in soils, sludges, sediments, surface water, and ground water. By harnessing the natural capabilities of plants, we can remove, degrade, or stabilize contaminants.

There are several ways in which plants are used to clean up or remediate contaminated sites. To remove pollutants from soil, sediment, and water, plants can breakdown or degrade the organic pollutants and stabilize metal contaminants by acting as filters or traps (Kong & Bitton, 2003).

13.7 Conclusion

An abrupt rise in plastic waste has become one of the most serious global environmental problems during the past five decades, and many strategies have been suggested to control the increasing levels of contaminants associated with plastic waste. Although many studies have focused on the fate, toxicity, and health problems of plastic waste contaminants, only a very few have investigated microbial remediation of contaminants using cutting-edge nanoscience. This review focuses on addressing the environmental problems caused by microplastics (MP) and nanoplastics (NP) particles in view of nanoscience. Test microplastics were not absorbed in either soil or tissue culture seedlings, even though plants can absorb small dye molecules. Microplastics breakdown slowly in the soil and it is possible that these breakdown products could be absorbed.

There are several ways in which plants are used to clean up or remediate contaminated sites. To remove pollutants from soil, sediment and/or water, and air, plants can break down or degrade organic pollutants or contain and stabilize inorganic contaminants by acting as filters or traps. The success of phytoremediation at a given site cannot always be attributed to just one of these mechanisms because a combination of mechanisms may be at work. Phytoremediation is a low-cost, solar-energy driven and natural cleanup technique, which are most useful at sites with shallow, low levels of contamination. They are useful for treating a wide variety of environmental contaminants and are effective with or, in some cases, in place of mechanical cleanup methods. Phytoremediation harnesses natural processes to assist in the clean-up of pollutants in the environment (Park et al., 2011). The mechanisms by which plants promote the removal of pollutants are varied, including uptake and concentration, transformation of pollutants, stabilization, and rhizosphere degradation, in which plants promote the growth of bacteria underground in the root zone that in turn break down pollutants. Phytoremediation is amenable to a variety of organic and inorganic compounds and may be applied either in situ or ex situ. In situ applications decrease soil disturbance and the possibility of contaminants spreading via air and water, reduce the amount of waste to be land filled (up to 95%), and are low-cost compared with other treatment methods. In addition to this, it is easy to implement and maintain, does not require the use of expensive equipment or highly specialized personnel, and is environmentally friendly and aesthetically pleasing to the public.

References

Ahmad, M., Hashimoto, Y., Moon, D. H., Lee, S. S., & Ok, Y. S. (2012). Immobilization of lead in a Korean military shooting range soil using eggshell waste: An integrated mechanistic approach. *Journal of Hazardous Materials, 4*, 392–401.

Allouzi, M. M. A., Tang, D. Y. Y., Chew, K. W., Rinklebe, J., Bolan, N., Allouzi, S. M. A., & Show, P. L. (2021). Micro (nano) plastic pollution: The ecological influence on soil-plant system and human health. *Science of the Total Environment, 20*(788), 147815.

Andrady, A. L. (2011). Microplastics in the marine environment. *Marine Pollution Bulletin, 62*(2011), 1596–1605. https://doi.org/10.1016/j.marpolbul.2011.05.030

Boisson, J., Ruttens, A., Mench, M., & Vangronsveld, J. (1999). Evaluation of hydroxyapatite as a metal immobilizing soil additive for the remediation of polluted soils. Part 1. Influence of hydroxyapatite on metal exchangeability in soil, plant growth and plant metal accumulation. *Environmental Pollution, 104*, 225–233.

Brennan, M. A., & Shelley, M. L. (1999). A model of the uptake, translocation, and accumulation of lead (Pb) by maize for the purpose of phytoextraction. *Ecological Engineering, 12*, 271–297.

Cunningham, S. D., & Ow, D. W. (1996). Promises and prospect of phytoremediation. *Plant Physiology, 110*, 715–719.

Cunningham, S. D., Anderson, T. A., Schwab, P. A., & Hsu, F. C. (1996). Phytoremediation of soils contaminated with organic pollutants. *Advances in Agronomy, 56*, 55–114.

EPA. (2000). *A citizen's guide to phytoremediation*. EPA 542-F-98-011. United States Environmental Protection Agency, p. 6. Available at: http://www.bugsatwork.com/XYCLONYX/EPA_GUIDES /PHYTO.PDF

Freitas, E. V., Nascimento, C. W., & Silva, W. M. (2014). Citric acid-assisted phytoextraction of lead in the field: The use of soil amendments. *Water, Air, and Soil Pollution, 225*, 1–9.

Guo, J. J., Huang, X. P., Xiang, L., Wang, Y. Z., Li, Y. W., Li, H., & Wong, M. H. (2020). Source, migration and toxicology of microplastics in soil. *Environment International, 137*, 105263.

International Union for Conservation of Nature annual report. (2017). International Union for Conservation of Nature annual report 2017, https://portals.iucn.org/library/node/47536, 1, IUCN-2018-007, (2017).

Kong, I. C., & Bitton, G. (2003). Correlation between heavy metal toxicity and metal fractions of contaminated soils in Korea. *Bulletin of Environmental Contamination and Toxicology, 70*, 557–565.

Li, Z., Zhou, M., & Lin, W. (2014). The research of nanoparticle and microparticle hydroxyapatite amendment in multiple heavy metals contaminated soil remediation. *Journal of Nanomaterials, 17*, 1–8. [Google Scholar] [CrossRef].

Miretzky, P., & Fernandez-Cirelli, A. (2008). Phosphates for Pb immobilization in soils: A review. *Environmental Chemistry Letters, 6*, 121–133.

Park, J. H., Lamb, D., Paneerselvam, P., Choppala, G., Bolan, N., & Chung, J. W. (2011). Role of organic amendments on enhanced bioremediation of heavy metal(loid) contaminated soils. *Journal of Hazardous Materials, 85*, 549–574.

Raskin, I., & Ensley, B. D. (2000). Recent developments for in situ treatment of metal contaminated soils. In *Phytoremediation of toxic metals: Using plants to clean up the environment*. John Wiley & Sons. Available at: http://clu-n.org/techfocus

Tanee, F. B. G., & Akonye, L. A. (2009). Effectiveness of Vigna unguiculata as a phytoremediation plant in the remediation of crude oil polluted soil for cassava (Manihot esculenta Crantz) cultivation. *Journal of Applied Sciences and Environmental Management, 13*, 43–47.

Tessier, A., Campbell, P. G. C., & Bisson, M. (1979). Sequential extraction procedure for the speciation of particulate trace metals. *Analytical Chemistry, 51*, 844–851.

Tokalıoğlu, Ş., Yılmaz, V., & Kartal, Ş. (2010). An assessment on metal sources by multivariate analysis and speciation of metals in soil samples using the BCR sequential extraction procedure. *Clean-Soil Air Water, 38*, 713–718.

United Nations Environment Programme Division of Technology, Industry, and Economics. United States Environmental Protection Agency (USEPA). (2000). *Introduction to phytoremediation*. EPA 600/R-99/107. U.S. Environmental Protection Agency, Office of Research and Development.

Wani, S. H., Sanghera, G. S., & Athokpam, H. (2012). Phytoremediation: Curing soil problems with crops. *African Journal of Agricultural Research, 7*, 3991–4002.

Wuana, R. A., Okieimen, F. E., & Imborvungu, J. A. (2010). Removal of heavy metals from contaminated soil using chelating organic acids. *International Journal of Environmental Science & Technology, 7*, 485–496.

Yadav, B. K., Siebel, M. A., & van Bruggen, J. J. A. (2011). Rhizofiltration of a Heavy Metal (Lead) Containing Wastewater Using the Wetland Plant Carex pendula. *Clean-Soil Air Water, 39*, 467–474. [Google Scholar] [CrossRef].

Zhuang, P., Ye, Z. H., Lan, C. Y., Xie, Z. W., & Hsu, W. S. (2005). Chemically assisted phytoextraction of heavy metal contaminated soils using three plant species. *Plant Soil, 276*, 153–162.

Chapter 14
Bacterial Remediation of Micro-Nanoplastics (MNPs): Contaminated Soil

Srinivas Jukuri and Saida Lavudi

Abstract According to the recent studies, approximately 380 million tons of plastic is being generated across the world per year and 90% of it is recycled, so that it is converted into a pollutant. The majority of plastic waste has been sent to landfills; therefore, the soil acts as a major sink for plastic wastes. During the process of plastic breakdown in the soil, the plastic debris will be changed into micro-nanoplastics (MNPs), which will have a negative impact on the flora and fauna in the ecosystem, including the human health. Hence, appropriate degradation methods are needed to overcome this issue. Microbial biodegradation is the best method and is considered to be a more profitable and more effective and a highly accepted method. The microorganisms which are responsible for the biodegradation are differing from one another and have their own optimal growth conditions in the soil. Many kinds of microorganisms are involved in the biodegradation of MNPs. Among these biodegrade microorganisms, the bacteria are easier to grow and degrade MNPs compared to others. The objectives of this chapter are (1) to summarize the bacterial degradation of MNPs in soil and (2) to list out various kinds of bacteria and enzymes, which are involved in the degradation of MNPs in the soil system.

Keywords Micro-nanoplastics · Soil · Plastics · Bacteria · Biofilm · Biodegradation

---

S. Jukuri · S. Lavudi (✉)
Centre for Biotechnology, Institute of Science and Technology, JNTUH, Hyderabad, India
e-mail: lavudisaida@jntuh.ac.in

N. R. Maddela et al. (eds.), *Micro and Nanoplastics in Soil*,
https://doi.org/10.1007/978-3-031-21195-9_14

14.1 Introduction

Plastics are organic polymers containing molecules composed of long carbon chains like back-bones formed during the polymerization (Koushal et al., 2014). They are made of carbon and hydrogen, with nitrogen, sulfur, and other various organic and inorganic materials derived from fossil fuels (Kumari & Murthy, 2013). Many of the same units (or mers) are connected together to form a long chain or polymer or macromolecules. Plastics are polymers that, when heated, become mobile and can be molded into required shapes. Plastic-derived materials can be pushed into any required shape because they are non-metallic compounds. Plastics are predominantly used in the packaging business, which includes industries such as food, pharmaceuticals, and cosmetics. Polyethylene (LDPE, MDPE, HDPE, LLDPE), polypropylene (PP), polystyrene (PS), polyvinyl chloride (PVC), polybutyrene tetraphthalate (PBT), and nylon are the most regularly used polymers in the industries. Since the last six decades, when the commercial production of plastics began, we are depending on plastic as an affordable, versatile, and durable material. The production is accelerating so rapidly that it has created 8.3 billion tons of plastic, and unfortunately over 90% of it is not being recycled. As of 2018, approximately 380 million tons of plastic is produced worldwide each year and to combat the problem of plastic waste, the strategies of reuse, reduction, and recycling are now widely adopted. However, this method is less effective, especially for plastics waste that has been mixed with other types of waste (Drzyzga & Prieto, 2018). So that the majority of plastic materials has been sent to landfills and yet we are still producing and consuming more plastic. The decomposition process of plastic polymers takes thousands of years, and the landfill plastics waste processing requires large space, and incineration plastics waste processing can produce toxic gases into the environment (Kumar et al., 2017). As a result, people commonly burn plastic debris to combat the accumulation of plastic waste in the environment; however, this activity pollutes the air. It emits hazardous substances like CO2 and dioxins into the atmosphere, which are causes of lung diseases and cancer (Kale et al., 2015). Plastic waste is a contaminant that pollutes the land, air, and water ecosystems, harming the biosphere including human beings (Soud, 2019; Sowmya et al., 2014). Micro and nanoplastics (MNPs) are pieces of any plastic material having a size less than 5 mm in length that form as a result of bigger plastic goods degrading in the environment due to natural processes such as weathering. In recent years, the MNPs are abundantly found in the sea, freshwater, terrestrial environment, and organisms. MNPs contamination is becoming a major issue, and it is considered to be the second-most important scientific topic in the study of environment and ecology. Microplastics are seen as a serious threat to terrestrial ecosystems, including the soil, which potentially holds more plastic than the seas (Hurley & Nizzetto, 2018). Microplastics were abundantly found in floodplain soils (Scheurer & Bigalke, 2018), coastal beach soils (Zhou et al., 2018), and farming soils (Liu et al., 2018). Microplastics entered in soil will get stored, translocated, cause erosion, degradation and leach the groundwater, and thus threaten organisms and further effect human health (Hurley

& Nizzetto, 2018). Microplastics accumulation can be influenced by soil biota. Microplastics can be consumed by soil fauna and transformed into smaller MNPs in their gizzard. Digging mammals, such as gophers and moles, can incidentally contribute to the further abrasion into nanoplastics and translocation of microplastics (Rillig et al., 2017). Microplastics pollution can have negative impact on organisms in soils. Plastics are manufactured with multiple types of chemical ingredients to enhance their quality, including plasticizers, stabilizers, flame retardants, and monomers (de Souza Machado et al., 2018). The chemical ingredients can be leached out during the life cycle of the product, especially in the soil environment. On the other hand, plastics can also absorb other toxicants such as metals, polychlorinated biphenyls (PCBs), polycyclic aromatic hydrocarbons (PAHs), and organochlorine pesticides (e.g. DDT, HCH) due to their hydrophobic surface. The annual plastic release into soil is approximately 4 to 23 times higher than that of the sea (Horton et al., 2017). The study of plastic pollution in the oceans preceded that of soil contamination (da Costa et al., 2016). The terrestrial environment can't cope with this amount of plastic polluting MNPs; thus, a proper method of processing plastic waste is necessary. The plastic degradation mainly comprises of the following three types: photodegradation, oxy-photodegradation, and biodegradation (Shah et al., 2008). Among the best methods until date is biodegradation, as it uses microbes to degrade plastic, which is advantageous and efficient as well as widely accepted. In biodegradation, several types of plastic are degraded by various microbes and decomposers, such as actinomycetes, algae, bacteria, fungi, and others (Agrawal & Singh, 2016). They have the ability to create enzymes (both intracellular and extracellular) that aid in the decomposition of polymeric polymers. Bacteria are easier than fungi to grow and break down polymeric materials because fungi require more stable conditions to develop and degrade (Amobonye et al., 2021). The aim of this chapter is to discuss the state of soil pollution and highlight the major knowledge points on the microplastics and nanoplastics (MNPs) degradation by using of bacteria in the soil environment. This will help to improve our knowledge on the exposure, effect, and risks of MNPs degradation in contaminated soil by bacteria.

14.2 Types of Most Commonly Used Plastics

There are different types of plastics, classified on the basis of their origin, chemical structure, and physical properties (Fig. 14.1).

14.2.1 Classification Based on the Origin of the Plastics

Based on the origin, the plastics are divided into three types.

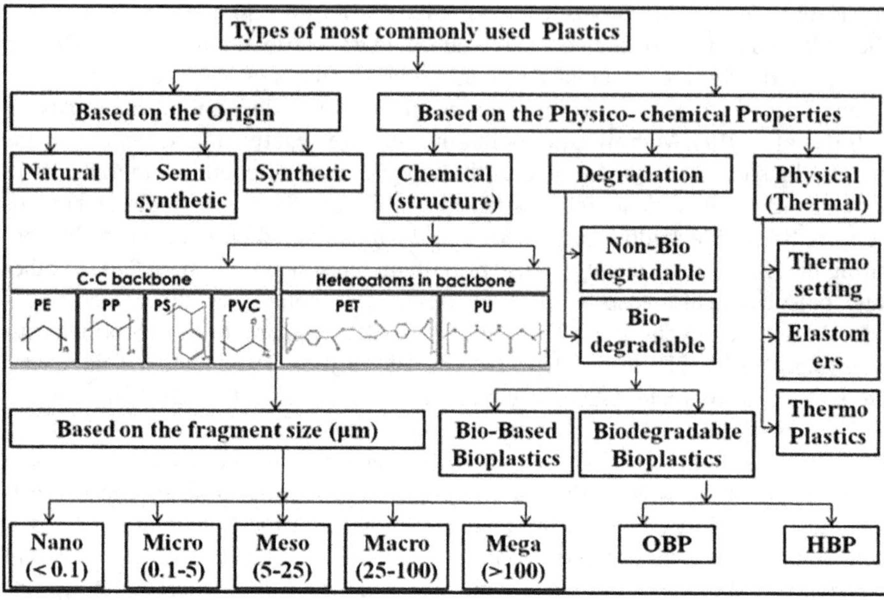

Fig. 14.1 Detailed classification of the most commonly used plastics

(a) *Natural plastics*:
Natural plastics are material, which can be moulded in its natural form (Tar, shellac, tortoiseshell, animal horn, cellulose, amber, and latex from tree sap).

(b) *Semi-synthetic plastics*:
Semi-synthetic plastics are chemically altered natural materials. Celluloid and vulcanized rubber were the first polymers, which are chemically modified from natural polymers, such as cellulose and latex.

(c) *Synthetic plastics*:
Synthetic polymers are synthesized entirely in the lab, usually by polymerizing monomers sourced from oil or gas, and plastics are made from them by adding different chemical ingredients. Bakelite is the first complete synthetic polymer. The use of synthetic plastics is widespread in the packaging of products, such as pharmaceuticals, food, cosmetics, chemicals, and detergents.

14.3 Classification Based on the Structure of the Atoms

Plastic polymers are classified into two groups based on their atomic structure (Kumar et al., 2013).

(a) *C–C backbone polymers*: C–C backbone polymers, including PE, PVC, PS, and PP, represent 77% of the total market share. (A) Polypropylene or PP (e.g., bottle caps, drinking straws, medicine bottles, car batteries, disposable syringes). (B) Polyvinyl chloride or PVC (e.g., bottles of juice, cling films, raincoats, visors, shoe soles, garden hoses, and electrical wiring pipes). (C) Polystyrene or PS (e.g., disposable cups, plates, trays, and cutlery, as well as packing materials and laboratory ware). (D) Polyethylene (PE): Polyethylene $((C_2H_4)_n)$ is the most common plastic used for packaging. PE is usually a mixture of similar polymers of ethylene, with various values of n. The commonly used PEs are as follows: (1) High-density polyethylene or HDPE (e.g., water bottles, trash, and retail bags). (2) Low-density polyethylene or LDPE (e.g., frozen food bags, squeezable bottles, flexible container lids). (3) Medium-density polyethylene or MDPE (e.g., gas pipes and fittings, sacks, shrink film, packaging film, carrier bags). (4) Linear low-density or LLDPE (e.g., cable coverings, toys, lids, buckets, containers, and pipe).

(b) *C–O backbone polymers or hetero atomic polymers*: C–C backbone polymers, including PET and PU, represent ~18% of the market share. (A) Polyethylene terephthalate or PET (e.g., soft drink, water and dressing bottles, peanut butter, and jam bars). (B) Polyurethane or PU (e.g., bedding, truck seating).

14.4 Classification on the Basis of Thermal Properties

Plastics can be classified into three kinds based on their thermal properties.

(a) *Thermosetting plastics*: Thermosets are hard and have a very tight-meshed, branched molecular structure. These plastics can withstand high temperatures and once hardened these cannot be reformed or recycled even with the application of heat. Thermosets are used, for example, to make light switches (e.g., bakelite, polyurethane, epoxy resin, vinyl ester resin, and vulcanized rubber).

(b) *Elastomers*: Elastomers also have a cross-linked structure and a looser mesh than thermosets, allowing for elasticity. Elastomers also cannot be reshaped with heat once they have been shaped (e.g., automobile tires).

(c) *Thermoplastics*: Thermoplastics usually have low melting points, which allow them to be remolded or recycled easily. They have a linear or branched molecular structure that determines their strength and thermal behavior; they are flexible at ordinary temperatures. At approx. 120–180 °C, thermoplastics become a pasty/liquid mass. The service temperature range for thermoplastics is lower than that of thermosets. Most plastics are thermoplastics, which are commonly used in packaging (e.g., polyvinyl chloride, polyethylene Polystyrene, Teflon, Acrylic, and Nylon are some of the thermoplastic materials).

14.5 Classification on the Basis of Degradability

14.5.1 Non-biodegradable Plastics

Non-biodegradable plastics also known as synthetic plastics are derived from petrochemicals and are very high molecular weight polymers. They do not degrade naturally and hence accumulate in environment.

14.5.2 Biodegradable Plastics

Biodegradable plastics are derived from natural substances such as components of algae, plants, and animals, which provide cellulose, starch, and protein needed for their production. They can easily be destroyed by UV radiation, water, enzymes, pH changes, and other factors. They are further divided into four groups

(i) *Bio-based bioplastics*: Plastics whose entire carbon content is produced from agricultural and forestry resources such as corn starch, soybean protein, and so on.

(ii) *Biodegradable bioplastics*: A biodegradable bioplastic is commonly made of renewable raw materials, microorganisms, petrochemicals, or a combination of all three. These plastics degrade completely by microorganisms into biogases and biomass (primarily carbon dioxide and water) without releasing harmful compounds. The use of biodegradable plastics is common in disposable items such as packaging, crockery, cutlery, medical devices, personal hygiene products, and foodservice containers. Several biodegradable bioplastics have been developed over the past few years, including polyhydroxyalkanoates (PHAs), polylactides, polycaprolactone, and polysaccharides. The polyhydroxyalkanoates (PHAs) were first observed in bacteria in 1888 by Martinus Beijerinck. There are two main types of biodegradable plastics.

   (a) *Oxo-biodegradable bioplastics (OBP)*: OBP is produced by mixing ordinary plastics with a little portion of fatty acid compounds obtained from transition metals.

   (b) *Hydro-biodegradable bioplastics (HBP)*: HBP is made from bio-based sources like corn, wheat, sugar cane, petroleum-based sources, or a combination of both. Both types of degradation begin with a chemical breakdown (oxidation and hydrolysis, respectively), followed by a biodegradation process. In both cases, degrading plastic emits CO2, but hydro-biodegradable plastics can emit methane as well. Examples for biodegradable plastics are polyglycolic acid (PGA), polyhydroxy butyrate (PHB), polylactic acid (PLA), polycaprolactone (PCL), poly hydroxyl alkanoates (PHA), polyhydroxyl valerate (PHV), polyvinyl alcohol (PVOH), and polyvinyl acetate (PVAc)

(iii) *Compostable bioplastics*: When composted, these bioplastics decompose at a similar rate as other compostable materials without leaving behind any toxic residues.
(iv) *Photodegradable bioplastics*: A photodegradable bioplastic is composed of light-sensitive groups attached to its backbone; therefore, prolonged exposure to UV light disintegrates their polymeric structure, making them prone to further degradation by microbes.

14.6 Classification Based on the Fragment Size

Plastic fragments are categorized as micro nanoplastics (MNPs) and micro, macro, and megaplastics based on their size in the environment. Micro and nano-sized plastic particles are produced by the physico-chemical breakdown of plastic waste and are referred to as micro-nanoplastics (MNPs). MNPs are split into two groups based on their source of origin: primary MNPs and secondary MNPs.

(i) *Primary MNPs*:
Primary MNPs are derived from household items, cosmetics, and polymeric raw materials from the plastics industry, such as polyethylene (PE), polystyrene (PS), and polypropylene (PE).The primary microplastics contain micro-beads in personal care products, tiny beads used for exfoliation, the abrasives in toothpastes, the plastic pellets used for grinding and polishing in industrial production, or the tiny debris originally produced in the manufacturing process (Wang et al., 2020).
(ii) *Secondary MNPs*:
Secondary MNPs are formed due to fragmentation of extensive plastic waste from exposure to abiotic factors such as temperature, UV radiation, and microbial degradation. Secondary MPs are mainly from the industrial and daily plastic goods discarded in the environment (e.g., bottles, packaging bags, boxes, clothing, various instruments, and production wastes) (Ammala et al., 2011).

Toxicological Effect of Micro-Nanoplastics (MNPs)
There are two types of pollutants transported together with MNPs that damage the ecosystem: The first are chemicals applied to plastics to increase their performance; the second is pollutants (chemical substances or pathogens) acquired and carried by MNPs from their surroundings in the continual transfer process in the environment (Yuan et al., 2020). When plastic ages in the ecosystem, it promotes the absorption of contaminants (such as heavy metal ions, POPs, and microbes) (Mao et al., 2020). Moreover, when soil MPs are becoming more abundant, the interaction between MPs and microorganisms becomes more frequent (Sangeetha Devi et al., 2015). MPs can be consumed or attached to organisms at various trophic levels in the soil, and then transported to organisms at higher trophic levels in the food chain, resulting in MP flow in the food web (Kumar Sen & Raut, 2015). The movements of MPs

in the food web can cause physical and chemical damages to organisms. Pollutants or pathogens spread by plastic particles will enter the food web and travel up in the food chain. Chlorinated plastic can release toxic soil, affecting the environment and groundwater. MNPs have been detected for the first time in human blood (Leslie et al., 2022), warning that the ubiquitous particles may be making their way into organs. According to a Dutch study, half of the blood samples showed traces of PET plastic, which is widely used to make drink bottles, and more than third contained polystyrene, which is widely used in disposable food containers and other products. According to the study, MNPs might have entered the body by a variety of means, including air, water, and food, as well as toothpaste, lip glosses, and tattoo ink. Methane gas, a major greenhouse gas generated during the decomposition process, affects significantly to global warming (Hester & Harrison, 2011).

14.7 Degradation of Plastics

There are two ways to degrade plastic waste. (1) Abiotic methods, (2) Biotic methods

14.7.1 Abiotic Methods

The degradation process of plastics is influenced by abiotic factors, which involve mechanical and chemical forces. The process will convert plastics into brittle materials, which leads to the formation of MNPs. The MW of the polymer is decreased during chemical fragmentation, but not during mechanical fragmentation. It is controlled by a number of factors, including polymer chain length, intramolecular forces, mechanical stability, polymer crystallinity, and plastic weight. Polymer degradation has been classified as follows based on the nature of the causing agents.

(i) *Photo-oxidative degradation*:

A photo-oxidation process, also known as ultraviolet degradation, degrades polymeric materials by exposing them to terrestrial light energy in combination with a chemical oxidizer, such as air. Certain plastics are naturally susceptible to photo-oxidation due to their structure and functional groups. As a result of adequate light energy input, these functional groups (chromospheres) cleave and produce free radicals, which is very similar to thermal oxidation. In essence, light energy accelerates the generation of free radicals, which initiate the degradation reaction. The photo-oxidation reaction reduces the molecular weight of a polymer by incorporating oxygen into its backbone structure as carbonyl groups. Its rate of initiation is very slow; once the plastic started to degrade, it propagates very fast. It is an environmentally friendly method, but it is quite expensive.

(ii) *Thermo-oxidative degradation*:

Thermal oxidation is the process by which polymeric materials are degraded by contacting a chemical oxidizer. Most polymers are susceptible to thermal oxidation, and it is by far the most common degradation process for plastics.

In oxidation, oxygen is added into the molecular structure of a polymer, creating a type of carbon–oxygen bond known as carbonyl functionality. The process of oxidation produces a permanent change in a plastic by shortening its chains by reducing its molecular weight. The oxidation is driven by the formation of free radicals within the plastic. In plastic formulations, free radicals can be unintentional byproducts of polymerization, as additives to formulations, or as contaminants. These free radicals are reactive and attack the covalent bonds in the polymer backbone. Polymer chains are cleaved through thermal oxidation, and the resulting shortened chains are terminated by oxygenated functional groups, such as carboxylic acids, esters, ketones, and aldehydes. In this method, oxygen is needed as well as heat (75–200 °C, a temperature higher than ambient). Various harmful gases are emitted into the environment at high temperatures. This approach is very quick, but that's not widely accepted.

(iii) *Hydrolytic degradation*:

The destruction of a polymeric material by contact with water, specifically hydrogen cations (H^+) or hydroxyl anions, is known as hydrolysis (OH^-). The degradation of plastic materials can be caused by immersion in water, condensation cycles, or exposure to steam. It can also be caused by interaction with acids (high H^+ concentration) or bases (high OH^- concentration), both of which can speed up the process significantly.

14.7.2 Biotic Methods

The microbial aspect of the synthetic plastic degradation is mostly due to the action of diverse microbial populations that have been identified as potential xenobiotics degraders based on their adaptability to and use these compounds as growth and energy substrates. These organisms use their diverse enzyme systems to break down polymers into intermediates that can then be absorbed and metabolized to meet their energy requirements. In this regard, the ability to biodegrade certain plastic polymers has recently been explored. Microbial degradation is a practical, clean, and affordable way to remediate MNPs contaminants. In this process, the plastic gets modified chemically, physically, and mechanically through surface degradation caused by diverse microbes and decomposer organisms such as actinomycetes, algae, bacteria, as well as fungi. As a result of microbial activity, various metabolic reaction pathways are involved in the conversion of organic molecules (MNPs) into biogas and residual biomass. The biodegradation of plastic waste is an efficient, profitable, and economically viable method. These bacteria can produce a variety of enzymes, both intracellular and extracellular, which can catalyze the degradation of plastic polymers into small and safe fragments (Agrawal & Singh, 2016). The use of microbial cells to break down plastic C–C linkages is considered to be more successful (Wei & Zimmermann, 2017). Microbial degradation is a specific enzymatic reaction. Certain enzymes are responsible for the breakdown of specific substrates (Adamcová & Vaverková, 2014).

14.8 Mechanism of Plastic Biodegradation

Abiotic degradation occurs before biodegradation and is triggered by thermal, hydrolytic, or UV light in the environment. Smaller polymer fragments are generated by abiotic breakdown can penetrate through the cell membrane and be biodegraded by enzymatic action inside microbial cells; nevertheless, some microorganisms secrete extracellular enzymes that can act on specific plastic polymers. The entire process of microbial degradation can be summarized in four essential stages (Fig. 14.2):

(i) *Biodeterioration or colonization (Adherence of microbes to the surface of polymer superficially)*: The first step of biodegradation is the biodeterioration that includes the combined action of microbial communities. Physico-chemical reactions lead to the incorporation of aquaphilic groups, which make the polymer more hydrophilic and reduce surface energy. It may allow the polymer's carbon to be used for microbial growth and development. Deterioration is a

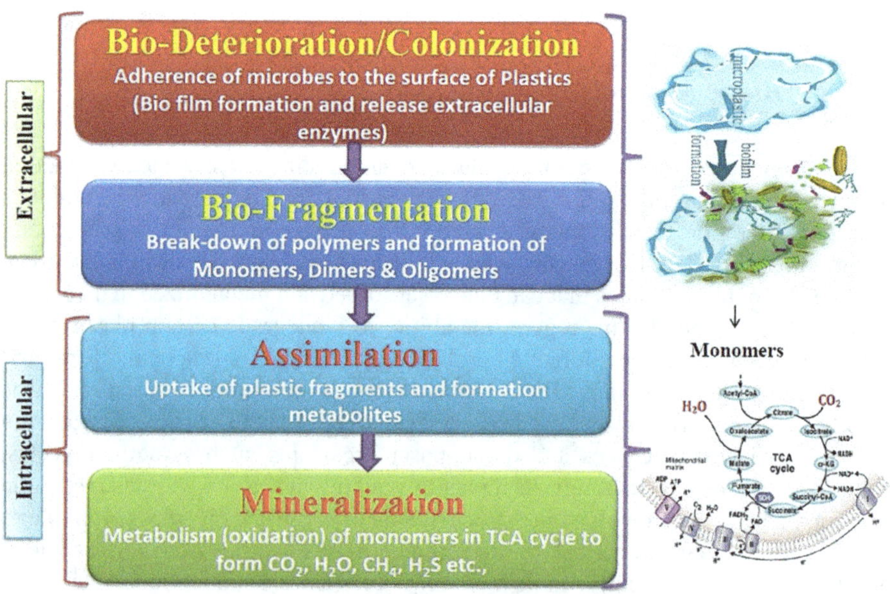

Fig. 14.2 Detailed mechanism of plastics biodegradation

type of surface degradation that affects a material's mechanical, physical, and chemical properties. This process will be accelerated by biofilms formed by microorganisms on the plastic surface. Biofilm is a colony of living organisms. Microbes attach to one another in a polymer matrix and colonize the surface of the material to form biofilms with the help of polysaccharides and proteins extracellular polymeric substances (EPS) are produced by themselves to break down the plastic surface. The EPS contains polysaccharides, proteins, and nucleic acids. The EPS penetrates into the surface pores of the plastic, causing them to expand. Microbes and bacteria have been enhanced in their ability to degrade plastic polymers, produce holes, and promote the physical deterioration of plastic polymers. Furthermore, the growth of biofilms on plastic surfaces supports the formation of different acid compounds (nitrous acid, nitric acid, sulfuric acid, citric, fumaric, gluconic, glutaric, glyoxylic, oxalic, and oxaloacetic acid) affecting the pH of plastic polymers and causing changes in the microstructures of the polymer, called chemical plastic deterioration.

(ii) *Biofragmentation or depolymerization (Exploitation of polymer as a food/carbon source to the microbes)*: Cleavage of the primary carbon chain takes place by catalytic agents (depolymerase enzymes), which are secreted by microorganisms and result in the formation of low molecular weight fragments such as oligomers, dimers, and monomers. Microbe-secreted extracellular and intracellular depolymerase enzymes play a significant role in the breakdown of plastic waste degradation. The released enzymes will break down complex polymers into smaller and simpler chains during the breakdown process. These decomposed small molecules will be easily dissolved in water and then absorbed by microorganisms through their semi-permeable cell membranes and utilized as carbon and energy sources.

(iii) *Assimilation*: Assimilation is the process of integrating molecules transported in the cytoplasm into the microbial metabolism to generate energy, biomass, vesicles, and numerous primary and secondary metabolites. Bacteria thus secrete some enzymes which played a significant part in the degradation process. The main end products of biodegradation of plastic in an aerobic environment are CO2, H2O, and biomass, whereas in anaerobic conditions, the main products are CO2, H2O, biomass, and CH4, while the main products of biodegradation of plastic in a sulfidogenic environment are H2S, CO2, and water.

(iv) *Mineralization*: Mineralization refers to the excretion of simple and different salts, and also complex metabolites that reach the extracellular surroundings. In this process, hazardous compounds are transformed into more environmentally friendly compounds. In mineralization, biodegradable materials or biomass are converted into gases, water, salt, minerals, and other residues. These gases include carbon dioxide, methane, and nitrogen. Mineralization will be completed when all biodegradable compounds have been consumed by microorganisms and all carbon has been converted to carbon dioxide.

Biodegradation is influenced by various factors, including polymer characteristics, organism type, and pretreatment method. The polymer characteristics such as its mobility, tactility, crystallinity, molecular weight, the type of functional groups and substituents present in its structure, and plasticizers or additives added to the polymer all play an important role in its degradation. Environmental conditions mediate the interaction between microbes and the degradative pathway during degradation. At commercial level, additives, antioxidants, and stabilizers get attached to the surface of polymers which may be proven harmful and susceptible to organisms in the environment and may also lead to slowing down of the speed of biodegradation process. The majority of plastics deteriorate at first on the surface, which is exposed and vulnerable to chemical or enzymatic attack. Therefore, degradation of microplastics proceeds faster than meso- and microplastics, as microplastics has a higher surface-to-volume ratio.

14.9 Plastics Biodegradation Bacteria

First Report of Plastic Degradation by Bacteria: For the first time, comparative degradation assay of lignin and paraffin's was studied due to action of bacteria (Fuhs, 1961) by growing bacteria on different kinds of alkenes as the only source of carbon. They further reported that bacteria can deteriorate only polymers with molecular weight up to 4800. Later, reports on plastic degradation by microbes started increasing significantly in the literature from various regions. Similarly, bacterial strains can degrade plastic polymeric substances in contaminated water or soil. Several studies have reported that plastics biodegradation by specialized bacteria can be a promising bioremediation strategy for contaminated ecosystems (Yoshida et al., 2016). Bacterial strains such as Pseudomonas spp., Bacillus spp., and Streptomyces spp. have exhibited high degradation efficiency against various plastic polymers (Li et al., 2020; Matjašič et al., 2021). In many cases, the plastics degradation rates by fungi exceed those achieved by bacterial strains (Muhonja et al., 2018). According to Amobonye et al. (2021), bacteria are easier to grow and degrade polymeric materials than fungi that need more stable conditions.

14.9.1 Plastic-Degrading Bacteria

Bacteria are considered to be the engine of the earth's nutrients, as they are responsible for the conversion and cycling of nutrients in the environment. Bacteria use the contaminants for their growth, nutrition, and reproduction. This is the main reason behind bacterial transformation of different contaminants which are organic in nature. Microorganisms get carbon (C) from Organic Carbon (OC). Carbon (C) is essential for bacteria and other microorganisms as it acts as a building block for new cell. Carbon (C) is also a source of energy utilized by the organisms (Mondal & Palit, 2019). Most of the identified bacteria belong to the phyla proteobacteria

(48%), firmicutes (37.4%) and actinobacteria (9.8%). Research has reported a wide range of plastic-degrading bacteria, summarized below.

(a) *C–C backbone plastic polymer degradation bacteria in soil*:

PE, PP, PVC, and PS are the four main types of synthetic plastics in the C–C backbone group. The polymer's structure renders it resistant to biodegradation. Furthermore, their short tenure in natural ecosystems (a few decades) is inadequate for nature to evolve new enzymatic systems that can degrade these synthetic polymers (Mueller, 2006).

(i) *PE, HDPE, and LDPE Biodegradation Bacteria*:

Polyethylene is a thermoplastic polymer made from ethylene gas and serves as a basis for multiple plastic products. Polyethylene is the most produced plastic in the world, which contains high hydrophobic level and high-molecular weight. The most commonly used PEs is LDPE and HDPE. Low-Density Polyethylene (LDPE) is the most extensively used packaging material, due to its outstanding mechanical qualities, water barrier capabilities, low cost, lightweight, and high energy effectiveness. HDPE is a denser version of polyethylene which is used to make water and drain pipes because of its rigidity and crystalline structure. In its natural form, it cannot be degraded easily by microorganisms. As early as the 1970s, Albertsson carried out experiments on microbial degradation of ^{14}C-labeled PE by using three different soil microbiotas as inocula (Albertsson, 1978). After that, Kawai et al. claimed that the upper limit of molecular weight for PE degradation by microorganisms was about 2000 Da based on the results of a numerical simulation (Kawai et al., 2004). *Actinobacter* sp. can partially break down lower molecular weight PE oligomers (MW = 600–800), whereas high molecular weight PE cannot be degraded. (Ghosh et al., 2013). In order to make it biodegradable, the crystallinity molecular weight and mechanical properties of the PE have to be modified. PE is activated by UV light at the beginning of the degradation process, which acts as an activator. As part of a similar study, PE was exposed to UV light as well as treated with nitric acid (Hasan et al., 2007). The pretreated polymer was applied to a microbial treatment. More than 20 bacterial genera have been shown to degrade different types of PE, those include various Gram-negative and Gram-positives species belonging to the genera *Pseudomonas, Ralstonia, tenotrophomonas, Klebsiella,* and *Acinetobactor* and *Rhodococcus, Staphylococcus, Streptococcus, Streptomyces,* and *Bacillus* (Danso et al., 2019). Majority of these bacterial strains can degrade the surface of PE and/or form a biofilm over it. In the process of biodegradation, the PE or paraffin molecules containing carbonyl group first get converted into an alcohol (containing –OH group) (Fig. 14.3) by a mono-oxygenase enzyme. The alcohol is then oxidized to an aldehydes (containing -CHO group) by alcohol dehydrogenase enzyme. An aldehydes dehydrogenase converts aldehydes to a fatty acid (containing -COOH group). This fatty acid then undergoes β-oxidation pathway inside cells (Hasan et al., 2007). Pseudomonas species has the unique ability to degrade and metabolize polymers with extracellular oxidative and/or hydrolytic enzymes, which facilitate uptake and degradation of polymer fragments and control the interaction between biofilms and polymer surfaces (Wilkes & Aristilde, 2017). *Brevibacillus*

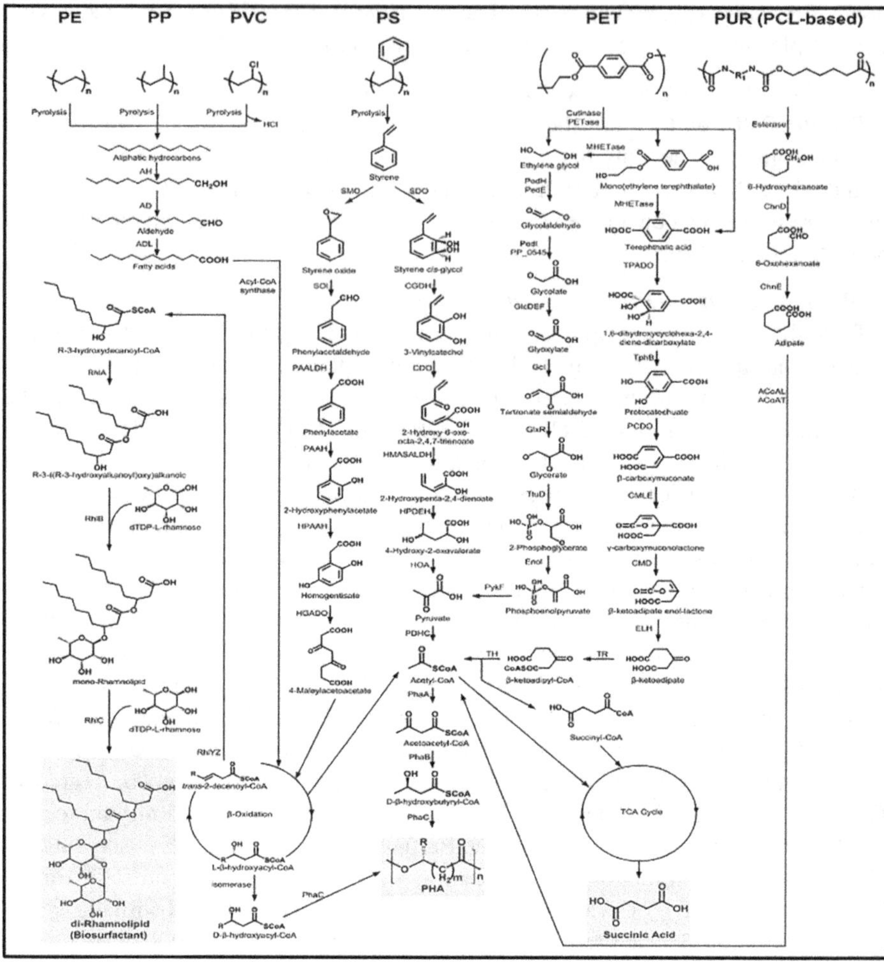

Fig. 14.3 Microbial biodegradation pathways of synthetic plastic material. (Adapted from Ru et al. (2020))

borstelensis, a thermophilic soil bacterium, utilizes BLDPE as the sole carbon and energy source, thus causing a reduction of 30% in the molecular weight of PE film after 30 days of incubation (Hadad et al., 2005). After thermal treatment, *Klebsiella pneumoniae* degraded the HDPE. This strain was able to adhere strongly to HDPE surfaces, leading to an increase in biofilm thickness while simultaneously decreasing the weight and tensile strength of the HDPE film by 18.4% and 60%, respectively, after 60 days (Awasthi et al., 2017). In the soil mixed with municipal waste, the decreasing order of degradation susceptibility of polymers was PE>>>LDPE>HDPE as determined by analyzing the weight loss of samples, changes in tensile strength, changes in FTIR, and bacterial activity in the soil (Orhan et al., 2004). Table 14.1 shows that the soil-isolated microbial strains can degrade PE, HDPE, and LDPE.

(ii) *Polypropylene (PP) biodegradation bacteria:*

PP is a thermoplastic polymer resin with a semi-crystalline structure, which is the second most commonly used plastic in the world. Most commercial PP is isotactic and has an intermediate level of crystallinity between that of LDPE and HDPE due to its durability and outstanding characteristics, it is used in a variety of applications to include packaging for consumer products, plastic moldings, plastic tubs, stationary folders, packaging materials non-absorbable sutures, diapers, automotive industry, and textiles. It can be degraded when exposed to ultraviolet UV light from the sun, and it can also be oxidized at high temperatures. Even though PP is a polyolefin, it has the same oxidative degradation susceptibility as PE. However, its substitution of methyl for hydrogen in the ß position allows it to be more resistant to microbial degradation. Microbial degradation of PP was firstly assessed by Cacciari et al. (1993) by cultures enriched from sandy soils containing PE wastes. These isolated bacterial communities from soil samples mixed with starch have been shown to be capable of degrading PP. Biodegradation of isotactic PP without any treatment is reported with one of the community designated as 3S among the four microbial communities (designated as 1S, 2S, 3S, and 6S) adapted to grow on starch containing PE obtained from enrichment culture. *Pseudomonas chlororaphis, Pseudomonasstuzeri,* and *Vibrio species* were identified in the community 3S. It is reported that UV-treated PP sample is more susceptible to degradation (Sameh et al., 2006). Pseudomonas and Bacillus bacterial species were isolated from the soil of a plastic-dumping site, could utilize PP as their carbon source for growth and degrade 0.05–5% of PP after incubation for 12 months (Arkatkar et al., 2010). A mixed consortium of four bacterial isolates from waste management landfills and sewage treatment plants could degrade the PP strips and pellets, lost 44.2–56.3% of their weight after 140 days (Skariyachan et al., 2018). *Bacillus Rhodococcus, Bacillus gottheilii* were isolated from mangrove sediments and also able to grow in aqueous synthetic media containing PP microplastics and resulting in a weight loss of 4.0–6.4% after 40 days (Auta et al., 2018). Helen AS et al. reported in 2017 that B. cereus had a PP degradation capacity of 0.003 grams per day and S.globispora had a PP degradation capacity of 0.002 g per day. Table 14.1 shows that the soil-isolated microbial strains can degrade the Polypropylene (PP)

(iii) *PVC biodegradation bacteria:*

PVC is a strong plastic that resists abrasion and chemicals. It also has low moisture absorption properties. There are a lot of studies about the thermal and photodegradation of PVC, but only a few studies on the biodegradation of this material. PVC is the primary synthetic plastic type with the highest percentage of plasticizers (up to 50%). Plasticized PVC is susceptible to microbial attack because plasticizers can be utilized as a carbon source by bacteria. Microorganisms degrading plasticized PVC just break down components of the plasticizer [e.g., bis (2-Ethylhexyl) phthalate, DEHP] rather than the backbone of PVC. The degradation of both PVC

Table 14.1 List of bacteria used in the degradation of various plastics

Type of plastic	*Bacteria*	References
PE	*Bacillus amylolyticus*	Patil (2018)
	Bacillus gottheilii	Auta et al. (2018); Yoshida et al. (2016)
	Ideonella sakaiensis	Palm et al. (2019)
	Bacillus subtilis	Patil (2018)
	Desulfotomaculum nigrifans	Begum et al. (2015)
	Lysinibacillus fusiformis *Bacillus cereus*	Shahnawaz et al. (2016)
	Pseudomonas alcaligenes	Begum et al. (2015)
	Pseudomonas fluorescens	Patil (2018)
	Pseudomonas putida	Patil (2018)
	Pseudomonas putida MTCC 2475	Saminathan et al. (2014)
	Streptomyces SSP2	Soud (2019)
	Streptomyces SSP4	Soud (2019)
	Sterptomyces SSP14	Soud (2019)
	Actinobacter ursingii	Hussein et al. (2015)
	Brevibacillus borstelensis	Mohanrasu et al. (2018)
	Pseudomonas spp.	Skariyachan et al. (2015)
	Acidobacteria, Bacteriodietes,	
	Actinobacteria; Diminish: Acidobacteria	Ren et al. (2020)
HDPE	*Ochrobacterum anthropi*	Riandi et al. (2017)
	Arthrobacter sp. *GMB5* and Pseudomonas sp. *GMB7*	
	Bacillus cereus *Brevibacillus borstelensis*	Muhonja et al. (2018)

(continued)

Table 14.1 (continued)

Type of plastic	*Bacteria*	References
LDPE	*Lysinibacillus macrolides* *Pseudomonas putida* *Cellulosimicrobium funkei*	Muhonja et al. (2018)
	Pantoea sp. *Enterobacter* sp.	Skariyachan et al. (2016)
	Actinobacter ursingii	Hussein et al. (2015)
	Alcanivorax borkumensis	Delacuvellerie et al. (2019)
	Streptomyces spp. *Pseudomonas spp.*	Deepika and Jaya (2015)
	Pseudomonas sp.	Tribedi and Sil (2013)
	Bacillus carbonipphilus	Shresta et al. (2019)
	Bacillus coagulans	Shresta et al. (2019)
	Bacillus licheniformis KC2-MRL	Jamil et al. (2017)
	Bacillus megaterium	Shresta et al. (2019)
	Bacillus nedei	Shresta et al. (2019)
	Bacillus smithii	Shresta et al. (2019)
	Bacillus sp. *KC3-MRL*	Jamil et al. (2017)
	Bacillus sporothermo-durans	Shresta et al. (2019)
	Bacillus weihenstephanensis	Mukherjee and Chatterjee (2014)
	Burkholderia cepacia	Mukherjee and Chatterjee (2014)
	Escherichia coli	Mukherjee and Chatterjee (2014)
	Pseudomonas aeruginosa	Riandi et al. (2017)
	Pseudomonas fluorescens	Patil (2018)
	Serratia sp. *KCI-MRL*	Jamil et al. (2017)
	Stenotropphomonas sp. *KC4-MRL*	Jamil et al. (2017)
	Streptomyces coelicoflavus NBRC 15399^{T}	Duddu et al. (2015)
	Streptomyces SSP2	Soud (2019)
	Streptomyces SSP4	Soud (2019)
	Sterptomyces SSP14	Soud (2019)
	Sphingobacterium moltivorum	Montazer et al. (2018)
PP	*Bacillus cereus*	Helen et al. (2017)
	Sporosacrina globispora	Helen et al. (2017)
	Bacillus Rhodococcus	Auta et al. (2018)
	Bacillus gottheilii	Auta et al. (2018)
PVC	*Chryseomicrobium imtechense; Lysinibacillus fusiformis; Acinetobacter calcoaceticus; Stenotrophomonas pavanii*	Latorre et al. (2012)
	Acanthopleurobacter pedis; Bacillus cereus; Pseudomonas otitidis; Bacillus aerius	Anwar et al. (2016)
	Pseudomonas citronellolis	Giacomucci et al. (2019)

(continued)

Type of plastic	*Bacteria*	References
PS	*Bacillus subtilis*	Asmita et al. (2015)
	Pseudomonas auroginosa	Asmita et al. (2015)
	Staphylococcus aureus	Asmita et al. (2015)
	Staphylococcus pyogenes	Asmita et al. (2015)
Polyester PU coating (including Impranil®)	*Alicycliphilus* sp.	Oceguera-Cervantes et al. (2007)
Polyester PU foam	*Arthrobacter calcoaceticus*	El-Sayed et al. (1996)
	Acinetobacter garnei	Howard et al. (2012)
	Arthrobacter globiformis	El-Sayed et al. (1996)
	Bacillus subtilis	Rowe and Howard (2002); Nakkabi et al. (2015)
	Bacillus pumilus	Nair and Kumar (2007)
	Commamonas acidovorans	Allen et al. (1999)
	Pseudomonas aeruginosa	El-Sayed et al. (1996)
	Pseudomonas cepacian	El-Sayed et al. (1996)
	Pseudomonas chlororaphis	Howard et al. (2001a)
	Pseudomonas fluorescens	Ruiz et al. (1999)
	Pseudomonas putida	El-Sayed et al. (1996); Peng et al. (2014)
	Alycycliphilus sp.	Pérez-Lara et al. (2016)
Thermoplastic polyester PU	*Pseudomonas aeruginosa*	Kay et al. (1991)
	Pseudomonas chlororaphis	Gautam et al. (2007)
	Arthrobacter sp. *Bacillus* sp.	Shah et al. (2008)
Thermoplastic polyether PU	*Comamonas acidovorans*	Akutsu et al. (1998); Nakajima-Kambe et al. (1997, 1995)
	Corynebacterium	Kay et al. (1991)
	Micrococcus sp.	Shah et al. (2008)
	Pseudomonas sp.	Shah et al. (2008)
	Pseudomonas aeruginosa	Shah et al. (2008, 2013, 2016); Fernandes et al. (2016)
	Staphylococcus epidermidis	Jansen et al. (1991)
PET	*Bacillus subtilis*	Asmita et al. (2015)
	Staphylococcus pyogenes	Asmita et al. (2015)
	Staphylococcus aureus	Asmita et al. (2015)

and plasticizers by microorganisms has not been observed so far. Therefore, we do not know what enzymes are responsible for the microbial degradation of PVC. Nevertheless, a number of bacterial varieties have been reported to be able to degrade the plasticized PVC, including those isolated from garden soil (Nakamiya et al., 2005; Giacomucci et al., 2019), landfill leachate, waste disposal sites (Latorre et al., 2012; Anwar et al., 2016), and marine environments (Kumari et al., 2019).

Table 14.1 shows a list of soil-isolated bacteria used in the degradation of Polyvinyl chloride (PVC).

(iv) *PS biodegradation bacteria:*

The PS polymer $(C_8H_8)_n$ is an aromatic polymer with a high molecular weight, which is made of monomer styrene. PS can be solid or foamed, while styrene monomer is liquid. The general purpose polystyrene (GPPS) is clear, rigid, and brittle. In many aspects of human life and industry, polystyrene is widely used due to its properties such as low cost, lightweight, ease of manufacture and versatility, thermal efficiency, durability, and water resistance. PS is used in the manufacture of disposable cups, packaging materials, and laboratory ware, as well as in certain electronic products. It is used for its lightweight, stiffness, and excellent thermal insulation. Polystyrene is extremely stable and difficult to degrade in the environment due to its hydrophobic nature, making it resistant to hydrolysis (Albertsson & Karlsson, 1993). Styrene, benzene, toluene, and acrolein are released when it is decomposed by thermal or chemical processes. There are limited publications on PS biodegradation, however, a few researchers have reported on the microbial decomposition of its monomer, styrene. There are several ways of styrene catabolism; however, a predominant pathway involves the oxidation of styrene to phenyl acetate, which is then converted via the TCA cycle. This pathway is shown in Fig. 14.2. According to Kaplan et al., PS breakdown is less than 1% after 90 days in farmed soils with a wide range of fungi, bacteria, and invertebrates, with no notable rise in degradation rate after this one time (Kaplan et al., 1979). Otake et al., on the other hand, observed that a PS sheet buried in soil for 32 years showed no signs of degradation (Otake et al., 1995). The *Rhodococcus ruber* has been demonstrated to create biofilms on PS and partially break ot down (Mor & Sivan, 2008). A biofilter made up of Brevibacillus sp. has been found to remove 3 kg of styrene in a day (Baggi et al., 1983). Oikawa et al. was isolated and identified *Pseudomonas* sp. and *Bacillus* sp. for styrene degradation, and also *Xanthomonas* sp. and *Sphingobacterium* sp. for PS decomposition by 16 S ribosomal DNA analyses from soil (Oikawa et al., 2003). Four microbial strains have been isolated from garden soil after 8 months of buried samples of PS and EPS solution (2%) in chloroform. They were identified as *Microbacterium* sp. NA23, *Paenibacillus urinalis* NA26, *Bacillus* sp. NB6, and *Pseudomonas aeruginosa* NB26. They were able to extract some carbon from the complex molecules of PS, but the process was very slow and did not cause any significant chemical changes on the surface (Atiq et al., 2010). The biodegradation of PS involved Gram-positive coccobacillus, Gram-negative cocci, Gram-negative rod-shaped bacillus, Gram-positive cocci (in clusters) in Garden soil, and Gram-negative cocci (in singles) in garbage soil with weight loss up to 30% (Asmita et al, 2015). Krueger et al. (2015) found a reduction in molecular mass of polystyrene sulfonate (PSS) by 50% within 20 days as a result of the activity of *Gloeophyllum trabeum* DSM 1398. *Citrobacter sedlakii*, *Enterobacter* sp., *Alcaligenes* sp., and *Brevundimonas diminuta* were isolated and identified by Sekhar et al. (2016), and the highest PS degradation rate was estimated to be 12.4% within 30 days of Enterobacter. Pseudomonas spp. and Bacillus spp. were reported to be able to breakdown high-impact PS (HIPS) film by Mohan et al. (2016). Bacillus spp., in particular, succeeded a reduction of plastic weight loss by 23% after 30 days. Table 14.1 shows list of soil-isolated bacteria used in the degradation of Poly styrene (PS)

(b) *C–O backbone plastic polymer degradation bacteria in soil*:

Two synthetic plastic polymers lie under the C–O backbone category, namely PU and PET. However, this type of plastic material can be hydrolyzed due to the presence of ester bonds

(i) *PU biodegradation bacteria:*

Polyurethanes (PUs) are an important branch of synthetic plastics belonging to the thermosetting group, which can be re-used for production. PUs can be broadly categorized as follows: flexible, semi-rigid, rigid, microcellular, viscoelastic, or thermoplastic urethanes. The polyurethanes industries were laid in the late 1930s with the discovery by German scientist Otto Bayer (Szycher, 1999). Since that time, scientists have been finding its use in an ever-increasing number of applications, and polyurethanes are now all around us, playing a vital role in many industries—from furniture to footwear, construction to cars, i.e., furniture coatings, adhesives, construction materials, flame retardants, fibers, paints, elastomeric parts, and synthetic skins are just a few examples. Polyurethanes have also been employed in a variety of biomedical applications, including vascular prostheses, prosthetic skin, pericardial patches, soft-tissue adhesive, drug delivery devices, and tissue engineering scaffolds (Young & Lovell, 1994). Now Polyurethanes (PU) represent almost 8% of produced plastics which place them as the sixth most used polymer in the world (Kemona & Piotrowska, 2020). PU is the condensation product of polyisocyanate and polyol having intramolecular urethane bonds (carbonate ester bond –NHCOO–) (Sauders & Frisch, 1964). There are two types of PU when it comes to biodegradation: polyester polyurethane (PSPU) and polyether polyurethane (PEPU) which have several applications in the industrial field. Microbial degradation (fungal) of PU was firstly reported by Darby and Kaplan in 1968 (Darby & Kaplan, 1968). In comparison to PEPU, PSPU was easier to biodegrade. Following that, numerous microorganisms were shown to be capable of degrading polyester PU. The potential of Staphylococcus epidermidis KH11 to break down polyether PU was examined by Jansen et al. (1991). As well as three esterases purified from *Pseudomonas chlororaphis* (Ruiz et al., 1999), the protease purified from *Pseudomonas fluorescens* (Vega et al., 1999), and a lipase purified from *B. subtilis* (Rowe & Howard, 2002) have the ability to degrade PSPU. In addition, they also cloned a gene named pulA from Pseudomonas fluorescens (Ruiz & Howard, 1999) and two genes, pue A and pue B, from Pseudomonas chlororaphis (Stern & Howard, 2000; Howard et al., 2001b). These genes encoded three different esterases involved in the microbial degradation of emulsified polyester PU by *Pseudomonas fluorescens* and *Pseudomonas chlororaphis*. The list of bacterial strains were degrading different kinds of polyurethane (Kemona & Piotrowska, 2020). Table 14.1 shows the list of soil-isolated bacteria used in the degradation of various polyurethanes (PUs).

(ii) *Polyethylene terephthalate (PET) biodegradation bacteria:*

PET, a synthetic polymer generated from crude oil, is today one of the most widely used plastics (Liu et al., 2019), contributing to more than 10% of the plastic market share (Carr et al., 2020). This polymer is made up of terephthalic acid (TPA) and ethylene glycol repeating units (EG) (Fig. 14.1). PET is convenient both in terms of manufacture and utility, as it is utilized in containers, films, and fibers, in addition to bottles, due to its lightweight, durability, and mold ability. It is resistant to biodegradation due to the polymer's backbone's high stability, as well as its crystallinity and surface hydrophobicity, which are some of the underlying elements that limit the natural breakdown of this plastic. *B.subtilis, S.aureus, and S.pyogenes* are considered as important PET and PS degrading bacteria (Asmita et al., 2015). *Ideonella sakaiensis* was also reported to degrade PET polymer (Yoshida et al., 2016; Oberbeckmann & Labrenz, 2020).Table 14.1 shows the list of soil-isolated bacteria used in the degradation of Polyethylene terephthalate (PET).

14.9.2 Plastic-Degrading Actinomycetes

Actinomycetes are a phylum of gram positive bacteria. They are prokaryotic organisms with a primitive unicellular organization. Actinomycetes are anaerobic microorganisms. On solid substrates, they have filamentous and branching growth patterns that resemble fungal mycelia. Their colonies, like myceliums, are large. Many genera of actinomycetes have aerial hyphae. Some actinomycetes have flagella and are motile. Actinomycetes can be found in both soil and water. Actinomycetes include the Streptomyces groups, Rhodococcus ruber, Actinomadura spp., and the thermophilic Thermo actinomycetes species have been isolated from different ecological zones and demonstrated to possess significant plastic biodegradative potentials (Auta et al., 2018; Jabloune et al., 2020). Their ability to produce a wide range of hydrolytic enzymes as well as other bioactive metabolites has previously been emphasized. These hydrolytic enzymes are one of the most important components in their ability to grow on a variety of plastic polymers and degrade the large molecular weight molecules into simpler ones. They can produce extracellular polymers such as dextrin, glycogen, levan, and N-acetyl glucosamine-rich slime polysaccharides which probably facilitate their attachment to plastic surfaces for subsequent microbial action. Biofilm formation has been found to be an important factor in the actinomycetal colonization of plastics, similar to the bacteria PET and other polymers including p-nitro phenyl esters, cutin, and suberin were found to be degraded by Streptomyces scabies, isolated from potatoes (Jabloune et al., 2020). Nocardiopsis sp., an endophytic actinomycetes isolated from the hibiscus, was similarly found to break down PE and fuel. The effectiveness of actinomycetal plastic degradation has also been highlighted in a microbial consortium with a substantial fraction of actinomycetal species degrading polyurethane and different chemical additives.

The plastics of poly (β-hydroxybutyrate) (PHB)-and poly (ε-caprolactone) (PCL) were degraded by aerobic microorganisms that persist in the natural environment.

Plastic depolymerizing microbes can be found over many kinds of material, including landfill leachate, compost, sewage sludge, forest soil, farm soil, paddy soil, weed field soil, roadside sand, and pond sediment (Nisida & Tokiwa, 1993). Actinomycetes strains *Streptomyces* genus and *Micromonospora* genus were isolated and screened for the capability to degrade poly (ethylene succinate) (PES), poly (ε-caprolactone) (PCL), and poly (β-hydroxybutyrate) (PHB) from the upstream and downstream regions of the Touchien River in Taiwan (Hoang et al., 2007). *Streptoverticillium kashmirense* AF1 can degrade a natural polymer; poly (3-hydroxybutyrate-co-3-hydroxyvalerate) (PHBV) was isolated from municipal sewage sludge by soil burial technique. Extracellular enzymes PHBV depolymerase secreted by *Streptoverticillium kashmirense* AF1 was purified and degrade PHBV film (Shah et al., 2007) *Actinomadura, Microbispora, Streptomyces, Thermo actinomyces,* and *Saccharomonospora* were thermophilic actinomycetes strains able to degrade poly (ethylene succinate) (PES), poly (ε-caprolactone) (PCL), and poly (β-hydroxybutyrate) (PHB). Thermophilic actinomycetes *Microbispora rosea, Excellospora japonica,* and *E. viridilutea* were able to degrade aliphatic polyester, poly (tetramethylene succinate) (100 mg PTMS film) (Jarerat & Tokiwa, 2001) *Rhodococcusruber* (C208) *Rhodococcus* sp. strain RHA1, strong polychlorinated biphenyl (PCB) degrader has diverse biphenyl/PCB degradative genes and harbors huge linear plasmids, including pRHL1 (1100 kb), pRHL2 (450 kb), and pRHL3 (330 kb). Linear plasmids of Rhodococcus sp. strain RHA1 having degradative genes such as bphB2, etbD2, etbC, bphDEF, bphC2, and bphC4 (Shimizu et al., 2001) *Amycolatopsis* strains, poly(L-lactide) degrader stain has the ability to absorb breakdown products such as poly lactic acids (Pranamuda et al., 1999). Polylactide (PLA)-degrading microorganisms are sparsely distributed in soil environments. Totally 34 different kinds of marine Actinomycetes isolates were discovered in marine soil. Five of the most common Actinomycetes cultures were tested for plastic degradation such as *Streptomyces* sp., *Pseudonocardia* sp., *Actinoplanes* sp., *Sporichthya* sp. Among them, *Streptomyces* sp. has shown significant reduction (20%) when compared to other tested organisms (Sathya et al., 2012). Table 14.2 summarizes Actinomycetes strains associated with various plastics biodegradation in the soil.

Table 14.2 List of Actinomycetes used in the degradation of various plastics

Types of plastic	Actinomycetes	Reference
Polyurethane	*Acinetobacter calcoaceticus, A. gerneri*	Howard et al. (2012)
Polyethylene	*Streptomyces* sp., *Sporichthya* sp., *Actinoplanes* sp.	Sathya et al. (2012)
Disposable plastic films	*Streptomyces* sp.	El-Shafei et al. (1998)
LDPE powder	*Streptomyces KU5, Streptomyces KU1, Streptomyces KU6*	Usha et al. (2011); Abraham et al. (2017)
HDPE	*Streptomyces* sp.	Farzi et al. (2017)
PET	*Streptomyces scabies* *Streptomyces* sp.	Jabloune et al. (2020); Farzi et al. (2019)

14.10 Plastic-Degrading Bacterial Enzymes

Plastic-degrading bacteria and other microorganisms are involved in plastic biodegradation by producing a variety of essential enzymes. This polymer biodegradation process involves two reactions: Hydrolysis, and Oxidation. Hydrolysis is the breakdown of polymers catalyzed by hydrolases enzymes, which are one of the most important aspects in their ability to grow on various polymers and degrade high molecular weight to simpler ones. Hydrolase enzymes catalyze the hydrolysis of esters, carbonates, amides, and glycosidic linkages to create monomers from various hydrolyzed polymers. Oxidation is a biodegradation process that is conducted by oxidoreductase enzymes. Meanwhile, oxidoreductase enzymes catalyze ethylene, carbonate, amide, urethane, and other oxidizing and reducing processes (Ganesh et al., 2017). Some polymer compounds cannot be degraded by certain enzymes, the other appropriate enzymes will work together to break down those compounds. This phenomenon is known as oxidation. Plastic biodegradation enzymes are classified into two broad categories, viz., extracellular and intracellular enzymes (Gu, 2003).

Extracellular Enzymes

These types of enzymes are involved in heterogonous reactions, as these act on the macromolecules at the surface of the solid plastic while it is in a liquid state (Chinaglia et al., 2018). Additionally, other groups of enzymes are involved in the surface functionalization of hydrophobic plastic surfaces, the degradation of the plastic metabolic intermediates into monomeric units, and the mineralization of the final monomeric intermediates.

Intracellular Enzymes

These enzymes convert intermediates into compounds that can be assimilated by microbes via aerobic and anaerobic processes (Pathak, 2017).

Enzyme technology has recently been investigated for the production, isolation, purification, and providing the enzymes for the degradation of plastics. These enzymes are non-toxic and biodegradable. In the last decade, a few polymer plastics chains (PE, PP, PS, and PVC) are subjected to degrade by a distinct group of enzymes as shown in Table 14.3. Many enzymes like esterases, protease, cutinase, and laccase have shown significant results in the breakdown of MNPs. A bacterium named *Ideonella sakaiensisis* can utilize PET as its primary carbon and energy source (Yoshida et al., 2016). By the presence of two active enzymes (PETase and MHET ase), this bacterium converts PET into its monomers terephthalic acid and ethylene glycol (Palm et al., 2019). Recent research on the enzymatic degradation of plastics has generated a lot of interest in protein/enzyme engineering to improve enzyme activity.

An engineered PETase mutant from *Ideonella sakaiensis* exhibits an increase in the three mutants (R61A, L88F, and I179F) by 1.4-fold, 2.1-fold, and 2.5-fold, in comparison to the wild type strain. It has been demonstrated that enzyme activity can be significantly improved by rational protein engineering and by modifying key hydrophobic grooves of substrate binding sites (Ma et al., 2018). Surprisingly, a

Table 14.3 List of plastic-degrading bacterial enzymes

Type of plastic polymer	Bacteria	Enzyme	Reference
PE	*Pseudomonas* sp. E4 expressed, in *Escherichia coli* BL21	Recombinant Alkane hydroxylases (AH)	Yoon et al. (2012)
	Pseudomonas aeruginosa E7	Alkane monooxygenase, Rubredoxin and Rubredoxin reductase	Jeon and Kim (2015)
	Rhodococcus ruber C208, *Bacillus cereus*	Laccase	Sowmya et al. (2014)
	Bacillus cereus	Manganese peroxidase	Sowmya et al. (2014)
Polyester PUR	*Comamonas acidovorans*	Polyurethane esterase	Akutsu et al. (1998)
	Pestalotiopsis microspore	Serine hydrolases	Russell et al. (2011)
	Pseudomonas chlororaphis	Putative Polyurethanases	Russell et al. (2011)
	Delftiaacidovorans *Comamonas acidovorans* *Bacillus subtilis*	Esterase	Nakajima-Kambe et al. (1997); Shah et al. (2013); Wei and Zimmermann (2017)
PET	*Thermobifidafusca* *Ideonella sakaiensis* *Bacillus gottheilii*	Cutinases Glycoside Hydrolases, PETase and MHETase	Auta et al. (2018); Palm et al. (2019); Ronkvist et al. (2009); Yoshida et al. (2016)
	Bacillus licheniformis *Bacillus subtilis* *Thermobifidafusca*	Carboxylesterases	Wei and Zimmermann (2017)
Low molecular weight poly lactic acid (PLA)	*Brevibacillus* sp. *Bacillus* sp.	Protease	Bhardwaj et al. (2012)

recent study found that protein-engineered enzymes were effective in degrading MNPs (Islam et al., 2019). According to the study, the degradation of the MNPs of PU has increased by about 6.7 times. These remarkable results indicate that protein/enzyme modification could be one of the approaches for more effectively removing MNPs. Immobilized enzyme techniques have recently been used to degrade MNPs in the environment (Shakerian et al., 2020). Bis phenol A (BPA), a monomer of polycarbonate plastics, is one of the most produced chemicals on the planet (Hacıosmanoğlu et al., 2019). Laccase enzyme was reported to be the most commonly used enzyme in immobilized systems to break down the BPA (Piao et al., 2019). When compared to free enzymes, immobilization of oxidative enzymes (laccase and horseradish peroxide) has shown high stability, durability, reusability, and cost-effectiveness (Shakerian et al., 2020). Hence, the combination of membranes and enzymes/microbial technology is expected to have a promising future in the degradation of MNPs from the soil and other eco systems.

14.11 Conclusions

The biodegradation of MNPs by using plastic-degrading bacteria is a viable and cost-effective plastic waste degradation technique that can be easily implemented in real-time to maintain the environmental quality of the soil caused by MNPs. This process has minimal or no side effects on the environment. The degradation of MNPs involves some intra and extracellular enzymes (Hydrolase and Oxidase), which are produced by bacteria. This enzymatic process breaks down the recalcitrant plastic polymers into microbial biomass and other environmentally safe compounds through several steps, including biodeterioration, depolymerization, assimilation, and mineralization. Optimizing the right environmental factors is the main factor to enhance the ability of bacteria to degrade plastics waste. Many advanced techniques like enzyme/protein engineering and enzymatic immobilization techniques have been developed to facilitate the biodegradation of MNPs.

References

Abraham, J., Ghosh, E., Mukherjee, P., & Gajendiran, A. (2017). Microbial degradation of low density polyethylene. *Environmental Progress & Sustainable Energy, 36*, 147–154.

Adamcová, D., & Vaverková, M. (2014). Degradation of biodegradable/degradable plastics in municipal solid-waste landfill. *Polish Journal of Environmental Studies, 23*(4), 1071–1078.

Agrawal, P., & Singh, R. K. (2016). Breaking down of polyethylene by *Pseudomonas* species. *International Journal of Scientific & Engineering Research, 7*(3), 124–127.

Akutsu, Y., Nakajima-Kambe, T., Nomura, N., & Nakahara, T. (1998). Purification and properties of a polyester polyurethane-degrading enzyme from Comamonasacidovorans TB-35. *Applied and Environmental Microbiology, 64*, 62–67.

Albertsson, A. C. (1978). Biodegradation of synthetic polymers. II. A limited microbial conversion of 14C in polyethylene to $^{14}CO_2$ by some soil fungi. *Journal of Applied Polymer Science, 22*, 3419–3433.

Albertsson, A. C., & Karlsson, S. (1993). Aspects of biodeterioration of inert and degradable polymers. *International Biodeterioration & Biodegradation, 31*, 161–170.

Allen, A. B., Hilliard, N. P., & Howard, G. T. (1999). Purification and characterization of a soluble polyurethane degrading enzyme from Comamonasacidovorans. *International Biodeterioration and Biodegradation, 43*, 37–41.

Ammala, A., Bateman, S., Dean, K., Petinakis, E., Sangwan, P., Wong, S., Yuan, Q., Yu, L., Patrick, C., & Leong, K. H. (2011). An overview of degradable and biodegradable Polyolefins. *Progress in Polymer Science, 36*(8), 1015–1049.

Amobonye, A., Bhagwat, P., Singh, S., & Pillai, S. (2021). Plastic biodegradation: Frontline microbes and their enzymes. *Science of the Total Environment, 759*, 143536.

Anwar, M. S., Kapri, A., Chaudhry, V., Mishra, A., Ansari, M. W., Souche, Y., et al. (2016). Response of indigenously developed bacterial consortia in progressive degradation of polyvinyl chloride. *Protoplasma, 253*, 1023–1032.

Arkatkar, A., Juwarkar, A. A., Bhaduri, S., Uppara, P. V., & Doble, M. (2010). Growth of *Pseudomonas* and *Bacillus* biofilms on pretreated polypropylene surface. *International Biodeterioration & Biodegradation, 64*, 530–536.

Asmita, K., Shubhamsingh, T., & Tejashree, S. (2015). Isolation of plastic degrading microorganisms from soil samples collected at various locations in Mumbai, India. *International Research of Environetmental Sciences, 4*(3), 77–85.

Atiq, N., Safia, A., & Ali, M. I. (2010). Isolation and identification of polystyrene biodegrading bacteria from soil. *African Journal of Microbiology Research, 4*, 1537–1541.

Auta, H. S., Emenike, C. U., Jayanthi, B., & Fauziah, S. H. (2018). Growth kinetics and biodeterioration of polypropylene microplastics by Bacillus sp. and Rhodococcus sp. isolated from mangrove sediment. *Marine Pollution Bulletin, 127*, 15–21.

Awasthi, S., Srivastava, P., Singh, P., Tiwary, D., & Mishra, P. K. (2017). Biodegradation of thermally treated high-density polyethylene (HDPE) by Klebsiella pneumoniae CH001. *3 Biotech, 7*, 332.

Baggi, G., Boga, M. M., & Catelani, D. (1983). (1983). Styrene catabolism by a strain of Pseudomonas fluorescens. *Systematic and Applied Microbiology, 4*, 141–147.

Begum, M. A., Varalakshmi, B., & Umamagheswari, K. (2015). Biodegradation of polythene bag using bacteria isolated from soil. *International Journal of Current Microbiology and Applied Sciences, 4*(11), 674–680.

Bhardwaj, H., Gupta, R., & Tiwari, A. (2012). Microbial population associated with plastic degradation. *Scientific Reports, 5*, 272–274.

Cacciari, I., Quatrini, P., Zirletta, G., Mincione, E., Vinciguerra, V., Lupattelli, P., et al. (1993). Isotactic polypropylene biodegradation by a microbial community: Physicochemical characterization of metabolites produced. *Applied and Environmental Microbiology, 59*, 3695–3700.

Carr, C. M., Clarke, D. J., & Dobson, A. D. W. (2020). Microbial polyethylene terephthalate hydrolases: Current and future perspectives. *Frontiers in Microbiology, 11*, 571265.

Chinaglia, S., Tosin, M., & Degli-Innocenti, F. (2018). Biodegradation rate of biodegradable plastics at molecular level. *Polymer Degradation and Stability, 147*, 237–244.

da Costa, J. P., Santos, P. S. M., Duarte, A. C., & Rocha-Santos, T. (2016). (Nano) plastics in the environment—Sources, fatesandeects. *Science of the Total Environment, 566–567*, 15–26.

Danso, D., Chow, J., & Streit, W. R. (2019). Plastics: Environmental and biotechnological perspectives on microbial degradation. *Applied and Environmental Microbiology, 85*, 1–14.

Darby, R. T., & Kaplan, A. M. (1968). Fungal susceptibility of polyurethanes. *Applied Microbiology, 16*, 900–905. https://doi.org/10.1128/aem.16.6.900-905.1968

de Souza Machado, A. A., Kloas, W., Zarfl, C., Hempel, S., & Rillig, M. C. (2018). Microplastics as an emerging threat to terrestrial ecosystems. *Global Change Biology, 24*(2018), 1405–1416.

Deepika, S., & Jaya, M. R. (2015). Biodegradation of low density polyethylene by microorganisms from garbage soil. *Journal of Experimental Biology and Agricultural Sciences, 3*, 15–21.

Delacuvellerie, A., Cyriaque, V., Gobert, S., Benali, S., & Wattiez, R. (2019). The plastisphere in marine ecosystem hosts potential specific microbial degraders including Alcanivorax borkumensis as a key player for the low density polyethylene degradation. *Journal of Hazardous Materials, 380*, 1–11.

Drzyzga, O., & Prieto, A. (2018). Plastic waste management, a matter for the community. *Microbial Biotechnology, 12*(1), 66–68.

Duddu, M. K., Tripura, K. L., Gantuku, G., & Divya, D. S. (2015). Biodegradation of low density polyethylene (LDPE) by a new biosurfactant-producing thermophilic Streptomyces coelicoflavus NBRC 15399T. *African Journal of Biotechnology, 14*(4), 327–340.

El-Sayed, A. H. M. M., Mahmoud, W. M., Davis, E. M., & Coughlin, R. W. (1996). Biodegradation of polyurethane coatings by hydrocarbon-degrading bacteria. *International Biodeterioration and Biodegradation, 37*, 69–79.

El-Shafei, H. A., El-Nasser, N. H. A., Kansoh, A. L., & Ali, A. M. (1998). Biodegradation of disposable polyethylene by fungi and Streptomyces species. *Polymer Degradation and Stability, 62*, 361–365.

Farzi, A., Dehnad, A., Shirzad, N., & Norouzifard, F. (2017). Biodegradation of high density polyethylene using Streptomyces species. *Journal of Coastal Life Medicine, 5*, 474–479.

Farzi, A., Dehnad, A., & Fotouhi, A. F. (2019). Biodegradation of polyethylene terephthalate waste using Streptomyces species and kinetic modeling of the process. *Biocatalysis and Agricultural Biotechnology, 17*, 25–31.

Fernandes, I. P., Barbosa, M., Amaral, J. S., Pinto, V., Rodrigues, J. L., Ferreira, M. J., & Barreiro, M. F. (2016). Biobased additives as biodegradability enhancers with application in TPU-based footwear components. *Journal of Renewable Materials, 4*, 47–56.

Fuhs, G. W. (1961). Der mikrobielleAbbau von Kohlenwasserstoffen. *Archiv fur Mikrobiologie, 39*, 374–422.

Ganesh, P., Dineshraj, D., & Yoganathan, K. (2017). Production and screening of depolymerasing enzymes by potential bacteria and fungi isolated from plastic waste dump yard sites. *International Journal of Applied Research, 3*(3), 693–695.

Gautam, R., Bassi, A. S., Yanful, E. K., & Cullen, E. (2007). Biodegradation of automotive waste polyester polyurethane foam using Pseudomonas chlororaphis ATCC55729. *International Biodeterioration and Biodegradation, 60*, 245–249.

Ghosh, S. K., Pal, S., & Ray, S. (2013). Study of microbes having potentiality for biodegradation of plastics. *Environmental Science and Pollution Research, 20*, 4339–4355.

Giacomucci, L., Raddadi, N., Soccio, M., Lotti, N., & Fava, F. (2019). Polyvinyl chloride biodegradation by Pseudomonas citronellolis and Bacillus flexus. *New Biotechnology, 52*, 35–41.

Gu, J. D. (2003). Microbiological deterioration and degradation of synthetic polymeric materials: Recent research advances. *International Biodeterioration and Biodegradation, 52*, 69–91.

Hacıosmanoğlu, G. G., Doğruel, T., Genç, S., Oner, E. T., & Can, Z. S. (2019). Adsorptive removal of 525 bisphenol a from aqueous solutions using phosphonatedlevan. *Journal of Hazardous Materials, 374*(526), 43–49.

Hadad, D., Geresh, S., & Sivan, A. (2005). Biodegradation of polyethylene by the thermophilic bacterium Brevibacillusborstelensis. *Journal of Applied Microbiology, 98*, 1093–1100.

Hasan, F., Shah, A. A., Hameed, A., & Ahmed, S. (2007). Synergistic effect of photo and chemical treatment on the rate of biodegradation of low density polyethylene by fusarium sp. AF4. *Journal of Applied Polymer Science, 105*, 1466–1470.

Helen, A. S., Uche, E. C., & Hamid, F. S. (2017). Screening of polyptopylene degradation potential of bacteria isolated from mangrove ecosystems in Peninsular Malaysia. *International Journal of Bioscience, Biochemistry and Bioinformatics, 7*(4), 245–251.

Hester, R. E., & Harrison, R. M. (2011). *Marine pollution and human health. Issues in environmental science and technology* (Vol. 33, pp. 84–85). Royal Society of Chemistry.

Hoang, K. C., Lee, C. Y., Tseng, M., & Chu, W. S. (2007). Polyester degrading actinomycetes isolated from the Touchien River of Taiwan. *World Journal of Microbiology and Biotechnology, 23*(2), 201–205.

Horton, A. A., Walton, A., Spurgeon, D. J., Lahive, E., & Svendsen, C. (2017). Microplastics in freshwater and terrestrial environments: Evaluating the current understanding to identify the knowledge gaps and future research priorities. *Science of the total environment, 586*(2017), 127–141.

Howard, G. T., Crother, B., & Vicknair, J. (2001a). Cloning, nucleotide sequencing and characterization of a polyurethanase gene (pueB) from pseudomonas chlororaphis. *International Biodeterioration and Biodegradation, 47*, 141–149.

Howard, G. T., Vicknair, J., & Mackie, R. I. (2001b). Sensitive plate assay for screening and detection of bacterial polyurethanase activity. *Letters in Applied Microbiology, 32*, 211–214. [CrossRef].

Howard, G. T., Norton, W. N., & Burks, T. (2012). Growth of Acinetobacter gerneri P7 on polyurethane and the purification and characterization of a polyurethane enzyme. *Biodegradation, 23*, 561–573.

Hurley, R. R., & Nizzetto, L. (2018). Fate and occurrence of micro(nano)plastics in soils: Knowledge gaps and possible risks. *Current Opinion in Environmental Science & Health, 1*, 6–11.

Hussein, A. A., Al-Mayaly, I. K., & Khudeir, S. H. (2015). Isolation, screening and identification of low density polyethylene (LDPE) degrading bacteria from contaminated soil with plastic wastes. *Mesopotamia Environmental Journal, 1*(4), 1–14.

Islam, S., Apitius, L., Jakob, F., & Schwaneberg, U. (2019). Targeting microplastic particles in the 541 void of diluted suspensions. *Environment International, 123*, 428–435.

Jamil, S. U. U., Zada, S., Khan, I., Sajjad, W., Rafiq, M., Shah, A. A., & Hasan, F. (2017). Biodegradation of polyethylene by bacterial strains isolated from Kashmir cave, Buner, Pakistan. *Journal of Cave and Karst Studies, 79*(1), 73–80.

Jansen, B., Schumacher-Perdreau, F., Peters, G., & Pulverer, G. (1991). Evidence for degradation of synthetic polyurethanes by Staphylococcus epidermidis. *Zentralblatt für Bakteriologie, 276*, 36–45.

Jarerat, A., & Tokiwa, Y. (2001). Degradation of poly (tetramethylene succinate) by thermophilic actinomycetes. *Biotechnology Letters, 23*(8), 647–651.

Jeon, H. J., & Kim, M. N. (2015). Functional analysis of alkane hydroxylase system derived from Pseudomonas aeruginosa E7 for low molecular weight polyethylene biodegradation. *International Biodeterioration and Biodegradation, 103*, 141–146.

Jabloune, R., Khalil, M., Ben Moussa, I. E., Simao-Beaunoir, A. M., Lerat, S., Brzezinski, R., & Beaulieu, C. (2020). Enzymatic degradation of p-Nitrophenyl Esters, Polyethylene Terephthalate, Cutin, and Suberin by Sub1, a Suberinase encoded by the plant pathogen streptomyces scabies. *Microbes Environments, 35*(1):ME19086. https://doi.org/10.1264/jsme2.ME19086. PMID: 32101840; PMCID: PMC7104285.

Kale, S. K., Deshmukh, A. G., Dudhare, M. S., & Patil, V. B. (2015). Microbial degradation of plastic: A review. *Journal of Biochemical Technology, 6*(2), 952–961.

Kaplan, D. L., Roy, H., & Jim, S. (1979). Biodegradation of polystyrene, poly(metnyl methacrylate), and phenol formaldehyde. *Applied and Environmental Microbiology, 38*, 551–553.

Kawai, F., Watanabe, M., Shibata, M., Yokoyama, S., Sudate, Y., & Hayashi, S. (2004). Comparative study on biodegradability of polyethylene wax by bacteria and fungi. *Polymer Degradation and Stability, 86*, 105–114.

Kay, M. J., Morton, L. H. G., & Prince, E. L. (1991). Bacterial degradation of polyester polyurethane. *International Biodeterioration, 27*, 205–222.

Kemona, A., & Piotrowska, M. (2020). Polyurethane recycling and disposal: Methods and prospects. *Polymers, 12*, 1752.

Koushal, V., Sharma, R., Sharma, M., Sharma, R., & Sharma, V. (2014). Plastics: Issues challenges and remediation. *International Journal of Waste Resources, 4*(1), 1–6.

Krueger, M. C., Hofmann, U., Moeder, M., & Schlosser, D. (2015). Potential of wood-rotting fungi to attack polystyrene sulfonate and its depolymerisation by Gloeophyllumtrabeum via hydroquinone-driven Fenton chemistry. *PLoS One, 10*, e0131773.

Kumar Sen, S., & Raut, S. (2015). Microbial degradation of low density polyethylene (LDPE): A review. *Journal of Environmental Chemical Engineering, 3*(1), 462–473.

Kumar, S., Das, M. L., Rebecca, J., & Sharmila, S. (2013). Isolation and identification of LDPE degrading fungi from municipal solid waste. *Journal of Chemical and Pharmaceutical Research, 5*(3), 78–81.

Kumar, V. R., Kanna, G. R., & Elumalai, S. (2017). Biodegradation of polyethylene by green photosynthetic microalgae. *Journal of Bioremediation & Biodegradation, 8*(1), 1–8. OMICS Publishing Group.

Kumari, P., & Murthy, N. S. (2013). A novel mathematical approach for optimization of plastic degradation. *International Journal of Engineering Trends and Technology, 4*(8), 3539–3542.

Kumari, A., Chaudhary, D. R., & Jha, B. (2019). Destabilization of polyethylene and polyvinylchloride structure by marine bacterial strain. *Environmental Science and Pollution Research International, 26*, 1507–1516.

Latorre, I., Hwang, S., & Montalvo-Rodriguez, R. (2012). Isolation and molecular identification of landfill bacteria capable of growing on di-(2-ethylhexyl) phthalate and deteriorating PVC materials. *Journal of Environmental Science and Health, Part A, 47*, 2254–2262.

Leslie, H. A., van Velzen, M. J. M., Brandsma, S. H., Vethaak, D., Garcia-Vallejo, J. J., & Lamoree, M. H. (2022). Discovery and quantification of plastic particle pollution in human blood. *Environment International, 163*, 107199. ISSN 0160-4120.

Li, J., Kim, H. R., Lee, H. M., Yu, H. C., Jeon, E., Lee, S., & Kim, D.-H. (2020). Rapid biodegradation of polyphenylene sulfide plastic beads by pseudomonas sp. *Science of the Total Environment, 720*, 137616.

Liu, M., Lu, S., Song, Y., Lei, L., Hu, J., Lv, W., Zhou, W., Cao, C., Shi, H., Yang, X., & He, D. (2018). Microplastic and mesoplastic pollution in farmland soils in suburbs of Shanghai, China. *Environmental Pollution, 242*(2018), 855–862.

Liu, C., Shi, C., Zhu, S., Wei, R., & Yin, C.-C. (2019). Structural and functional characterization of polyethylene terephthalate hydrolase from Ideonellasakaiensis. *Biochemical and Biophysical Research Communications, 508*, 289–294.

Ma, Y., Yao, M., Li, B., Ding, M., He, B., Chen, S., Zhou, X., & Yuan, Y. (2018). Enhanced poly(ethylene terephthalate) hydrolase activity by protein engineering. *Engineering, 4*, 888–893.

Mao, R., Lang, M., Yu, X., Wu, R., Yang, X., & Guo, X. (2020). Aging mechanism of microplastics with UV irradiation and its effects on the adsorption of heavy metals. *Journal of Hazardous Materials, 393*, 122515.

Matjašič, T., Simčič, T., Medvešček, N., Bajt, O., Dreo, T., & Mori, N. (2021). Critical evaluation of biodegradation studies on synthetic plastics through a systematic literature review. *Science of the Total Environment, 752*, 141959.

Mohan, A. J., Sekhar, V. C., Bhaskar, T., & Nampoothiri, K. M. (2016). Microbial assisted high impact polystyrene (HIPS) degradation. *Bioresource Technology, 213*, 204–207.

Mondal, S., & Palit, D. (2019). Effective role of microorganism in waste management and environmental sustainability. In M. Jhariya, A. Banerjee, R. Meena, & D. Yadav (Eds.), *Sustainable agriculture, forest and environmental management*. Springer. https://doi.org/10.1007/978-981-13-6830-1_14

Mor, R., & Sivan, A. (2008). Biofilm formation and partial biodegradation of polystyrene by the actinomycetes Rhodococcusruber: Biodegradation of polystyrene. *Biodegradation, 19*, 851–858.

Mueller, R. J. (2006). Biological degradation of synthetic polyesters—Enzymes as potential catalysts for polyester recycling. *Process Biochemistry, 41*, 2124–2128.

Muhonja, C. N., Makonde, H., Magoma, G., & Imbuga, M. (2018). Biodegradability of polyethylene by bacteria and fungi from Dandora dumpsite Nairobi-Kenya. *PLoS One, 13*(7), e0198446.

Mukherjee, S., & Chatterjee, S. (2014). A Comparative study of commercially available plastic carry bag biodegradation by microorganisms isolated from hydrocarbon effluent enriched soil. *International Journal of Current Microbiology and Applied Sciences, 3*(5), 318–325.

Nair, S., & Kumar, P. (2007). Molecular characterization of a lipase-producing Bacillus pumilus strain (NMSN-1d) utilizing colloidal water-dispersible polyurethane. *World Journal of Microbiology and Biotechnology, 23*, 1441–1449.

Nakajima-Kambe, T., Onuma, F., Kimpara, N., & Nakahara, T. (1995). Isolation and characterization of a bacterium which utilizes polyester polyurethane as a sole carbon and nitrogen source. *FEMS Microbiology Letters, 129*, 39–42.

Nakajima-Kambe, T., Onuma, F., Akutsu, Y., & Nakahara, T. (1997). Determination of the polyester polyurethane breakdown products and distribution of the polyurethane degrading enzyme of Comamonasacidovorans strain TB-35. *Journal of Fermentation and Bioengineering, 83*, 456–460.

Nakamiya, K., Hashimoto, S., Ito, H., Edmonds, J. S., Yasuhara, A., & Morita, M. (2005). Microbial treatment of bis (2-ethylhexyl) phthalate in polyvinyl chloride with isolated bacteria. *Journal of Bioscience and Bioengineering, 99*, 115–119.

Nakkabi, A., Sadiki, M., Fahim, M., Ittobane, N., IbnsoudaKoraichi, S., Barkai, H., & El abed, S. (2015). Biodegradation of poly(ester urethane)s by Bacillus subtilis. *International Journal of Environmental Research, 9*, 157–162.

Nisida, H., & Tokiwa, Y. (1993). Distribution of poly (β-hydroxybutyrate) and poly (ε-caprolacton) aerobic degrading microorganisms in different environments. *Journal of Environmental Polymer Degradation, 1*, 227–233.

Oberbeckmann, S., & Labrenz, M. (2020). Marine microbial assemblages on microplastics: Diversity, adaptation, and role in degradation. Annual review of marine. *Science, 12*(1), 209–232.

Oceguera-Cervantes, A., Carrillo-García, A., López, N., Bolaños-Nuñez, S., Cruz-Gómez, M. J., Wacher, C., & Loza-Tavera, H. (2007). Characterization of the polyurethanolytic activity of two Alicycliphilus sp. strains able to degrade polyurethane and N-methylpyrrolidone. *Applied and Environmental Microbiology, 73*, 6214–6223.

Oikawa, E., Linn, K. T., & Endo, T. (2003). Isolation and characterization of polystyrene degrading microorganisms for zero emission treatment of expanded polystyrene. *Environmental Engineering Research, 40*, 373–379.

Orhan, Y., Hrenovic, J., & Buyukgungor, H. (2004). Biodegradation of plastic compost bags under controlled soil conditions. *Acta Chimica Slovenica, 51*, 579–588.

Otake, Y., Kobayashi, T., & Asabe, H. (1995). Biodegradation of low-density polyethylene, polystyrene, polyvinyl chloride, and urea formaldehyde resin buried under soil for over 32 years. *Journal of Applied Polymer Science, 56*, 1789–1796.

Palm, G. J., Reisky, L., Böttcher, D., Müller, H., Michels, E. A., Walczak, M. C., Berndt, L., Weiss, M. S., Bornscheuer, U. T., & Weber, G. (2019). Structure of the plastic-degrading Ideonellasakaiensis MHETase bound to a substrate. *Nature Communications, 10*, 1–10.

Pathak, V. M. (2017). Review on the current status of polymer degradation: A microbial approach. *Bioresources and Bioprocessing, 4*, 15.

Patil, R. C. (2018). Screening and characterization of plastic degrading bacteria from garbage soil. *British Journal of Environmental Sciences, 6*(4), 33–40.

Peng, Y. H., Shih, Y. H., Lai, Y. C., Liu, Y. Z., Liu, Y. T., & Lin, N. C. (2014). Degradation of polyurethane by bacterium isolated from soil and assessment of polyurethanolytic activity of a pseudomonas putida strain. *Environmental Science and Pollution Research, 21*, 9529–9537.

Pérez-Lara, L. F., Vargas-Suárez, M., Lõpez-Castillo, N. N., Cruz-Gõmez, M. J., & Loza-Tavera, H. (2016). Preliminary study on the biodegradation of adipate/phthalate polyester polyurethanes of commercial-type by Alicycliphilus sp. BQ8. *Journal of Applied Polymer Science, 133*, 1–9.

Piao, M., Zou, D., Ren, X., Gao, S., Qin, C., & Piao, Y. (2019). High efficiency biotransformation of bisphenol A in a fluidized bed reactor using stabilized laccase in porous silica. *Enzyme and Microbial Technology, 126*, 1–8.

Pranamuda, H., Chollakup, R., & Tokiwa, Y. (1999). Degradation of polycarbonate by a polyester-degrading strain, Amycolatopsis sp. strain HT-6. *Applied and Environmental Microbiology, 65*(9), 4220–4222.

Ren, X., Tang, J., Liu, X., & Liu, Q. (2020). Effects of microplastics on greenhouse gas emissions and the microbial community in fertilized soil. *Environmental Pollution*. (Barking, Essex: 1987), *256*, 113347.

Riandi, M. I., Kawuri, R., & Sudirga, S. K. (2017). Potential of pseudomonas sp. and Ochrobacterum sp. isolated from various soil sample as degrading bacteria of high density polyethylene (HDPE) and low density polyethylene (LDPE) plastic. *Symbiosis Journal of Biological Science, 5*(2), 58–63.

Rillig, M. C., Ziersch, L., & Hempel, S. (2017). Microplastic transport in soil by earthworms. *Scientific Reports, 7*(2017), 1362.

Ronkvist, A. M., Xie, W., Lu, W., & Gross, R. A. (2009). Cutinase-catalyzed hydrolysis of poly (ethylene terephthalate). *Macromolecules, 42*, 5128–5138.

Rowe, L., & Howard, G. T. (2002). Growth of Bacillus subtilis on polyurethane and the purification and characterization of a polyurethanaselipase enzyme. *International Biodeterioration & Biodegradation, 50*, 33–40.

Ru, J., Huo, Y., & Yang, Y. (2020). Microbial degradation and valorization of plastic wastes. *Frontiers in Microbiology, 11*, 442.

Ruiz, C., & Howard, G. T. (1999). Nucleotide sequencing of a polyurethanase gene (pulA) from Pseudomonas fluorescens. *International Biodeterioration & Biodegradation, 44*, 127–131.

Ruiz, C., Main, T., Hilliard, N. P., & Howard, G. T. (1999). Purification and characterization of two polyurethanase enzymes from Pseudomonas chlororaphis. *International Biodeterioration & Biodegradation, 43*, 43–47.

Russell, J. R., Huang, J., Anand, P., Kucera, K., Sandoval, A. G., Dantzler, K. W., Hickman, D., Jee, J., Kimovec, F. M., & Koppstein, D. (2011). Biodegradation of polyester polyurethaneby endophytic fungi. *Applied and Environmental Microbiology, 77*, 6076–6084.

Sameh, A. S., Alariqi Kumar, A. P., Rao, B. S. M., & Singh, R. P. (2006). Biodegradation of γ-sterilized biomedical polyolefins under composting and fungal culture environments. *Journal of Polmer Degradation and Stability, 91*, 1105–1116.

Saminathan, P., Sripriya, A., Nalini, K., Sivakumar, T., & Thangapandian, V. (2014). Biodegradation of plastics by Pseudomonas putida isolated from garden soil samples. *Journal of Advanced Botany and Zoology, 1*(3), 1–4.

Sangeetha Devi, R., Rajesh Kannan, V., Nivas, D., Kannan, K., Chandru, S., & Robert Antony, A. (2015). Biodegradation of HDPE by aspergillus Spp. from marine ecosystem of gulf of Mannar, India. *Marine Pollution Bulletin, 96*(1–2), 32–40.

Sathya, R., Ushadevi, T., & Panneerselvam, A. (2012). Plastic degrading actinomycetes isolated from mangrove sediments. *International Journal of Current Research, 4*(10), 001–003.

Sauders, J. H., & Frisch, K. C. (1964). *Polyurethanes: Chemistry and technology, part II technology*. Inter-Science Publishers.

Scheurer, M., & Bigalke, M. (2018). Microplastics in Swiss floodplain soils. *Environmental Science & Technology, 52*(2018), 3591–3598.

Sekhar, V. C., Nampoothiri, K. M., Mohan, A. J., Nair, N. R., Bhaskar, T., & Pandey, A. (2016). Microbial degradation of high impact polystyrene (HIPS), an e-plastic with decabromodiphenyl oxide and antimony trioxide. *Journal of Hazardous Materials, 318*, 347–354.

Shah, A. A., Hasan, F., Hameed, A., & Ahmed, S. (2007). Isolation and characterization of poly (3-hydroxybutyrate-co-3-hydroxyvalerate) degrading actinomycetes and purification of PHBV depolymerase from newly isolated Streptoverticilliumkashmirense AF1. *Annals of Microbiology, 57*(4), 583–588.

Shah, A. A., Hasan, F., Akhter, J. I., Hameed, A., & Ahmed, S. (2008). Degradation of polyurethane by novel bacterial consortium isolated from soil. *Annals of Microbiology, 58*, 381–386. [CrossRef].

Shah, Z., Hasan, F., Krumholz, L., Atkas, D., & Shah, A. A. (2013). Degradation of polyester polyurethane by newly isolated Pseudomonas aeruginosa strain MZA-85 and analysis of degradation products by GC-MS. *International Biodeterioration and Biodegradation, 77*, 114–122.

Shah, Z., Gulzar, M., Hasan, F., & Shah, A. A. (2016). Degradation of polyester polyurethane by an indigenously developed consortium of Pseudomonas and Bacillus species isolated from soil. *Polymer Degradation and Stability, 134*, 349–356.

Shahnawaz, M., Sangale, M. K., & Ade, A. B. (2016). Rhizosphere of Avicennia marina (Forsk.) Vierh. As a landmark for polythene degrading bacteria. *Environmental Science and Pollution Research, 23*, 14621–14635.

Shakerian, F., Zhao, J., & Li, S.-P. (2020). Recent development in the application of immobilized oxidative enzymes for bioremediation of hazardous micro pollutants–a review. *Chemosphere, 239*, 124716.

Shimizu, S., Kobayashi, H., Masai, E., & Fukuda, M. (2001). Characterization of the 450-kb linear plasmid in a polychlorinated biphenyl degrader, Rhodococcus sp. strain RHA1. *Applied and Environmental Microbiology, 67*(5), 2021–2028.

Shresta, J. K., Joshi, D. R., Regmi, P., & Badahit, G. (2019). Isolation and identification of low density polyethylene (LDPE) degrading bacillus spp. from a soil of landfill site. *Acta Scientific Microbiology, 2*(4), 30–34.

Skariyachan, S., Megha, M., Kini, M. N., Mukund, K. M., Rizvi, A., & Vasist, K. (2015). Selection and screening of microbial consortia for efficient and ecofriendly degradation of plastic garbage collected from urban and rural areas of Bangalore, India. *Environmental Monitoring and Assessment, 187*(1), 4174.

Skariyachan, S., Manjunatha, V., Sultana, S., Jois, C., Bai, V., & Vasist, K. S. (2016). Novel bacterial consortia isolated from plastic garbage processing areas demonstrated enhanced degradation for low density polyethylene. *Environmental Science and Pollution Research International, 23*(18), 18307–18319.

Skariyachan, S., Patil, A. A., Shankar, A., Manjunath, M., Bachappanavar, N., & Kiran, S. (2018). Enhanced polymer degradation of polyethylene and polypropylene by novel thermophilic consortia of Brevibacillussps. and Aneurinibacillus sp. screened from waste management landfills and sewage treatment plants. *Polymer Degradation and Stability, 149*, 52–68.

Soud, S. A. (2019). Biodegradation of polyethylene LDPE plastic waste using locally isolated Streptomyces sp. *Journal of Pharmaceutical Sciences and Research, 11*(4), 1333–1339.

Sowmya, H., Ramalingappa, V., Krishnappa, M., & Thippeswamy, B. (2014). Low density polyethylene degrading fungi isolated from local dumpsite of Shivamogga district. *International Journal of Biological Research, 2*(2), 39–43.

Stern, R. V., & Howard, G. T. (2000). The polyester polyurethanase gene (pueA) from Pseudomonas chlororaphis encodes a lipase. *FEMS Microbiology Letters, 185*, 163–168.

Szycher, M. (1999). *Szycher's handbook of polyurethanes*. CRC Press.

Tribedi, P., & Sil, A. K. (2013). Low-density polyethylene degradation by Pseudomonas sp. AKS2 biofilm. *Environmental Science and Pollution Research, 20*, 4146–4153.

Usha, R., Sangeetha, T., & Palaniswamy, M. (2011). Screening of Polyethylene degrading microorganisms from garbage soil. *Libyan Agriculture Research Center Journal International, 2*, 200–204.

Vega, R. E., Main, T., & Howard, G. T. (1999). Cloning and expression in Escherichia coli of a polyurethane-degrading enzyme from Pseudomonas fluorescens. *International Biodeterioration & Biodegradation, 43*, 49–55.

Wang, W., Ge, J., Yu, X., & Li, H. (2020). Environmental fate and impacts of microplastics in soil ecosystems: Progress and perspective. *Science of the Total Environment, 708*, 134841.

Wei, R., & Zimmermann, W. (2017). Microbial enzymes for the recycling of recalcitrant petroleum-based plastics: How far are we? *Microbial Biotechnology, 10*(6), 1308–1322.

Wilkes, R. A., & Aristilde, L. (2017). Degradation and metabolism of synthetic plastics and associated products by Pseudomonas sp. capabilities and challenges. *Journal of Applied Microbiology, 123*, 582–593.

Yoon, M. G., Jeon, H. J., & Kim, M. N. (2012). Biodegradation of polyethylene by a soil bacterium and AlkB cloned recombinant cell. *Journal of Bioremediation & Biodegradation, 3*, 1–8.

Yoshida, S., Hiraga, K., Takehana, T., Taniguchi, I., Yamaji, H., Maeda, J., Toyohara, K., Miyamoto, K., Kimura, Y., & Oda, K. (2016). A bacterium that degrades and assimilates poly(ethylene terephthalate). *Science, 351*(6278), 1196–1199.

Young, R. J., & Lovell, P. A. (Eds.). (1994). *Introduction to polymers* (2nd ed.). Chapman & Hall.

Yuan, J., Ma, J., Sun, Y., Zhou, T., Zhao, Y., & Yu, F. (2020). Microbial degradation and other environmental aspects of microplastics/plastics. *Science of the Total Environment, 715*, 136968.

Zhou, Q., Zhang, H., Fu, C., Zhou, Y., Dai, Z., Li, Y., Tu, C., & Luo, Y. (2018). The distribution andmorphology of microplastics in coastal soils adjacent to the Bohai Sea and the Yellow Sea. *Geoderma, 322*(2018), 201–208.

Chapter 15
Mycoremediation of Micro-/Nanoplastics-Contaminated Soils

Chittari Amaravathi Sneha Latha Varma, Malay Sahasrabudhe, Sneha Ganguly, and Maheswara Reddy Mallu

Abstract Plastic pollution has increased by nearly 200-fold in the last 50 years. There is a plethora of literature existing on the clear negative implications of the non-biodegradable material on land and in oceans. There are sophisticated waste management systems available; however, instead of simply managing it, a complete degradation is required so that they do not adversely affect the environment as some plastics are said to leach chemicals into the soil thereby causing toxicity. This, when ingested, can alter genetic constitution and have fatal effects in both flora and fauna. Fungi has exhibited its ability to degrade certain forms of plastics, especially the filamentous ones by using it as a carbon source and using a wide sequence of methods for its degradation. This review talks widely about the variety of fungal species that degrade plastics, the enzymes produced by the fungal strains to bring about biodegradation and the different methods used for the detection of the fungal degradation of plastics. It is important not only study the rate but also the mechanism of degradation for future studies. It is crucial for researchers as well agencies alike across the globe to discuss and consider available methods and alternative species of fungal strains and their enzymatic counterparts to combat this menace on the terrestrial front.

Keywords Mycoremediation · Microplastics · Nanoplastics · Soil microflora · Biodegradation · In situ · Plastics · Physicochemical degradation

Chittari Amaravathi Sneha Latha Varma · M. Sahasrabudhe · S. Ganguly
Excelra Knowledge Solutions Pvt Ltd, Hyderabad, Telangana, India

M. R. Mallu (✉)
Department of Biotechnology, Koneru Lakshmaiah Education Foundation, Guntur, Andhra Pradesh, India
e-mail: mahesh_bt@kluniversity.in

N. R. Maddela et al. (eds.), *Micro and Nanoplastics in Soil*,
https://doi.org/10.1007/978-3-031-21195-9_15

15.1 Introduction

Plastics, in simple words, are petroleum products, which are durable, moldable, and cheap; they affordable commercial products used in a variety of industrial and home applications which interact minimally with the natural ecosystem and thus are hard to utilize as a mineral source by most of the flora and fauna in the ecosystem (Shah et al., 2008). Today, in a normal household, you can see more plastic than metal or wood, both dissimilated into the ecosystem. Even though natural plastics can be dated back to centuries, full synthetic plastic was made in the early twentieth century (American Chemical Society National Historic Chemical Landmarks, 2015).

The abundance of plastic can be verified by collecting it from the loneliest places of the world. In 2016, it was predicted that the output of plastic material will be doubled in the next 10 years (MacLeod et al., 2021). Then COVID happened. Due to lack of information, a sudden spike in hygiene awareness caused a fold increase in demand for single-use plastic. Excessive Personal Protective Equipment (PPE) kits, masks, and other safety equipment made of plastics are now contaminating our water bodies, which will complete its life cycle by reaching into soil or oceans (Patrício Silva et al., 2021).

Microplastics (MPs) and nanoplastics (NPs) have become widespread both in aquatic and terrestrial environments as a consequence of their huge build-up in urban waste systems. As defined by European Food Safety Authority, MPs are microscopic particles of plastic having a diameter ranging between 0.1 and 5000 μm, whereas NPs has a diameter ranging between 0.001 and 0.1 μm. Despite the fact that yearly plastic releases into soil are around 4–23 times larger than those into aquatic waters, and MPs and NPs detected in soil are around four times greater levels than in aquatic systems, studies into the oceanic plastic pollution has taken precedence, with terrestrial NPs and MPs receiving less attention.

The end of the plastic life cycle is even recorded in human placenta in early 2021, in human blood and live lung cells in early 2022 (Ragusa et al., 2021; Leslie et al., 2022; Jenner et al., 2022). This would convey the gravity of the matter. There are many attempts at promoting biodegradable plastics, replacing polyethylene (PE) films by mulch films in the agricultural industry in developed countries. But, as the transition is difficult, it is even worse that there is no ready plan to utilize or degrade the non-degradable plastics we have been using for centuries. For many recent years, scientists have been finding several ways to degrade plastics in landfill, oceans, etc., through various means such as Ultraviolet (UV) irradiation, ozone treatment, photolytic damage, microbial degradation, etc. Although it may be comparatively faster than biotic degradation, these abiotic degradation techniques are hard to implement in situ.

Concurrently there is research happening on biotic degradation in oceanic, soil, and arctic environments. Many naturally occurring fungi and bacteria are a point of interest worldwide. As one can depict fungi as an octopus of soil (due to their hyphae and ability to attach to anything), they are amusing creatures if you believe. As they are omnipresent, one has no problem using them as in situ degraders. Humans are too aware of this biological family but we have been successfully transforming them to fulfill our needs. Many common plant pathogens are identified as promising hits for plastic biodegradation. Like, *Aspergillus spp., Candida spp., Penicillium spp., Alternaria spp.*, list is unending (Ghosh et al., 2013). As this family hosts one of the most robust molecular technologies, and their minimalistic needs are often observed by their abundance in any extreme conditions, it is ideal for industrial applications.

As we know, there is a clear relation between an organism, its capabilities, and the enzymes produced by it. The extent at which any organism causes degradation of plastic lies in the enzymes and their ability to catalyze the process. However, there are very few studies evaluating enzymatic interactions between plastic. However, many enzymes such as glycosidase, cutinase, lipase, serine hydrolase, and magnesium peroxidase (MnP) are identified as biological catalysts (Bhardwaj et al., 2013).

In this chapter, we summarized knowledge gathered from different research in the field of plastic mycoremediation about various aspects like known fungi who can degrade diverse types of plastics, enzymes responsible in the work, methodologies of plastic degradation including both biotic and abiotic and different means to assess those.

15.2 Plastics

Thermoplastics and thermosets are the two main families of synthetic plastics depending on their thermal characteristics. Thermoplastics are synthetic polymers that do not alter chemical composition when reheated, allowing them to be remodeled after melting (Amobonye et al., 2021). Acrylonitrile butadiene styrene (ABS), PE, polyamide (PA), polyimide (PI), polypropylene (PP), poly-methyl methacrylate (PMMA), polystyrene (PS), polyvinyl chloride (PVC), polytetrafluoroethylene (PFE), and polyvinylidene chloride (PVDC) are a few examples (Li et al., 2019; Makhlouf et al., 2016). The thermoplastics' carbon backbone makes them difficult to degrade and resistant to hydrolytic cleavage. Unlike thermoplastics, thermosets cannot be remodeled through melting because the chemical changes caused by heat are irreversible. Additionally, thermoset backbones are heteroatomic and extensively cross-linked, making them more prone to hydrolytic cleavage. Polyurethane (PU), polyester polyethylene terephthalate (PET), epoxy resins, acrylic resins, silicone, and vinyl resins are all common thermosets (Jog, 1995; Ray & Cooney, 2018).

15.2.1 Types of Plastic Degradation

As we already know that plastics are difficult-to-degrade chemically produced polymers. Plastic deterioration is often a slow process that is regulated by a variety of environmental elements including temperature, sun energy, air humidity, polymer characteristics and moisture in it, pH, and biological components. Many different mechanisms for plastic breakdown are studied. Various types of plastic degradation and their end products are shown in Fig. 15.1 (Modified from Krzan et al., 2006).

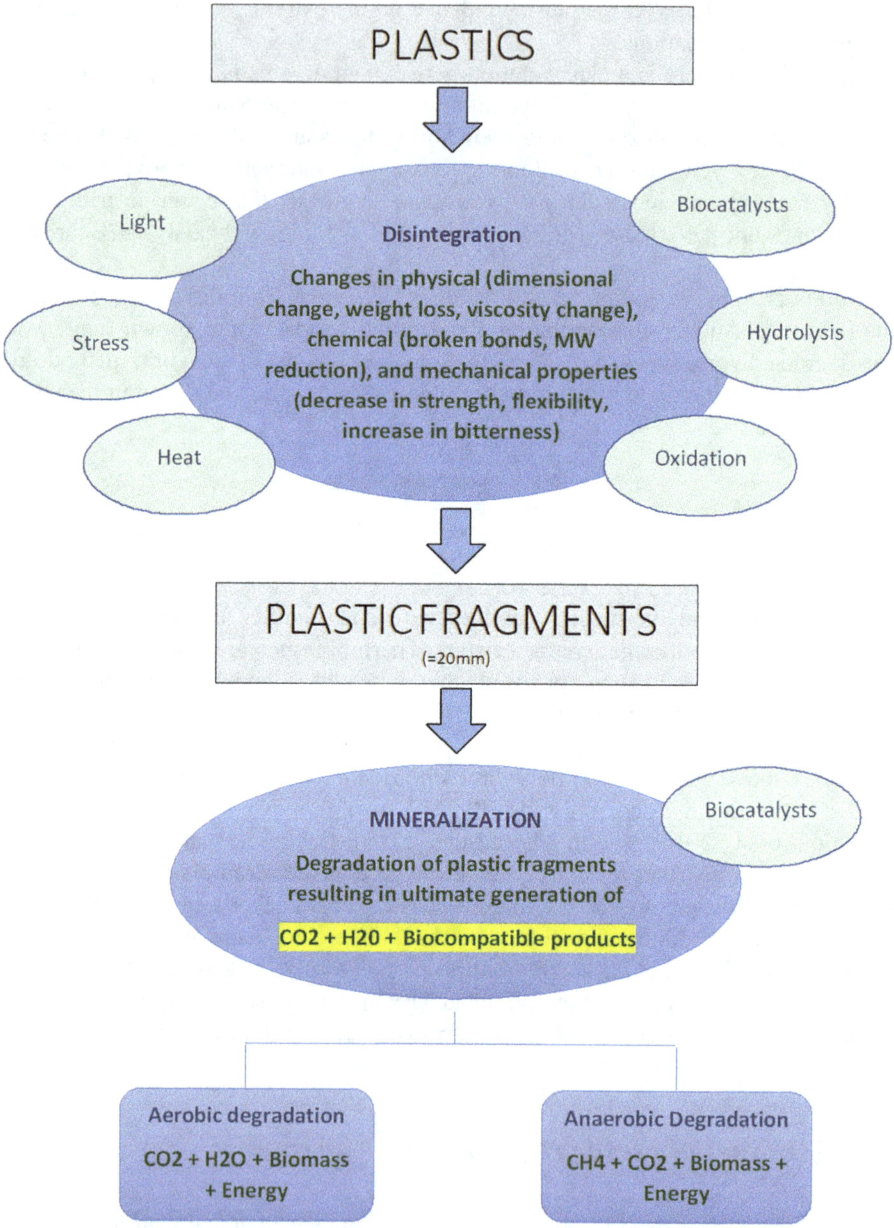

Fig. 15.1 Various types of plastic degradation and the end products. (Modified from Krzan et al., 2006)

Thermal Degradation

High temperature breaks larger, more complex polymers into smaller yet simpler ones. When oxygen is used, it is called thermal oxidation. Usually visible light in the range of 400–760 nm can start degradation of plastics. When they are exposed to infrared radiation, thermal oxidation occurs (Pospíšil & Nešpůrek, 1997). This can cause change in malleability and other alterations in physical appearance.

Photodegradation

High intensity photon particles bombard a long chain polymer, converting it into smaller particles. Solar radiations are entrapped by these polymers; when oxygen interacts to help in their breakage, it becomes photooxidation (Iram et al., 2019).

Ozone Degradation

Polymers when exposed to ozone break down into ketones, lactones, esters, etc.

Mechanochemical Degradation

Ultrasonic waves, due to mechanical stress break down larger polymers into simpler ones. External stimulus, such as heat and electricity, triggers the reaction.

Catalytic Degradation

Transformation of polymers into hydrocarbons to produce oils and gases is yet another method of great interest. After the pyrolysis of plastics, the overall quality of end product is improved. This can be utilized on a much larger scale to manage larger quantities of plastics. Different types of catalysts used are Pt-Mo, Pt-Co, transition metal catalysts such as Cr, Ni, Mb, and Co.

Computational Biology

Computational biology and bioinformatics store a lot of data related to deoxyribonucleic acid (DNA), ribonucleic acid (RNA) of microbes, the folding pattern of their enzymes. They are eventually being used in certain research studies to find out if a certain microbial population from a certain area are capable of degrading plastics. Databases such as MetaRouter (Pazos, 2004), BSD (Urbance, 2003), PAHbase (Kessner et al., 2008), MetaboLights (Haug et al., 2013), BioRadBase (Reena et al., 2012), KBase (Arkin et al., 2018), BiofOmics (Lourenço et al., 2012) are already in use for biodegradation study of plastics. Some of these tools are not only used for prediction of biodegradation studies but also give necessary information on toxicity of chemicals formed due to degradation process. For example, metagenomic analysis is used to identify genes that can degrade PET. With the aid of various computational biology tools, we can develop native structures of the novel plastic-degrading enzymes that can be modelled when their actual structure is absent from literature or yet to be designed (Skariyachan et al., 2022; Arora & Bae, 2014).

Out of all these methods, microbial degradation is the easiest, most cost effective, eco-friendly method used for plastic breakdown. In India, approximately 2500 tons of plastic is generated annually but owing to their extremely slow degradation process, it becomes difficult to eliminate them without harming the ecosystem. So, microbes which use carbon as the source breaks down plastic in varying degrees, this can be potentially developed as an effective method in dealing with plastics.

15.3 Bioremediation

Bioremediation is a broad phrase that refers to the use of a biological entity to clean up pollution, whereas biological recycling refers to the use of biological techniques to recover useful goods from trash. Depolymerization of waste petro-plastics either by microbes or by enzymes into monomers for recycling, or mineralization into carbon dioxide, water, and fresh biomass, with simultaneous generation of higher value bio products is a promising technique (Grima et al., 2000; Montazer et al., 2019; Montazer et al., 2020). Both techniques, i.e., bioremediation and biological recycling, are based on polymer biodegradation (Wierckx et al., 2015). Microbial degradation of plastic waste has sparked a lot of curiosity. Biodegradation is a chemical decomposition process caused by living forms (bacteria, fungi, and actinomycetes). Bond cleavage, chemical transformation, and the synthesis of new functional groups are all part of the process (Prabhat et al., 2013). Biodegradation is described as degradation caused by changes in surface characteristics or mechanical strength, microbial degradation using enzymes, backbone chain breaking, and consequent decrease in the average polymeric molecular weight (Singh & Sharma, 2008). Plastics can be biodegraded in either aerobic or anaerobic environment. CO_2 and H_2O are produced during aerobic decomposition, whereas CH_4 along with CO_2 and H_2O are the end products of anaerobic decomposition (Arkatkar et al., 2009). The microbe utilized, the kind of pre-treatment employed, and the plastic properties all have an impact on plastic biodegradation.

Plastic biodegradation entails the microorganism excreting extracellular enzymes, enzyme attaches to the surface of the plastic, and hydrolysis of plastic polymer to intermediates, such as oligomers and monomers, which are then ingested by microbes as a source of carbon, releasing CO_2 (Lucas et al., 2008). A single enzyme from microbiological, vegetal, or mammalian origin can also accomplish degradation (Marchant et al., 1987; Tokiwa et al., 1988; Brzeska et al., 2015). Due to their relatively low cost and minimal environmental impact, bioremediation and biological recycling are appealing methods for combating plastic pollution (Azubuike et al., 2016). By using microbial biotechnology to identify and genetically engineer these plastic-degrading microbes and/or enzymes, it will be possible to enhance recycling of plastic and thus minimize the plastic pollution in the environment by absorption of plastic waste into carbon sources or breakdown of plastic waste into useful alkane products (Mohanan et al., 2020).

Steps in Plastic Biodegradation

Biodeterioration occurs when the surface properties of the plastic changes. This affects their physical as well as chemical properties. Environmental conditions such as wind, oxygen, moisture also affect the rate of biodeterioration (Vivi et al., 2019). Initially the microbes aim to reduce the durability of the said polymer by colonizing its surface; biofilm forming microbes are useful in this regard. They cause localized swelling leading to plastic becoming brittle; microbes use them as carbon sources and continue to proliferate (Amobonye et al., 2021).

Bio Fragmentation is the next step in degradation of polymers. Microbial enzymes, mostly oxygenases as well as hydrolytic enzymes, based on the presence of oxygen aerobically or anaerobically, degrade plastics into monomers and subsequent oxidation of these monomers. The hydrolytic enzymatic reactions occur via endo or exo-attacks. In the endo-reaction, the degraded monomers are easily assimilated into the microbes, whereas in the exo method the degraded monomers require further degradation to be assimilated by the said microbes (Amobonye et al., 2021).

Assimilation is when monomers are integrated into the cell membrane and within the microbe. Some of them undergo transformation to be transported outside the cell for use by other microbes and continue with degradation (Zeenat et al., 2021).

It is theorized that this method also has active and passive forms of transport involved. For example, DG17 facilitated passive transport at higher concentrations and active transport at lower concentrations (Amobonye et al., 2021; Hua et al., 2013).

Mineralization is when monomers from the plastic enter into the cell via cell membrane, and they are used by the cell for energy production, thus increasing its own biomass. These monomers are degraded into simpler products such as water, carbon dioxide, methane, nitrogen. Mineralization can occur either aerobically or anaerobically in the presence of a variety of enzymes. Intermediates formed during this process can be channeled through a number of different mechanistic processes (Amobonye et al., 2021; Ho et al., 2018).

15.4 Importance of Mycoremediation

Fungi's involvement in bioremediation has become more widely recognized in recent years (Singh, 2006; Jafari et al., 2013). Several authors have emphasized fungi's potential to breakdown or alter harmful substances, particularly biotrophic and saprotrophic basidiomycetes (Baldrian, 2008; Spina et al., 2018). Mycoremediation is a biodegradation approach that uses fungus to remove hazardous substances from the environment. It can be done using filamentous fungi such as molds and also by employing macrofungi such as mushrooms (Chatterjee et al., 2017; Ali et al., 2017) where both these classes have enzymes that can degrade a wide range of contaminants (Purnomo et al., 2013; Kulshreshtha et al., 2013).

The PU biodegradation is being researched more and more since it is one of the most polluting polymers, with 18 Mt. generated in 2016 (Furtwengler et al., 2017; Reddy et al., 2006). Because of the hydrolysable ester linkages, polyester-based PU is more degradable than polyether-based PU (Cregut et al., 2013). Due to their superior enzyme systems and abiotic properties related to filament formation, filamentous fungi are said to be more efficient at degrading PU than bacteria (Barratt et al., 2003; Lucas et al., 2008). Numerous types of PU materials have been already characterized as susceptible to fungal biodegradation, including polyether urethane foam (Álvarez-Barragán et al., 2016) and thermoplastic polyester (Khan et al., 2017b). Fungi have been found to be superior degraders of PE and PU than bacteria in general (Pathak, 2017; Muhonja et al., 2018). Polymers including PP, PVC, PET, polyesters, and MPs have also been studied for fungal breakdown. *Gliocladium, Spicaria spp., Penicillium spp., Geomyces pannorum, Cunninghamella, Phoma spp., Aspergillus, Alternaria solani, Mucor, Fusarium solani,* and *Mortierella* are among the fungal species that have previously been implicated in the plastic polymer biodegradation (Konduri et al., 2010; Constantin et al., 2012; Sowmya et al., 2015; Muhonja et al., 2018; Zhang et al., 2020).

Fungi are a wide and complex kingdom of eukaryotic creatures that can be found in a variety of habitats, and some have adapted to thrive in both aquatic and terrestrial environments even in harsh conditions (Raghukumar, 2017). Filamentous fungi are the most widely classified fungi, and they have a remarkable capability to adjust to changing surroundings and to endure a wide range of contaminants. They can degrade contaminants and utilize their chemical components to develop or make them available to other microbes. As a result, filamentous fungus plays a significant role in the breakdown and mineralization of a wide range of contaminants by accelerating key chemical reactions (Črešnar & Petrič, 2011).

Various fungal species have been emphasized for their potential to decompose diverse plastic polymers depending on their capability to use these polymers as their main source of energy and carbon. In this context, a diverse range of fungal strains was shown to break down plastics, spanning various classes, ecologies, and morphologies. According to recent research, the *Aspergillus* genus is the most important fungus group involved in the biodegradation of manmade plastics. Three *Aspergillus* species that have been isolated from diverse terrestrial environments, which include *Aspergillus fumigatus* (Osman et al., 2018), *Aspergillus clavatus* (Gajendiran et al., 2016), and *Aspergillus niger* (Usman et al., 2020), have been proven to breakdown PU, PE, and PP, respectively.

Endophytic fungi obtained from a variety of plants when grown in submerged as well as solid-state fermentation were found to degrade PU to diverse degrees (Russell et al., 2011). Interestingly, in contrary to most research that have concentrated on the potency of pure cultures, PE (Sowmya et al., 2015) and PU (Cosgrove et al., 2007) have found to be degraded by some fungal consortium in a synergistic manner. Fungal enzymes, particularly depolymerases, were emphasized in many of this research, as in all biological processes. Moreover, these enzymes' broad specificity, which enables them to degrade a variety of polymers, is important (Rodrigues et al., 2020). The potential of fungal hyphae to generate hydrophobins for improved hyphal adhesion to hydrophobic substrates, along with their dispersion and penetrative capability, has been found to be an important feature in their early colonization prior to eventual depolymerization (Sánchez, 2020).

Aspergillus flavus has already been shown to degrade a polyester PU as a sole source of carbon (Mathur & Prasad, 2012); however, *A. oryzae* has never been reported to degrade PU. Hydrolytic urethane bond cleavage is responsible for PU biodegradation (Nakajima-Kambe et al., 1999). Enzymatic hydrolysis of ester bonds is carried out by few fungi and bacteria in order to biodegrade polyester-polyurethane (Nakajima-Kambe et al., 1999; Howard, 2002). Fungi such as *Aureobasidium pullulans, Curvularia senegalensis, Cladosporium spp.,* and

Fusarium solani have all been found to breakdown polyester-polyurethane (Crabbe et al., 1994).

Aspergillus, Fusarium, Acremonium, Phanerochaete, and *Penicillium* are only a few of the PE-degrading fungus that have recently been discovered (Restrepo-Flórez et al., 2014). PE degradation has been found to be carried out significantly by fungi such as *Pleurotus ostreatus* and *Aspergillus flavus* (Gómez-Méndez et al., 2018; Zhang et al., 2020). It has been reported that fungal enzymes can shorten PE chains, allowing a metabolic pathway for PE degradation to be postulated (Restrepo-Flórez et al., 2014). Fungi such as *Fusarium* and *Aspergillus* have been found to depolymerize PE following pre-treatments such as thermal and/or UV treatments, which make the polymer's carbon chains biodegradable (Ammala et al., 2011). El-Shafei et al. (1998) reported on the potential of *A. flavus* culture to degrade PE plastic bags.

Low density polyethylene (LDPE) degrading capabilities have also been observed in *Aspergillus terreus, A. niger, Aureobasidium pullulans, Scopulariopsis brevicaulis, Paecilomyces varioti, Penicillium ochrochloron, Penicillium funiculosum,* and *Trichoderma viride* (Ojha et al., 2017). Ability of *A. flavus MCP5* and *A. flavus MMP10* to biodegrade LDPE has been demonstrated by Kunlere et al. (2019) revealing that *A. flavus* can utilize LDPE as source of carbon and nitrogen without the presence of any additives.

There have been very few investigations on biodegradation of PP. PP has been observed to be degraded by the fungus *Aspergillus niger* (Cacciari et al., 1993).

Whenever exposed to nutrient (carbon, nitrogen, or sulphur) restricting conditions, white rot fungi have been found to biodegrade PVC of low molecular weight (Kirbaş et al., 1999). The white rot fungi's ability to convert organic pollutants to CO_2 has been found to be reliant on a non-stereoselective and nonspecific lignin-degrading mechanism. *Trichocladium sp.* and *Chaetomium sp.* grew faster using PVC as a source of carbon and cellulose being present along, implying that PVC and cellulose co-metabolize (Kaczmarek & Bajer, 2007). In addition, plasticizers and other chemicals in plastics promote fungal growth by providing nutrients.

Krueger et al. (2015) and Paço et al. (2017) revealed that the ascomycetes *Gloeophyllum trabeum* and *Zalerion maritimum* can decompose MPs. list of fungi responsible for degradation of various polymers are given in Table 15.1.

15.5 Important Fungal Strains Involved in Bioremediation

As discussed in the earlier part of this chapter, there are many types of plastic degradation. But these methods are good for industrial purposes only. But when we are evaluating promising methods for degradation in the field, we cannot achieve the goal with methods requiring big machinery or isolation of plastic in the closed chamber or work in biologically harsh conditions. At that time, one needs methods which will achieve one's goal with no extra setup or no extra instruments and treatments. Some of the important fungal strains along with their habitat and percentages of plastic degradation are discussed in this section.

Table 15.1 List of fungi responsible for degradation of various polymers

Plastic polymer type	Microorganism	References
Polyethylene (PE)	*Aspergillus fumigatus, Alternaria alternate, Curvularia lunata, Penicillium simplicissimum, Phanerochaete chrysosporium, Fusarium sp, Trametes versicolor, Aspergillus nomius, A. terreus, A. flavus, A. sydowii, Acremonium, Trichoderma viride, Pleurotus ostreatus, Zalerion maritimum* (PE microplastics)	Vishnu et al. (2012); Sowmya et al. (2015); Munir et al. (2018); Sangale et al. (2019); Paço et al. (2017); Gómez-Méndez et al. (2018); Zhang et al. (2020); Restrepo-Flórez et al. (2014)
Low density polyethylene (LDPE)	*Trichoderma viride, Aspergillus oryzae, A. niger, A. nomius, A. flavus, A. terreus, A. caespitosus, A. japonicas, A. versicolor, A. fumigatus, Paecilomyces variotii, Phialophora alba, Eupenicillium hirayamae, Alternaria alternate, Fusarium solani, Mucor sp., Penicillium sp.*	Munir et al. (2018); Muhonja et al. (2018); Deepika and Jaya (2015); Ameen et al. (2015); Jyoti and Gupta (2014); Das and Kumar (2014); Sindujaa et al. (2011); Zahra et al. (2010)
High density polyethylene (HDPE)	*Aspergillus terreus*	Balasubramanian et al. (2014)
Polylactic acid (PLA)	*F. moniliforme, A. fumigates* and *Thermomyces lanuginosus.*	Torres et al. (1996); Karamanlioglu et al. (2014)
Polycaprolactone (PCL)	*Fusarium solani, Aspergillus niger, A. flavus, A. fumigatus, Penicillium funiculosum, Chaetomium globosum,* and *Fusarium sp*	Pathak (2017)
Poly-vinyl alcohol (PVA)	*Fusarium, Phanerochaete chrysosporium, Aspergillus, Galactomyces geotrichum, Fimetariella rabenhorsti, Trichosporon laibachii, Fusarium oxysporum, G. geotrichum*	Lipşa et al. (2015)
Polyurethane (PU)	*Chaetomium globosum, Curvularia senegalensis, Aspergillus terreus, A. flavus, Fusarium solani, Cladosporium sp, Aureobasidium pullulans, Exophiala jeanselmei*	Howard (2012); Owen et al. (1996); Mathur and Prasad (2012)
Polyvinyl chloride (PVC)	*Phanerochaete chrysosporium, Aspergillus niger, A. flavus, Lentinus tigrinus, Aspergillus sydowii, Aspergillus fumigates, Trichocladium sp, White rot fungi, Chaetomium globosum, Cochliobolus sp., Engyodontium album*	Ali et al. (2014); Kirbaş et al. (1999); Kaczmarek and Bajer (2007); Vivi et al. (2019); Zhang et al. (2020); Sumathi et al. (2016); Jeyakumar et al. (2013)
Polyhydroxybutyrate (PHB)	*Aspergillus spp, Penicillium*	Altaee et al. (2016)
Polystyrene (PS)	*Curvularia sp., Cephalosporium sp., Gloeophyllum trabeum* (Polystyrene Sulfonate-microplastic), *Mucor spp*	Motta et al. (2009); Krueger et al. (2015); Chaudhary and Vijayakumar (2020)

(continued)

Table 15.1 (continued)

Plastic polymer type	Microorganism	References
Polypropylene (PP)	*Aspergillus niger, Phanerochaete chrysosporium*	Cacciari et al. (1993); Jeyakumar et al. (2013)
Polyamide (PA)	*Fusarium sp.*	
Polyester polyurethane	*Cladosporium cladosporioides, Penicillium griseofulvum, Xepiculopsis graminea, Leptosphaeria sp., Monascus sp., Monascus sanguineus, Monascus ruber, Aspergillus tubingensis, Pestalotiopsis microspora*	Brunner et al. (2018); El-Morsy (2017); Khan et al. (2017b); Russell et al. (2011)
Polyester polyethylene terephthalate (PET)	*Penicillium funiculosum, Penicillium citrinum, Fusarium solani, Thermomyces (formerly Humicola) insolens, Penicillium sp.*	Sepperumal et al. (2013); Nowak et al. (2011); Ronkvist et al. (2009); Liebminger et al. (2007)

15.5.1 Aspergillus flavus

Aspergillus flavus, a dominant fungus in the rhizosphere, is a major phytopathogen which holds the capability to infect both pre- and post-harvest. This pathogen is also known to infect animals including humans and causes a disease called aspergillosis which is reported in immunocompromised patients. First described by Florentine priest and mycologist P. A. Micheli in early eighteenth century, this fungus is fully sequenced now with size of 36.8 Mb (Mega base pairs or one million base pair) holds 55 secondary metabolite clusters and 12,197 genes (Amaike & Keller, 2011).

Aspergillus species are one of the majorly evaluated organisms for plastic biodegradability. Reports are that *A. flavus* can degrade around 2% of plastic film in a week's time. Though plastic comes in many forms such as poly-uracil, PE, PVC, nylon, and many more, *A. flavus* is still responsible for degrading major types of plastics. This degradation capability is dependent on several external factors. For instance, reports indicate that biodegradability is better off solid media, which means soil can provide crucial substratum for *A. flavus* to grow. It is also said to be nutrient dependent. Right now, there is very few research published which can give extended account on various minerals and their impact on biodegradation, but data says in vitro capacity of *A. flavus* to biodegrade is better on mineral salt medium (MSM) media than normal Potato dextrose agar (PDA) media. It is also reported that magnesium salt can accelerate biodegradation whereas zinc inhibits it. Reason is unknown but it can be due to the metabolic profile of the chemical reaction (Khan et al., 2020).

A. flavus attaches itself to the hydrophobic surface of the plastic which is a major obstacle for many other microorganisms. Mycelium growth on the surface then secrets some enzymes which carry on hydrolytic and oxidative reactions (Kunlere et al., 2019) leading to hydrophilicity of the surface. This can be seen by loss of surface integrity. This then leads to either growth of other organisms which otherwise were unable to grow on the hydrophobic surface or rapid growth of *A. flavus*.

Some genes reported to be upregulated belong to laccase-like multicopper oxidase. Fourier-Transform Infrared Spectroscopy (FTIR) studies show release of diverse groups formed in the process such as hydroxyl, ether, and carbonyl (Zhang et al., 2020). These groups vary with distinct types of plastic. Like in PU, methylene dianiline (MDA) is released which is a precursor of PU and produced globally. It is also said that biodegradation is population density (Accinelli et al., 2020) dependent which seems a logical statement.

A. flavus can use plastic as carbon and nitrogen source which can be extrapolated to real-life situations which means that it has the ability to degrade plastic in low mineral value soil. Other proteins which help in degradation are laccases, MnP, and lignin peroxidases (LiP). The pathway to degrade plastic is same as fatty acid metabolism ending in acetyl co-enzyme A which in turn enters glycolysis.

Like all the positive studies, it is also believed that increased activity of *A. flavus* can lead to aflatoxicity in soil (Pramila & Ramesh, 2011). Even though there is not much evidence of this, urgent work is needed to isolate and study plastic biodegradable microorganisms from an industrial point of view.

15.5.2 Fusarium solani

Fusarium solani is a complex of 26 different closely related fungi (Sabino, 2021). They are phytopathogens. It is the primary infesting organism in saffron fields of Jammu and Kashmir region of India (Bazoobandi et al., 2020). A whopping 50% of cases reported of fusariosis are caused by this species. Fusariosis is a human disease caused in immunodeficient people, who have neutrophil deficiency. They are also pathogens of chondrichthyans, a class of fish.

Fusarium solani is one of the major contestants who degrade polycaprolactone (PCL). Its cutinase enzyme is active specially against PCL. It can also degrade LDPE at the rate of 1.6% a week (Das et al., 2018). The enzyme is pH and nutrient dependent. Optimum pH is around 9–10 and any simple carbon source can inhibit production of an enzyme (Murphy et al., 1996), which suggests that the degradation is a by-product of growth but cannot degrade plastics. Also, the catalytic active site of *F. solani* is inferior to *A. fumigatus*. Which means *A. fumigatus* can degrade PCL more efficiently (Ping et al., 2017).

In conclusion, *F. solani* is not an excellent choice for plastic degradation if one is aiming for industrial application.

15.5.3 Aspergillus niger

Like *Aspergillus flavus, Aspergillus niger* is also a dominant fungus in the rhizosphere. It is a filamentous ascomycete. It has industrial importance due to its ability to produce excessive amounts of citric acid efficiently. Coming from the same family of *Aspergillus flavus*, *A. niger* is also a human pathogen. It causes otomycosis, an infection of the ear which can even cause damage to the ear canal. The Genome Sequence Project is in progress for *A. niger*. In fact, three independent genome projects are in progress in the laboratories of US Department of Energy, Integrated genomics (US company), and DSM corporation, a Dutch multinational, respectively. It is estimated to have a genome size of 35.5–38.5 Megabases (Mb), which is divided among eight chromosomes. *A. niger* has the ability to produce a wide range of hydrolytic and oxidative enzymes, which are most sought-after characteristics for biodegradation of polymers (Baker, 2006).

Aspergillus niger, not as efficient as *A. flavus*, can degrade around 0.87% LDPE per week (Esmaeili et al., 2013) but is reported to degrade much more types of plastics than *A. flavus* such as PVC, PU, PE, and PCL. But it fluctuates based on the type of plastic. There is a drastic change in types of enzymes produced when grown on plastics. Laccase is anonymously reported to be the most prominent enzyme upregulated than control conditions. According to research, the carbonyl group is present after a degraded sample. Some studies also observed aromatic compounds within the fungi body. Alkene and aliphatic groups are also observed in FTIR spectra after degradation. Tensile strength, which describes the maximum load material can carry, was also declined after degradation (Khruengsai et al., 2021).

Aspergillus niger, a dominant rhizosphere habitant can be considered as a contestant for biodegradation of plastics. But the scope of research varies. So further detailed investigations on the ability of *A. niger* is required. Research on PCL and PVC degradation is important due to their abundance in the rhizosphere for various civil purposes.

15.5.4 Candida antarctica

Candida antarctica is also known as *Pseudozyma antarctica*. It is one of the yeasts, utilized globally at industrial scale as of the day. Its enzyme lipase is one of the most efficient and is being exploited at global scale (Johnson & Echavarri-Erasun, 2011).

C. antarctica is studied for degradation on specialty plastics such as PCL and PET. PCL is a resin added to the manufacturing of PU. It provides additional strength to PU. Lipase enzymes from *C. antarctica* can degrade several types (based on molecular weight) of PCL. It is recorded to degrade in two phases. Initial phase shows logarithmic degradation. Second phase shows a steady biodegradation rate. Even though *C. antarctica* can degrade PCL effectively, it does not change major components of the plastic, which is caprolactone. It may be suggested that fungi are able to internalize caprolactone which in turn is used for metabolism (Ma et al., 2020).

*C. antarctic*a can also mulch films. Although they are biodegradable, they still are not actively degraded in situ conditions. Whereas *C. antarctica* decreases the tensile strength by 20% (Sameshima-Yamashita et al., 2019). It even makes it impossible to separate mulch films, which compared to control are easily collected.

15.5.5 Phanerochaete chrysosporium

The basidiomycete *Phanerochaete chrysosporium* is categorized as a white root fungus because of the physical alterations it causes in the wood and is one of the best lignin degraders (Mäkelä et al., 2021). It has even become the standard laboratory model for the same purpose (Couto et al., 1998). Extracellular enzymes are produced by the fungus when it is grown under certain conditions. These enzymes were linked to lignin degradation as reported in 1983 (Tien & Kirk, 1983). Ligninolytic enzymes are a class of enzymes engaged in the breakdown of resistant natural and manufactured polymeric substances such as lignin and PVC (Paszczynski & Crawford, 1995). Because it has most studied multiple ligninolytic enzymes such as MnP, LiP, glyoxal oxidases, and polyphenol oxidase laccase, the fungus has a complex mechanism for degradation of organic material (Nunes & Malmlöf, 2018; Messerschmidt, 2010). These enzymes can oxidize a wide spectrum of harmful chemical molecules, transforming them to harmless products or CO_2 and H_2O (Bumpus & Aust, 1987).

Phanerochaete chrysosporium is one of the majorly studied fungi on plastic bioremediation. It is indeed able to degrade a variety of plastics. It is reported to degrade a massive 13% of LDPE in a week's time. FTIR shows it produces groups such as ester, aldehydes, carboxylic acids during the remediation (Orhan & Büyükgüngör, 2000). It is even reported to degrade ethylene vinyl alcohol (EVOH), a specialty polymer (Carolina et al., 2004). It is a great barrier for gas, moisture, and flavor; hence, it is being used in the packaging industry of perishable goods. It can be imagined how hard it is to degrade EVOH as moisture and gas exchange is key for growth of any organism. *P. chrysosporium* can degrade EVOH by deploying its robust enzyme LiP. It forms the C=C group, claimed in the FTIR study. It also produces alcohol studied by another researcher (Mejía et al., 2001). It can also degrade PVC which is used in civil works. It can even use PVC as a main carbon source. FTIR shows a noble peak in the region 2350–2370 (Ali et al., 2014). Researchers report that it can aggressively degrade bisphenol A (BPA), an additive used in different plastic consumer products (Wang et al., 2022). Adversities caused by BPA can be pointed to the fact that it is now banned from certain products in the United States and the EU such as baby feeding bottles and many others.

15.5.6 Aspergillus japonicus

Aspergillus japonicus is found in soil as well as an endophyte. It is a metabolically active strain which secretes various anticancer alkaloids, flavonoids, phenols which are major classes of secondary metabolites (Chandran et al., 2020; Nadumane et al., 2016) and enzymes such as lipases used in lignin breakdown, in excessive amounts, one should not be surprised when it was reported that *A. japonicus* can sustain on phenolics which are by-products of lignin metabolism (Ferrari et al., 2021). It participates in the phosphorus cycle in soil (Singer et al., 2019).

Moreover, it looks biased when three species of *Aspergillus* family make the list of biodegradables, but *A. japonicus* is simply amusing. It has immense potential besides the mycoremediation of plastics, which qualifies it to be in the list. There are very scanty reports present on the ability of *A. japonicus* to degrade plastic, but the one present are promising. It shows 3% degradation of LDPE in a week (Raaman et al., 2012) where A. flavus manages to degrade 2% only. Additionally, when studying the growth pattern, it is found to be heavily populated on the inhabitable surface of PE sheets (Annamalai & Nallamuthu, 2011).

Even though there is not much technical data on its biodegradation ability, we should not be surprised if we find it being used in industrial scale for biodegradation processes of plastic in the next decade or so.

15.5.7 Cladosporium cladosporioides

Cladosporium cladosporioides is an endophyte. Being an endophyte, it is a characteristic of them to produce crucial enzymes for degradation of plants' carbohydrate structures such as xylase, cellulose, and pectin (Gohain et al., 2020). It is also responsible for infections of the lower respiratory system, even the central nervous system (Dubey, 2016; Gholami-Shabani et al., 2018).

Cladosporium cladosporioides is a very rarely studied microorganism. Yet it is a very efficient degrader described by some researchers. It can degrade almost 44% PS-PU varnish also called impranil. Even though it does not come under the term plastic exactly, studies can be conducted on plastics as there is none yet. It transforms impranil to alcohols and hexane diisocyanate (Álvarez-Barragán et al., 2016).

15.5.8 Alternaria solani

Alternaria solani is very weakly studied, mostly dominated by research on finding cure or restricting its disease-causing ability towards tomato and potato (Ng, 2013; Petrov et al., 2021) both commercially, unbelievably valuable plants. It is reported that it can degrade PS-PU composites to an extent of 21–23% weight loss in a matter of one week (Ibrahim et al., 2011). It causes 24% decrease in tensile strength. It forms an N-H bond (Ibrahim et al., 2009).

15.5.9 Trichoderma viride

A mold, *Trichoderma viride*, is both fungi and fungicide. It is even used as a biocontrol agent in the fertilizer industry (Deepa & Sreenivasa, 2019). It produces substantial number of extracellular enzymes, usually characteristic of most fungi. *T. viride* gives tough competition to *A. niger* in the production of large-scale citric acid (Moresi & Parente, 1999).

Trichoderma viride is a weak biodegrader of plastics. There are very few reports of this organism degrading plastics. It can degrade PE films in the range of 0.3–0.7%. But it degraded almost 11% compared to 8% by *A. nomius* (Munir et al., 2018). This was achieved by its laccase enzyme (Johnnie et al., 2021).

15.5.10 Chaetomium globosum

C. globosum is a very versatile fungus. It secrets many secondary metabolites such as polyketides (Oikawa, 2010). It also poses some therapeutic effects towards human diseases such as rheumatoid arthritis, even cancer (Abdel-Azeem et al., 2019; Yashavantha Rao et al., 2021).

Chaetomium globosum cannot degrade plastics in efficient ways. There is scanty nature of reports on this fungus. It can degrade PE by around half a percent in a week (Saxena et al., 2022). Promising that it can degrade both PCL and PVC, i.e., both hydrophilic and hydrophobic, respectively. It can use them as a carbon source (Vivi et al., 2019) very few others hold this capacity indeed. Even though it cannot degrade it efficiently, it can serve as a good host. It can also degrade polyester, a polymer used in several types of fabrics around the world (Kim & Rhee, 2003).

15.6 Factors Influencing Mycoremediation

Biodegradation by fungi is known to be a time-consuming process and is influenced by several factors. Soil's physical-chemical parameters such as temperature, pH, content of H_2O, and potential to carry out redox reactions have a substantial effect on microbial development and, as a result, on the efficacy of a biodegradation process, in general (Bosco & Mollea, 2019).

Temperature influences the rate of polymer breakdown by fungus. The metabolic rate of fungus increases as soil temperature rises, resulting in faster plastic degradation. As a result, the rate at which fungi degrade plastic changes according to the season. The fungal proliferation and rate of plastic breakdown are both influenced by pH. Fungi grow faster at acidic and basic pH levels than at neutral pH. *Penicillium* grew best at a basic pH, while *Aspergillus* grew best at an acidic pH (Pawar, 2015).

Factors influencing metabolism of microbes such as micro- and macro-nutrients availability, type of pollutant and its concentration, along with their toxicity, bioavailability, and mobility will impact the bioremediation activities of microbes (Lukić et al., 2017). Although soil normally contains sufficient nutrients for microbe's development, nutrients can indeed be provided in a functional form that donates electrons to accelerate the biodegradation procedure (Omokhagbor Adams et al., 2020). The genetic properties of the microbe, particularly the intracellular and extracellular enzymatic machinery, have a major role in the microbial degradation of a harmful substance (Chanda et al., 2016). The concentration of pollutant has direct impact on microbial action: a range of harmful effects can be produced on different microbial population due to higher concentration, whilst a lower pollutant concentration may not be adequate to stimulate production of degradative enzymes. Filamentous fungi capable of establishing a large network of mycelium and producing a large number of nonspecific enzymes are more resistant to high concentrations of pollutant than bacteria (Harms et al., 2011). Degradative enzymes can also be synthesized at low concentrations of pollutant due to their poor substrate selectivity. The intracellular metabolic processes part of fungal bioremediation is strikingly comparable to those which control fungal secondary metabolism, particularly that which control synthesis of mycotoxins (Chanda et al., 2016). Mycotoxin- producing filamentous fungi such as *Penicillium* and *Aspergillus spp.* can breakdown endocrine-disrupting chemicals (EDCs), the prominent pharmaceutical contaminants (Harms et al., 2011; Esteban et al., 2014), which bacteria cannot degrade effectively. Biodegradation employing fungi of soils contaminated with aged polycyclic aromatic hydrocarbons (PAHs) established that bioavailability of pollutant is one among the most critical parameters that may be adjusted to enhance and expedite biodegradation (Leonardi et al., 2007). Various publications (Prakash, 2017; Shekhar et al., 2015) have described the potential of fungi to chemically change or impact bioavailability of pollutant by producing biosurfactants. Biosurfactant production has been observed in *Aspergillus* and *Penicillium spp.* (Gao et al., 2011; Lin et al., 2003).

The process of degradation is heavily influenced by the physical and chemical properties of plastic, such as density and its molecular weight; complexity of the structure; available functional groups; toughness and the form of polymer. The polymer size also impacts the rate of degradation (Hari, 2019). The enormous molecular weight and very hydrophobic surfaces of these polymers contribute to their great resistance to passing through the cell wall (Mohanan et al., 2020). Low molecular weight polymers break down quickly. The capacity to degrade is also affected by the melting point. When the melting point rises, polymer degradation reduces. The degree of crystalline and amorphous forms, as well as the existence of strong C-C bonds, which are resistant to enzyme attack, can influence degradability. Polymer solubility in an organic solvent is required for polymer degradation in solution. The activity of enzymes in the polymer biodegradation is mostly determined by the solvent characteristics, which rises with polarity and declines with viscosity.

15.7 Diverse Enzymes Involved in Mycoremediation

Enzymes are biological catalysts that are used in a variety of industries that take part in a process, operate on a specific substrate, and speed up transformation of the substrate into useful products. To carry out their metabolism, microbes produce a variety of enzymes. Microorganisms rely heavily on enzymes for biodegradation as well. At least two types of enzymes are employed in the biological breakdown of polymers throughout the depolymerization process: extracellular and intracellular depolymerases (Gu et al., 2000). The first and foremost step in the degradation process by microbes is the production of biofilm on the surface of polymer in order for microorganisms to optimally utilize non-soluble substrates through enzymatic activities. The degradation or unzipping of chain of certain polymers occur via two mechanisms: biological hydrolysis and biological oxidation. Because large enzymes are unable to permeate the polymer, breakdown happens primarily on the surface. Extracellular enzymes released by microbes breakdown bulky complex polymers with high molecular weight into smaller molecules with short chains, such as oligomeric, dimeric, and/or monomeric units, tiny enough to pass through the microbes' semi-permeable outer membranes and to be absorbed into the microbial cells. Once inside the cells, these breakdown products use traditional catabolic pathways to produce energy and reduce power for cellular development (Mohanan et al., 2020; Frazer, 1994; Hamilton et al., 1995). Figure 15.2 gives an overview of degradation by microbial enzymes (Modified from Alshehrei, 2017). For example, CO_2 and acids are the end products of aliphatic polycarbonate biodegradation (Patel et al., 2013).

Fungi are heterotrophic creatures that obtain their food by taking nutrients from the environment around them. They secrete digestive enzymes outside their hyphae via exocytosis, which degrade complex organic compounds into simpler molecules in order to consume them back, releasing out carbon dioxide and water in aerobic conditions (along with methane in anaerobic environments) when the substrate is mineralized (Pathak, 2017). Fungi offer a variety of unique methods for combating a wide range of complex chemicals, some of which are pollutants and harmful (Olicón-Hernández et al., 2017). These methods include a robust enzymatic machinery, adsorption capabilities, and the creation of endogenous biosurfactants (i.e., hydrophobins), all of which allow them to employ plastic polymers as carbon and electron sources, giving cellular machinery and source of energy, respectively.

Enzymatic system of fungi consists of intracellular and extracellular enzyme machinery which work together and catalyze a wide range of reactions, making them ideal for the degradation of pollutants (e.g., plastic polymers). Jeon et al. (2016); Olicón-Hernández et al. (2017); Schwartz et al. (2018) found that the intracellular enzyme machinery functions as an detoxifying system for intrinsic harmful metabolites and external toxic pollutants and is important for fungal adaptability (Morel et al., 2009). Phase 1 enzymes: Cytochrome P450 (CYP) epoxidases and Phase 2 enzymes: glutathione transferases constitute this intracellular enzymatic system. CYP epoxidases, or monooxidases, are oxidoreductases having heme-thiolate in their structure, activated by reduced form of heme iron acts on a variety of substrates in a regio- and stereoselective way and transfers 1 atom to the substrate from O_2. NAD(P)H is required as a donor of electrons for hydroxylation, sulfoxidation, epoxidation, and dealkylation (Mäkelä et al., 2015). Reduced glutathione is utilized to carry out catalysis of the nucleophilic attack reaction of an electrophile such as carbon, sulphur, or nitrogen in non-polar molecules by glutathione transferases, which are found in several cellular compartments. Electrophilic substrates

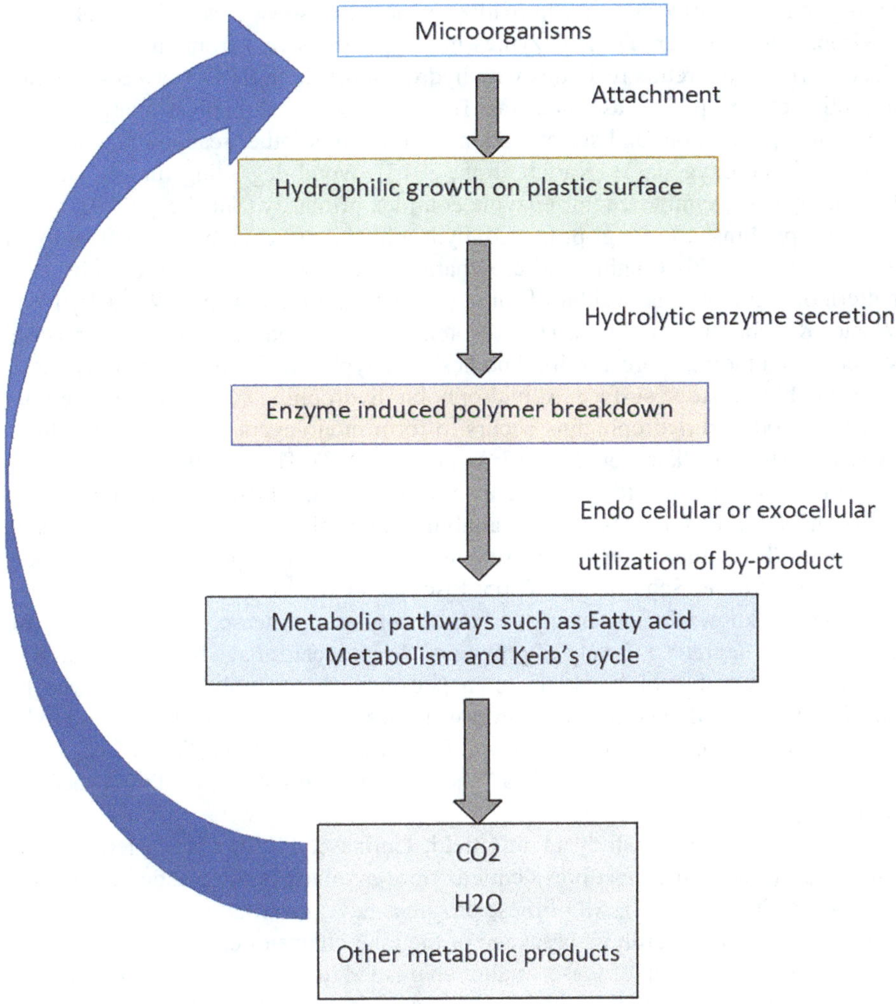

Fig. 15.2 Overview of degradation by microbial enzymes. (Modified from Alshehrei, 2017)

become a little more soluble in H_2O when coupled with glutathione (GSH). The extracellular enzymatic system comprises of a hydrolytic system producing hydrolases and is an exceptional and non-specific oxidative system, which are engaged in the degradation of polysaccharides and complex substances (e.g., lignin), respectively (Sánchez, 2009). Since it enhances the solubility of contaminants and diminishes their bioaccumulation potential (Olicón-Hernández et al., 2017), hydroxylation might be regarded as a biotransformation strategy for bioremediation procedures. Unspecific oxidoreductases include nonspecific peroxygenases, class II peroxidases (versatile peroxidase (VP), LiP and MnP), dye decolorizing peroxidases (DyPs), and laccases which all together make up the non-specific oxidative system. The non-specific oxidative system can oxidize a variety of substrates and is an effective environmental cleaner. These enzymes move electrons (e^-) from substrates to O_2 (laccases) through redox reactions with hydrogen peroxide (H_2O_2) as a co-substrate accepting e^- (dye peroxidases and class II peroxidases) or through aromatic peroxygenation, epoxidation, and sulfoxidation, among some other reaction mechanisms (unspecific peroxygenase) (Karich et al., 2017). Wood-degrading fungus, such as basidiomycetes, synthesize this enzyme complex primarily (Sánchez, 2009).

Hydrophobins are fungi-produced hydrophobic proteins having a length of 70–350 amino acids roughly and are characterized by the presence of conserved pattern of eight cysteine residues forming four disulphide linkages (Wessels, 1996; Wösten & Scholtmeijer, 2015). These proteins help filamentous fungi develop aerial structures including spores, fruiting bodies, and hyphae, as well as attach hyphae to hydrophobic substrate surface. At hydrophobic-hydrophilic surfaces, self-assembly of fungi produced hydrophobins occurs to form monolayers of amphipathic films (Wessels, 1996; Kulkarni et al., 2017; Wu et al., 2017). These proteins are useful in bioremediation because they behave as biosurfactants, which improve mobility of substrate while also increasing bioavailability. Fibroblasts on the surface of plastic can benefit from a hydrophobin coating as it improves their proliferation and morphology (Hektor & Scholtmeijer, 2005; Piscitelli et al., 2017).

Fungi are known for producing cellulase, amylase, protease, laccase, and lipase enzymes that degrade a variety of polymers. Ascomycetes have been discovered to be the most capable of breaking down petroleum-based polymers, followed by basidiomycetes and zygomycetes. Several studies have shown that fungi can biodegrade and biodeteriorate (break down in characteristics) plastics. Only a few research, however, have looked at the enzymes involved in this processes (Sánchez, 2020).

According to recent study in this field, Cutinase, PETase (an Esterase), and Lipase are three of the most prevalent enzymes involved in the plastic degradation (Tokiwa & Suzuki, 1977). All of these enzymes have a similar effect on the plastic polymer, initiating hydrolytic breakage of the long chains of carbon into small carbon chains, assimilation of these smaller chains for further breakdown by enzymes in the microbial cell and metabolic product release (Tokiwa & Suzuki, 1977). However, "Hydrolases", i.e., the third class of enzymes, are the primary enzymes involved in the degradation of plastics found in the environment (Müller et al., 2005) and includes enzymes that have been identified as degraders of plastic polymers. This class of enzymes participate in a chemical reaction that causes the substrate's chemical bonds to break down in the presence of H_2O resulting in the breakup of a larger substrate molecule into smaller subunits (Tokiwa et al., 1976). The following reaction represents the general reaction of hydrolases:

$$X - Y + H_2O \rightarrow X - OH + Y - H.$$

Long carbon chains are cleaved in two steps by the hydrolases involved in plastic polymer degradation. All of the plastics in the environment are hydrophobic. As a first step in the interaction between enzyme and plastic polymer, extracellular enzymes synthesized by different microbes cling to the plastic surface via hydrophobic interactions. Many hydrolases have a hydrophobic cleft near the active site that can accept hydrophobic groups in the polymer, making the enzyme more accessible to the polymer (Wilkes & Aristilde, 2017). The enzyme's active site participates in the hydrolytic breakdown of long polymeric chains into smaller oligomers, dimers, or monomers, which can be collected by the microbe and utilized as a source of carbon in the second step of the reaction (Barth et al., 2015).

When *Pestalotiopsis microspora* was cultivated on polyester polyurethane (impranil) as its only carbon source, Russell et al. (2011) found that it had higher serine as well as cysteine hydrolase activity. Cutinases (EC 3.1.1.74), carboxylesterases (EC 3.1.1.1), and lipases (EC 3.1.1.3) are typical serine hydrolases implicated in plastic polymers breakdown (Danso et al., 2019).

15.7.1 Cutinases

Cutinases (EC 3.1.1.74) are a type of esterase enzymes that has gained popularity due to their capacity to carry out hydrolysis of high molecular weight polyesters (Chen et al., 2013). Cutinases are carboxylic ester hydrolases derived from plant pathogenic fungi, such as *Fusarium solani pisi* (Kolattukudy, 1981; Heredia, 2003). Because the enzyme can hydrolyze the cutin present in the cuticular layer of leaves or the suberin present in the bark under natural circumstances, it has attracted attention for its phyto pathogenicity. Cutinase is a serine esterase with a catalytic triad of Ser-His-Asp/Glu that is very similar to lipase. With the exception of a recently discovered cutinase from *Trichoderma reesei* (Roussel et al., 2014), most cutinases vary from lipases, in that their active regions are uncovered. Both cutinase and lipase can hydrolyze the ester link between -OH and -COOH groups despite their structural differences (Shi et al., 2019). Cutinase's catalytic region is exposed to the solvent and is positioned at one end of the protein ellipsoid, enclosed by a loop and additional hydrophobic amino acids (Jelsch et al., 1998). The serine's job is to engage a transacylation process with the substrate and produce an acyl-enzyme intermediate, which is subsequently hydrolyzed to produce the final product.

Cutinases have been found, cloned, and studied for their activity on polymeric substrates from a variety of microbiological sources to date (Muñoz & Bailey, 1998; Griswold et al., 2003; Carvalho et al., 1999; Baker et al., 2012; van Gemeren et al., 1996; Liu et al., 2009). Cutinases have hydrolytic activity against a variety of esters, including synthetic esters such para-nitrophenol ester, PET, polyvinyl acetate, and PCL (Melo et al., 1995; Egmond, 2000), making them potential for bioplastic biodegradation. Cutinases and their homologues have demonstrated the most potential for PET hydrolysis among the microbial polyester hydrolases that have been described (Kawai et al., 2019). When PET films with low crystallinity (7%) were utilized as substrates, Ronkvist et al. (2009) discovered that *Fusarium solani* and *Thermomyces* (previously *Humicola*) *insolens* exhibited cutinase catalytic activities. *Fusarium solani*'s cutinase has been overexpressed in *Pichia pastoris* to breakdown polybutylene succinate (PBS) plastic (Hu et al., 2016). This recombinant cutinase was capable of accelerating the process of degradation, and the PBS films were totally degraded within just 6 hours (Nikolic & Djonlagic, 2001). The recombinant enzyme degraded both the crystalline and amorphous structure of PBS, according to the analysis of films after being degraded by enzyme (Maeda et al., 2005). Many different forms of biodegradable polymers have been proven to be degraded by cutinases (Masaki et al., 2005). High molecular weight polylactic acid (PLA) based polymeric material was degraded by a cutinase-like enzyme isolated from *Cryptococcus sp. Strain S-2* (van Gemeren et al., 1998). 0.8 g/mL concentration of this cutinase-like enzyme required only 60 hours to breakdown PLA, which is significantly faster than other known PLA-degrading enzymes (Oda et al., 2000). Cutinase enzyme from *Humicola insolens* (HiC) was used to degrade PBF (poly (1,4-butylene-2,5-furandicarboxylate) and PBTF (poly (1,4-butylene 2,5-thiophenecarboxylate) films. Cutinase is a PCL degradation enzyme produced by *Aspergillus oryzae* and *Fusarium solani* (Liu et al., 2009).

15.7.2 Lipases and Esterases

These enzymes are members of the α/β hydrolase superfamily, which perform the catalysis of the hydrolysis and formation of ester bonds (Sharma et al., 2017, 2018a, b, c). Long-chain acylglycerols (C8) are hydrolyzed by lipases (EC3.1.1.3), whereas short-chain fatty esters (C8) are hydrolyzed by esterases (EC3.1.1.1).

Lipase produced by *Rhizopus delemar* was used to hydrolyze polybutylene terephthalate, PLA, and PET, whereas lipase from *Chromobacterium viscosum* and *Pseudomonas sp.*, hydrolyzed PE (Müller et al., 2001). Poly bisphenol-A (PBPA) carbonate has been reported to be degraded by lipases which include *Candida Rugosa*, Lipolase, Hog pancreas, and Novozyme (Sivalingam & Madras, 2004). PLA polymer and its nanomaterials augmented with nanocrystals of cellulose were also degraded by *Candida rugosa*'s lipase and *Tritirachium album*'s proteinase K enzymes (Kaushal et al., 2021). In the hydrolysis of polybutylene succinate-co-adipate (PBSA) and PBS, a lipase enzyme isolated from *Cryptococcus sp. MTCC 5455*, which was cultivated on leftover agricultural waste, showed a very good potential (Thirunavukarasu et al., 2008). Within 16 hours for PBSA and 72 hours for PBS, the plastic materials were completely degraded (Thirunavukarasu et al., 2016). The poly (butylene succinate-co-hexamethylene succinate) (P (BS-co-HS)) copolymer was degraded using *Candida rugosa*'s lipase. According to the findings, the higher a copolymer's HS level, the more vulnerable it is to lipase enzyme attack. P (BS-co-HS) copolymer degradation was studied using weight loss as a measure (Pereira et al., 2001).

Degradation of PU by *esterase* from *Penicillium griseofulvum* and *Xepiculopsis graminea* is a two-step mechanism: hydrophobic enzyme attachment onto the surface of PU after which PU ester linkages are hydrolyzed (Brunner et al., 2018). In enzyme experiments of PU, the *Alternaria* and *Penicillium* strains had similar esterase activities, however the *Aspergillus* strain had a much reduced esterase activity (Magnin et al., 2019). Liebminger et al. (2007) isolated PET pellet hydrolyzing polyesterase from *Penicillium citrinum*. Despite the fact that several bacteria and fungi have been shown to degrade PS, the primary enzymes that catalyze the first depolymerization of the polymers is yet to be discovered. PS has been demonstrated to be broken down by an extracellular esterase of *Lentinus tigrinus* (Tahir et al., 2013).

15.7.3 Peroxidases and Laccases

Hydroxylases, peroxidases (LiP and MnP), laccases, and reductases have all been identified as PE degrading biocatalysts from bacterial, actinomycetes, and fungal sources in recent years (Wei & Zimmermann, 2017). Multiple isoenzymes of manganese and lignin peroxidase have been found to be synthesized by white rot fungi such as Basidiomycetes, *Trametes versicolor, P. chrysosporium,* and *Pleurotus ostreatus*. Two lignin-degrading fungus, *Trametes versicolor* and *Phanerochaete chrysosporium*, synthesized MnP (EC 1.11.1.13) and LiP (EC 1.11.1.14) enzymes with MnP being the primary enzyme responsible for the PE degradation (de Albuquerque et al., 2019; Iiyoshi et al., 1998). Even though the particular mechanism involved in the PE degradation via enzymes has not been revealed, these enzymes are involved in the degradation process carrying out oxidation at the terminal, chain cleavage, and metabolization of fatty acid (Albertsson et al., 1987). By cleaving plastic polymers into water-soluble short chains, these enzymes generate non-specific and highly reactive free radicals, which promote their transit through microbe membranes for intracellular breakdown (Khatoon et al., 2019).

Fungal peroxidases produce oxidants that start the oxidation of the substrate outside the cell (Deshmukh et al., 2016). Peroxidases (E.C. 1.11.1.7) are class II peroxidases (Hofrichter et al., 2010) which carry out the catalysis of redox reactions by converting various chemicals into oxidized or polymerized end products using the free radical mechanism employing hydrogen peroxide as electron acceptor. The principal high-redox class II peroxidases in fungi are LiPs, VPs, and MnPs, as previously documented. Peroxidase's prosthetic group is made up of a protein-bound heme, which is normally linked to a histidine amino acid residue that serves as a proximal ligand (Hofrichter, 2002). When heme peroxidase activates H_2O_2, it produces intermediate molecules with a high valence that can abstract electrons from a variety of substrates.

DyPs responsible for hydrolytic and oxidative actions on non-phenolic and phenolic organic substrates can also be produced by fungi (Lauber et al., 2017). Heme-thiolate peroxidases (HTPs) are enzymes that transmit peroxide oxygen to substrate from hydrogen peroxide or R-COOH; this category also comprises aromatic or unspecific peroxygenases (APOs or UPOs) and chloroperoxidases (CPOs). Aromatic peroxygenation, hydroxylation, or double-bond epoxidation of aliphatic molecules allow UPOs to particularly act on heterogeneous substrates (Hofrichter et al., 2010).

Ameen et al. (2015) discovered that a group of ascomycete strains may degrade LDPE by using MnP, laccase, and LiP enzymes. Fungal and bacterial sources of nylon-degrading enzymes have been discovered (Nomura et al., 2001; Yamano et al., 2019). A MnP from the strain *IZU-154* of white-rot fungus has been demonstrated to peel the Nylon 6's surface and cause intense horizontal fissures to appear in the polymer (Deguchi et al., 1998).

Laccases (EC 1.10.3.2) are benzenediol oxygen reductases belonging to the family multi-copper oxidases (MCOs) found in a variety of bacteria, fungi, lichens, plants, and insects, with each laccase from different species having distinct sequences and catalytic properties (Arregui et al., 2019; Giardina et al., 2010; Riva, 2006; Mayer, 2002; Santhanam et al., 2011; Mate & Alcalde, 2015). Laccases with poor substrate selectivity oxidizes a variety of substrates, most commonly aromatic amines (phenylenediamines) and substituted phenols (para-, ortho-, amino-phenols) by initiating cleavage in aromatic ring via transferring 4 e^- from substrate to O_2 and produce free radicals (Yang et al., 2015; Lassouane et al., 2019). Chemically unstable products and predominantly produced free radicals frequently initiate domino reactions resulting in complicated chemical transformations with biological significance. Laccase-mediator systems (LMSs) increase the range of substrates to be acted upon by laccases by affecting the chain of electron transfer (Senthivelan et al., 2016). Laccases primarily found in lignin-degrading fungi can also degrade non-aromatic substrates (Mayer, 2002). Laccase is produced by fungi such as *Trametes versicolor, Streptomyces, Pleurotus ostreatus,* and *T. pubescens*, which breakdown PE (Sivan, 2011). Laccase-like multicopper oxidases (LMCOs) and laccases isolated from *A. flavus strain PEDX3* were also examined for their potential to degrade PE. Low molecular weight PVC is degraded under laboratory conditions by white rot fungi *Cochliobolus sp*. It is known to produce laccase which uses the plastic as a carbon source (Sumathi et al., 2016).

15.7.4 Dehydrogenase

Dehydrogenase and oxidase produced by *Aspergillus fumigatus* cleaves the polymer backbone (Mollasalehi, 2013). PU dehydrogenase is produced by *Pestalotiopsis microspora* and is responsible for PU degradation (Tokiwa et al., 2009). Endopolyurethenases hydrolyze PU at random sites, while exopolyurethenases removes successive monomeric and dimeric units from the chain terminals (Howard et al., 2001).

15.7.5 Depolymerase

A Polyhydroxybutyrate (PHB) depolymerase has been examined to see if it can degrade PHB, polyethylene succinate (PES) and PBS (Jung et al., 2018). PHB depolymerase was isolated from the fungus *Aspergillus fumigatus* and then purified (Kasuya et al., 1998). PHB depolymerases, which break down PHB, are produced by *Fusarium solani* (Shivakumar, 2013). Because it has two substrate binding sites, PHB depolymerase has quite a high substrate selectivity and enhances enzyme adsorption (Roohi & Kuddus, 2018).

15.7.6 Protease

Serine protease produced by *Arthrobotrys oligospora* degrades PLA (Hari, 2019). Protease is more potent than esterase in degrading PU. On PU, a combination of protease and esterase proved to be quite successful (Ozsagiroglu et al., 2012). Acid protease has been found to be produced by *Aspergillus, Rhizopus, Penicillium, Mucor, Thermoascus, Humicola,* and *Thermomyces* (Souza et al., 2015). Based on a protein's amino acid sequence, proteases can either perform restricted proteolysis (break down particular peptide bonds) or unlimited proteolysis (break an entire peptide to individual amino acids) (Gilan & Sivan, 2013).

15.7.7 Urease and Papain

Medical polyester polyurethane is degraded by proteolytic enzymes urease and papain. Hydroxyl and free amine groups generated upon the hydrolysis of urea and urethane bonds allowed papain to breakdown the polymer.

15.7.8 Plastic Degradation by Enzyme Consortium

Cutinases, lipases, esterases, laccases, proteases, peroxidases, and ureases, among other enzymes from bacterial and fungal origins, have been demonstrated to degrade PU (Magnin et al., 2020). El-Morsy (2017), for example, showed that *Monascus sp.* strains produce protease, esterase, and lipase in the biodegradation of PU. Esterase, urease, and protease enzymes found in fungi have been found to breakdown ester-type PU.

Enzymatic PET hydrolysis can take place in one of two ways either by modifying surface of polyester fibers or by polymer hydrolysis. Hydrolases such as lipases, carboxylesterases, cutinases, and proteases are used to enzymatically modify the surface of PET polymers. Lipases from *Candida antarctica* (Vertommen et al., 2005) and cutinases from *Fusarium solani* (Alisch-Mark et al., 2006; O'Neill et al., 2007), *Aspergillus oryzae, Penicillium citrinum* (Liebminger et al., 2007), and *Humicola insolens* (Ronkvist et al., 2009) have been investigated for their ability to hydrophilize the surface of PET fibers.

PCL polymer is also degraded by lipase and esterase (Tokiwa et al., 2009).

Cutinase and lipase are esterase enzymes that belong to the hydrolases category of enzymes. A combination of lipase and cutinase was recently found to be effective in the biological breakdown of the plastic PCL (Liu et al., 2019). End-to-end fusion method was used to create the fusion system of the two enzymes to finally overexpress in *Pichia pastoris* (Aris et al., 2016). PCL breakdown with Lip–Cut fusion enzyme was more faster after 6 hours than with these two enzymes alone (Khan et al., 2017a). List of fungi along with their enzymes involved in plastic degradation is given in Table 15.2.

15.8 Methods of Analysis of Plastic Degradation

The most significant aspect of plastic biodegradation is knowing the by-products generated by the activity of the microorganisms used to degrade the plastic after it has been degraded using various methods. The importance and efficiency of the procedure can be proposed only after the degradation products are known. Some of the common methods used to determine plastic degradation are as follows:

15.8.1 Gravimetric Measurement

Gravimetric measurement of weight loss of plastic due to biodegradation is the most commonly used method in assessment. Especially when the plastic is exposed to environmental factors such as water, wind, heat, and microbiota. It is, however, difficult to ascertain the extent of fungal degradation of plastic, as other microorganisms such as bacteria are also present. However, in the laboratory, it is easier to study the effect of fungal degradation on a plastic object, either as a single strain or in multiples. One can also take note of the time taken, if any other products are formed during degradation by the fungi. Fungal isolates recovered from municipal solid waste revealed that *Fusarium oxysporum, Aspergillus fumigatus, Lasiodiplodia crassispora, Aspergillus niger, Penicillium sp.*, and *Trichoderma harzianum* were also able to cause weight loss of PE as well as PU (Zeghal et al., 2021; Raghavendra et al., 2016).

This method requires long incubation, followed by post incubation treatment. Often fungal degradation is accompanied by formation of biofilm, which must be removed. This can affect weight loss in plastic. Furthermore, it is not very clear whether the fungal degradation of plastic is photochemical or even physiological.

Table 15.2 List of fungi along with their enzymes involved in plastic degradation

Microorganism	Enzyme	Plastic polymer	References
Pestalotiopsis microspora	Serine hydrolase	Polyester polyurethane (impranil)	Russell et al. (2011)
Fusarium solani, Thermomyces (previously Humicola) insolens	Cutinase	PET films	Ronkvist et al. (2009)
Fusarium solani	Cutinase	PBS	Hu et al. (2016)
Fusarium solani's cutinase overexpressed in Pichia pastoris	Recombinant cutinase	PBS	Hu et al. (2016); Nikolic and Djonlagic (2001)
Cryptococcus sp. Strain S-2	Cutinase-like enzyme	PLA	van Gemeren et al. (1998)
Humicola insolens	Cutinase	PBF and PBTF	Gigli et al. (2019)
Aspergillus oryzae, Fusarium solani	Cutinase	PCL	Liu et al. (2009)
Rhizopus delemar	Lipase	Polybutylene terephthalate, PLA, PET	Müller et al. (2001); Masaki et al. (2005)
Rhizopus delemar, R. arrhizus, Penicillium	Lipase	PCL, Polyethylene adipate, PBS	Tokiwa et al. (2009)
Chromobacterium viscosum, Pseudomonas sp.	Lipase	PE	Müller et al. (2001)
Candida rugosa	Lipase	PBPA carbonate, PLA polymer, and its nanomaterials augmented with nanocrystals of cellulose, P (BS-co-HS),	Sivalingam and Madras (2004); Kaushal et al. (2021); Pereira et al. (2001); Hegyesi et al. (2019)
Tritirachium album	Proteinase K	PLA polymer and its nanomaterials augmented with nanocrystals of cellulose, Benzoyl peroxide-reinforced PBS/PLA blend films	Kaushal et al. (2021); Hegyesi et al. (2019); Hu et al. (2018)
Cryptococcus sp. MTCC 5455	Lipase	PBSA and PBS	Thirunavukarasu et al. (2008)
Penicillium griseofulvum, Xepiculopsis graminea, Alternaria sp., Penicillium sp., Aspergillus sp.	Esterase	PU	Brunner et al. (2018); Magnin et al. (2019)
Penicillium citrinum	Polyesterase	PET	Liebminger et al. (2007)

(continued)

Table 15.2 (continued)

Microorganism	Enzyme	Plastic polymer	References
Lentinus tigrinus	Esterase	PS	Tahir et al. (2013)
Candida antarctica	Esterase	Biodegradable plastic	Sameshima-Yamashita et al. (2019)
Monascus sp., M. ruber, M. sanguineus	Esterase	PU	El-Morsy (2017)
Phanerochaete chrysosporium	Manganese peroxidase, Lignin peroxidase, Laccase	PVC	Paszczynski and Crawford (1995)
White-rot fungus strain IZU-154	Manganese peroxidase	Nylon 6's surface	Deguchi et al. (1998)
Trametes versicolor, Streptomyces, Pleurotus ostreatus, T. pubescens, A. flavus strain PEDX3	Laccase	PE	Sivan (2011)
Cochliobolus sp.	Laccase	Low molecular weight PVC	Sumathi et al. (2016)
Aspergillus fumigatus	Dehydrogenase and oxidase	Polymer backbone	Mollasalehi (2013)
Pestalotiopsis microspora	Polyurethane dehydrogenase	PU	Tokiwa et al. (2009)
Aspergillus fumigatus, Fusarium solani	PHB depolymerase	PHB	Kasuya et al. (1998); Shivakumar (2013)
Arthrobotrys oligospora	Serine protease	PLA	Hari (2019)
Monascus sp.	Protease, esterase, and lipase	PU	El-Morsy (2017)
Candida antarctica	Lipase	PET fibers	Vertommen et al. (2005)
Fusarium solani, Aspergillus oryzae, Penicillium citrinum, Humicola insolens	Cutinase	PET fibers	Alisch-Mark et al. (2006); O'Neill et al. (2007); Liebminger et al. (2007); Ronkvist et al. (2009)
Expressed in *Pichia pastoris*	Lip–Cut fusion enzyme	PCL	Liu et al. (2019); Aris et al. (2016); Khan et al. (2017a)

Then arises the presence of additives in plastic which degrade, due to fungal action, but not the rest of the object. Such a case can cause false weight loss signals. If plastic serves as the only carbon source for the fungi, then we can observe growth in the fungal biomass followed by a corresponding weight loss in plastics.

15.8.2 Clearance Zone Formation

Another method used to study plastic degradation is clearance zone formation. Plates containing agar and solubilized plastic are inoculated with fungi. The said fungi derive carbon from plastic in the plate and forms a clear zone around itself, proving that it can degrade plastic. These halos are called clearance zone. Attempts to use PE and clearance zone formation have been successful for *Aspergillus niger* and *Aspergillus flavus* (Deepika & Jaya, 2015). Although they can be used for visual observation, it is difficult to prove whether carbon source is taken from agar or from the plastic.

15.8.3 Scanning Electron Microscopy (SEM)

SEM sends a high intensity electron beam on to the surface of plastics in order to ascertain the topography and to check for any dents or microbes, and if it is degrading the plastic surface. SEM is used to evaluate colonization of the plastic surface. SEM has been used to study growth of *C. tenuissimum* hyphae within PE-PU (Álvarez-Barragán et al., 2016). It can be done without any chemical treatment; however, it does not allow identification of the strain as well as depth of the cracks.

15.8.4 Fourier Transform Infrared Spectroscopy (FTIR)

FTIR is used to obtain information on the chemical identity of most polymers. It enables semi-quantification of changes of the configuration of the plastic when it undergoes oxidation. Measuring carbonyl index helps in determining degree of carbonylation, which is found in plastic degradation, such as PU (Filip, 1979), PE (Paço et al., 2017), PS (Tian et al., 2017), and PP (Sheik et al., 2015). PE degradation has been evaluated by FTIR in co-cultures of bacteria and fungi, *Lysinibacillusxylanilyticus* and *A. niger*, isolated from soil (Esmaeili et al., 2013). This method is fairly straightforward, but its values are changed due to formation of biofilms on plastics. Thus, biofilm needs to be removed by chemical treatment such as by hydrogen peroxide (Löder & Gerdts, 2015).

15.8.5 Carbon Labeled Polymers

Radiolabeled carbon complexes can be incorporated into plastics to study its fungal degradation, both qualitatively as well as quantitatively. Since radioisotopes are incredibly sensitive, incredibly small degradation rates can also be detected by the radiolabels. Detection using radiolabeled compound requires proper laboratory and sufficient protection for the personnel, which makes it inconvenient to work with.

15.8.6 Other Methods

Additional methods to evaluate plastic degradation include the following techniques:

(i) Measuring O_2 consumption or CO_2 production via respiration, which can provide ample information on the progress of degradation. It does not tell us whether the degradation occurs due to aerobic or anaerobic respiration.
(ii) Thermogravimetric Analysis (TGA) helps in determining the thermal stability of a polymer, which can potentially indicate its degradation.
(iii) Strength/integrity of the plastics can be determined based on tensile resistance alterations of plastic, which will decrease as a function of degradation.
(iv) Differential scanning calorimetric (DSC) analysis measures and analyzes the thermal properties such as glass transition temperature of synthetic polymers.
(v) High-Temperature Gel Permeation Chromatography (HT-GPC) gives detailed information on the molecular weight distribution of the plastic. A decrease in the molecular weight of the polymer proves that microbial degradation occurred on account of chain cleavage.

15.9 Some Studies Undertaken for Microbial Degradation of Plastics

As observed by University of Lagos, an experiment was conducted on water sachet to check degradation of low density polythene using bacteria and fungi. *Pseudomonas* and *Aspergillus* were used. They were collected from the university dumpsite, and the change in weight of the plastic over a period of 60 days was observed. There was a definite decrease in weight and the loss in weight was higher in *Aspergillus* as compared to *Pseudomonas sp.* Two microorganisms had synergistic effect upon one another, and their resultant degrading effect on plastic is higher as compared to the either. What causes the microorganisms to recognize each other's chemical pattern and work in concert to bring about the reaction is to be studied. This can shed some light on genes acting from outside the organism acting in tandem with the action taken up by the other gene (Ogunbayo et al., 2019).

The research team, composed of Manisha Sangale (Savitribai Phule Pune University) and Md.Shahnawaz (CSIR, Indian Institute of Integrative Medicine, Jammu), collected samples from 12 different locations such as Kerala, Karnataka, Maharashtra, Goa, and Gujarat. They collected the plastic wastes primarily from dumping sites with growing mangroves surrounded by marine water. From the soil samples, they isolated fungal strains *Aspergillus terreus* and Aspergillus *sydowii* which are efficient polythene degraders. The *Aspergillus sydowii* fungi strain degraded about 94% polythene. According to the researchers, the study was carried out in vitro, there is a need to conduct it in vivo and then it can be introduced on a massive scale for plastic bioremediation.

Mangrove Soil Microbes
Similar studies were undertaken to study biodegradation of plastic cups against polythene by mangrove soil. They were incubated in different settings for 2, 4, 6, 9 months. Among the microbes, *Aspergillus* species recorded the most, contributing to the weight loss of the biomass within a month. Whereas among bacteria, *Pseudomonas* recorded the highest weight loss (Kathiresan, 2003)

Thus, mangrove soils are good source of microbes that can degrade plastics. Whether it be the temperature or salinity of the soil, it is hitherto unknown as to what helps to grow the certain population of microbes and fungi that can be used for the above process.

15.10 Conclusion and Future Considerations

Plastic waste is growing at an alarming rate which needs to be curtailed. Plastic waste can be recycled as well as degraded. Among them, bioremediation has become an innovative area of research as it is eco-friendly and cost-effective. Because of the remarkable capability of fungi to adjust to changing surroundings and to endure a wide range of contaminants, mycoremediation is grabbing a lot of interest these days. This review discussed copious amounts of information on different soil-borne fungi and the different classes of enzymes for plastic degradation. Some were able to degrade several types of plastic such as PE, PU, PET, PVC, PS, PP, PLA, and PCL, whereas some were able to degrade only PU films. Genus *Aspergillus* was the major versatile group found to be able to degrade a wide variety of plastic waste efficiently. Enzymes are the backbone for any biocatalytic reaction, in this case, any biodegradation process. Different classes of enzymes such as cutinases, lipases, esterases, peroxidases, laccases, depolymerases, dehydrogenases, protease, urease, and papain were found to be involved in plastic degradation. Furthermore, a bifunctional system that combines two enzymes that can break down more than one type of plastic in a single experiment, extending the range of degradability of plastic. PCL breakdown with Lip–Cut fusion enzyme was much faster, after 6 hours, than with these two enzymes alone.

Understanding the molecular enzymatic mechanisms and the factors that influence the degradation process are the two important aspects to be researched in order to develop efficient and effective degradation techniques. However, there are other aspects that need to be worked on too. One of the drawbacks of some of the plastic degradation studies is that they use commercial models such as Impranil to represent PU. But Impranil is not a perfect PU model which makes the studies not that accurate for realistic applications.

Most significantly, the technique of isolating fungal species producing plastic-degrading enzymes must be standardized and made considerably more effective, because the pure reproducing populations of these microbes may be recognized in the shortest possible time. In order to reduce the time required to breakdown the plastic polymer, standardization of this first stage is very much crucial which can increase the amount of enzyme produced by pure colonies. Also, manipulating the functional group and/or structure of plastic degrading enzyme's active site by employing, for example, site-directed mutagenesis can improve the enzyme stability so that they can easily accommodate plastic polymers and function even in harsh conditions. With the help of microbial biotechnology, identified fungi and their enzymes responsible for bioremediation can be grown in large scale to breakdown plastic waste in enormous amounts. Moreover, these fungi and/or their enzymes can be cloned into other robust microbes to develop recombinant microbes and enzymes for future use.

References

Abdel-Azeem, A. M., Abdel-Azeem, M. A., & Khalil, W. F. (2019). Endophytic fungi as a new source of antirheumatoid metabolites. In *Bioactive food as dietary interventions for arthritis and related inflammatory diseases* (pp. 355–384). Elsevier.

Accinelli, C., Abbas, H. K., Bruno, V., Nissen, L., Vicari, A., Bellaloui, N., Little, N. S., & Thomas Shier, W. (2020). Persistence in soil of microplastic films from ultra-thin compostable plastic bags and implications on soil Aspergillus flavus population. *Waste Management, 113*, 312–318. https://doi.org/10.1016/j.wasman.2020.06.011

Albertsson, A.-C., Andersson, S. O., & Karlsson, S. (1987). The mechanism of biodegradation of polyethylene. *Polymer Degradation and Stability, 18*(1), 73–87. https://doi.org/10.1016/0141-3910(87)90084-X

Ali, M. I., Ahmed, S., Robson, G., Javed, I., Ali, N., Atiq, N., & Hameed, A. (2014). Isolation and molecular characterization of polyvinyl chloride (PVC) plastic degrading fungal isolates: Fungal degradation of PVC plastics. *Journal of Basic Microbiology, 54*(1), 18–27. https://doi.org/10.1002/jobm.201200496

Ali, A., Guo, D., Mahar, A., Wang, P., Shen, F., Li, R., & Zhang, Z. (2017). Mycoremediation of potentially toxic trace elements—A biological tool for soil cleanup: A review. *Pedosphere, 27*(2), 205–222. https://doi.org/10.1016/S1002-0160(17)60311-4

Alisch-Mark, M., Herrmann, A., & Zimmermann, W. (2006). Increase of the hydrophilicity of polyethylene terephthalate fibres by hydrolases from Thermomonospora fusca and Fusarium solani f. sp. pisi. *Biotechnology Letters, 28*(10), 681–685. https://doi.org/10.1007/s10529-006-9041-7

Alshehrei, F. (2017). Biodegradation of synthetic and natural plastic by microorganisms. *Journal of Applied & Environmental Microbiology, 5*(1), 8–19.

Altaee, N., El-Hiti, G. A., Fahdil, A., Sudesh, K., & Yousif, E. (2016). Biodegradation of different formulations of polyhydroxybutyrate films in soil. *Springerplus, 5*(1), 762. https://doi.org/10.1186/s40064-016-2480-2

Álvarez-Barragán, J., Domínguez-Malfavón, L., Vargas-Suárez, M., González-Hernández, R., Aguilar-Osorio, G., & Loza-Tavera, H. (2016). Biodegradative activities of selected environmental fungi on a polyester polyurethane varnish and polyether polyurethane foams. *Applied and Environmental Microbiology, 82*(17), 5225–5235. https://doi.org/10.1128/AEM.01344-16

Amaike, S., & Keller, N. P. (2011). Aspergillus flavus. *Annual Review of Phytopathology, 49*(1), 107–133. https://doi.org/10.1146/annurev-phyto-072910-095221

Ameen, F., Moslem, M., Hadi, S., & Al-Sabri, A. E. (2015). Biodegradation of low density polyethylene (Ldpe) by mangrove fungi from the red sea coast. *Progress in Rubber, Plastics and Recycling Technology, 31*(2), 125–143. https://doi.org/10.1177/147776061503100204

American Chemical Society National Historic Chemical Landmarks. (2015). *Bakelite: The world's first synthetic plastic*. Retrieved 23 February 2015. Accessed 10 March 2022.

Ammala, A., Bateman, S., Dean, K., Petinakis, E., Sangwan, P., Wong, S., Yuan, Q., Yu, L., Patrick, C., & Leong, K. H. (2011). An overview of degradable and biodegradable polyolefins. *Progress in Polymer Science, 36*(8), 1015–1049. https://doi.org/10.1016/j.progpolymsci.2010.12.002

Amobonye, A., Bhagwat, P., Singh, S., & Pillai, S. (2021). Plastic biodegradation: Frontline microbes and their enzymes. *Science of the Total Environment, 759*, 143536. https://doi.org/10.1016/j.scitotenv.2020.143536

Annamalai, J., & Nallamuthu, T. (2011). Assessment of potential plastic degrading fungi isolated from soil buried plastic pieces at Coovamriver bank. *Research in Environment and Life Sciences, 4*(3), 125–128.

Aris, M. H., Annuar, M. S. M., & Ling, T. C. (2016). Lipase-mediated degradation of poly-ε-caprolactone in toluene: Behavior and its action mechanism. *Polymer Degradation and Stability, 133*, 182–191. https://doi.org/10.1016/j.polymdegradstab.2016.08.015

Arkatkar, A., Arutchelvi, J., Sudhakar, M., Bhaduri, S., Uppara, P. V., & Doble, M. (2009). Approaches to enhance the biodegradation of polyolefins. *The Open Environmental Engineering Journal, 2*(1), 68–80. https://doi.org/10.2174/1874829500902010068

Arkin, A. P., Cottingham, R. W., Henry, C. S., Harris, N. L., Stevens, R. L., Maslov, S., Dehal, P., Ware, D., Perez, F., Canon, S., Sneddon, M. W., Henderson, M. L., Riehl, W. J., Murphy-Olson, D., Chan, S. Y., Kamimura, R. T., Kumari, S., Drake, M. M., Brettin, T. S., Glass, E. M., Chivian, D., Gunter, D., Weston, D. J., Allen, B. H., Baumohl, J., Best, A. A., Bowen, B., Brenner, S. E., Bun, C. C., Chandonia, J.-M., Chia, J.-M., Colasanti, R., Conrad, N., Davis, J. J., Davison, B. H., DeJongh, M., Devoid, S., Dietrich, E., Dubchak, I., Edirisinghe, J. N., Fang, G., Faria, J. P., Frybarger, P. M., Gerlach, W., Gerstein, M., Greiner, A., Gurtowski, J., Haun, H. L., He, F., Jain, R., Joachimiak, M. P., Keegan, K. P., Kondo, S., Kumar, V., Land, M. L., Meyer, F., Mills, M., Novichkov, P. S., Oh, T., Olsen, G. J., Olson, R., Parrello, B., Pasternak, S., Pearson, E., Poon, S. S., Price, G. A., Ramakrishnan, S., Ranjan, P., Ronald, P. C., Schatz, M. C., Seaver, S. M. D., Shukla, M., Sutormin, R. A., Syed, M. H., Thomason, J., Tintle, N. L., Wang, D., Xia, F., Yoo, H., Yoo, S., & Yu, D. (2018). KBase: The United States department of energy systems biology knowledgebase. *Nature Biotechnology, 36*(7), 566–569. https://doi.org/10.1038/nbt.4163

Arora, P. K., & Bae, H. (2014). Integration of bioinformatics to biodegradation. *Biological Procedures Online, 16*(1), 8. https://doi.org/10.1186/1480-9222-16-8

Arregui, L., Ayala, M., Gómez-Gil, X., Gutiérrez-Soto, G., Hernández-Luna, C. E., Herrera de los Santos, M., Levin, L., Rojo-Domínguez, A., Romero-Martínez, D., Saparrat, M. C. N., Trujillo-Roldán, M. A., & Valdez-Cruz, N. A. (2019). Laccases: Structure, function, and potential application in water bioremediation. *Microbial Cell Factories, 18*(1), 200. https://doi.org/10.1186/s12934-019-1248-0

Azubuike, C. C., Chikere, C. B., & Okpokwasili, G. C. (2016). Bioremediation techniques–classification based on site of application: Principles, advantages, limitations and prospects. *World Journal of Microbiology and Biotechnology, 32*(11), 180. https://doi.org/10.1007/s11274-016-2137-x

Baker, S. E. (2006). *Aspergillus niger* genomics: Past, present and into the future. *Medical Mycology, 44*(s1), 17–21. https://doi.org/10.1080/13693780600921037

Baker, P. J., Poultney, C., Liu, Z., Gross, R., & Montclare, J. K. (2012). Identification and comparison of cutinases for synthetic polyester degradation. *Applied Microbiology and Biotechnology, 93*(1), 229–240. https://doi.org/10.1007/s00253-011-3402-4

Balasubramanian, V., Natarajan, K., Rajeshkannan, V., & Perumal, P. (2014). Enhancement of in vitro high-density polyethylene (HDPE) degradation by physical, chemical, and biological treatments. *Environmental Science and Pollution Research, 21*(21), 12549–12562. https://doi.org/10.1007/s11356-014-3191-2

Baldrian, P. (2008). Wood-inhabiting ligninolytic basidiomycetes in soils: Ecology and constraints for applicability in bioremediation. *Fungal Ecology, 1*(1), 4–12. https://doi.org/10.1016/j.funeco.2008.02.001

Barratt, S. R., Ennos, A. R., Greenhalgh, M., Robson, G. D., & Handley, P. S. (2003). Fungi are the predominant micro-organisms responsible for degradation of soil-buried polyester polyurethane over a range of soil water holding capacities. *Journal of Applied Microbiology, 95*(1), 78–85. https://doi.org/10.1046/j.1365-2672.2003.01961.x

Barth, M., Oeser, T., Wei, R., Then, J., Schmidt, J., & Zimmermann, W. (2015). Effect of hydrolysis products on the enzymatic degradation of polyethylene terephthalate nanoparticles by a polyester hydrolase from Thermobifidafusca. *Biochemical Engineering Journal, 93*, 222–228. https://doi.org/10.1016/j.bej.2014.10.012

Bazoobandi, M., Rahimi, H., & Karimi-Shahri, M.-R. (2020). Saffron crop protection. In *Saffron* (pp. 169–185). Elsevier.

Bhardwaj, H., Gupta, R., & Tiwari, A. (2013). Communities of microbial enzymes associated with biodegradation of plastics. *Journal of Polymers and the Environment, 21*(2), 575–579. https://doi.org/10.1007/s10924-012-0456-z

Bosco, F., & Mollea, C. (2019). Mycoremediation in soil. In H. Saldarriaga-Noreña, M. Alfonso Murillo-Tovar, R. Farooq, R. Dongre, & S. Riaz (Eds.), *Environmental chemistry and recent pollution control approaches*. IntechOpen.

Brunner, I., Fischer, M., Rüthi, J., Stierli, B., & Frey, B. (2018). Ability of fungi isolated from plastic debris floating in the shoreline of a lake to degrade plastics. *PLoS One, 13*(8), e0202047. https://doi.org/10.1371/journal.pone.0202047

Brzeska, J., Heimowska, A., Sikorska, W., Jasińska-Walc, L., Kowalczuk, M., & Rutkowska, M. (2015). Chemical and enzymatic hydrolysis of polyurethane/polylactide blends. *International Journal of Polymer Science, 2015*, 1–8. https://doi.org/10.1155/2015/795985

Bumpus, J. A., & Aust, S. D. (1987). Biodegradation of environmental pollutants by the white rot fungus Phanerochaete chrysosporium: Involvement of the lignin degrading system. *BioEssays, 6*(4), 166–170. https://doi.org/10.1002/bies.950060405

Cacciari, I., Quatrini, P., Zirletta, G., Mincione, E., Vinciguerra, V., Lupattelli, P., & GiovannozziSermanni, G. (1993). Isotactic polypropylene biodegradation by a microbial community: Physicochemical characterization of metabolites produced. *Applied and Environmental Microbiology, 59*(11), 3695–3700. https://doi.org/10.1128/aem.59.11.3695-3700.1993

Carolina, A. E., Amanda Inés, M. G., & Betty Lucy, L. O. (2004). Poly (vinylalcohol-co-ethylene) biodegradation on semi solid fermentation by Phanerochaete chrysosporium. *Acta Farmaceutica Bonaerense, 23*(2), 123–128.

Carvalho, C. M. L., Aires-Barros, M. R., & Cabral, J. M. S. (1999). Cutinase: From molecular level to bioprocess development. *Biotechnology and Bioengineering, 66*(1), 17–34. https://doi.org/10.1002/(SICI)1097-0290(1999)66:1<17::AID-BIT2>3.0.CO;2-F

Chanda, A., Gummadidala, P. M., & Gomaa, O. M. (2016). Mycoremediation with mycotoxin producers: A critical perspective. *Applied Microbiology and Biotechnology, 100*(1), 17–29. https://doi.org/10.1007/s00253-015-7032-0

Chandran, V., Shaji, H., & Mathew, L. (2020). Endophytic microbial influence on plant stress responses. In *Microbial endophytes* (pp. 161–193). Elsevier.

Chatterjee, S., Sarma, M. K., Deb, U., Steinhauser, G., Walther, C., & Gupta, D. K. (2017). Mushrooms: From nutrition to mycoremediation. *Environmental Science and Pollution Research, 24*(24), 19480–19493. https://doi.org/10.1007/s11356-017-9826-3

Chaudhary, A. K., & Vijayakumar, R. P. (2020). Studies on biological degradation of polystyrene by pure fungal cultures. *Environment, Development and Sustainability, 22*(5), 4495–4508. https://doi.org/10.1007/s10668-019-00394-5

Chen, S., Su, L., Chen, J., & Wu, J. (2013). Cutinase: Characteristics, preparation, and application. *Biotechnology Advances, 31*(8), 1754–1767. https://doi.org/10.1016/j.biotechadv.2013.09.005

Constantin, M., Iuliana, R., Vasilescu, G., Arsene, M. L., & Luiza, J. (2012). Colonization and degradation of polyethylene composites by fungal strains isolated. *Scientific Bulletin. Series F. Biotechnologies, 16*, 109–112.

Cosgrove, L., McGeechan, P. L., Robson, G. D., & Handley, P. S. (2007). Fungal communities associated with degradation of polyester polyurethane in soil. *Applied and Environmental Microbiology, 73*(18), 5817–5824. https://doi.org/10.1128/AEM.01083-07

Couto, S. R., Longo, M. A., Cameselle, C., & Sanromán, A. (1998). Influence of some inducers on activity of ligninolytic enzymes from corncob cultures of Phanerochaete chrysosporium in semi-solid-state conditions. In *Progress in biotechnology* (pp. 703–708). Elsevier.

Crabbe, J. R., Campbell, J. R., Thompson, L., Walz, S. L., & Schultz, W. W. (1994). Biodegradation of a colloidal ester-based polyurethane by soil fungi. *International Biodeterioration & Biodegradation, 33*(2), 103–113. https://doi.org/10.1016/0964-8305(94)90030-2

Cregut, M., Bedas, M., Durand, M.-J., & Thouand, G. (2013). New insights into polyurethane biodegradation and realistic prospects for the development of a sustainable waste recycling process. *Biotechnology Advances, 31*(8), 1634–1647. https://doi.org/10.1016/j.biotechadv.2013.08.011

Črešnar, B., & Petrič, Š. (2011). Cytochrome P450 enzymes in the fungal kingdom. *Biochimica et Biophysica Acta (BBA) – Proteins and Proteomics, 1814*(1), 29–35. https://doi.org/10.1016/j.bbapap.2010.06.020

Danso, D., Chow, J., & Streit, W. R. (2019). Plastics: Environmental and biotechnological perspectives on microbial degradation. *Applied and Environmental Microbiology, 85*(19), e01095–e01019. https://doi.org/10.1128/AEM.01095-19

Das, M. P., & Kumar, S. (2014). Microbial deterioration of low density polyethylene by Aspergillus and Fusarium sp. *International Journal of ChemTech Research, 6*(1), 299–305.

Das, M. P., Kumar, S., & Das, J. (2018). Fungal-mediated deterioration and biodegradation study of low-density polyethylene (LDPE) isolated from municipal dump yard in Chennai, India. *Energy, Ecology and Environment, 3*(4), 229–236. https://doi.org/10.1007/s40974-018-0085-z

de Albuquerque, T. L., Silva, J. d. S., de Macedo, A. C., Gonçalves, L. R. B., & Rocha, M. V. P. (2019). Biotechnological strategies for the lignin-based biorefinery valorization. In *Reference module in chemistry, molecular sciences and chemical engineering* (p. B9780124095472146000). Elsevier.

Deepa, N., & Sreenivasa, M. Y. (2019). Biocontrol strategies for effective management of phytopathogenic fungi associated with cereals. In *New and future developments in microbial biotechnology and bioengineering* (pp. 177–189). Elsevier.

Deepika, S., & Jaya, M. R. (2015). Biodegradation of low density polyethylene by microorganisms from garbage soil. *Journal of Experimental Biology and Agricultural Sciences, 3*, 1–5.

Deguchi, T., Kitaoka, Y., Kakezawa, M., & Nishida, T. (1998). Purification and characterization of a nylon-degrading enzyme. *Applied and Environmental Microbiology, 64*(4), 1366–1371. https://doi.org/10.1128/AEM.64.4.1366-1371.1998

Deshmukh, R., Khardenavis, A. A., & Purohit, H. J. (2016). Diverse metabolic capacities of fungi for bioremediation. *Indian Journal of Microbiology, 56*(3), 247–264. https://doi.org/10.1007/s12088-016-0584-6

Dubey, T. (2016). Indoor air pollution due to mycoflora causing acute lower respiratory infections. In *The microbiology of respiratory system infections* (pp. 95–112). Elsevier.

Egmond, M. (2000). Fusarium solani pisi cutinase. *Biochimie, 82*(11), 1015–1021. https://doi.org/10.1016/S0300-9084(00)01183-4

El-Morsy, E. (2017). Biodegradative activities of fungal isolates from plastic contaminated soils. *Mycosphere, 8*(8), 1071–1087. https://doi.org/10.5943/mycosphere/8/8/13

El-Shafei, H. A., Abd El-Nasser, N. H., Kansoh, A. L., & Ali, A. M. (1998). Biodegradation of disposable polyethylene by fungi and Streptomyces species. *Polymer Degradation and Stability, 62*(2), 361–365. https://doi.org/10.1016/S0141-3910(98)00019-6

Esmaeili, A., Pourbabaee, A. A., Alikhani, H. A., Shabani, F., & Esmaeili, E. (2013). Biodegradation of low-density polyethylene (LDPE) by mixed culture of Lysinibacillusxylanilyticus and Aspergillus niger in soil. *PLoS One, 8*(9), e71720. https://doi.org/10.1371/journal.pone.0071720

Esteban, S., Gorga, M., Petrovic, M., González-Alonso, S., Barceló, D., & Valcárcel, Y. (2014). Analysis and occurrence of endocrine-disrupting compounds and estrogenic activity in the surface waters of Central Spain. *Science of the Total Environment, 466–467*, 939–951. https://doi.org/10.1016/j.scitotenv.2013.07.101

Ferrari, R., Gautier, V., & Silar, P. (2021). Lignin degradation by ascomycetes. In *Advances in botanical research* (pp. 77–113). Elsevier.

Filip, Z. (1979). Polyurethane as the sole nutrient source for Aspergillus niger and Cladosporium herbarum. *European Journal of Applied Microbiology and Biotechnology, 7*(3), 277–280. https://doi.org/10.1007/BF00498022

Frazer, A. C. (1994). O-demethylation and other transformations of aromatic compounds by acetogenic bacteria. In H. L. Drake (Ed.), *Acetogenesis* (pp. 445–483). Springer.

Furtwengler, P., Perrin, R., Redl, A., & Avérous, L. (2017). Synthesis and characterization of polyurethane foams derived of fully renewable polyester polyols from sorbitol. *European Polymer Journal, 97*, 319–327. https://doi.org/10.1016/j.eurpolymj.2017.10.020

Gajendiran, A., Krishnamoorthy, S., & Abraham, J. (2016). Microbial degradation of low-density polyethylene (LDPE) by Aspergillus clavatus strain JASK1 isolated from landfill soil. *3 Biotech, 6*(1), 52. https://doi.org/10.1007/s13205-016-0394-x

Gao, Y., Li, Q., Ling, W., & Zhu, X. (2011). Arbuscular mycorrhizal phytoremediation of soils contaminated with phenanthrene and pyrene. *Journal of Hazardous Materials, 185*(2–3), 703–709. https://doi.org/10.1016/j.jhazmat.2010.09.076

Gholami-Shabani, M., Zamani, S., Moosa, H., Shams-Ghahfarokhi, M., Jamzivar, F., & Razzaghi-Abyaneh, M. (2018). Recent advances in fungal infections of the central nervous system: From etiology to diagnosis and management. In *The microbiology of central nervous system infections* (pp. 215–259). Elsevier.

Ghosh, S. K., Pal, S., & Ray, S. (2013). Study of microbes having potentiality for biodegradation of plastics. *Environmental Science and Pollution Research, 20*(7), 4339–4355. https://doi.org/10.1007/s11356-013-1706-x

Giardina, P., Faraco, V., Pezzella, C., Piscitelli, A., Vanhulle, S., & Sannia, G. (2010). Laccases: A never-ending story. *Cellular and Molecular Life Sciences, 67*(3), 369–385. https://doi.org/10.1007/s00018-009-0169-1

Gigli, M., Quartinello, F., Soccio, M., Pellis, A., Lotti, N., Guebitz, G. M., Licoccia, S., & Munari, A. (2019). Enzymatic hydrolysis of poly(1,4-butylene 2,5-thiophenedicarboxylate) (PBTF) and poly(1,4-butylene 2,5-furandicarboxylate) (PBF) films: A comparison of mechanisms. *Environment International, 130*, 104852. https://doi.org/10.1016/j.envint.2019.05.046

Gilan, I., & Sivan, A. (2013). Effect of proteases on biofilm formation of the plastic-degrading actinomycete *Rhodococcus ruber* C208. *FEMS Microbiology Letters, 342*(1), 18–23. https://doi.org/10.1111/1574-6968.12114

Gohain, A., Manpoong, C., Linggi, B., Saikia, R., & De Mandal, S. (2020). Microbial endophytes of plants: Diversity, benefits, and their interaction with host. In *Recent advancements in microbial diversity* (pp. 19–36).

Gómez-Méndez, L. D., Moreno-Bayona, D. A., Poutou-Piñales, R. A., Salcedo-Reyes, J. C., Pedroza-Rodríguez, A. M., Vargas, A., & Bogoya, J. M. (2018). Biodeterioration of plasma pretreated LDPE sheets by Pleurotus ostreatus. *PLoS One, 13*(9), e0203786. https://doi.org/10.1371/journal.pone.0203786

Grima, S., Bellon-Maurel, V., Feuilloley, P., & Silvestre, F. (2000). Aerobic biodegradation of polymers in solid-state conditions: A Review of environmental and physicochemical parameter settings in laboratory simulations. *Journal of Polymers and the Environment, 8*(4), 183–195. https://doi.org/10.1023/A:1015297727244

Griswold, K. E., Mahmood, N. A., Iverson, B. L., & Georgiou, G. (2003). Effects of codon usage versus putative 5′-mRNA structure on the expression of Fusarium solani cutinase in the Escherichia coli cytoplasm. *Protein Expression and Purification, 27*(1), 134–142. https://doi.org/10.1016/S1046-5928(02)00578-8

Gu, J. D., Ford, T. E., Mitton, D. B., & Mitchell, R. (2000). Microbial degradation and deterioration of polymeric materials. In Review (Ed.), *The Uhlig corrosion handbook* (2nd ed., p. 439). Wiley.

Hamilton, J. D., Reinert, K. H., Hagan, J. V., & Lord, W. V. (1995). Polymers as solid waste in municipal landfills. *Journal of the Air & Waste Management Association, 45*(4), 247–251. https://doi.org/10.1080/10473289.1995.10467364

Hari, S. (2019). Review on effect of fungi on plastic degradation. *JRAR, 6*(1), 261–265.

Harms, H., Schlosser, D., & Wick, L. Y. (2011). Untapped potential: Exploiting fungi in bioremediation of hazardous chemicals. *Nature Reviews. Microbiology, 9*(3), 177–192. https://doi.org/10.1038/nrmicro2519

Haug, K., Salek, R. M., Conesa, P., Hastings, J., de Matos, P., Rijnbeek, M., Mahendraker, T., Williams, M., Neumann, S., Rocca-Serra, P., Maguire, E., González-Beltrán, A., Sansone, S.-A., Griffin, J. L., & Steinbeck, C. (2013). MetaboLights—An open-access general-purpose repository for metabolomics studies and associated meta-data. *Nucleic Acids Research, 41*(D1), D781–D786. https://doi.org/10.1093/nar/gks1004

Hegyesi, N., Zhang, Y., Kohári, A., Polyák, P., Sui, X., & Pukánszky, B. (2019). Enzymatic degradation of PLA/cellulose nanocrystal composites. *Industrial Crops and Products, 141*, 111799. https://doi.org/10.1016/j.indcrop.2019.111799

Hektor, H. J., & Scholtmeijer, K. (2005). Hydrophobins: Proteins with potential. *Current Opinion in Biotechnology, 16*(4), 434–439. https://doi.org/10.1016/j.copbio.2005.05.004

Heredia, A. (2003). Biophysical and biochemical characteristics of cutin, a plant barrier biopolymer. *Biochimica et Biophysica Acta (BBA) – General Subjects, 1620*(1–3), 1–7. https://doi.org/10.1016/S0304-4165(02)00510-X

Ho, B. T., Roberts, T. K., & Lucas, S. (2018). An overview on biodegradation of polystyrene and modified polystyrene: The microbial approach. *Critical Reviews in Biotechnology, 38*(2), 308–320. https://doi.org/10.1080/07388551.2017.1355293

Hofrichter, M. (2002). Review: Lignin conversion by manganese peroxidase (MnP). *Enzyme and Microbial Technology, 30*(4), 454–466. https://doi.org/10.1016/S0141-0229(01)00528-2

Hofrichter, M., Ullrich, R., Pecyna, M. J., Liers, C., & Lundell, T. (2010). New and classic families of secreted fungal heme peroxidases. *Applied Microbiology and Biotechnology, 87*(3), 871–897. https://doi.org/10.1007/s00253-010-2633-0

Howard, G. T. (2002). Biodegradation of polyurethane: A review. *International Biodeterioration & Biodegradation, 49*(4), 245–252. https://doi.org/10.1016/S0964-8305(02)00051-3

Howard, G. T. (2012). Polyurethane biodegradation. In S. N. Singh (Ed.), *Microbial degradation of xenobiotics* (pp. 371–394). Springer.

Howard, G. T., Vicknair, J., & Mackie, R. I. (2001). Sensitive plate assay for screening and detection of bacterial polyurethanase activity: Detection of polyurethanase activity. *Letters in Applied Microbiology, 32*(3), 211–214. https://doi.org/10.1046/j.1472-765x.2001.00887.x

Hu, X., Gao, Z., Wang, Z., Su, T., Yang, L., & Li, P. (2016). Enzymatic degradation of poly(butylene succinate) by cutinase cloned from fusarium solani. *Polymer Degradation and Stability, 134*, 211–219. https://doi.org/10.1016/j.polymdegradstab.2016.10.012

Hu, X., Su, T., Li, P., & Wang, Z. (2018). Blending modification of PBS/PLA and its enzymatic degradation. *Polymer Bulletin, 75*(2), 533–546. https://doi.org/10.1007/s00289-017-2054-7

Hua, F., Wang, H. Q., Li, Y., & Zhao, Y. C. (2013). Trans-membrane transport of n-octadecane by pseudomonas sp. DG17. *Journal of Microbiology, 51*(6), 791–799. https://doi.org/10.1007/s12275-013-3259-6

Ibrahim, N. I., Anwar, M., Khalid, M. H., Ismail, M. S., Hamzah, M. M., & Toshiaki, N. K. (2009). Polyester-polyurethane biodegradation by Alternaria Solani, isolated from Northern Jordan. *Advances in Environmental Biology, 3*(2), 162–170.

Ibrahim, I. N., Maraqa, A., Hameed, K. M., Saadoun, I. M., & Maswadeh, H. M. (2011). Assessment of potential plastic-degrading fungi in Jordanian habitats. *Turkish Journal of Biology, 35*(5), 551–557. https://doi.org/10.3906/biy-0901-9

Iiyoshi, Y., Tsutsumi, Y., & Nishida, T. (1998). Polyethylene degradation by lignin-degrading fungi and manganese peroxidase. *Journal of Wood Science, 44*(3), 222–229. https://doi.org/10.1007/BF00521967

Iram, D., Riaz, R., & Iqbal, R. (2019). Usage of potential micro-organisms for degradation of plastics. Open. *Journal of Environmental Biology, 4*(1), 007–015. https://doi.org/10.17352/ojeb.000010

Jafari, M., Danesh, Y. R., Goltapeh, E. M., & Varma, A. (2013). Bioremediation and genetically modified organisms. In E. M. Goltapeh, Y. R. Danesh, & A. Varma (Eds.), *Fungi as Bioremediators* (pp. 433–451). Springer.

Jelsch, C., Longhi, S., & Cambillau, C. (1998). Packing forces in nine crystal forms of cutinase. *Proteins, 31*(3), 320–333.

Jenner, L. C., Rotchell, J. M., Bennett, R. T., Cowen, M., Tentzeris, V., & Sadofsky, L. R. (2022). Detection of microplastics in human lung tissue using μFTIR spectroscopy. *Science of the Total Environment, 831*, 154907. https://doi.org/10.1016/j.scitotenv.2022.154907

Jeon, H., Durairaj, P., Lee, D., Ahsan, M. M., & Yun, H. (2016). Improved NADPH regeneration for fungal cytochrome P450 monooxygenase by co-expressing bacterial glucose dehydrogenase in resting-cell biotransformation of recombinant yeast. *Journal of Microbiology and Biotechnology, 26*(12), 2076–2086. https://doi.org/10.4014/jmb.1605.05090

Jeyakumar, D., Chirsteen, J., & Doble, M. (2013). Synergistic effects of pretreatment and blending on fungi mediated biodegradation of polypropylenes. *Bioresource Technology, 148*, 78–85. https://doi.org/10.1016/j.biortech.2013.08.074

Jog, J. P. (1995). Crystallization of polyethyleneterephthalate. *Journal of Macromolecular Science, Part C: Polymer Reviews, 35*(3), 531–553. https://doi.org/10.1080/15321799508014598

Johnnie, D. A., Issac, R., & Prabha, M. L. (2021). Bio efficacy assay of laccase isolated and characterized from Trichoderma viride in biodegradation of low density polyethylene (LDPE) and textile industrial effluent dyes. *Journal of Pure and Applied Microbiology, 15*(1), 410–420. https://doi.org/10.22207/JPAM.15.1.38

Johnson, E. A., & Echavarri-Erasun, C. (2011). Yeast biotechnology. In *The yeasts* (pp. 21–44). Elsevier.

Jung, H.-W., Yang, M.-K., & Su, R.-C. (2018). Purification, characterization, and gene cloning of an Aspergillus fumigatus polyhydroxybutyrate depolymerase used for degradation of polyhydroxybutyrate, polyethylene succinate, and polybutylene succinate. *Polymer Degradation and Stability, 154*, 186–194. https://doi.org/10.1016/j.polymdegradstab.2018.06.002

Jyoti, S., & Gupta, K. C. (2014). Screening and identification of low density polyethylene (LDPE) degrading soil fungi isolated from polythene polluted sites around Gwalior city (MP). *International Journal of Current Microbiology and Applied Sciences, 3*(6), 443–448.

Kaczmarek, H., & Bajer, K. (2007). Biodegradation of plasticized poly(vinyl chloride) containing cellulose. *Journal of Polymer Science: Part B: Polymer Physics, 45*, 903–918.

Karamanlioglu, M., Houlden, A., & Robson, G. D. (2014). Isolation and characterisation of fungal communities associated with degradation and growth on the surface of poly(lactic) acid (PLA) in soil and compost. *International Biodeterioration & Biodegradation, 95*, 301–310. https://doi.org/10.1016/j.ibiod.2014.09.006

Karich, A., Ullrich, R., Scheibner, K., & Hofrichter, M. (2017). Fungal unspecific peroxygenases oxidize the majority of organic EPA priority pollutants. *Frontiers in Microbiology, 8*, 1463. https://doi.org/10.3389/fmicb.2017.01463

Kasuya, K., Takagi, K., Ishiwatari, S., Yoshida, Y., & Doi, Y. (1998). Biodegradabilities of various aliphatic polyesters in natural waters. *Polymer Degradation and Stability, 59*(1–3), 327–332. https://doi.org/10.1016/S0141-3910(97)00155-9

Kathiresan, K. (2003). Polythene and plastics-degrading microbes from the mangrove soil. *Revista de biologia tropical, 51*(3–4), 629–633.

Kaushal, J., Khatri, M., & Arya, S. K. (2021). Recent insight into enzymatic degradation of plastics prevalent in the environment: A mini review. *Cleaner Engineering and Technology, 2*, 100083. https://doi.org/10.1016/j.clet.2021.100083

Kawai, F., Kawabata, T., & Oda, M. (2019). Current knowledge on enzymatic PET degradation and its possible application to waste stream management and other fields. *Applied Microbiology and Biotechnology, 103*(11), 4253–4268. https://doi.org/10.1007/s00253-019-09717-y

Kessner, D., Chambers, M., Burke, R., Agus, D., & Mallick, P. (2008). ProteoWizard: Open source software for rapid proteomics tools development. *Bioinformatics, 24*(21), 2534–2536. https://doi.org/10.1093/bioinformatics/btn323

Khan, I., Ray Dutta, J., & Ganesan, R. (2017a). Lactobacillus sps. lipase mediated poly (ε-caprolactone) degradation. *International Journal of Biological Macromolecules, 95*, 126–131. https://doi.org/10.1016/j.ijbiomac.2016.11.040

Khan, S., Nadir, S., Shah, Z. U., Shah, A. A., Karunarathna, S. C., Xu, J., Khan, A., Munir, S., & Hasan, F. (2017b). Biodegradation of polyester polyurethane by Aspergillus tubingensis. *Environmental Pollution, 225*, 469–480. https://doi.org/10.1016/j.envpol.2017.03.012

Khan, S., Nadir, S., Dong, Y., Schaefer, D. A., Mortimer, P. E., Gui, H., Khan, A., Yu, M., Iqbal, S., Sheng, J., & Xu, J. (2020). Biodegradation of polyester polyurethane by *Aspergillus flavus* G10. *Bioengineering* 241.

Khatoon, N., Jamal, A., & Ali, M. I. (2019). Lignin peroxidase isoenzyme: A novel approach to biodegrade the toxic synthetic polymer waste. *Environmental Technology, 40*(11), 1366–1375. https://doi.org/10.1080/09593330.2017.1422550

Khruengsai, S., Sripahco, T., & Pripdeevech, P. (2021). Low-density polyethylene film biodegradation potential by fungal species from Thailand. *Journal of Fungi, 7*(8), 594. https://doi.org/10.3390/jof7080594

Kim, D. Y., & Rhee, Y. H. (2003). Biodegradation of microbial and synthetic polyesters by fungi. *Applied Microbiology and Biotechnology, 61*(4), 300–308. https://doi.org/10.1007/s00253-002-1205-3

Kirbaş, Z., Keskin, N., & Güner, A. (1999). Biodegradation of polyvinylchloride (PVC) by white rot fungi. *Bulletin of Environmental Contamination and Toxicology, 63*(3), 335–342. https://doi.org/10.1007/s001289900985

Kolattukudy, P. E. (1981). Structure, biosynthesis, and biodegradation of Cutin and Suberin. *Annual Review of Plant Physiology, 32*(1), 539–567. https://doi.org/10.1146/annurev.pp.32.060181.002543

Konduri, M. K., Anupam, K. S., Vivek, J. S., Kumar, D. B. R., & Narasu, M. L. (2010). Synergistic effect of chemical and photo treatment on the rate of biodegradation of high density polyethylene by indigenous fungal isolates. *International Journal of Biotechnology & Biochemistry, 6*(2), 157–175.

Krueger, M. C., Hofmann, U., Moeder, M., & Schlosser, D. (2015). Potential of wood-rotting fungi to attack polystyrene sulfonate and its depolymerisation by Gloeophyllum trabeum via hydroquinone-driven Fenton chemistry. *PLoS One, 10*(7), e0131773. https://doi.org/10.1371/journal.pone.0131773

Krzan, A., Hemjinda, S., Miertus, S., Corti, A., & Chiellini, E. (2006). Standardization and certification in the area of environmentally degradable plastics. *Polymer Degradation and Stability, 91*(12), 2819–2833. https://doi.org/10.1016/j.polymdegradstab.2006.04.034

Kulkarni, S., Nene, S., & Joshi, K. (2017). Production of Hydrophobins from fungi. *Process Biochemistry, 61*, 1–11. https://doi.org/10.1016/j.procbio.2017.06.012

Kulshreshtha, S., Mathur, N., & Bhatnagar, P. (2013). Mycoremediation of paper, pulp and cardboard industrial wastes and pollutants. In E. M. Goltapeh, Y. R. Danesh, & A. Varma (Eds.), *Fungi as Bioremediators* (pp. 77–116). Springer.

Kunlere, I. O., Fagade, O. E., & Nwadike, B. I. (2019). Biodegradation of low density polyethylene (LDPE) by certain indigenous bacteria and fungi. *International Journal of Environmental Studies, 76*(3), 428–440. https://doi.org/10.1080/00207233.2019.1579586

Lassouane, F., Aït-Amar, H., Amrani, S., & Rodriguez-Couto, S. (2019). A promising laccase immobilization approach for bisphenol a removal from aqueous solutions. *Bioresource Technology, 271*, 360–367. https://doi.org/10.1016/j.biortech.2018.09.129

Lauber, C., Schwarz, T., Nguyen, Q. K., Lorenz, P., Lochnit, G., & Zorn, H. (2017). Identification, heterologous expression and characterization of a dye-decolorizing peroxidase of Pleurotus sapidus. *AMB Express, 7*(1), 164. https://doi.org/10.1186/s13568-017-0463-5

Leonardi, V., Šašek, V., Petruccioli, M., D'Annibale, A., Erbanová, P., & Cajthaml, T. (2007). Bioavailability modification and fungal biodegradation of PAHs in aged industrial soils. *International Biodeterioration & Biodegradation, 60*(3), 165–170. https://doi.org/10.1016/j.ibiod.2007.02.004

Leslie, H. A., van Velzen, M. J. M., Brandsma, S. H., Vethaak, A. D., Garcia-Vallejo, J. J., & Lamoree, M. H. (2022). Discovery and quantification of plastic particle pollution in human blood. *Environment International, 163*, 107199. https://doi.org/10.1016/j.envint.2022.107199

Li, D., Zhou, L., Wang, X., He, L., & Yang, X. (2019). Effect of crystallinity of polyethylene with different densities on breakdown strength and conductance property. *Materials, 12*(11), 1746. https://doi.org/10.3390/ma12111746

Liebminger, S., Eberl, A., Sousa, F., Heumann, S., Fischer-Colbrie, G., Cavaco-Paulo, A., & Guebitz, G. M. (2007). Hydrolysis of PET and bis-(benzoyloxyethyl) terephthalate with a new polyesterase from Penicillium citrinum. *Biocatalysis and Biotransformation, 25*(2–4), 171–177. https://doi.org/10.1080/10242420701379734

Lin, W., Brauers, G., Ebel, R., Wray, V., Berg, A., & Sudarsono, P. P. (2003). Novel Chromone derivatives from the fungus Aspergillus versicolor isolated from the marine sponge Xestospongia exigua. *Journal of Natural Products, 66*(1), 57–61. https://doi.org/10.1021/np020196b

Lipşa, R., Tudorachi, N., Grigoraş, A., Vasile, C., & Grădinariu, P. (2015). Study on poly (vinyl alcohol) copolymers biodegradation. *Memoirs of the Scientific Sections of the Romanian Academy, 38*(1), 7–25.

Liu, Z., Gosser, Y., Baker, P. J., Ravee, Y., Lu, Z., Alemu, G., Li, H., Butterfoss, G. L., Kong, X.-P., Gross, R., & Montclare, J. K. (2009). Structural and functional studies of Aspergillus oryzae cutinase: Enhanced thermostability and hydrolytic activity of synthetic ester and polyester degradation. *Journal of the American Chemical Society, 131*(43), 15711–15716. https://doi.org/10.1021/ja9046697

Liu, M., Zhang, T., Long, L., Zhang, R., & Ding, S. (2019). Efficient enzymatic degradation of poly (ε-caprolactone) by an engineered bifunctional lipase-cutinase. *Polymer Degradation and Stability, 160*, 120–125. https://doi.org/10.1016/j.polymdegradstab.2018.12.020

Löder, M. G. J., & Gerdts, G. (2015). Methodology used for the detection and identification of microplastics—A critical appraisal. In M. Bergmann, L. Gutow, & M. Klages (Eds.), *Marine anthropogenic litter* (pp. 201–227). Springer. https://doi.org/10.1007/978-3-319-16510-3

Lourenço, A., Ferreira, A., Veiga, N., Machado, I., Pereira, M. O., & Azevedo, N. F. (2012). BiofOmics: A web platform for the systematic and standardized collection of high-throughput biofilm data. *PLoS One, 7*(6), e39960. https://doi.org/10.1371/journal.pone.0039960

Lucas, N., Bienaime, C., Belloy, C., Queneudec, M., Silvestre, F., & Nava-Saucedo, J.-E. (2008). Polymer biodegradation: Mechanisms and estimation techniques – A review. *Chemosphere, 73*(4), 429–442. https://doi.org/10.1016/j.chemosphere.2008.06.064

Lukić, B., Panico, A., Huguenot, D., Fabbricino, M., van Hullebusch, E. D., & Esposito, G. (2017). A review on the efficiency of landfarming integrated with composting as a soil remediation treatment. *Environmental Technology Reviews, 6*(1), 94–116. https://doi.org/10.1080/21622515.2017.1310310

Ma, Q., Shi, K., Su, T., & Wang, Z. (2020). Biodegradation of Polycaprolactone (PCL) with different molecular weights by Candida antarctica lipase. *Journal of Polymers and the Environment, 28*(11), 2947–2955. https://doi.org/10.1007/s10924-020-01826-4

MacLeod, M., Arp, H. P. H., Tekman, M. B., & Jahnke, A. (2021). The global threat from plastic pollution. *Science, 373*(6550), 61–65. https://doi.org/10.1126/science.abg5433

Maeda, H., Yamagata, Y., Abe, K., Hasegawa, F., Machida, M., Ishioka, R., Gomi, K., & Nakajima, T. (2005). Purification and characterization of a biodegradable plastic-degrading enzyme from Aspergillus oryzae. *Applied Microbiology and Biotechnology, 67*(6), 778–788. https://doi.org/10.1007/s00253-004-1853-6

Magnin, A., Hoornaert, L., Pollet, E., Laurichesse, S., Phalip, V., & Avérous, L. (2019). Isolation and characterization of different promising fungi for biological waste management of polyurethanes. *Microbial Biotechnology, 12*(3), 544–555. https://doi.org/10.1111/1751-7915.13346

Magnin, A., Pollet, E., Phalip, V., & Avérous, L. (2020). Evaluation of biological degradation of polyurethanes. *Biotechnology Advances, 39*, 107457. https://doi.org/10.1016/j.biotechadv.2019.107457

Mäkelä, M. R., Marinović, M., Nousiainen, P., Liwanag, A. J. M., Benoit, I., Sipilä, J., Hatakka, A., de Vries, R. P., & Hildén, K. S. (2015). Aromatic metabolism of filamentous fungi in relation to the presence of aromatic compounds in plant biomass. In *Advances in applied microbiology* (pp. 63–137). Elsevier.

Mäkelä, M. R., Hildén, K. S., & Kuuskeri, J. (2021). Fungal lignin-modifying peroxidases and H2O2-producing enzymes. In *Encyclopedia of mycology* (pp. 247–259). Elsevier.

Makhlouf, A., Satha, H., Frihi, D., Gherib, S., & Seguela, R. (2016). Optimization of the crystallinity of polypropylene/submicronic-talc composites: The role of filler ratio and cooling rate. *Express Polymer Letters, 10*(3), 237–247. https://doi.org/10.3144/expresspolymlett.2016.22

Marchant, R. E., Zhao, Q., Anderson, J. M., & Hiltner, A. (1987). Degradation of a poly(ether urethane urea) elastomer: Infra-red and XPS studies. *Polymer, 28*(12), 2032–2039. https://doi.org/10.1016/0032-3861(87)90037-1

Masaki, K., Kamini, N. R., Ikeda, H., & Iefuji, H. (2005). Cutinase-like enzyme from the yeast *Cryptococcus* sp. strain S-2 hydrolyzes polylactic acid and other biodegradable plastics. *Applied and Environmental Microbiology, 71*(11), 7548–7550. https://doi.org/10.1128/AEM.71.11.7548-7550.2005

Mate, D. M., & Alcalde, M. (2015). Laccase engineering: From rational design to directed evolution. *Biotechnology Advances, 33*(1), 25–40. https://doi.org/10.1016/j.biotechadv.2014.12.007

Mathur, G., & Prasad, R. (2012). Degradation of polyurethane by Aspergillus flavus (ITCC 6051) isolated from soil. *Applied Biochemistry and Biotechnology, 167*(6), 1595–1602. https://doi.org/10.1007/s12010-012-9572-4

Mayer, A. (2002). Laccase: New functions for an old enzyme. *Phytochemistry, 60*(6), 551–565. https://doi.org/10.1016/S0031-9422(02)00171-1

Mejía, G. A. I., López, O. B. L., & Sierra, L. (2001). Biodegradation of poly(vinylalcohol co-ethylene) with the fungus phanerochaete chrysosporium. *Materials Research Innovations, 4*(2–3), 148–154. https://doi.org/10.1007/s100190000085

Melo, E. P., Airesbarros, M. R., & Cabral, J. M. S. (1995). Triglyceride hydrolysis and stability of a recombinant cutinase from Fusarium solani in AOT-iso-octane reversed micelles. *Applied Biochemistry and Biotechnology, 50*(1), 45–56. https://doi.org/10.1007/BF02788039

Messerschmidt, A. (2010). Copper Metalloenzymes. In *Comprehensive natural products II* (pp. 489–545). Elsevier.

Mohanan, N., Montazer, Z., Sharma, P. K., & Levin, D. B. (2020). Microbial and enzymatic degradation of synthetic plastics. *Frontiers in Microbiology, 11*, 580709. https://doi.org/10.3389/fmicb.2020.580709

Mollasalehi, S. (2013). *Fungal biodegradation of polyvinyl alcohol in soil and compost environments*. The University of Manchester.

Montazer, Z., Habibi Najafi, M. B., & Levin, D. B. (2019). Microbial degradation of low-density polyethylene and synthesis of polyhydroxyalkanoate polymers. *Canadian Journal of Microbiology, 65*(3), 224–234. https://doi.org/10.1139/cjm-2018-0335

Montazer, Z., Habibi Najafi, M. B., & Levin, D. B. (2020). Challenges with verifying microbial degradation of polyethylene. *Polymers, 12*(1), 123. https://doi.org/10.3390/polym12010123

Morel, M., Ngadin, A. A., Droux, M., Jacquot, J.-P., & Gelhaye, E. (2009). The fungal glutathione S-transferase system. Evidence of new classes in the wood-degrading basidiomycete Phanerochaete chrysosporium. *Cellular and Molecular Life Sciences, 66*(23), 3711–3725. https://doi.org/10.1007/s00018-009-0104-5

Moresi, M., & Parente, E. (1999). Fermentation (industrial)|production of organic acids. In *Encyclopedia of food microbiology* (pp. 705–717). Elsevier.

Motta, O., Proto, A., De Carlo, F., De Caro, F., Santoro, E., Brunetti, L., & Capunzo, M. (2009). Utilization of chemically oxidized polystyrene as co-substrate by filamentous fungi. *International Journal of Hygiene and Environmental Health, 212*(1), 61–66. https://doi.org/10.1016/j.ijheh.2007.09.014

Muhonja, C. N., Makonde, H., Magoma, G., & Imbuga, M. (2018). Biodegradability of polyethylene by bacteria and fungi from Dandora dumpsite Nairobi-Kenya. *PLoS One, 13*(7), e0198446. https://doi.org/10.1371/journal.pone.0198446

Müller, R.-J., Kleeberg, I., & Deckwer, W.-D. (2001). Biodegradation of polyesters containing aromatic constituents. *Journal of Biotechnology, 86*(2), 87–95. https://doi.org/10.1016/S0168-1656(00)00407-7

Müller, R.-J., Schrader, H., Profe, J., Dresler, K., & Deckwer, W.-D. (2005). Enzymatic degradation of poly(ethylene terephthalate): Rapid hydrolyse using a hydrolase from T. fusca. *Macromolecular Rapid Communications, 26*(17), 1400–1405. https://doi.org/10.1002/marc.200500410

Munir, E., Harefa, R. S. M., Priyani, N., & Suryanto, D. (2018). Plastic degrading fungi Trichoderma viride and Aspergillus nomius isolated from local landfill soil in Medan. *IOP Conference Series: Earth and Environmental Science, 126*, 012145. https://doi.org/10.1088/1755-1315/126/1/012145

Muñoz, C. I., & Bailey, A. M. (1998). A cutinase-encoding gene from Phytophthora capsici isolated by differential-display RT-PCR. *Current Genetics, 33*(3), 225–230. https://doi.org/10.1007/s002940050330

Murphy, C. A., Cameron, J. A., Huang, S. J., & Vinopal, R. T. (1996). Fusarium polycaprolactone depolymerase is cutinase. *Applied and Environmental Microbiology, 62*(2), 456–460. https://doi.org/10.1128/aem.62.2.456-460.1996

Nadumane, V. K., Venkatachalam, P., & Gajaraj, B. (2016). Aspergillus applications in cancer research. In *New and future developments in microbial biotechnology and bioengineering* (pp. 243–255). Elsevier.

Nakajima-Kambe, T., Shigeno-Akutsu, Y., Nomura, N., Onuma, F., & Nakahara, T. (1999). Microbial degradation of polyurethane, polyester polyurethanes and polyether polyurethanes. *Applied Microbiology and Biotechnology, 51*(2), 134–140. https://doi.org/10.1007/s002530051373

Ng, T. B. (2013). Antifungal and antibacterial peptides of fungal origin. In *Handbook of biologically active peptides* (pp. 157–161). Elsevier.

Nikolic, M. S., & Djonlagic, J. (2001). Synthesis and characterization of biodegradable poly(butylene succinate-co-butylene adipate)s. *Polymer Degradation and Stability, 74*(2), 263–270. https://doi.org/10.1016/S0141-3910(01)00156-2

Nomura, N., Deguchi, T., Shigeno-Akutsu, Y., Nakajima-Kambe, T., & Nakahara, T. (2001). Gene structures and catalytic mechanisms of microbial enzymes able to blodegrade the synthetic solid polymers nylon and polyester polyurethaoe. *Biotechnology and Genetic Engineering Reviews, 18*(1), 125–147. https://doi.org/10.1080/02648725.2001.10648011

Nowak, B., Pająk, J., Łabużek, S., Rymarz, G., & Talik, E. (2011). Biodegradation of poly (ethylene terephthalate) modified with polyester Bionolle® by Penicillium funiculosum. *Polimery, 56*(1), 35–44.

Nunes, C. S., & Malmlöf, K. (2018). Enzymatic decontamination of antimicrobials, phenols, heavy metals, pesticides, polycyclic aromatic hydrocarbons, dyes, and animal waste. In *Enzymes in human and animal nutrition* (pp. 331–359). Elsevier.

O'Neill, A., Araújo, R., Casal, M., Guebitz, G., & Cavaco-Paulo, A. (2007). Effect of the agitation on the adsorption and hydrolytic efficiency of cutinases on polyethylene terephthalate fibres. *Enzyme and Microbial Technology, 40*(7), 1801–1805. https://doi.org/10.1016/j.enzmictec.2007.02.012

Oda, Y., Yonetsu, A., Urakami, T., & Tonomura, K. (2000). Degradation of polylactide by commercial proteases. *Journal of Polymers and the Environment, 8*(1), 29–32. https://doi.org/10.1023/A:1010120128048

Ogunbayo, A. O., Olanipekun, O. O., & Adamu, I. A. (2019). Preliminary studies on the microbial degradation of plastic waste using Aspergillus niger and Pseudomonas sp. *Journal of Environmental Protection, 10*(05), 625–631. https://doi.org/10.4236/jep.2019.105037

Oikawa, H. (2010). Diels–Alderases. In *Comprehensive natural products II* (pp. 277–314). Elsevier.

Ojha, N., Pradhan, N., Singh, S., Barla, A., Shrivastava, A., Khatua, P., Rai, V., & Bose, S. (2017). Evaluation of HDPE and LDPE degradation by fungus, implemented by statistical optimization. *Scientific Reports, 7*(1), 39515. https://doi.org/10.1038/srep39515

Olicón-Hernández, D. R., González-López, J., & Aranda, E. (2017). Overview on the biochemical potential of filamentous fungi to degrade pharmaceutical compounds. *Frontiers in Microbiology, 8*, 1792. https://doi.org/10.3389/fmicb.2017.01792

Omokhagbor Adams, G., Tawari Fufeyin, P., Eruke Okoro, S., & Ehinomen, I. (2020). Bioremediation, biostimulation and bioaugmention: A review. *Journal of Environmental Bioremediation & Biodegradation, 3*(1), 28–39. https://doi.org/10.12691/ijebb-3-1-5

Orhan, Y., & Büyükgüngör, H. (2000). Enhancement of biodegradability of disposable polyethylene in controlled biological soil. *International Biodeterioration & Biodegradation, 45*(1–2), 49–55. https://doi.org/10.1016/S0964-8305(00)00048-2

Osman, M., Satti, S. M., Luqman, A., Hasan, F., Shah, Z., & Shah, A. A. (2018). Degradation of polyester polyurethane by Aspergillus sp. strain S45 isolated from soil. *Journal of Polymers and the Environment, 26*(1), 301–310. https://doi.org/10.1007/s10924-017-0954-0

Owen, S., Otani, T., Masaoka, S., & Ohe, T. (1996). The biodegradation of low-molecular-weight urethane compounds by a strain of *Exophiala jeanselmei*. *Bioscience, Biotechnology, and Biochemistry, 60*(2), 244–248. https://doi.org/10.1271/bbb.60.244

Ozsagiroglu, E., Iyisan, B., & Guvenilir, Y. A. (2012). Biodegradation and characterization studies of different kinds of polyurethanes with several enzyme solutions. *Polish Journal of Environmental Studies, 21*(6), 1777–1782.

Paço, A., Duarte, K., da Costa, J. P., Santos, P. S. M., Pereira, R., Pereira, M. E., Freitas, A. C., Duarte, A. C., & Rocha-Santos, T. A. P. (2017). Biodegradation of polyethylene microplastics by the marine fungus Zalerion maritimum. *Science of the Total Environment, 586*, 10–15. https://doi.org/10.1016/j.scitotenv.2017.02.017

Paszczynski, A., & Crawford, R. L. (1995). Potential for bioremediation of xenobiotic compounds by the white-rot fungus Phanerochaete chrysosporium. *Biotechnology Progress, 11*(4), 368–379. https://doi.org/10.1021/bp00034a002

Patel, C., Yadav, S., Rahi, S., & Dave, A. (2013). Studies on biodiversity of fungal endophytes of indigenous monocotaceous and dicotaceous plants and evaluation of their enzymatic potentialities. *International Journal of Scientific and Research Publications, 3*(7), 2250–3153.

Pathak, V. M., & Navneet. (2017). Review on the current status of polymer degradation: A microbial approach. *Bioresources and Bioprocessing, 4*(1), 15. https://doi.org/10.1186/s40643-017-0145-9

Patrício Silva, A. L., Prata, J. C., Walker, T. R., Duarte, A. C., Ouyang, W., Barcelò, D., & Rocha-Santos, T. (2021). Increased plastic pollution due to COVID-19 pandemic: Challenges and recommendations. *Chemical Engineering Journal, 405*, 126683. https://doi.org/10.1016/j.cej.2020.126683

Pawar, R. M. (2015). The effect of soil pH on bioremediation of polycyclic aromatic hydrocarbons (PAHS). *Journal of Bioremediation & Biodegradation, 6*(3), 291–304.

Pazos, F. (2004). MetaRouter: Bioinformatics for bioremediation. *Nucleic Acids Research, 33*(Database issue), D588–D592. https://doi.org/10.1093/nar/gki068

Pereira, E. B., De Castro, H. F., De Moraes, F. F., & Zanin, G. M. (2001). Kinetic studies of lipase from Candida rugosa a comparative study between free and immobilized enzyme onto porous chitosan beads. *Applied Biochemistry and Biotechnology, 91–93*(1–9), 739–752. https://doi.org/10.1385/ABAB:91-93:1-9:739

Petrov, N., Stoyanova, M., & Gaur, R. K. (2021). Ecological methods to control viral damages in tomatoes. In *Plant virus-host interaction* (pp. 469–488). Elsevier.

Ping, L.-F., Chen, X.-Y., Yuan, X.-L., Zhang, M., Chai, Y.-J., & Shan, S.-D. (2017). Application and comparison in biosynthesis and biodegradation by Fusarium solani and Aspergillus fumigatus cutinases. *International Journal of Biological Macromolecules, 104*, 1238–1245. https://doi.org/10.1016/j.ijbiomac.2017.06.118

Piscitelli, A., Cicatiello, P., Gravagnuolo, A. M., Sorrentino, I., Pezzella, C., & Giardina, P. (2017). Applications of functional amyloids from fungi: Surface modification by class I Hydrophobins. *Biomolecules, 7*(4), 45. https://doi.org/10.3390/biom7030045

Pospíšil, J., & Nešpůrek, S. (1997). Highlights in chemistry and physics of polymer stabilization. *Macromolecular Symposia, 115*(1), 143–163. https://doi.org/10.1002/masy.19971150110

Prabhat, S., Bhattacharyya, S., Vishal, V., Kalyan, R. K., Vijai, K., Pandey, K. N., & Singh, M. (2013). Studies on isolation and identification of active microorganisms during degradation of polyethylene/starch film. *International Research Journal of Environmental Sciences, 2*(9), 83–85.

Prakash, V. (2017). Mycoremediation of environmental pollutants. *International Journal of ChemTech Research, 10*(3), 149–155.

Pramila, R., & Ramesh, K. V. (2011). Biodegradation of low density polyethylene (LDPE) by fungi isolated from municipal landfill area. *Journal of Microbiology and Biotechnology Research, 1*(4), 131–136.

Purnomo, A. S., Mori, T., Putra, S. R., & Kondo, R. (2013). Biotransformation of heptachlor and heptachlor epoxide by white-rot fungus Pleurotus ostreatus. *International Biodeterioration & Biodegradation, 82*, 40–44. https://doi.org/10.1016/j.ibiod.2013.02.013

Raaman, N., Rajitha, N., Jayshree, A., & Jegadeesh, R. (2012). Biodegradation of plastic by Aspergillus spp. isolated from polythene polluted sites around Chennai. *Journal of Academia and Industrial Research, 1*(6), 313–316.

Raghavendra, V. B., Uzma, M., Govindappa, M., Vasantha, R. A., & Lokesh, S. (2016). Screening and identification of polyurethane (PU) and low density polyethylene (LDPE) degrading soil fungi isolated from municipal solid waste. *International Journal of Current Research, 8*, 34753–34761.

Raghukumar, S. (2017). *Fungi in coastal and oceanic marine ecosystems: Marine fungi* (1st ed.). Springer International Publishing: Imprint: Springer.

Ragusa, A., Svelato, A., Santacroce, C., Catalano, P., Notarstefano, V., Carnevali, O., Papa, F., Rongioletti, M. C. A., Baiocco, F., Draghi, S., D'Amore, E., Rinaldo, D., Matta, M., & Giorgini, E. (2021). Plasticenta: First evidence of microplastics in human placenta. *Environment International, 146*, 106274. https://doi.org/10.1016/j.envint.2020.106274

Ray, S., & Cooney, R. P. (2018). Thermal degradation of polymer and polymer composites. In *Handbook of environmental degradation of materials* (pp. 185–206). Elsevier.

Reddy, M. S., Basha, S., Adimurthy, S., & Ramachandraiah, G. (2006). Description of the small plastics fragments in marine sediments along the Alang-Sosiya ship-breaking yard, India. *Estuarine, Coastal and Shelf Science, 68*(3–4), 656–660. https://doi.org/10.1016/j.ecss.2006.03.018

Reena, R., Majhi, M., Arya, A., Kumar, R., & Kumar, A. (2012). BioRadBase: A database for bioremediation of radioactive waste. *African Journal of Biotechnology, 11*(35), 8718–8721. https://doi.org/10.5897/AJB12.020

Restrepo-Flórez, J.-M., Bassi, A., & Thompson, M. R. (2014). Microbial degradation and deterioration of polyethylene – A review. *International Biodeterioration & Biodegradation, 88*, 83–90. https://doi.org/10.1016/j.ibiod.2013.12.014

Riva, S. (2006). Laccases: Blue enzymes for green chemistry. *Trends in Biotechnology, 24*(5), 219–226. https://doi.org/10.1016/j.tibtech.2006.03.006

Rodrigues, M., da Luz, J., de Cássia Soares da Silva, M., Ferreira dos Santos, L., & Catarina Megumi Kasuya, M. (2020). Plastics polymers degradation by fungi. In M. Blumenberg, M. Shaaban, & A. Elgaml (Eds.), *Microorganisms*. IntechOpen.

Ronkvist, Å. M., Xie, W., Lu, W., & Gross, R. A. (2009). Cutinase-catalyzed hydrolysis of poly(ethylene terephthalate). *Macromolecules, 42*(14), 5128–5138. https://doi.org/10.1021/ma9005318

Roohi, Z. M. R., & Kuddus, M. (2018). PHB (poly-β-hydroxybutyrate) and its enzymatic degradation. *Polymers for Advanced Technologies, 29*(1), 30–40. https://doi.org/10.1002/pat.4126

Roussel, A., Amara, S., Nyyssölä, A., Mateos-Diaz, E., Blangy, S., Kontkanen, H., Westerholm-Parvinen, A., Carrière, F., & Cambillau, C. (2014). A cutinase from Trichoderma reesei with a lid-covered active site and kinetic properties of true lipases. *Journal of Molecular Biology, 426*(22), 3757–3772. https://doi.org/10.1016/j.jmb.2014.09.003

Russell, J. R., Huang, J., Anand, P., Kucera, K., Sandoval, A. G., Dantzler, K. W., Hickman, D., Jee, J., Kimovec, F. M., Koppstein, D., Marks, D. H., Mittermiller, P. A., Núñez, S. J., Santiago, M., Townes, M. A., Vishnevetsky, M., Williams, N. E., Vargas, M. P. N., Boulanger, L.-A., Bascom-Slack, C., & Strobel, S. A. (2011). Biodegradation of polyester polyurethane by endophytic fungi. *Applied and Environmental Microbiology, 77*(17), 6076–6084. https://doi.org/10.1128/AEM.00521-11

Sabino, R. (2021). Exposure to fungi in health care facilities. In *Encyclopedia of mycology* (pp. 1–10). Elsevier.

Sameshima-Yamashita, Y., Ueda, H., Koitabashi, M., & Kitamoto, H. (2019). Pretreatment with an esterase from the yeast Pseudozyma antarctica accelerates biodegradation of plastic mulch film in soil under laboratory conditions. *Journal of Bioscience and Bioengineering, 127*(1), 93–98. https://doi.org/10.1016/j.jbiosc.2018.06.011

Sánchez, C. (2009). Lignocellulosic residues: Biodegradation and bioconversion by fungi. *Biotechnology Advances, 27*(2), 185–194. https://doi.org/10.1016/j.biotechadv.2008.11.001

Sánchez, C. (2020). Fungal potential for the degradation of petroleum-based polymers: An overview of macro- and microplastics biodegradation. *Biotechnology Advances, 40*, 107501. https://doi.org/10.1016/j.biotechadv.2019.107501

Sangale, M. K., Mohd, S., & Ade, A. B. (2019). Potential of fungi isolated from the dumping sites mangrove rhizosphere soil to degrade polythene. *Scientific Reports, 9*(1), 5390. https://doi.org/10.1038/s41598-019-41448-y

Santhanam, N., Vivanco, J. M., Decker, S. R., & Reardon, K. F. (2011). Expression of industrially relevant laccases: Prokaryotic style. *Trends in Biotechnology, 29*(10), 480–489. https://doi.org/10.1016/j.tibtech.2011.04.005

Saxena, A., Jain, S., & Pareek, A. (2022). Estimation of possible biodegradation of polythene by fungal isolates growing on polythene debris. *Pollution, 8*(2), 567. https://doi.org/10.22059/poll.2021.331370.1195

Schwartz, M., Perrot, T., Aubert, E., Dumarçay, S., Favier, F., Gérardin, P., Morel-Rouhier, M., Mulliert, G., Saiag, F., Didierjean, C., & Gelhaye, E. (2018). Molecular recognition of wood polyphenols by phase II detoxification enzymes of the white rot Trametes versicolor. *Scientific Reports, 8*(1), 8472. https://doi.org/10.1038/s41598-018-26601-3

Senthivelan, T., Kanagaraj, J., & Panda, R. C. (2016). Recent trends in fungal laccase for various industrial applications: An eco-friendly approach – A review. *Biotechnology and Bioprocess Engineering, 21*(1), 19–38. https://doi.org/10.1007/s12257-015-0278-7

Sepperumal, U., Markandan, M., & Palraja, I. (2013). Micromorphological and chemical changes during biodegradation of polyethylene terephthalate (PET) by Penicillium sp. *Journal of Microbiology Biotechnology Research, 3*(4), 47–53.

Shah, A. A., Hasan, F., Hameed, A., & Ahmed, S. (2008). Biological degradation of plastics: A comprehensive review. *Biotechnology Advances, 26*(3), 246–265. https://doi.org/10.1016/j.biotechadv.2007.12.005

Sharma, T., Sharma, A., & Kanwar, S. S. (2017). An overview on esterases: Structure, classification, sources and their application. *Recent Advances in Biotechnology, 2*, 216–229.

Sharma, A., Sharma, T., Meena, K. R., Kumar, A., & Kanwar, S. S. (2018a). High throughput synthesis of ethyl pyruvate by employing superparamagnetic iron nanoparticles-bound esterase. *Process Biochemistry, 71*, 109–117. https://doi.org/10.1016/j.procbio.2018.05.004

Sharma, A., Sharma, T., & Kanwar, S. S. (2018b). Entrapment of Streptomyces sp. lipase in cupric-beads for methyl salicylate synthesis. *Trends in Carbohydrate Research, 10*, 21–29.

Sharma, A., Meena, K. R., & Kanwar, S. S. (2018c). Microbial thermo-tolerant and solvent-tolerant lipases and their applications. *Current Research Microbiology, 2*, 1–15.

Sheik, S., Chandrashekar, K. R., Swaroop, K., & Somashekarappa, H. M. (2015). Biodegradation of gamma irradiated low density polyethylene and polypropylene by endophytic fungi. *International Biodeterioration & Biodegradation, 105*, 21–29. https://doi.org/10.1016/j.ibiod.2015.08.006

Shekhar, S., Sundaramanickam, A., & Balasubramanian, T. (2015). Biosurfactant producing microbes and their potential applications: A Review. *Critical Reviews in Environmental Science and Technology, 45*(14), 1522–1554. https://doi.org/10.1080/10643389.2014.955631

Shi, K., Su, T., & Wang, Z. (2019). Comparison of poly(butylene succinate) biodegradation by fusarium solani cutinase and Candida antarctica lipase. *Polymer Degradation and Stability, 164*, 55–60. https://doi.org/10.1016/j.polymdegradstab.2019.04.005

Shivakumar, S. (2013). Poly- β-hydroxybutyrate (PHB) Depolymerase from *Fusarium solani* Thom. *Journal of Chemistry, 2013*, 1–9. https://doi.org/10.1155/2013/406386

Sindujaa, P., Padmapriya, M., Pramila, R., & Ramesh, K. V. (2011). Bio-degradation of low density polyethylene (LDPE) by fungi isolated from marine water. *Research Journal of Biological Sciences, 6*(4), 141–145.

Singer, S. D., Foroud, N. A., & Laurie, J. D. (2019). Molecular improvement of grain: Target traits for a changing world. In *Encyclopedia of food security and sustainability* (pp. 545–555). Elsevier.

Singh, H. (2006). *Mycoremediation*. John Wiley & Sons.

Singh, B., & Sharma, N. (2008). Mechanistic implications of plastic degradation. *Polymer Degradation and Stability, 93*(3), 561–584. https://doi.org/10.1016/j.polymdegradstab.2007.11.008

Sivalingam, G., & Madras, G. (2004). Dynamics of lipase catalyzed enzymatic degradation of poly(bisphenol-A carbonate). *Journal of Applied Polymer Science, 91*(4), 2391–2396. https://doi.org/10.1002/app.13394

Sivan, A. (2011). New perspectives in plastic biodegradation. *Current Opinion in Biotechnology, 22*(3), 422–426. https://doi.org/10.1016/j.copbio.2011.01.013

Skariyachan, S., Taskeen, N., Kishore, A. P., & Krishna, B. V. (2022). Recent advances in plastic degradation – From microbial consortia-based methods to data sciences and computational biology driven approaches. *Journal of Hazardous Materials, 426*, 128086. https://doi.org/10.1016/j.jhazmat.2021.128086

Souza, P. M. d., Bittencourt, M. L. d. A., Caprara, C. C., Freitas, M. d., Almeida, R. P. C. d., Silveira, D., Fonseca, Y. M., Ferreira Filho, E. X., Pessoa Junior, A., & Magalhães, P. O. (2015). A biotechnology perspective of fungal proteases. *Brazilian Journal of Microbiology, 46*(2), 337–346. https://doi.org/10.1590/S1517-838246220140359

Sowmya, H. V., Ramalingappa, B., Nayanashree, G., Thippeswamy, B., & Krishnappa, M. (2015). Polyethylene degradation by fungal consortium. *International Journal of Environmental Research, 9*(3), 823–830. https://doi.org/10.22059/ijer.2015.969

Spina, F., Cecchi, G., Landinez-Torres, A., Pecoraro, L., Russo, F., Wu, B., Cai, L., Liu, X. Z., Tosi, S., Varese, G. C., Zotti, M., & Persiani, A. M. (2018). Fungi as a toolbox for sustainable bioremediation of pesticides in soil and water. *Plant Biosystems – An International Journal Dealing with all Aspects of Plant Biology, 152*(3), 474–488. https://doi.org/10.1080/11263504.2018.1445130

Sumathi, T., Viswanath, B., Sri Lakshmi, A., & SaiGopal, D. V. R. (2016). Production of laccase by *Cochliobolus* sp. isolated from plastic dumped soils and their ability to degrade low molecular weight PVC. *Biochemistry Research International, 2016*, 1–10. https://doi.org/10.1155/2016/9519527

Tahir, L., Ishtiaq Ali, M., Zia, M., Atiq, N., Hasan, F., & Ahmed, S. (2013). Production and characterization of esterase in Lantinus tigrinus for degradation of polystyrene. *Polish Journal of Microbiology, 62*(1), 101–108.

Thirunavukarasu, K., Edwinoliver, N. G., DuraiAnbarasan, S., Gowthaman, M. K., Iefuji, H., & Kamini, N. R. (2008). Removal of triglyceride soil from fabrics by a novel lipase from Cryptococcus sp. S-2. *Process Biochemistry, 43*(7), 701–706. https://doi.org/10.1016/j.procbio.2008.02.011

Thirunavukarasu, K., Purushothaman, S., Sridevi, J., Aarthy, M., Gowthaman, M. K., Nakajima-Kambe, T., & Kamini, N. R. (2016). Degradation of poly(butylene succinate) and poly(butylene succinate-co-butylene adipate) by a lipase from yeast Cryptococcus sp. grown on agro-industrial residues. *International Biodeterioration & Biodegradation, 110*, 99–107. https://doi.org/10.1016/j.ibiod.2016.03.005

Tian, L., Kolvenbach, B., Corvini, N., Wang, S., Tavanaie, N., Wang, L., Ma, Y., Scheu, S., Corvini, P. F.-X., & Ji, R. (2017). Mineralisation of 14C-labelled polystyrene plastics by Penicillium variabile after ozonation pre-treatment. *New Biotechnology, 38*, 101–105. https://doi.org/10.1016/j.nbt.2016.07.008

Tien, M., & Kirk, T. K. (1983). Lignin-degrading enzyme from the hymenomycete Phanerochaete chrysosporium burds. *Science, 221*(4611), 661–663. https://doi.org/10.1126/science.221.4611.661

Tokiwa, Y., & Suzuki, T. (1977). Hydrolysis of polyesters by lipases. *Nature, 270*(5632), 76–78. https://doi.org/10.1038/270076a0

Tokiwa, Y., Ando, T., & Suzuki, T. (1976). Degradation of polycaprolactone by a fungus. *Journal of Fermentation Technology, 54*, 603–608.

Tokiwa, Y., Suzuki, T., & Takeda, K. (1988). Two types of lipases in hydrolysis of polyester. *Agricultural and Biological Chemistry, 52*(8), 1937–1943. https://doi.org/10.1080/00021369.1988.10868966

Tokiwa, Y., Calabia, B., Ugwu, C., & Aiba, S. (2009). Biodegradability of plastics. *International Journal of Molecular Sciences, 10*(9), 3722–3742. https://doi.org/10.3390/ijms10093722

Torres, A., Li, S. M., Roussos, S., & Vert, M. (1996). Screening of microorganisms for biodegradation of poly(lactic-acid) and lactic acid-containing polymers. *Applied and Environmental Microbiology, 62*(7), 2393–2397. https://doi.org/10.1128/aem.62.7.2393-2397.1996

Urbance, J. W. (2003). BSD: The biodegradative strain database. *Nucleic Acids Research, 31*(1), 152–155. https://doi.org/10.1093/nar/gkg032

Usman, M. A., Momohjimoh, I., & Usman, A. O. (2020). Mechanical, physical and biodegradability performances of treated and untreated groundnut shell powder recycled polypropylene composites. *Materials Research Express, 7*(3), 035302. https://doi.org/10.1088/2053-1591/ab750e

van Gemeren, I. A., Beijersbergen, A., Musters, W., Gouka, R. J., van den Hondel, C. A. M. J. J., Verrips, C. T., & van Gemeren, I. A. (1996). The effect of pre- and pro-sequences and multicopy integration on heterologous expression of the Fusarium solani pisi cutinase gene in Aspergillus awamori. *Applied Microbiology and Biotechnology, 45*(6), 755–763. https://doi.org/10.1007/s002530050759

van Gemeren, I. A., Beijersbergen, A., van den Hondel, C. A. M. J. J., & Verrips, C. T. (1998). Expression and secretion of defined cutinase variants by *Aspergillus awamori*. *Applied and Environmental Microbiology, 64*(8), 2794–2799. https://doi.org/10.1128/AEM.64.8.2794-2799.1998

Vertommen, M. A. M. E., Nierstrasz, V. A., Veer, M. v. d., & Warmoeskerken, M. M. C. G. (2005). Enzymatic surface modification of poly(ethylene terephthalate). *Journal of Biotechnology, 120*(4), 376–386. https://doi.org/10.1016/j.jbiotec.2005.06.015

Vishnu, S., Manish, D., & Seema, B. (2012). Microbial degradation of polyethylene (low density) by Aspergillius fumigatus and Penicillium sp. *Asian Journal of Experimental Biological Sciences, 3*(3), 498–501.

Vivi, V. K., Martins-Franchetti, S. M., & Attili-Angelis, D. (2019). Biodegradation of PCL and PVC: Chaetomium globosum (ATCC 16021) activity. *Folia Microbiologica, 64*(1), 1–7. https://doi.org/10.1007/s12223-018-0621-4

Wang, J., Xie, Y., Hou, J., Zhou, X., Chen, J., Yao, C., Zhang, Y., & Li, Y. (2022). Biodegradation of bisphenol A by alginate immobilized Phanerochaete chrysosporium beads: Continuous cyclic treatment and degradation pathway analysis. *Biochemical Engineering Journal, 177*, 108212. https://doi.org/10.1016/j.bej.2021.108212

Wei, R., & Zimmermann, W. (2017). Microbial enzymes for the recycling of recalcitrant petroleum-based plastics: How far are we? *Microbial Biotechnology, 10*(6), 1308–1322. https://doi.org/10.1111/1751-7915.12710

Wessels, J. G. H. (1996). Fungal hydrophobins: Proteins that function at an interface. *Trends in Plant Science, 1*(1), 9–15. https://doi.org/10.1016/S1360-1385(96)80017-3

Wierckx, N., Prieto, M. A., Pomposiello, P., Lorenzo, V., O'Connor, K., & Blank, L. M. (2015). Plastic waste as a novel substrate for industrial biotechnology. *Microbial Biotechnology, 8*(6), 900–903. https://doi.org/10.1111/1751-7915.12312

Wilkes, R. A., & Aristilde, L. (2017). Degradation and metabolism of synthetic plastics and associated products by *Pseudomonas* sp.: Capabilities and challenges. *Journal of Applied Microbiology, 123*(3), 582–593. https://doi.org/10.1111/jam.13472

Wösten, H. A. B., & Scholtmeijer, K. (2015). Applications of hydrophobins: Current state and perspectives. *Applied Microbiology and Biotechnology, 99*(4), 1587–1597. https://doi.org/10.1007/s00253-014-6319-x

Wu, Y., Li, J., Yang, H., & Shin, H.-J. (2017). Fungal and mushroom hydrophobins: A review. *Journal of Mushroom, 15*(1), 1–7. https://doi.org/10.14480/JM.2017.15.1.1

Yamano, N., Kawasaki, N., Ida, S., & Nakayama, A. (2019). Biodegradation of polyamide 4 in seawater. *Polymer Degradation and Stability, 166*, 230–236. https://doi.org/10.1016/j.polymdegradstab.2019.05.032

Yang, J., Yang, X., Lin, Y., Ng, T. B., Lin, J., & Ye, X. (2015). Laccase-catalyzed decolorization of malachite green: Performance optimization and degradation mechanism. *PLoS One, 10*(5), e0127714. https://doi.org/10.1371/journal.pone.0127714

Yashavantha Rao, H. C., Sruthi, D., Kamalraj, S., Parthasarathy, R., & Jayabaskaran, C. (2021). Endophytic fungi as a potential source of cytotoxic drugs: A fungal solution to cancer. In *Volatiles and metabolites of microbes* (pp. 305–323). Elsevier.

Zahra, S., Abbas, S. S., Mahsa, M.-T., & Mohsen, N. (2010). Biodegradation of low-density polyethylene (LDPE) by isolated fungi in solid waste medium. *Waste Management, 30*(3), 396–401. https://doi.org/10.1016/j.wasman.2009.09.027

Zeenat, E. A., Bukhari, D. A., Shamim, S., & Rehman, A. (2021). Plastics degradation by microbes: A sustainable approach. *Journal of King Saud University – Science, 33*(6), 101538. https://doi.org/10.1016/j.jksus.2021.101538

Zeghal, E., Vaksmaa, A., Vielfaure, H., Boekhout, T., & Niemann, H. (2021). The potential role of marine fungi in plastic degradation – A Review. *Frontiers in Marine Science, 8*, 738877. https://doi.org/10.3389/fmars.2021.738877

Zhang, J., Gao, D., Li, Q., Zhao, Y., Li, L., Lin, H., Bi, Q., & Zhao, Y. (2020). Biodegradation of polyethylene microplastic particles by the fungus Aspergillus flavus from the guts of wax moth Galleria mellonella. *Science of the Total Environment, 704*, 135931. https://doi.org/10.1016/j.scitotenv.2019.135931

Chapter 16
Emerging Techniques for the Mitigation of Micro and Nanoplastics in Soil

Chilumoju Sai Pavan, Palakeerti Srinivas Kumar, Nalam Renuka Satya Sree, and T. Mohammad Munawar

Abstract For many years, global plastic pollution has been a severe concern, and micro (nano) plastics (MNPs) have attracted the attention of researchers all over the world. Because MNPs can exhibit toxicological and interact with potentially toxic elements (PTEs) in the environment, soil toxicity can occur. Despite the fact that MNPs may accumulate in plant roots and have deleterious impacts on terrestrial environments, their impact on soil systems and plant crops has been overlooked. Human and animal use of MNP-contaminated plants or fruits will eventually result in health problems. Because identifying and measuring MNPs in diverse soil samples is difficult, knowledge of their fate, environmental, and ecological effects in terrestrial ecosystems are limited. As a result, it is critical to use an innovative strategy to remove MNPs from the natural environment. Microbial remediation is regarded to be a greener option amongst the many MNPs remediation processes. Enzymatic processes, substrates and co-substrates concentration, temperature, pH, oxidative stress, and other biotic and abiotic variables all impact the microbial breakdown of plastics. As a result, it is critical to understand the fundamental routes that microorganisms use to consume plastic particles as their only carbon source for growth and development. The benefits and downsides of different MNP remediation methods, such as enzymatic, advanced molecular, and biomembrane technologies, in stimulating the bioremediation of MNPs from diverse environmental compartments, as well as future research prospects, were discussed in this review.

Keywords Microbial remediation · Enzymatic processes · micro (nano) plastics (MNPs) · membrane technologies

C. S. Pavan · P. S. Kumar · N. R. S. Sree
Centre for Biotechnology, Institute of science & Technology, Jawaharlal Nehru Technological University, Hyderabad, Telangana, India

T. M. Munawar (✉)
Department of Biological and Chemical Engineering, Mekelle Institute of Technology, Mekelle University, Mekelle, Ethiopia

N. R. Maddela et al. (eds.), *Micro and Nanoplastics in Soil*,
https://doi.org/10.1007/978-3-031-21195-9_16

16.1 Introduction

Plastics, as a conveniently available, cost-effective, and handy material, are widely employed in industries and in our daily lives all over the world, and they provide a great deal of convenience to mankind. Along with its practical utility, it has recently been identified as a global environmental threat (Ali et al., 2021a, 2021b; Cheng et al., 2021). Microplastics (MPs) and nanoplastics (NPs) are solid polymer particles that are water-insoluble and created for several applications (Mancia et al., 2021). MPs and NPs have been found in a variety of environmental matrixes (Meng et al., 2019; Depledge et al., 2013). Plastic garbage pollutes the environment and will not disintegrate in the natural environment for at least 100 years (Ricardo et al., 2021; Tiwari et al., 2020). Nanoplastics (NPs) have a smaller size (100 nm), hence they have a greater influence on live tissue than MPs. NPs may easily penetrate through cell membranes, causing harm to cells and tissues (Wu et al., 2021). Plastics have a recovery rate of no more than 5%. Furthermore, plastics that we refer to as 'white pollution' will break into smaller particles known as MNPs as a result of different physical, chemical, and biological processes. Micro-plastics are plastic pieces with a diameter of less than 5 mm (Kumar et al., 2020a). MNPs are more difficult to manage and remove than macroparticle plastics, but they have a bigger influence on living beings (Kumar et al., 2021). MNPs interact and react with organic and inorganic pollutants in the environment, and they can even assist them in gaining access to environmental compartments (Sridharan et al., 2021a, 2021b). MNPs can be used as both a substrate for collecting and a vector for conveying organic and inorganic pollutants as a result of these interactions (Horton et al., 2017; Ricardo et al., 2021). As a result, MNPs help other contaminants bioaccumulate in the environment.

Although MNPs may accumulate in the plant root system and have negative effects on terrestrial habitats, their effects on soil systems and plant crops have been disregarded. Since MNPs accumulate in the soil and interact with PTEs and organic contaminants, they have a substantial influence on the soil-plant system (Chai et al., 2020). MNPs can impact plant development and chlorophyll concentration. Furthermore, MNPs can interact with soil organic detritus and remain in the soil for hundreds of years, affecting the soil's physio-chemical characteristics and polluting the groundwater (Wahl et al., 2020).

The presence of MNPs in the natural environment prompts microorganisms' adaptive mechanisms to cope with the negative effects of MNPs (Yang et al., 2020). Microbes respond to environmental stress in a variety of ways, including raising or lowering their growth rate, metabolic rate, and metabolic rate (Fuke et al., 2021; Mishra et al., 2021; Kumar et al., 2021) and to prevent environmental stress, novel microbial bioproducts are being synthesized (Guan & Liu, 2020; Gupta & Thakur, 2016; Kumar et al., 2018c). These microbial enzymes are involved not only in physiological responses but also in the microbial breakdown of environmental pollutants such as MPs. For example, microbial-assisted enzymatic degradation of plastic polymers break polymers down into monomers, which may be utilised by microorganisms as a source of carbon and energy. Furthermore, bacterial-aided composting decreased polyhydroxyalkanoates (PHAs) and polyethylene (PE) MNP abundance by 13–29% (Sun et al., 2021).

In comparison to the aquatic environment, the soil is the biggest reservoir of MNPs. Plastics can combine with soil aggregates, allowing for long-term storage. Storage, transfer and erosion, degradation, and finally, leaching to groundwater are all recognised as MNP fate mechanisms in the soil (Hurley & Nizzetto, 2018). According to preliminary findings, microplastics have a deleterious influence on the reproduction, growth, and death of many soil-dwelling earthworms (Rodríguez-Seijo et al., 2017; Cao et al., 2017; Huerta et al., 2016). Furthermore, when released from plastic, chemicals such as plasticizing agents, especially phthalates, may affect soil-dwelling organisms. Many additives are only weakly integrated with the polymer structure of many plastics, and hence may be washed off. Bis(2-Ethylhexyl) phthalate, for example, has been observed to suppress soil microbiological activity (Wang et al., 2016). They may have carcinogenic, mutagenic, and endocrine-disrupting effects making them dangerous soil pollutants (Wang et al., 2013; Fu & Du, 2011; Magdouli et al., 2013). Phthalates, once released into the environment, maybe taken up by plants (Sun et al., 2015), entering the food chain and posing a health risk to humans (Hauser & Calafat, 2005). MNPs with greater densities will stay at the surface, migrate deeper into the soil, transport and pollute groundwater, transfer into plants, and affect the entire food chain, whereas MNPs with lower densities will stay at the surface and maybe moved by wind and water erosion (Wu et al., 2019). Floods, accumulation, and other causes can bury MNPs over time, speeding up the preservation process. Although factors like microbial populations and pH can influence the preservation process, other techniques, such as tilling, can bring buried particles to the surface. Scientists are trying to come up with mathematical models to help them comprehend the destiny of MNPs in the environment because it is still unknown how much MNPs exist in the environment. As a result, if plastic is not removed from the soil, detrimental impacts on soil living organisms, soil fertility, and human health cannot be ruled out.

In theory, removing plastic pollution from soil may be challenging. Plastic, once introduced into the environment, has shown to be tenacious, accumulating in water and sediments (e.g., Barnes et al., 2009). Other long-lasting compounds in soil, such as black carbon, can last hundreds of years (e.g., Czimczik & Masiello, 2007). Physical and chemical methods, such as micro/nanofiltration in sewage treatment plants (STPs), are now used to remove MNPs from the terrestrial environment (Dey et al., 2021; Sun et al., 2019).

16.2 Impacts of MNPs

16.2.1 Impacts of MNPs on Plants

The size and texture of MNPs have an impact on their mobility in the soil. Soil cracking, tilling, ploughing, ditching, and bioturbation were all common methods of transportation (Khalid et al., 2020). According to Zhang et al. (2020), the roots were the most damaged section of the plant, followed by leaves, shoots, and finally the stem. Roots absorbed and adsorbed MNPs from the soil and water. Some of them may get in through lateral root cracks. In comparison to grass and plants that grow above the ground, root plants have a greater concentration of MNPs (Eggen et al., 2012). M Ps from air deposition infect leaves, and these particles attach to them. Smaller NPs have been proven to be more hazardous than bigger NPs because they can pass through plant biological membrane (Zhang et al., 2020). Because of the presence of MNPs in the soil, certain plant's roots experience a slowdown in growth. Roots were shown to be physically injured when exposed to sharp-edged MNPs (Iqbal et al., 2020). Some microfibers entangled juvenile roots, preventing them from growing properly. It was unclear whether the decrease in root development was caused by a change in soil structure or by the hazardous components in MNPs (Khalid et al., 2020).

Seed germination was inhibited by MNP blockage, which slowed or stopped the intake of water by the seed pores. NPs obstruct cell wall pores that transport nutrients and water, resulting in a decrease in root and plant development. MNPs may be carried through the vascular system from roots to stems and leaves, and they can absorb additional pollutants and PTEs that are harmful to plants'health (Ebere et al., 2019). The kind of plant and several environmental parameters such as leaf area, waxy leaves, trichomes, and the type, size, and density of MNP particles all influenced the accumulation and deposition of MNPs on leaves from the atmosphere. MNPs are transported hundreds of kilometres by the wind from metropolitan areas to woods, which, due to their low density, can act as a sink for MNPs. MNPs enter the soil by leaf wash off or leaf litter from precipitation (Bi et al., 2020).

16.2.2 Impacts of MNPs on Microorganisms and Mammals

MNPs can also be transmitted down the food chain to humans. To give an example, MNPs are transferred down the food chain from algae to zooplankton, then fish, and finally humans in the aquatic environment. As a result, from the standpoint of the soil environment, it is envisaged that people would be able to consume these MNPs through the ingestion of MNP-contaminated edible plants (Ebere et al., 2019). MNPs, particularly nano-sized particles, can be taken up by plant roots and transferred to edible sections of the plant. MNPs will enter the food chain if that specific plant is ingested by animals and humans (Rillig et al., 2019). Humans are thought to ingest 80 g of MNPs each day from a variety of sources, with MNPs likely to be found in the intestines. Once swallowed, they can cause obstructions in the gastrointestinal tract, as well as a decline in energy. Its collection and distribution patterns in the body were determined by the amount consumed and the size of the particles. Polyethersulfone (PES), PET, PP, PE, PVC, and polyamide (PA) are commonly found in food and can have a harmful influence on human health. Because of the added chemicals or by absorbing PTEs from the environment, these minute plastic particles may be hazardous and carcinogenic. In humans and rats, PVC and PS generated hazardous compounds that related to cancer and reproductive problems (Karbalaei et al., 2018).

16.3 Micro-Nano Plastics' Biological and Toxicological Effects

MNPs'ecotoxicity in organisms is a constant hazard to the ecosystem. Because most of their surfaces are negatively charged (Bradney et al., 2019), MNPs can attract contaminants with positive charges. Furthermore, bacteria and mammals may absorb them. MNPs can cause aberrant metabolic and immunological responses in living creatures, both of which can be harmful to their health (Allouzi et al., 2021). Pollution with MNPs alters the physico-chemical properties of soil. Toxic additives used in the manufacture of plastics, as well as contaminants carried on the surface of MPs, penetrate the soil ecosystem with MPs, causing harm to the soil's microbial habitat (Guo et al., 2020). MNPs, as well as the hazardous substances they release, can alter a range of soil properties, including pH, conductivity, texture, nutrients including ammonium nitrogen, and organic carbon (OC) (OC) (de Souza Machado et al., 2017; Kim et al., 2020). Plastic pieces become more mobile as their size decreases. MNPs enter the plant system after being absorbed or adsorbed by the roots of plants in the soil (Rillig, 2020; Rillig et al., 2019; Wu et al., 2021). Plants are influenced by MNPs in the following order: roots, leaves, buds, and stems (Zhang et al., 2020). MNPs hanging in the air settle on the surface of the leaves of the plants'aboveground section (Sridharan et al., 2021b). For example, nanoparticles may readily infiltrate plants and pass through their cell walls and membranes (de Souza Machado et al., 2017). MNPs are more abundant in mature, bigger trees and arboreal plants, which may be due to their larger and sturdy root systems (Khalid et al., 2020). Boots et al. (2019) found that the build-up of MNPs has a negative influence on plant photosynthesis rates and that each plant has a varied reaction to MNP pollution.

16.4 Analytical Methods

16.4.1 Available Analytical Methods

Various methods for detecting (micro-)plastics in water and sediments have been developed, but none have been assessed for identifying and quantifying synthetic polymers in soil to our knowledge. Organic matter in soils ranges from 0.02% (desert soils) to over 100% (surface litter, bog soils), and it interacts with soil minerals and other elements in a complicated way (e.g., Tisdall & Oades, 2006). For preliminary sorting of plastic sizes, the soil sample should be crushed and sieved. It is important to note that soils are usually sieved at a size of <2 mm (according to the highest size limit of sand; Böttcher, 1996), even to <1 mm in certain countries, such as Russia. This sieved, so-called fine-earth fraction is commonly referred to in all normal soil studies. When used on plastic analyses, however, this technique will eliminate microplastics and bigger objects. We recommend sieving soil samples of <5 mm and <1 mm for plastic tests to ensure comparability with results from freshwater and marine research. We recommend sieving soil samples of <5 mm and <1 mm for plastic analysis (according to the definition of macroplastic and large and small microplastic as defined by the European Marine Strategy Framework Directive).

The mineral phase of soils may be easily extracted after sifting utilising density fractionation methods that have been developed for sediment analysis. Different density solutions, such as NaCl, ZnCl2, or NaI, have been utilised for sediments (e.g. Thompson et al., 2004; Liebezeit & Dubaish, 2012; Dekiff et al., 2014. In soils, sodium polytungstate (3Na2WO4 9WO3 H2O) has been the density solution of choice since it allows not only to separate free particulate organic matter (and hence also plastic) but also SOM in various organo-mineral complexes (e.g. Golchin et al., 1994; Christensen, 1996). It is unclear if and to what extent plastic wastes are entrapped inside soil aggregates in the latter case. Such attachments should be removed before density fractionation, for example, by ultrasonic treatment. Ultrasonic treatment of 60 JmL^{-1} has been demonstrated to be adequate for dispersion of soil macroaggregates (>250 m); with greater ultrasonic energy input, particulate plant materials can be disturbed (Amelung & Zech, 1999; Kaiser & Berhe, 2014). Löder and Gerdts (2015) found that ultrasonic treatment of sediments resulted in the development of secondary microplastics. It needs to be seen whether such treatment affects the size and characteristics of plastic in soil.

Other organic moieties, such as SOM, should be eliminated to ensure accurate identification and quantification of plastic in soil, according to sedimentary research methodologies. Because SOM has average densities between 1.0 and 1.4 g cm^{-3}, similar to that of various plastic kinds such as Polyethylenterephthalat (PET) and Nylon, simple density fractionations will fail to distinguish SOM from plastic components in soil. However, because huge portions of SOM are refractory, removing enough of it without breaking tiny plastic polymers is difficult. Acidic, alkaline, or oxidising treatments, as well as enzymatic digestion, are used to remove organic materials from sediments, water, and biological samples. Enzymatic digestion has shown to be effective in removing organic waste from sediments (Hidalgo-Ruz et al., 2012); however, its application for SOM stabilised by minerals may be questioned, but it must be studied at the very least. Electrostatic separation was also used, however, it failed to separate plastics 'from a wider range of natural sediments and organic materials' (Hidalgo-Ruz et al., 2012). For biological samples, alkaline treatments, such as NaOH or KOH, are effective without damaging the plastic (Cole et al., 2014; Dehaut et al., 2016). Controlled hot acid digestion may also be utilised as a start-up for black carbon studies, no black carbon particles will be eliminated (Glaser et al., 1998; Kappenberg et al., 2016). In soil research, the most frequent approach for removing organic materials is to use H2O2, for example, in texture analysis (Böttcher, Jet al., 1995). Plastic analyses in sediments were also subjected to such oxidising procedures (e.g. Imhof et al., 2013). Many polymers, although not all, are resistant to hydrogen peroxide treatment. In interaction with H2O2, polyethylene (PE) and polypropylene are degraded. Mintenig et al. (2016) devised a potential multistep approach for removing organic matter from wastewater samples that include enzymatic digestion in conjunction with brief H2O2 treatment; nevertheless, the appropriateness for soil samples has to be proven.

Fuller and Gautam (2016) used pressured fluid extraction to establish a progressive extraction technique for plastic <30 m from soil and urban garbage samples. This approach works well for a variety of plastics, including PE, PVC, PP, and others, and is thus one of the most promising ways for separating plastics from soil samples. The approach has the drawback of not being able to remove bigger plastic particles (>30 μm), resulting in major sections of typical microplastics being missed (<1 or 5 mm). Furthermore, there is a chance that the extraction could change particle shape, which might make physical characterization difficult, such as source tracking (Fuller & Gautam, 2016).

In conclusion, several promising tools for isolating plastic materials from sediments have been tested, all of which have potential limitations for soils, implying that more research is needed to develop a standard protocol for isolating plastics from the soil, ideally at low cost and without affecting plastic properties.

Once plastic has been isolated as a particle or tracked by a microscope or laser beam, visual identification, FT-IR or Raman spectroscopy, as well as pyrolyse GC–MS may be used to identify and quantify it in environmental samples. The effectiveness of eliminating interfering SOM may still be a factor in the success of these approaches once applied to soil, although this has yet to be determined. Each of these approaches has its own set of drawbacks. When used for sediment analysis, visual sorting has previously been criticised for having error rates of 20–70% (Hidalgo-Ruz et al., 2012; Eriksen et al., 2013). The auto-fluorescence of SOM, on the other hand, significantly limits the application of Raman spectroscopy to soils (Kruse et al., 2015; Löder & Gerdts, 2015).

This behaviour can be reduced by utilising low-energy lasers with wavelengths of >1000 nm, however, this results in a reduction in the polymer's signal. FT-IR may be more suited, at least when the materials are dry; otherwise, the presence of water suppresses IR signals. However, to ascribe the IR signals unambiguously to plastic and not to any other molecule in soil, adhering SOM must be removed. Although Zubris and Richards (2005) employed polarised light microscopy to detect synthetic fibres in soil samples, the utility of this technique for identifying additional plastics and distinguishing between plastic kinds has to be proven. The use of thermal desorption gas chromatography-mass spectrometry (TDS-GC–MS), as done by Dümichen et al. (2015) and Dümichen et al. (2017), might be a relatively unique way to solve these issues. To date, this approach has only been able to identify PE, PP, and polystyrene (PS) in complex environmental matrices such as soil or ferment leftovers from biogas plants, except PE in standardised laboratory studies. As a result, the authors stated that further study is needed to make it easier to quantify distinct plastic kinds in environmental samples such as soil. Overall, it appears that analysing plastic in the soil is more difficult when the soil is rich in SOM or when considerable components of the SOM are stable and difficult to extract. Soil types with high SOM content, such as Histosols (bog soils) with at least >30% SOM and Podzols with organic soil layers, should be included in method tests. Chernozems, and to a lesser extent Anthrosols (e.g. paddy soils), as well as organic-rich Fluvisols, are examples of soils rich in SOM that is refractory or reacts strongly with minerals (e.g. organic-rich riparian soils, such as the German Tschernitza), Fertile soils that have recently received organic matter additions, such as compost, sewage sludge, or solid manure, and vertisols (due to clays potentially sticking to plastic). In soils with extremely low SOM levels, such as desert soils (less than 1% SOM) or most subsoils, known analytical techniques from sediment analysis are likely well suited for plastic analysis.

16.5 Microbial Degradation of Micro-Nano Plastics

Microbial-assisted decomposition of plastic pieces results in MNP degeneration, which is a green strategy. It is easier to regulate changes in plastic pollution since microbial degradation procedures are significantly dependant on both biotic and abiotic parameters such as pH, temperature, oxidative strain, and so on. The entire degradation/elimination of MNPs may be anticipated utilising cutting-edge technology, such as omics, by employing plastic pieces as the only carbon source for the development of bacteria (Knott et al., 2020; Tiwari et al., 2020). Although the employment of microorganisms in MNP biodegradation is a promising method, this technology is still in its infancy and attracting attention owing to its sluggish speed, partial mineralisation, and undiscovered degradation mechanism (Ru et al., 2020; Silva et al., 2018; Anastopoulos & Pashalidis, 2021).

16.5.1 Mechanism of Micro-Nano Plastics Biodegradation

Microbe-driven MP degradation is influenced by the chemical structure and molecular weight of the MPs, as well as the kinds of bacteria and other environmental factors. The biodegradability of MNPs is affected by several characteristics such as density, functional group types, and bioavailability, as well as plasticizers or chemical additives added during production (Yuan et al., 2020). A variety of metabolic processes are involved in the microbial breakdown of MNPs. Microbial MNP degradation includes biodeterioration (changes in polymer size, shape, and chemical characteristics), biofragmentation, biosynthesis, and mineralization (Tiwari et al., 2020). Bacterial hydrolase (extracellular) enzymes can convert highly complicated compounds into polymers, which may subsequently be transformed into hydroxyl acid monomers (Kamrannejad et al., 2014; Lear et al., 2021).

16.6 Recent Advancements in the Breakdown of Micro-Nano Plastics

The hydrophobic characteristic of particulate plastic, as well as HMW and longer polymeric chain length, are crucial qualities that make it resistant to biodegradation (Sridharan et al., 2021a). The organic compound's high molecular weight (HMW) makes it difficult to transfer it over the microbial cell membrane, hence depolymerisation is necessary (Oberbeckmann & Labrenz, 2020). Various sophisticated approaches have lately been advocated to improve the biodegradation of plastic particles, such as enzymatic/enzyme-assisted degradation (Priya et al., 2021), advanced molecular tools and technologies (Sudhakar et al., 2007; Purohit et al., 2020), membrane bioreactor (MBR)-assisted remediation (Poerio et al., 2019), nano-technologies based remediation (Uheida et al., 2020), and so on.

16.6.1 Membrane and Enzyme Technology

Polypropylene, polyethylene and polyethylene terephthalate microplastics, were the most common microplastics found in the water system. The load of PET microplastics in wastewater treatment facilities accounts for more than 79% of all plastics present. Untreated microplastic fibres and particles are released into the aquatic ecosystem by WWTPs (wastewater treatment plants) (Lares et al., 2018). As a result, wastewater treatment plants have become one of the major sources of environmental micro-nano plastic contamination. A few studies have shown that standard treatment methods may remove microplastics from freshwater and drinking water. Filtration using granular activated carbon and sand, coagulation-flocculation, precipitation, and sedimentation are all its part (Pivokonsky et al., 2018). The creation of a substantial volume of chemical sludge, which is linked with many difficulties such as increased turbidity and reduced disposal capability, is the primary downside of these traditional treatment methods (Zinicovscaia, 2016). Membrane technology is currently, widely used in wastewater treatment plants (WWTPs) to purify water, and it is divided into three categories: ultrafiltration, nanofiltration, and reverse osmosis (Baker, 2012).Though membrane technology is effective in removing MNPs from water bodies, further degradation and use of these harmful substances are necessary for a pollution-free environment. Enzyme technology may be used to produce, isolate, purify, and provide enzymes for the destruction of plastics, which is interesting. These enzymes are biodegradable and non-toxic. A specific set of enzymes has been found to breakdown a few polymer polymers chains (PE, PP, PS, and PVC) in the previous decade. Esterases, proteases, cutinases, and laccase are just a few of the enzymes that have shown promise in the breakdown of plastics.

Bacterium *Ideonella sakaiensis* may use polyethylene terephthalate (PET) as its major energy and carbon source (Yoshida et al., 2016). PET is converted into its monomers terephthalic acid and ethylene glycol by two active enzymes (PETase and MHETase) in this bacterium (Palm et al., 2019). Although a few enzymes and their effects on plastic breakdown are well established, the problem is yet to be resolved. This is due to a lack of methods to increase the efficiency and commercial manufacturing of plastic degrading enzymes. Recent research on the enzymatic destruction of plastics has sparked a lot of interest in protein/enzyme modification to improve enzyme activity. In comparison to the wild type strain, a designed PETase mutant from *Ideonella sakaiensis* shows a 1.4-fold, 2.1-fold, and 2.5-fold rise in the three mutants (R61A, L88F, 303, and I179F). This research demonstrates that rational protein engineering and altering crucial hydrophobic grooves of substrate binding sites might boost enzyme performance (Ma et al., 2018). Surprisingly, a recent study found that protein-engineered enzymes are effective in degrading microplastics (Islam et al., 2019). According to the report, polymer-binding peptides, such as anchor peptides, which act as a binding tool for synthetic polymers, start the breakdown of polyurethane microplastics. The enhanced breakdown efficiency of the protein designed enzyme, Tachystatin A2 (anchor peptide) of polyurethane as compared to Tcur1278-T (wild type) is responsible for the observed drop in the shelf life of polyurethane from 41.8 to 6.2 h, or roughly 6.7 folds (Islam et al., 2019). These surprising findings show that protein/enzyme modification might be one of the ways for more effectively removing micro-nano plastics.

Chitinases are enzymes produced by a variety of bacteria, including *Achromobacter, Flavobacterium, Micrococcus, Pseudomonas, and Vibrio species*. They break down the polymer by hydrolysing it. The bacteria *Pseudomonas* sp. has been shown to destroy MPs particles, however, the exact method is uncertain. Further investigation found that the chitinase enzyme is used in the enzymatic breakdown of polymeric polymers (Rogers et al., 2020). MPs with larger fragment sizes have a harder time entering microbial cells across the cell membrane. As a result, before being absorbed and biodegraded within the cell, it must be broken down into tiny bits. Microbial enzymes and microorganisms break down MPs using a method that often involves hydrolysis, which is a common degradation process. MP biodegradation is triggered by enzymatic hydrolysis and depolymerisation of MP polymers. Without the help of microbial enzymes, MPs cannot be broken down (Yuan et al., 2020). Microorganisms degraded HMW polymers through a variety of mechanisms, including the use of MNPs as a source of carbon for growth and development (Othman et al., 2021).

Polystyrene is most commonly found in the food packaging industry. Polystyrene degradation has been extensively studied. This pollutant has been reported to be degraded by a variety of species. Polystyrene breakdown has been linked to some enzymes (styrene monooxygenase, styrene oxide isomerase, phenylacetaldehyde dehydrogenase, and phenylacetyl coenzyme A ligase), with acetyl-CoA as the final monomer used in the TCA cycle. Serine hydrolase is another enzyme that can breakdown polystyrene, although nothing is known about its mechanism (Othman et al., 2021).

16.6.2 Enzyme-Assisted Degradation

It is self-evident that once plastic garbage is put into the natural environment, as a result of its interaction with the ecosystem, it may disintegrate or degrade. Abiotic and biotic variables interact in a synergistic manner Kumar et al., 2020a). Microbial-assisted biodegradation of plastic waste has proven to be a more environmentally friendly and successful method. A microbial variety of genes, proteins, enzymes, and their roles in various metabolic pathways have been identified as important in altering plastic polymers and enabling the depolymerisation process (Priya et al., 2021; Yuan et al., 2020) Fig. 16.1. Cutinases, lipases, esterases, carboxylesterases, and other microbial enzymes have been shown to change and/or breakdown a variety of plastic pieces (Zhang et al., 2020) Table 16.1. Several oxygenases, including monooxygenases and dioxygenases, have also been found to help in the enzymatic oxidation of synthetic plastics (Jaiswal et al., 2019; Xu et al., 2019). The oxidation increases the plastic polymer's hydrophilic property, which encourages microbe colonization and the release of different plastic degrading enzymes such as esterases, hydrolases, and lipases (Kim et al., 2020; Onda et al., 2020; Puglisi et al., 2019). Laccase, a Cu binding enzyme isolated from *R. ruber* and *A. flavus*, is widely recognized for its role in PE biodegradation (Zhang et al., 2020; Priya et al., 2021).

Similarly, the biodegradation and use of PE oligomers by *R. rhodochrous* bacteria were investigated utilising specialized carrier proteins such as ATP binding cassettes (ABC) or major facilitator superfamily members. Similarly, *R. rhodochrous* bacteria were tested for their ability to biodegrade and utilize PE oligomers using

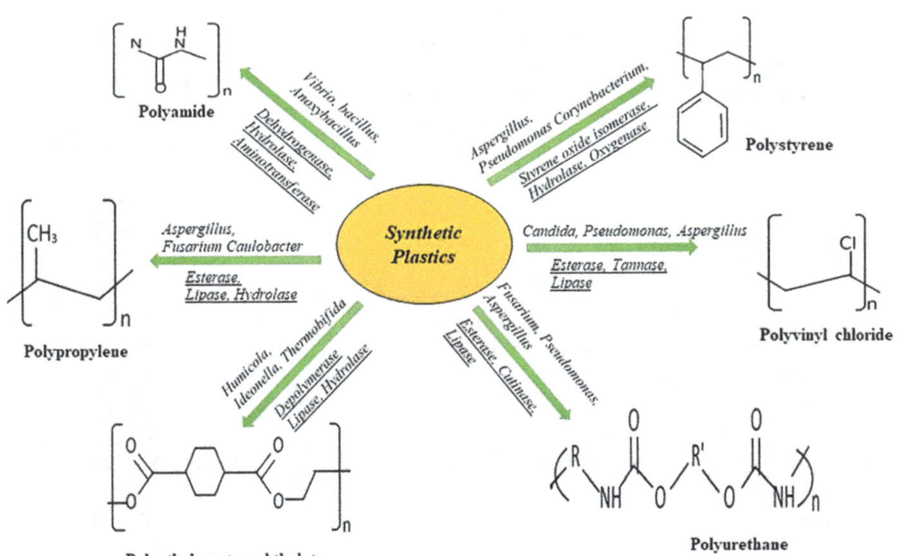

Fig. 16.1 Depolymerization of synthetic plastics by microorganisms and their enzymes. (Modified from Zhou et al. 2021)

Table 16.1 Degradation of plastic polymers by enzymes

Microorganisms	Enzymes	Plastic polymer	References
Pseudophormidium sp. *Bacillus subtilis; B. flexus; P. stutzeri Alcaligenes; Pseudomonas, Vibrio*	– – – –	PP	Urbanek et al. (2018) Arkatkar et al. (2009) Cacciari et al. (1993)
P. fluorescence; Rhodococcus Equi	Protease; aryl acylamidase	PUR	Howard et al. (2001), Purohit et al. (2020)
P. chlororaphis, P. protegens BC212	Lipase		Danso et al. (2019), Hung et al. (2016)
Pestalotiopsis microspore	Serine hydrolase		Russell et al. (2011)
Pseudomonas chlororaphis	polyurethenase		Howard and Blake (1998), Zheng et al. (2005)
Rhodococcus ruber	Laccase	PE	Santo et al. (2013)
Penicillium simplicissimum	lipase		Yamada-Onodera et al. (2001)
Phanerochaete chrysosporium	Manganese peroxidase		Shimao (2001)
(Bacillus, Micrococcus, Nocardia, Pseudomonas, Rhodococcus)	Styrene oxide Isomerase; Phenylacetaldehyde dehydrogenase	PS	
Microbial consortia PS P. putida AJ, P. putida CA-3	Styrene monooxygenase Alkane hydroxylase		Danso et al. (2019), Jacquin et al. (2019) Danko et al. (2004), O'Leary et al. (2005)
Pseudomonas sp.	Lipase	PET	Jacquin et al. (2019), Lewin et al. (2016)
Thermobifida fusca/ Thermomonospora fusca	Cutinase; lipase		Müller et al. (2005)
Humicola sp.	Cutinase		Danso et al. (2019)
Ideonella sakaiensis	MHETase; PETase		Yoshida et al. (2016)
Fusarium sp.	Cutinase		O'Neill et al. (2007)

(continued)

Table 16.1 (continued)

Microorganisms	Enzymes	Plastic polymer	References
Polyporus versicolor; Pleurotus sajor caju; Thermomonospora fusca	–	PVC	Kleeberg et al. (1998), Purohit et al. (2020)
Alteromonadaceae (Alteromonas); Cellvibrionaceae; Oceanospirillaceae; Aestuariicela	–		Jacquin et al. (2019), Danso et al. (2019)
Trametes versicolor	Laccase	Polycaprolactone	Fujisawa et al. (2001)
White-rot fungus IZU-154, Amycolatopsis sp.	Manganese peroxidase		
Agromyces sp.	Nylon hydrolase	Nylon	Negoro et al. (2012)

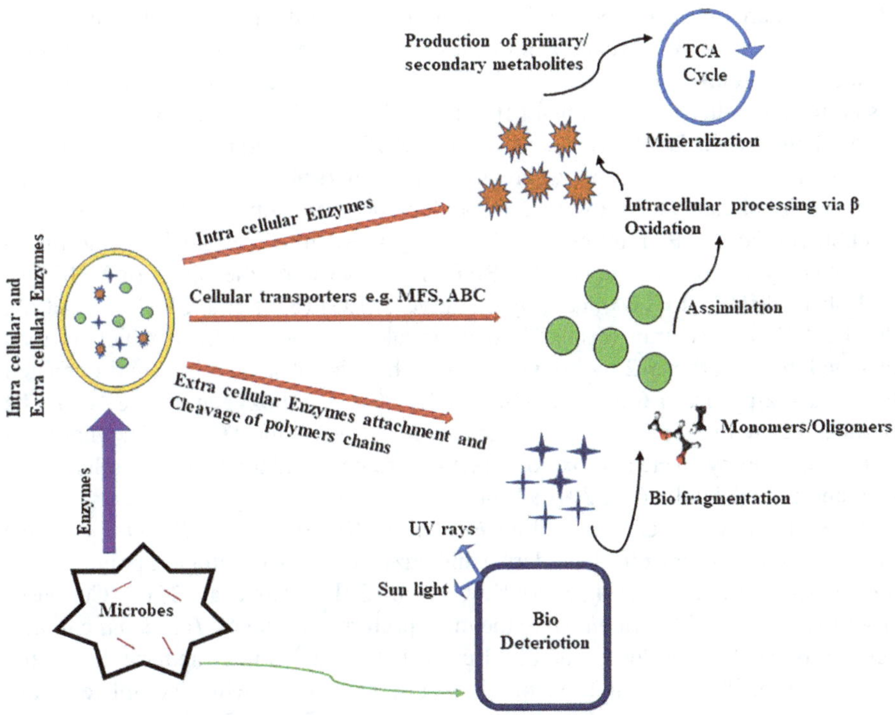

Fig. 16.2 A schematic representation of bio-mineralization of plastic waste. (Modified from Zhou et al. 2021)

specific carrier proteins such as ATP binding cassettes (ABC) or major facilitator superfamily (MFS) (Eyheraguibel et al., 2017; Gravouil et al., 2017). Gravouil et al., 2017 studied the bacteria's growth in PE-supplemented media as well as their enzymatic expression. The microorganisms consumed PE through a series of events that began with the engagement of acetyl CoA and succinyl CoA in the TCA cycle, followed by the creation of energy currency in the form of nicotinamide adenine dinucleotide hydrogen (NADH). The produced energy packet is then used to produce adenosine triphosphate (ATP) via the electron transport chain (ETC), as well as CO2 and H2O as by-products in the PE mineralisation process (Fig. 16.2). The involvement of certain genes, enzymes, and transporter proteins in PE biodegradation was disclosed in the fundamental results (Gravouil et al., 2017; Kumari & Chaudhary, 2020). Specific genes, such as alkane hydroxylase (alkB) from the *Pseudomonas* sp. E4 strain, allowed the PE to be degraded up to 28.6% in 80 days. In addition, to test the effectiveness of the alkB gene, the *Escherichia coli BL21* strain was chosen as the host for the production of the gene, which resulted in 19.3% enzymatic breakdown of the PE (Llorente-García et al., 2020; Yoon & Jeon, 2012). Hydrolases are produced by a variety of marine and soil bacteria, and their significance in the enzymatic breakdown of plastic wastes has been documented (Kawai et al., 2019; Tourova et al., 2020). PET hydrolase from *Ideonella sakaiensis* was tested for its ability to breakdown MHET and found to have similar activity as tannases (Palm et al., 2019). When exposed to PET as a carbon source, the *bacteria Ideonella sakaiensis* was able to create PETase (hydrolase) and MHETase. These two biological catalysts allow PET to be degraded into simpler chemicals such as terephthalic acid (TPA), mono(2-hydroxyethyl) terephthalate (MHET), and bis(2-hydroxyethyl) terephthalate (BHT) (BHET). Furthermore, the action of MHETase hydrolyses MHET to TPA and ethylene glycol (EG) (Kim et al., 2020; Xu et al., 2019). TPA is also transformed into protocatechuic acid (PCA), which is then changed into 4 carboxy-2 hydroxymuconic, which is subsequently dehydrogenated to produce 2-pyrone-4-6-dicarboxylic acid by PCA 3, 4 dioxygenase (PCA34). The produced 2-pyrone-4,6-dicarboxylic acid is employed in the TCA cycle, where it is transformed to pyruvate and oxaloacetate before being mineralised and released as CO_2 and H_2O (Mahdi et al., 2016; Tourova et al., 2020).

Several fungus species, including *Fusarium, Humicola, and Penicillium,* have been identified as PET bio-degraders using enzymes such as cutinase, polyesterase, and hydrolase (da Costa et al., 2020; Kawai et al., 2019; Palm et al., 2019). Cutinases from *Fusarium and Humicola* were the most preferred enzymes. *Humicola* cutinase activity is insufficient due to the build-up of MHET as an intermediary during the breakdown of PET; thus, lipase from C. Antarctica is used, which completely converts MHET to TPA (Carniel et al., 2016; Moharir & Kumar, 2018). PET esterases, like PET hydrolases, have been found to aid in the hydrolysis of bis (benzoyloxyethyl)-terephthalate and polycaprolactone (Hajighasemi et al., 2018; Nabi et al., 2020). Luu et al. (2013) found that styrene monooxygenase (extracted from the P. putida F1 strain) aided PS breakdown by oxidation with styrene epoxide. Monooxygenase produces phenylacetaldehyde from styrene epoxide in the second phase of oxidation, which is then converted to phenylacetic acid (PAA). Furthermore, several enzymes convert PAA to phenylacetyl-CoA, which then enters the TCA cycle as acetyl-CoA and succinyl-CoA. The bacterial strain *P. putida CA-3* was used to breakdown PS via the Phenylacetyl-CoA catabolic pathway. This method makes use of the catabolic operon, which makes it easier for P. putida CA-3 to use PS and produce PHAs (O'Leary et al., 2005).

Enzymatic degradation of synthetic plastic variations, such as PUR, has been widely documented. The bacterial strain *Comamonas acidovorans TB-35* has been used to biodegrade PUR via an esterase termed pudA (Yuan et al., 2020). Furthermore, various fungal species have been examined and found to degrade PUR via enzymes such as lipase, esterases, and hydrolases, including *Penicillium chrysogenum, Aspergillus fumigates, Fusarium solani, Candida ethanolica, and Candida rugosa* (Jenkins et al., 2019; Kalita et al., 2020; Vanleeuw et al., 2019). Various *Arthrobacter* bacterial species have been shown to produce hydrolases and aminotransferases capable of degrading nylon oligomers. The presence of the nylD1 and nylE1 genes, which encode 6-aminohexanoate aminotransferase and adipate semialdehyde dehydrogenase, in both, that mutually facilitated the metabolization of 6-aminohexanoate and adipate semialdehyde to adipate semialdehyde and adipate, is confirmed by the whole genome sequence of this organism (Gatz-Schrupp et al., 2020; Yuan et al., 2020).

16.6.3 Advance Molecular Technologies

The emergence of strong microbial strains as a result of advances in metabolic engineering and synthetic biology resulted in better biotransformation potency and more environmentally friendly recycling of synthetic polymers. Genetic modification tactics are powerful tools for changing microorganisms' fundamental features and increasing their ability to digest plastic wastes (Gu, 2021). Furthermore, the systems biology technique used a variety of omics strategies, such as genomics, metabolomics, proteomics, transcriptomics, and proteomics, to enhance the monitoring of environmental pollutants degradation (Basu et al., 2018; Kumar et al., 2017, 2018a, 2018b, 2019) (Fig. 16.3).

Various metabolic engineering techniques have recently been devised and implemented, either alone or in conjunction with genetically designed constructs that have been shown to digest resistant pollutants (Kumar et al., 2020b; Taha et al., 2021). Systems metabolomic engineering has emerged as a key technique for assisting the success of modified microorganisms by enhancing cellular development and increasing plastic breakdown efficiency (Yang et al., 2017). The advancement of gene editing methods and tools (TALENs, CRISPR/Cas9) leads to an increase in the plastic degradation potency of bacteria (Gaj et al., 2013; Priya et al., 2021). These methods may be used to introduce genes encoding of different plastic degrading enzymes, such as PETase, esterase, depolymerase, laccase, and others, into the genomes of bacteria (Gaj et al., 2013; Tofa et al., 2019).

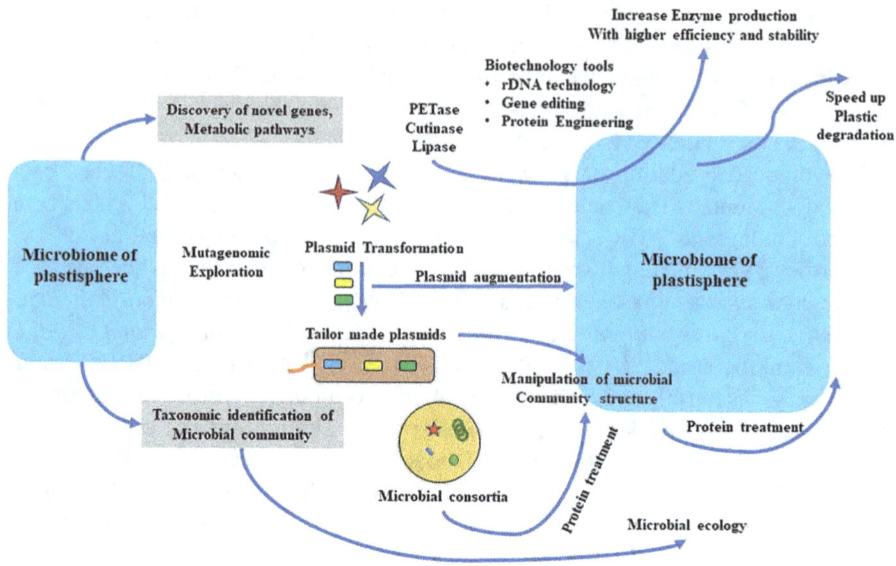

Fig. 16.3 Application of advanced molecular technologies/genetic engineering in biodegradation of plastic waste. (Modified from Zhou et al. 2021)

Several studies have found that the biodegradation of plastic garbage by wild bacteria is slower than that of engineered constructs (Gu, 2021). In comparison to their wild equivalent, modified enzyme construct cutinase was found to lower the degradation time of PUR 41.8 to 6.2 h (Islam et al., 2019). The improved plastic biodegradation potential of modified marine microorganisms' consortia has been proven by Syranidou et al. (2019). Plastic polymers may be depolymerized using modified plastic degrader microorganisms. Despite some victories related to microbe genetic engineering at lab sizes, the majority of modified microorganisms have had dismal field results. The better understanding of this information, the better will be the biodegradation of plastic polymers. Bioinformatics has emerged as a powerful technique for improving the biodegradation of pollutants like MNPs (Purohit et al., 2020). To assess biodegradation, many databases linked to biodegradation processes and toxicities have been produced. Few well-known databases, such as UM-BBD, MetaCyc, and BioCyc, give useful information on microbial metabolic pathways, microbial genes and enzymes, and their complicated enzymatic activities, which aid in the biodegradation of recalibrates compounds like plastic polymers (Priya et al., 2021; Tourova et al., 2020).

These computational approaches are particularly useful for not only identifying and researching the relevant enzymes but also anticipating the biodegradation pathways of previously unknown dangerous compounds. This bioinformatics strategy undoubtedly offered a collaborative platform on which metabolic engineering and synthetic biology might work together to develop a revolutionary approach to plastic biodegradation (Ali et al., 2021a, 2021b). However, the unavailability, inaccessibility, and validity of bioinformatics and experimental data are major limitations that should be considered in future investigations. Synthetic biology, particularly 'omic' studies, as well as computational biology and high-throughput sequencing, have continually aided in the understanding of microbial–plastic–sphere interactions and consequent polymer breakdown (Bouhajja et al., 2016; Wagner & Lambert, 2017). Designing a metabolic pathway for the biodegradation of synthetic polymers is also an important aspect of synthetic biology (Purohit et al., 2020; Wang et al., 2020). Nonetheless, there remain gaps in our understanding of the various categories of synthetic polymer degrader microorganisms and the enzymes that are responsible for their degradation. As a result, future research should focus on the characterization and identification of a strong polymer degrading bacterium and its enzymes. To overcome the stumbling block in the field of plastic polymer biodegradation, extensive study in environmental microbiology and biotechnology, gene modification, and protein engineering is necessary. Combinatorial approaches to biodegradation of plastic polymers, such as the integration of bioinformatics tools, metabolic engineering, genetics, molecular biology, and system biology, may soon give ground-breaking insight.

16.7 Bio-Membrane Technology

Bio-membrane technology, also known as membrane bioreactor (MBR) technology, is a system in which a biological catalyst, such as bacteria or enzymes, or both, is coupled to a partition mechanism that is controlled by a film-based system, such as microfiltration or ultrafiltration (Dey et al., 2021; Xiao et al., 2018). MBR is now widely regarded as a developing technology for the treatment of industrial and municipal waste. MBR is also widely used in the fields of food, medicine, biorefinery, and biodiesel production (Judd, 2015). The MBR is used in the treatment of MPs to lower the complexity of the MPs contaminated medium by biodegrading organic materials (OMs), resulting in enhanced MP degradation. When a stream of pre-treated WW reaches the bioreactor, the OMs are biodegraded there. For the removal method, the remaining treated liquid is urged in a semi-cross flow-filtering set-up and concentrated in the retentate flow (Poerio et al., 2019). Talvitie et al. (2017b) has compared the effectiveness of MBR with a few different WW treatment technologies for MP separation, including disc filters, quicksand filtering, and dissolved air flotation. In comparison to the technologies listed above, an MBR technology demonstrated high removal effectiveness of MPs (99%) as well as enhanced final effluent quality and fewer treatment stages. Talvitie et al. (2017b) presented a study on MP separation using Rapid Sand Filters (RSF), Reverse Osmosis (RO), Dissolved Air Flotation (DAF), and Membrane Bioreactors (MBR). According to the findings of these experiments, MBR technology was the most successful in treating MPs contaminated WW (Talvitie et al. 2017a). Although MPs from WW might be sorted out by MBR during WWT, MPs treatment technologies that aid in the removal of MPs from existing WWTPs are currently in development.

MBRs have been shown to be efficient at degrading organic pollutants in MNPs. For example, MBR has been examined in the lab alone or in conjunction with secondary sludge in the biodegradation of phthalate esters, starting from various synthetic or natural WW (Camacho-Muñoz et al., 2012), paper mill WW (Yoshida et al., 2016), MSW leachate (Boonyaroj et al., 2012), and so on. In WWTP, MBR showed a 70%t increase in DEHP removal compared to standard treatment technology (3%) and an even greater increase (83 percent) when combined with primary adsorption. When MBR was combined with primary anaerobic digestion and RO filtering, complete MPs biodegradation was accomplished (Balabanic et al., 2012). Nonetheless, the physico-chemical features of pollutants, as well as operating circumstances such as initial feeding rate and concentration, hydraulic retention time (HRT), and so on have a strong influence on the MBR's degrading effectiveness. The isolated bacteria species, *Ideonella sakaiensis*, can use PET as a carbon source (Yoshida et al., 2016), which might be useful shortly in connection with MBR. This bacteria in particular has a unique enzymatic system that effectively converts PET into less hazardous monomeric versions like TPA and EG. Dawson et al. (2018) found that when Antarctic Krill (Euphasia superba) feed on MP, it shrinks from 31.5 to <1 μm in size. The participation of a complicated enzymatic process in the size reduction of MPs by Antarctic Krill was discovered after an in-depth examination. These enzymes will easily combine with the MBR soon, and it will surely biodegrade the MP, as Barth et al. (2015) verified for PET biodegradation.

16.8 Conclusion

Micro-nano plastic pollution, amongst others, is recognized as a serious contaminant that lasts longer in the environment. Microbial-mediated MNP remediation can provide a platform for degrading or remediating MNPs via enzyme activity on polymers. Microbes' role in the degradation of MNPs is currently poorly understood. Furthermore, future challenges include the use of microbial enzymes to aid in the destruction and reformation of plastics for improved biodegradation. More significantly, future studies should focus on identifying organisms capable of acting on a wide range of plastics, as well as high molecular weight polymers. After polymers degrade, monomers, dimers, and oligomers are generated, which can be utilized to create new materials. Because identifying and isolating highly active and functioning enzymes is difficult, the metagenomics technique might help find non-cultivated microbes, thereby assisting in the development of function-based tests. Though an integrated function of the membrane, enzyme, nanoparticle, and metagenomics in the remediation of MNPs is promising, it is critical to investigate 'micro-nano plastics active' or 'micro-nano plastics degrading enzymes' and their process. Extensive study in this area should result in a significant reduction in global micro-nano-sized plastic pollution, as well as improved health for future generations. In addition, this study covers a wide range of MNP remediation techniques, including enzymatic, bio-membrane, advanced molecular, and nanoparticle technologies. Although numerous study investigations have been completed to mitigate MNPs by bioremediation and so minimize their ecotoxicological repercussions, further studies encompassing the features of MNP pollution, their effects, and mitigation techniques are still needed.

References

Ali, I., Ding, T., Peng, C., Naz, I., Sun, H., Li, J., & Liu, J.-F. (2021a). Micro- and Nanoplastics in wastewater treatment plants: Occurrence, removal, fate, impacts and remediation technologies – A critical review. *Chemical Engineering Journal, 423*. https://doi.org/10.1016/j.cej.2021.130205

Ali, S., Elsamahy, T., Altohamy, R., Zhu, D., Mahmoud, Y., Koutra, E., Metwally, M., Kornaros, M., & Sun, J. (2021b). Plastic wastes biodegradation: Mechanisms, challenges and future prospects. *Science of The Total Environment, 780*, 146590. https://doi.org/10.1016/j.scitotenv.2021.146590

Allouzi, M., Tang, D., Wayne, K., Chew, Rinklebe, J., Bolan, N., Allouzi, S., & Show, P.-L. (2021). Micro (Nano) plastic pollution: The ecological influence on soil-plant system and human health. *Science of The Total Environment, 788*, 147815. https://doi.org/10.1016/j.scitotenv.2021.147815

Amelung, W., & Zech, W. M. (1999). Minimisation of organic matter disruption during particle-size fractionation of grassland epipedons. *Geoderma, 92*, 73–85. https://doi.org/10.1016/S0016-7061(99)00023-3

Anastopoulos, I., & Pashalidis, I. (2021). Single-use surgical face masks, as a potential source of microplastics: Do they act as pollutant carriers? *Journal of Molecular Liquids, 326*. https://doi.org/10.1016/j.molliq.2020.115247

Arkatkar, A., Arutchelvi, J., Bhaduri, S., Uppara, P., & Doble, M. (2009). Degradation of unpretreated and thermally pretreated polypropylene by soil consortia. *International Biodeterioration & Biodegradation, 63*, 106–111. https://doi.org/10.1016/j.ibiod.2008.06.005.

Baker, R. W. (2012). *Membrane Technology and Applications*. Wiley.

Balabanic, D., Hermosilla, D., Merayo, N., Klemenčič, K., Aleksandra, & Blanco, A. (2012). Comparison of different wastewater treatments for removal of selected endocrine-disruptors from paper mill wastewaters. *Journal of Environmental Science and Health Part A, 47*, 1350–1363. https://doi.org/10.1080/10934529.2012.672301

Barnes, D., Galgani, F., Thompson, R., & Barlaz, M. (2009). Accumulation and fragmentation of plastic debris in global environments. *Philosophical Transactions of the Royal Society of London Series B, Biological Sciences, 364*, 1985–1998. https://doi.org/10.1098/rstb.2008.0205

Barth, M., Wei, R., Oeser, T., Then, J., Schmidt, J., Wohlgemuth, F., & Zimmermann, W. (2015). Enzymatic hydrolysis of polyethylene terephthalate films in an ultrafiltration membrane reactor. *Journal of Membrane Science, 494*, 182–187. https://doi.org/10.1016/j.memsci.2015.07.030

Basu, S., Rabara, R., Negi, S., & Shukla, P. (2018). Engineering PGPMOs through gene editing and systems biology: A solution for phytoremediation? *Trends in Biotechnology, 36*. https://doi.org/10.1016/j.tibtech.2018.01.011

Bi, M., He, Q., & Chen, Y. (2020). What roles are terrestrial plants playing in global microplastic cycling? *Environmental Science & Technology*. XXXX. https://doi.org/10.1021/acs.est.0c01009

Boonyaroj, V., Chiemchaisri, C., Chiemchaisri, W., Theepharaksapan, S., & Yamamoto, K. (2012). Toxic organic micro-pollutants removal mechanisms in long-term operated membrane bioreactor treating municipal solid waste leachate. *Bioresource Technology, 113*, 174–180. https://doi.org/10.1016/j.biortech.2011.12.127

Boots, B., Russell, C., & Green, D. (2019). Effects of microplastics in soil ecosystems: Above and below ground. *Environmental Science & Technology, 53*. https://doi.org/10.1021/acs.est.9b03304

Böttcher, J., Schlichting, E., Blume, H.-P., & Stahr, K. (1996). Bodenkundliches Praktikum - Eine Einführung in pedologisches Arbeiten für Ökologen, insbesondere Land- und Forstwirte und für Geowissenschaftler. 2., neubearbeitete Auflage. Blackwell Wissenschafts-Verlag, Berl. Zeitschrift für Pflanzenernährung und Bodenkunde. 159. https://doi.org/10.1002/jpln.1996.3581590315.

Bouhajja, E., Agathos, S., & George, I. (2016). Metagenomics: Probing pollutant fate in natural and engineered ecosystems. *Biotechnology Advances, 34*. https://doi.org/10.1016/j.biotechadv.2016.10.006

Bradney, L., Wijesekara, H., Palansooriya, K. N., Obadamudalige, N., Bolan, N., Ok, Y. S., Rinklebe, J., Kim, K.-H., & Kirkham, M. B. (2019). Particulate plastics as a vector for toxic trace-element uptake by aquatic and terrestrial organisms and human health risk. *Environment International, 131*, 104937. https://doi.org/10.1016/j.envint.2019.104937

Cacciari, I., Quatrini, P., Zirletta, G., Mincione, E., Vinciguerra, V., Lupattelli, P., & Sermanni, G. (1993). Isotactic polypropylene biodegradation by a microbial community: Physicochemical characterization of metabolites produced. *Applied and environmental microbiology, 59*. 3695–700. https://doi.org/10.1128/AEM.59.11.3695-3700.1993

Camacho-Muñoz, M., Martín, J., Santos, J., Alonso, E., Aparicio, I., De la Torre, T., Rodriguez, C., & Malfeito, J. (2012). Effectiveness of three configurations of membrane bioreactors on the removal of priority and emergent organic compounds from wastewater: Comparison with conventional wastewater treatments. *Journal of environmental monitoring : JEM, 14*, 1428–1436. https://doi.org/10.1039/c2em00007e

Cao, D., Wang, X., Luo, X., & Liu, G. (2017). Effects of polystyrene microplastics on the fitness of earthworms in an agricultural soil. *IOP Conference Series: Earth and Environmental Science, 61*, 012148. https://doi.org/10.1088/1755-1315/61/1/012148

Carniel, A., Valoni, É., Junior, J., Gomes, A., & Castro, A. (2016). Lipase from Candida antarctica (CALB) and cutinase from Humicola insolens act synergistically for PET hydrolysis to terephthalic acid. *Process Biochemistry, 59*. https://doi.org/10.1016/j.procbio.2016.07.023

Chai, B., Wei, Q., She, Y., Lu, G., Dang, Z., & Yin, H. (2020). Soil microplastic pollution in an e-waste dismantling zone of China. *Waste Management (New York, N.Y.), 118*, 291–301. https://doi.org/10.1016/j.wasman.2020.08.048

Cheng, Y. L., Kim, J. G., Kim, H.-B., Choi, J., Tsang, Y. F., & Baek, K. (2021). Occurrence and removal of microplastics in wastewater treatment plants and drinking water purification facilities: A review. *Chemical Engineering Journal, 410*, 128381. https://doi.org/10.1016/j.cej.2020.128381

Christensen, B. T. (1996). Carbon in primary and secondary organomineral complexes. In *Structure and organic matter storage in agricultural soils* (pp. 97–165).

Cole, M., Webb, H., Lindeque, P., Fileman, E., Halsband, C., & Galloway, T. (2014). Isolation of microplastics in biota-rich seawater samples and marine organisms. *Scientific Reports, 4*, 4528. https://doi.org/10.1038/srep04528

Costa, A., Lopes, V., Vidal, L., Nicaud, J.-M., Castro, A., & Coelho, M. A. (2020). Poly(ethylene terephthalate) (PET) degradation by Yarrowia lipolytica: Investigations on cell growth, enzyme production and monomers consumption. *Process Biochemistry, 95*. https://doi.org/10.1016/j.procbio.2020.04.001

Czimczik, C., & Masiello, C. (2007). Controls on black carbon storage in soils. *Global Biogeochemical Cycles - GLOBAL BIOGEOCHEM CYCLE., 21*. https://doi.org/10.1029/2006GB002798

Dawson, A., King, C., Townsend, K., King, R., Huston, W., Nash, B., & Susan. (2018). Turning microplastics into nanoplastics through digestive fragmentation by Antarctic krill. *Nature Communications, 9*. https://doi.org/10.1038/s41467-018-03465-9

Danso, D., Chow, J., & Streit, W. (2019). Plastics: Microbial Degradation, Environmental and Biotechnological Perspectives. *Applied and Environmental Microbiology, 85*. https://doi.org/10.1128/AEM.01095-19

de Souza Machado, A., Kloas, W., Zarfl, C., Hempel, S., & Rillig, M. (2017). Microplastics as an emerging threat to terrestrial ecosystems. *Global Change Biology., 24*. https://doi.org/10.1111/gcb.14020

Dehaut, A., Cassone, A.-L., Frère, L., Hermabessiere, L., Himber, C., Rinnert, E., Riviere, G., Lambert, C., Soudant, P., Huvet, A., Duflos, G., & Paul-Pont, I. (2016). Microplastics in seafood: Benchmark protocol for their extraction and characterization. *Environmental Pollution, 215*, 223–233. https://doi.org/10.1016/j.envpol.2016.05.018

Dekiff, J., Remy, D., Klasmeier, J., & Fries, E. (2014). Occurrence and spatial distribution of microplastics in sediments from Norderney. *Environmental Pollution (Barking, Essex : 1987), 186C*, 248–256. https://doi.org/10.1016/j.envpol.2013.11.019

Depledge, M., Galgani, F., Panti, C., Caliani, I., Casini, S., & Fossi, M. (2013). Plastic litter in the sea. *Marine Environmental Research, 92*. https://doi.org/10.1016/j.marenvres.2013.10.002

Dey, T. K., Uddin, M., & Jamal, M. (2021). Detection and removal of microplastics in wastewater: Evolution and impact. *Environmental Science and Pollution Research, 28*. https://doi.org/10.1007/s11356-021-12943-5

Dümichen, E., Barthel, A.-K., Braun, U., Bannick, C., Brand, K., Jekel, M., & Senz, R. (2015). Analysis of polyethylene microplastics in environmental samples, using a thermal decomposition method. *Water Research, 85*, 451–457. https://doi.org/10.1016/j.watres.2015.09.002

Dümichen, E., Eisentraut, P., Bannick, C., Barthel, A.-K., Senz, R., & Braun, U. (2017). Fast identification of microplastics in complex environmental samples by a thermal degradation method. *Chemosphere, 174*. https://doi.org/10.1016/j.chemosphere.2017.02.010

Danko, A., Luo, M., Bagwell, C., Brigmon, R., & Freedman, D. (2004). Involvement of Linear Plasmids in Aerobic Biodegradation of Vinyl Chloride. *Applied and environmental microbiology, 70*, 6092–7. https://doi.org/10.1128/AEM.70.10.6092-6097.2004

Ebere, E., & Wirnkor, V., & Evelyn Ngozi, V. (2019). Uptake of microplastics by plant: A reason to worry or to be happy?

Eggen, T., Heimstad, E., Stuanes, A., & Norli, H. (2012). Uptake and translocation of organophosphates and other emerging contaminants in food and forage crops. *Environmental Science and Pollution Research International, 20*. https://doi.org/10.1007/s11356-012-1363-5

Eriksen, M., Mason, S., Wilson, S., Box, C., Zellers, A., Edwards, W., Farley, H., & Amato, S. (2013). Microplastic pollution in the surface waters of the Laurentian Great Lakes. *Marine Pollution Bulletin., 77*, 177–182. https://doi.org/10.1016/j.marpolbul.2013.10.007

Eyheraguibel, B., Traikia, M., Fontanella, S., Sancelme, M., Bonhomme, S., Fromageot, D., Lemaire, J., Lauranson, G., Lacoste, J., & Delort, A.-M. (2017). Characterization of oxidized oligomers from polyethylene films by mass spectrometry and NMR spectroscopy before and after biodegradation by a Rhodococcus rhodochrous strain. *Chemosphere, 184*. https://doi.org/10.1016/j.chemosphere.2017.05.137

Fu, X., & Du, Q. (2011). Uptake of Di-(2-ethylhexyl) phthalate of vegetables from plastic film greenhouses. *Journal of Agricultural and Food Chemistry, 59*, 11585–11588. https://doi.org/10.1021/jf203502e

Fuke, P., Thangaraju, M., Kumar, M., Sawarkar, A., Pandey, A., & Singh, L. (2021). Role of microbial diversity to influence the growth and environmental remediation capacity of bamboo: A review. *Industrial Crops and Products, 167*. https://doi.org/10.1016/j.indcrop.2021.113567

Fuller, S., & Gautam, A. (2016). A procedure for measuring microplastics using pressurized fluid extraction. *Environmental Science & Technology, 50*. https://doi.org/10.1021/acs.est.6b00816

Gaj, T., Gersbach, C., & Barbas, C. (2013). ZFN, TALEN, and CRISPR/Cas-based methods for genome engineering. *Trends in Biotechnology, 31*. https://doi.org/10.1016/j.tibtech.2013.04.004

Gatz-Schrupp, J., Deckard, P., Hufford, B., Ly, S., Tupa, P., & Masuda, H. (2020). Isolation and genomic analysis of 11-aminoundecanoic acid-degrading bacterium Pseudomonas sp. JG-B from nylon 11 enrichment culture. *Journal of Genomics, 8*, 16–20. https://doi.org/10.7150/jgen.42095

Glaser, B., Haumaier, L., Guggenberger, G., & Zech, W. (1998). Black carbon in soils: The use of benzenecarboxylic acids as specific markers. *Organic Geochemistry, 29*, 811–819. https://doi.org/10.1016/S0146-6380(98)00194-6

Golchin, A., Oades, J. M., Skjemstad, J. O., & Clarke, P. (1994). Soil structure and carbon cycling. *Australian Journal of Soil Research - AUST J SOIL RES, 32*. https://doi.org/10.1071/SR9941043

Gravouil, K., Clément, R., Colas, S., Hélye, R., Kadri, L., Bourdeau, L., Moumen, B., Mercier, A., & Ferreira, T. (2017). Transcriptomics and Lipidomics of the environmental strain Rhodococcus ruber point out consumption pathways and potential metabolic bottlenecks for polyethylene degradation. *Environmental Science & Technology, 51*. https://doi.org/10.1021/acs.est.7b00846

Gu, J. D. (2021). Biodegradability of plastics: The issues, recent advances, and future perspectives. *Environmental Science and Pollution Research, 28*(2), 1278–1282.

Guan, N., & Liu, L. (2020). Microbial response to acid stress: Mechanisms and applications. *Applied Microbiology and Biotechnology, 104*, 1–15. https://doi.org/10.1007/s00253-019-10226-1

Guo, J., Huang, X.-P., Xiang, L., Wang, Y.-Z., Li, Y.-W., Li, H., Cai, Q.-Y., Mo, C.-H., & Wong, M. (2020). Source, migration and toxicology of microplastics in soil. *Environment International, 137*, 105263. https://doi.org/10.1016/j.envint.2019.105263

Gupta, J., & Thakur, I. (2016). Production and optimization of Polyhydroxyalkanoate from oleaginous bacteria Bacillus Sp. ISTC1. *Research & Reviews: Journal of Microbiology and Biotechnology, 5*, 80–89.

Hajighasemi, M., Tchigvintsev, A., Nocek, B., Flick, R., Popovic, A., Hai, T., Anna, K., Brown, G., Xu, X., Cui, H., Anstett, J., Chernikova, T., Bruls, T., Le Paslier, D., Yakimov, M., Joachimiak, A., Golyshina, O., Savtchenko, A., Golyshin, P., & Yakunin, A. (2018). Screening and characterization of novel Polyesterases from environmental metagenomes with high hydrolytic activity against synthetic polyesters. *Environmental Science & Technology, 52*. https://doi.org/10.1021/acs.est.8b04252

Hauser, R., & Calafat, A. M. (2005). Phthalates and human health. *Occupational and Environmental Medicine, 62*, 806–818. https://doi.org/10.1136/oem.2004.017590

Hidalgo-Ruz, V., Gutow, L., Thompson, R., & Thiel, M. (2012). Microplastics in the marine environment: A review of the methods used for identification and quantification. *Environmental Science & Technology, 46*, 3060–3075. https://doi.org/10.1021/es2031505

Horton, A., Walton, A., Spurgeon, D., Lahive, E., & Svendsen, C. (2017). Microplastics in freshwater and terrestrial environments: Evaluating the current understanding to identify the knowledge gaps and future research priorities. *Science of the Total Environment, 586*. https://doi.org/10.1016/j.scitotenv.2017.01.190

Howard, G., Crother, B., & Vicknair, J. (2001). Cloning, nucleotide sequencing and characterization of a polyurethanase gene (pueB) from Pseudomonas chlororaphis. *International Biodeterioration & Biodegradation, 47*, 141–149. https://doi.org/10.1016/S0964-8305(01)00042-7

Huerta, E., Gertsen, H., Gooren, H. P. A., Peters, P., Salánki, T., Ploeg, M., Besseling, E., Koelmans, A., & Geissen, V. (2016). Microplastics in the terrestrial ecosystem: Implications for Lumbricus terrestris (Oligochaeta, Lumbricidae). *Environmental Science & Technology, 50*. https://doi.org/10.1021/acs.est.5b05478

Hurley, R., & Nizzetto, L. (2018). Fate and occurrence of micro(nano)plastics in soils: Knowledge gaps and possible risks. *Current Opinion in Environmental Science & Health, 1*, 6–11. https://doi.org/10.1016/j.coesh.2017.10.006

Hung, C-S., Zingarelli, S., Nadeau, L., Biffinger, J., Drake, C., & Crouch, A., & Barlow, D., Russell, J., & Goodson, W. (2016). Carbon Catabolite Repression and Impranil Polyurethane Degradation in Pseudomonas protegens Strain Pf-5. *Applied and Environmental Microbiology, 82*, AEM.01448-16. https://doi.org/10.1128/AEM.01448-16

Howard, G., & Blake, R. (1998). Growth of Pseudomonas fluorescens on a polyester–polyurethane and the purification and characterization of a polyurethanase–protease enzyme. *International Biodeterioration & Biodegradation, 42*, 213–220. https://doi.org/10.1016/S0964-8305(98)00051-1

Imhof, H., Ivleva, N., Schmid, J., Niessner, R., & Laforsch, C. (2013). Contamination of beach sediments of a subalpine lake with microplastic particles. *Current Biology, 23*, R867–R868. https://doi.org/10.1016/j.cub.2013.09.001

Iqbal, S., Xu, J., Allen, S., Khan, S., Nadir, S., Arif, M., & Yasmeen, T. (2020). Unraveling consequences of soil micro- and nano-plastic pollution on soil-plant system: Implications for nitrogen (N) cycling and soil microbial activity. *Chemosphere, 127578*. https://doi.org/10.1016/j.chemosphere.2020.127578

Islam, S., Apitius, L., Jakob, F., & Schwaneberg, U. (2019). Targeting microplastic particles in the void of diluted suspensions. *Environment International, 123*, 428–435. https://doi.org/10.1016/j.envint.2018.12.029

Jaiswal, S., Sharma, B., & Shukla, P. (2019). Integrated approaches in microbial degradation of plastics. *Environmental Technology & Innovation, 17*, 100567. https://doi.org/10.1016/j.eti.2019.100567

Jenkins, S., Quer, A., Fonseca, C., & Varrone, C. (2019). Microbial degradation of plastics: New plastic degraders. *Mixed Cultures and Engineering Strategies*. https://doi.org/10.1002/9781119592129.ch12

Judd, S. (2015). The status of industrial and municipal effluent treatment with membrane bioreactor technology. *Chemical Engineering Journal, 305*. https://doi.org/10.1016/j.cej.2015.08.141

Jacquin, J,. Cheng, J., Odobel, C., Pandin, C., Conan, P., Pujo-Pay, M., Barbe, V., Meistertzheim, A.-L. Ghiglione, J-F. (2019). *Microbial Ecotoxicology of Marine Plastic Debris: A Review on Colonization and Biodegradation by the "Plastisphere". Frontiers in Microbiology, 10*, 865. https://doi.org/10.3389/fmicb.2019.00865

Kaiser, M., & Berhe, A. (2014). How does sonication affect the mineral and organic constituents of soil aggregates? —A review. *Journal of Plant Nutrition and Soil Science, 177*. https://doi.org/10.1002/jpln.201300339

Kalita, N. K., Kalamdhad, A., & Katiyar, V. (2020). Recent trends and advances in the biodegradation of conventional plastics. https://doi.org/10.1007/978-981-15-1251-3_17

Kamrannejad, M., Hasanzadeh, A., Nosoudi, N., Mai, L., & Babaluo, A. (2014). Photocatalytic degradation of polypropylene/TiO2 Nano-composites. *Materials Research*. https://doi.org/10.1590/1516-1439.267214

Kappenberg, A., Braun, M., Lehndorff, E., & Amelung, W. (2016). Black carbon assessment using benzene polycarboxylic acids: Limitations for organic-rich matrices. *Organic Geochemistry, 94*. https://doi.org/10.1016/j.orggeochem.2016.01.009

Karbalaei, S., Hanachi, P., Walker, T., & Cole, M. (2018). Occurrence, sources, human health impacts and mitigation of microplastic pollution. *Environmental Science and Pollution Research, 25*, 36046–36063. https://doi.org/10.1007/s11356-018-3508-7

Kawai, F., Kawabata, T., & Oda, M. (2019). Current knowledge on enzymatic PET degradation and its possible application to waste stream management and other fields. *Applied Microbiology and Biotechnology, 103*. https://doi.org/10.1007/s00253-019-09717-y

Khalid, N., Aqeel, M., & Noman, A. (2020). Microplastics could be a threat to plants in terrestrial systems directly or indirectly. *Environmental Pollution, 267*, 115653. https://doi.org/10.1016/j.envpol.2020.115653

Kim, H., Lee, H., Yu, H., Jeon, E., Lee, S., Li, J., & Kim, D.-H. (2020). Biodegradation of polystyrene by Pseudomonas sp. isolated from the gut of Superworms (Larvae of Zophobas atratus). *Environmental Science & Technology*, XXXX. https://doi.org/10.1021/acs.est.0c01495

Kleeberg, I., Hetz, C., Kroppenstedt, R., Müller, R.-J., & Deckwer, W.-D. (1998). Biodegradation of Aliphatic-Aromatic Copolyesters by Thermomonospora fusca and Other Thermophilic Compost Isolates. *Applied and environmental microbiology, 64*, 1731–5. https://doi.org/10.1128/AEM.64.5.1731-1735.1998

Knott, B., Erickson, E., Allen, M., Gado, J., Graham, R., Kearns, F., Pardo, I., Topuzlu, E., Anderson, J., Austin, H., Dominick, G., Johnson, C., Rorrer, N., Szostkiewicz, C., Copié, V., Payne, C., Woodcock, H., Donohoe, B., Beckham, G., & Mcgeehan, J. (2020). Characterization and engineering of a two-enzyme system for plastics depolymerization. *Proceedings of the National Academy of Sciences, 117*, 25476.

Kruse, J., Kanwischer, M., Amelung, W., Baum, C., Bol, R., Kühn, O., Lewandowski, H., Niederberger, J., Oelmann, Y., Rüger, C., Santner, J., Siebers, M., Siebers, N., Spohn, M., Vestergren, J., Vogts, A., & Leinweber, P. (2015). Innovative methods in soil phosphorus research: A review. *Journal of Plant Nutrition and Soil Science, 178*, 43.

Kumar, M., Morya, R., Gnansounou, E., Larroche, C., & Thakur, I. S. (2017). Characterization of carbon dioxide concentrating chemolithotrophic bacterium serratia sp. ISTD04 for production of biodiesel. *Bioresource Technology, 243*, 893–897. https://doi.org/10.1016/j.biortech.2017.07.067

Kumar, M., Kumar, M., Pandey, A., & Thakur, I. (2019). Genomic analysis of carbon dioxide sequestering bacterium for exopolysaccharides production. *Scientific Reports, 9*, 4270.

Kumar, M., Kumar, M., Verma, S., Gazara, R., Pandey, A., & Verma, P. (2018a). Genomic and proteomic analysis of lignin degrading and polyhydroxyalkanoate accumulating β-proteobacterium Pandoraea sp. ISTKB. *Biotechnology for Biofuels, 11*, 154.

Kumar, M., Sundaram, S., Gnansounou, E., & Larroche, C. (2018b). Carbon dioxide capture, storage and production of biofuel and biomaterials by bacteria: A review. *Bioresource Technology, 247*, 1059–1068.

Kumar, M., Varjani, S., Gnansounou, E., Ravindran, S., & Thakur, I. (2018c). Sequestration and utilization of carbon dioxide by chemical and biological methods for biofuels and biomaterials by chemoautotrophs: Opportunities and challenges. *Bioresource Technology, 256*, 478–490.

Kumar, M., Xiong, X., He, M., Tsang, D., Gupta, J., Khan, E., Harrad, S., Hou, D., Ok, Y. S., & Bolan, N. (2020a). Microplastics as pollutants in agricultural soils. *Environmental Pollution, 265*, 114980. https://doi.org/10.1016/j.envpol.2020.114980

Kumar, M., You, S., Beiyuan, J., Luo, G., Gupta, J., Kumar, S., Singh, L., Zhang, S., & Tsang, D. (2020b). Lignin valorization by bacterial genus Pseudomonas: State-of-the-art review and prospects. *Bioresource Technology*. https://doi.org/10.1016/j.biortech.2020.124412

Kumar, R., Sharma, P., Manna, C., & Jain, M. (2021). Abundance, interaction, ingestion, ecological concerns, and mitigation policies of microplastic pollution in riverine ecosystem: A review. *Science ofThe Total Environment, 782*, 146695. https://doi.org/10.1016/j.scitotenv.2021.146695

Kumari, A., & Chaudhary, D. (2020). Engineered microbes and evolving plastic bioremediation technology. In *Bioremediation of pollutants*. https://doi.org/10.1016/B978-0-12-819025-8.00021-1

Lares, V., Ncibi, M. C., Sillanpää, M., & Sillanpää, M. (2018). Occurrence, identification and removal of microplastic particles and fibers in conventional activated sludge process and advanced MBR technology. *Water Research, 133*, 236–246. https://doi.org/10.1016/j.watres.2018.01.049

Lear, G., Kingsbury, J., Franchini, S., Gambarini, V., Maday, S., Wallbank, J., Weaver, L., & Pantos, O. (2021). Plastics and the microbiome: Impacts and solutions. *Environmental Microbiome, 16*. https://doi.org/10.1186/s40793-020-00371-w

Lewin, A., Strand, Trine, H. T., Klinkenberg, G., Kotlar, H., Valla, S., Drablos, F., & Wentzel, A. (2016). Discovery and Characterization of a Thermostable Esterase from an Oil Reservoir Metagenome. *Advances in Enzyme Research, 04*, 68–86. https://doi.org/10.4236/aer.2016.42008

Liebezeit, G., & Dubaish, F. (2012). Microplastics in beaches of the East Frisian Islands Spiekeroog and Kachelotplate. *Bulletin of Environmental Contamination and Toxicology, 89*, 213–217. https://doi.org/10.1007/s00128-012-0642-7

Llorente-García, B., Hernandez Lopez, J., Zaldivar-Cadena, A. A., Siligardi, C., & Cedillo-González, E. (2020). First insights into photocatalytic degradation of HDPE and LDPE microplastics by a Mesoporous N–TiO2 coating: Effect of size and shape of microplastics. *Coatings, 10*, 658. https://doi.org/10.3390/coatings10070658

Löder, M., & Gerdts, G. (2015). Methodology used for the detection and identification of microplastics—A critical appraisal. In *Marine anthropogenic litter* (pp. 201–227). Springer. https://doi.org/10.1007/978-3-319-16510-3_8

Luu, R., Schneider, B., Ho, C., Nesteryuk, V., Ngwesse, S., Liu, X., Parales, J., Ditty, J., & Parales, R. (2013). Taxis of pseudomonas putida F1 toward phenylacetic acid is mediated by the energy taxis receptor Aer2. *Applied and Environmental Microbiology, 79*. https://doi.org/10.1128/AEM.03895-12

Ma, Y., Yao, M., Li, B., Ding, M., He, B., Chen, S., Zhou, X., & Yuan, Y. (2018). Enhanced poly(ethylene terephthalate) hydrolase activity by protein engineering. *Engineering, 4*. https://doi.org/10.1016/j.eng.2018.09.007

Magdouli, S., Daghrir, R., Brar, K., Drogui, P., & Tyagi, R. D. (2013). Di 2-ethylhexylphtalate in the aquatic and terrestrial environment: A critical review. *Journal of Environmental Management, 127*, 36–49. https://doi.org/10.1016/j.jenvman.2013.04.013

Mahdi, M., Ameen, R., & Ibrahim, H. (2016). *International Journal of Advanced Research in Biological Sciences Study on Degradation of Nylon 6 by thermophilic bacteria Anoxybacillus rupiensis Ir3 (JQ912241)*, 200–209.

Mancia, A., Chenet, T., Bono, G., Geraci, M., Vaccaro, C., Munari, C., Mistri, M., Cavazzini, A., & Pasti, L. (2021). Corrigendum "Adverse effects of plastic ingestion on the Mediterranean small-spotted catshark (Scyliorhinus canicula)" [Mar. Environ. Res. 155 March 2020 104876]. *Marine Environmental Research, 169*, 105325. https://doi.org/10.1016/j.marenvres.2021.105325

Meng, Y., Kelly, F., & Wright, S. (2019). Advances and challenges of microplastic pollution in freshwater ecosystems: A UK perspective. *Environmental Pollution, 256*, 113445. https://doi.org/10.1016/j.envpol.2019.113445

Mintenig, S., Int-Veen, I., Löder, M., Primpke, S., & Gerdts, G. (2016). Identification of microplastic in effluents of waste water treatment plants using focal plane array-based micro-Fourier-transform infrared imaging. *Water Research, 108*. https://doi.org/10.1016/j.watres.2016.11.015

Mishra, A., Kumar, M., Bolan, N., Kapley, A., Kumar, R., & Singh, L. (2021). Multidimensional approaches of biogas production and up-gradation: Opportunities and challenges. *Bioresource Technology, 338*. https://doi.org/10.1016/j.biortech.2021.125514

Moharir, R., & Kumar, S. (2018). Challenges associated with plastic waste disposal and allied microbial routes for its effective degradation: A comprehensive review. *Journal of Cleaner Production, 208*. https://doi.org/10.1016/j.jclepro.2018.10.059

Müller, R.-J., Schrader, H., Profe, J., Dresler, K. & Deckwer, W.-D.. (2005). Enzymatic Degradation of Poly(ethylene terephthalate): *Rapid Hydrolyse using a Hydrolase from T. fusca. Macromolecular Rapid Communications, 26*, 1400–1405. https://doi.org/10.1002/marc.200500410

Nabi, I., Bacha, A.-U.-R., Kejian, L., Cheng, H., Wang, T., Liu, Y., Ajmal, S., Yang, Y., Feng, Y., & Zhang, L. (2020). *Complete photocatalytic mineralization of microplastic on TiO2 nanoparticle film. 101326*. https://doi.org/10.1016/j.isci.2020.101326

Nishida, T., Fujisawa, M., & Hirai, H. (2001). Degradation of Polyethylene and Nylon-66 by the Laccase-Mediator System. *Journal of Polymers and the Environment, 9*. https://doi.org/10.1023/A:1020472426516

Negoro, S., Shibata, N., Tanaka, Y., Yasuhira, K., Shibata, H., Hashimoto, H., Lee, Y.-H., Oshima, S., Santa, R., Mochiji, K., Goto, Y., Ikegami, T., Nagai, K., Kato, D.-I., Takeo, M., & Higuchi, Y. (2011). Three-dimensional Structure of Nylon Hydrolase and Mechanism of Nylon-6 Hydrolysis. *The Journal of biological chemistry, 287*, 5079–90. https://doi.org/10.1074/jbc.M111.321992

O'Leary, N., O Connor, K., Ward, P., Goff, M., & Dobson, A. (2005). Genetic characterization of accumulation of Polyhydroxyalkanoate from Styrene in Pseudomonas putida CA-3. *Applied and Environmental Microbiology, 71*, 4380–4387. https://doi.org/10.1128/AEM.71.8.4380-4387.2005

Oberbeckmann, S., & Labrenz, M. (2020). Marine microbial assemblages on microplastics: Diversity, adaptation, and role in degradation. *Annual Review of Marine Science, 12*. https://doi.org/10.1146/annurev-marine-010419-010633

Onda, D. F., Gomez, N. C., Purganan, D. J., Tolentino, M. P., Bitalac, J. M., Calpito, J., Perez, J., & Viernes, A. (2020). Marine microbes and plastic debris: Research status and opportunities in the Philippines. *Philippine Journal of Science, 149*, 89–100.

Othman, A., Abu Hasan, H., Muhamad, M., Ismail, N. I., & Abdullah, S. (2021). Microbial degradation of microplastics by enzymatic processes: A review. *Environmental Chemistry Letters, 19*, 1–17. https://doi.org/10.1007/s10311-021-01197-9

O'Neill, A., Araújo, R., Casal, M. Guebitz, G., & Cavaco-Paulo, A. (2007). *Effect of the agitation on the adsorption and hydrolytic efficiency of cutinases on polyethylene terephthalate fibres. Enzyme and Microbial Technology, 40*, 1801–1805. https://doi.org/10.1016/j.enzmictec.2007.02.012

Palm, G., Reisky, L., Böttcher, D., Müller, H., Michels, E., Walczak, M., Berndt, L., Weiss, M., Bornscheuer, U., & Weber, G. (2019). Structure of the plastic-degrading Ideonella sakaiensis MHETase bound to a substrate. *Nature Communications, 10*. https://doi.org/10.1038/s41467-019-09326-3

Pivokonsky, M., Cermakova, L., Novotna, K., Peer, P., Cajthaml, T., & Janda, V. (2018). Occurrence of microplastics in raw and treated drinking water. *Science of The Total Environment, 643*. https://doi.org/10.1016/j.scitotenv.2018.08.102

Poerio, T., Piacentini, E., & Mazzei, R. (2019). Membrane processes for micro-plastic removal. *Molecules, 24*, 4148.

Priya, A., Dutta, K., & Daverey, A. (2021). A comprehensive biotechnological and molecular insight into plastic degradation by microbial community. *Journal of Chemical Technology & Biotechnology*. https://doi.org/10.1002/jctb.6675

Puglisi, E., Romaniello, F., Galletti, S., Boccaleri, E., Frache, A., & Cocconcelli, P. S. (2019). Selective bacterial colonization processes on polyethylene waste samples in an abandoned landfill site. *Scientific Reports, 9*. https://doi.org/10.1038/s41598-019-50740-w

Purohit, J., Chattopadhyay, A., & Teli, B. (2020). Metagenomic exploration of plastic degrading microbes for biotechnological application. *Current Genomics, 21*. https://doi.org/10.2174/1389202921999200525155711

Ricardo, I., Alberto, E., da Silva Júnior, A., Macuvele, D. L., Padoin, N., Soares, C., Riella, H., Starling, V. M., & Clara, M. (2021). A critical review on microplastics, interaction with organic and inorganic pollutants, impacts and effectiveness of advanced oxidation processes applied for their removal from aqueous matrices. *Chemical Engineering Journal, 424*. https://doi.org/10.1016/j.cej.2021.130282

Rillig, M. C. (2020). Plastic and plants. *Nature Sustainability, 3*(11), 887–888.

Rillig, M., Lehmann, A., Machado, A., & Yang, G. (2019). Microplastic effects on plants. *New Phytologist, 223*. https://doi.org/10.1111/nph.15794

Rodríguez-Seijo, A., Lourenço, J., Rocha-Santos, T. A. P., Da Costa, J., Duarte, A., Vala, H., & Pereira, R. (2017). Histopathological and molecular effects of microplastics in Eisenia andrei Bouché. *Environmental Pollution, 220, Part A*, 495–503. https://doi.org/10.1016/j.envpol.2016.09.092

Rogers, K., Carreres-Calabuig, J. A., Gorokhova, E., & Posth, N. (2020). Micro-by-micro interactions: How microorganisms influence the fate of marine microplastics. *Limnology and Oceanography Letters, 5*. https://doi.org/10.1002/lol2.10136

Ru, J., Huo, Y.-X., & Yang, Y. (2020). Microbial degradation and valorization of plastic wastes. *Frontiers in Microbiology, 11*, 442. https://doi.org/10.3389/fmicb.2020.00442

Russell, J., & Huang, J., Anand, P., Kucera, K., & Sandoval, A., Dantzler, K., & Hickman, D., Jee, J., Kimovec, F., Koppstein, D., Marks, D., Mittermiller, P., Núñez, S., Santiago, M., & Townes, M., & Lin, M., Williams, N., Mario, P., Nunez, B. L-A & Strobel, S. (2011). Biodegradation of Polyester Polyurethane by Endophytic Fungi. *Applied and environmental microbiology, 77*, 6076–84. https://doi.org/10.1128/AEM.00521-11

Silva, A., Costa, M., & Duarte, A. (2018). Biotechnology advances for dealing with environmental pollution by micro(nano)plastics: Lessons on theory and practices. *Current Opinion in Environmental Science & Health, 1*, 30–35. https://doi.org/10.1016/j.coesh.2017.10.005

Sridharan, S., Kumar, M., Bolan, N., Singh, L., Kumar, S., Kumar, R., & You, S. (2021a). Are microplastics destabilizing the global network of terrestrial and aquatic ecosystem services? *Environmental Research, 198*, 111243. https://doi.org/10.1016/j.envres.2021.111243

Sridharan, S., Kumar, M., Singh, L., Bolan, N., & Saha, M. (2021b). Microplastics as an emerging source of particulate air pollution: A critical review. *Journal of Hazardous Materials, 418*, 126245. https://doi.org/10.1016/j.jhazmat.2021.126245

Sudhakar, M., Priyadarshini, C., Doble, M., Murthy, S., & Venkatesan, R. (2007). Marine bacteria mediated degradation of nylon 66 and 6. *International Biodeterioration & Biodegradation, 60*, 144. https://doi.org/10.1016/j.ibiod.2007.02.002

Sun, J., Wu, X., & Gan, J. (2015). Uptake and metabolism of phthalate esters by edible plants. *Environmental Science & Technology, 49*. https://doi.org/10.1021/acs.est.5b01233

Sun, J., Dai, X., Wang, Q., van Loosdrecht, M., & Ni, B.-J. (2019). Microplastics in wastewater treatment plants: Detection, occurrence and removal. *Water Research, 152*. https://doi.org/10.1016/j.watres.2018.12.050

Sun, Y., Ren, X., Rene, E., Wang, Z., Zhou, L., Zhang, Z., & Wang, Q. (2021). The degradation performance of different microplastics and their effect on microbial community during composting process. *Bioresource Technology, 332*, 125133. https://doi.org/10.1016/j.biortech.2021.125133

Syranidou, E., Karkanorachaki, K., Amorotti, F., Avgeropoulos, A., Kolvenbach, B., Zhou, N.-Y., Fava, F., Corvini, P., & Kalogerakis, N. (2019). Biodegradation of mixture of plastic films by tailored marine consortia. *Journal of Hazardous Materials, 375*. https://doi.org/10.1016/j.jhazmat.2019.04.078

Santo, M., Weitsman, R., & Sivan, A. (2013). The role of the copper-binding enzyme – laccase – in the biodegradation of polyethylene by the actinomycete Rhodococcus ruber. *International Biodeterioration & Biodegradation, 84*, 204–210. https://doi.org/10.1016/j.ibiod.2012.03.001

Shimao, M. (2001). Biodegradation of plastics. *Current opinion in biotechnology, 12*(3), 242–247. https://doi.org/10.1016/s0958-1669(00)00206-8

Taha, Z., Md Amin, R., Anuar, S., Nasser, A., & Sohaimi, E. (2021). Microplastics in seawater and zooplankton: A case study from Terengganu estuary and offshore waters, Malaysia. *Science of The Total Environment, 786*, 147466. https://doi.org/10.1016/j.scitotenv.2021.147466

Talvitie, J., Mikola, A., Koistinen, A., & Setälä, O. (2017a). Solutions to microplastic pollution – Removal of microplastics from wastewater effluent with advanced wastewater treatment technologies. *Water Research, 123*. https://doi.org/10.1016/j.watres.2017.07.005

Talvitie, J., Mikola, A., Setälä, O., Heinonen, M., & Koistinen, A. (2017b). How well is microlitter purified from wastewater? – A detailed study on the stepwise removal of microlitter in a tertiary level wastewater treatment plant. *Water Research, 109*. https://doi.org/10.1016/j.watres.2016.11.046

Thompson, R., Olsen, Y., Mitchell, R., Davis, A., Rowland, S., John, A., Mcgonigle, D. F., & Russell, A. (2004). Lost at sea: Where is all the plastic? *Science (New York, N.Y.), 304*, 838. https://doi.org/10.1126/science.1094559

Tisdall, J., & Oades, J. (2006). Organic matter and water-stable aggregates in soils. *Journal of Soil Science, 33*, 141–163. https://doi.org/10.1111/j.1365-2389.1982.tb01755.x

Tiwari, N., Santhiya, D., & Sharma, J. (2020). Microbial remediation of micro-nano plastics: Current knowledge and future trends. *Environmental Pollution, 265*, 115044. https://doi.org/10.1016/j.envpol.2020.115044

Tofa, T., Ye, F., Laxman, K., & Dutta, J. (2019). Enhanced visible light photodegradation of microplastic fragments with Plasmonic platinum/zinc oxide nanorod photocatalysts. *Catalysts, 9*, 819. https://doi.org/10.3390/catal9100819

Tourova, T., Sokolova, D., Nazina, T., Grouzdev, D., & Laptev, A. (2020). Phylogenetic diversity of microbial communities from the surface of polyethylene terephthalate materials exposed to different water environments. *Microbiology, 89*, 96–106. https://doi.org/10.1134/S0026261720010154

Uheida, A., Giraldo Mejaí, H., Abdel-Rehim, M., Hamd, W., & Dutta, J. (2020). Visible light photocatalytic degradation of polypropylene microplastics in a continuous waterflow system. *Journal of Hazardous Materials, 406*. https://doi.org/10.1016/j.jhazmat.2020.124299

Urbanek, A., Rymowicz, W., & Mirończuk, A. (2018). Degradation of plastics and plastic-degrading bacteria in cold marine habitats. *Applied Microbiology and Biotechnology, 102*. https://doi.org/10.1007/s00253-018-9195-y

Vanleeuw, E., Winderickx, S., Thevissen, K., Lagrain, B., Dusselier, M., Cammue, B., & Sels, B. (2019). Substrate-specificity of Candida rugosa Lipase and its industrial application. *ACS Sustainable Chemistry & Engineering, 7*. https://doi.org/10.1021/acssuschemeng.9b03257

Wagner, M., & Lambert, S. (2017). *Freshwater microplastics: Emerging environmental contaminants?* https://doi.org/10.1007/978-3-319-61615-5

Wahl, A., Juge, C., Davranche, M., El Hadri, H., Grassl, B., Reynaud, S., & Gigault, J. (2020). Nanoplastic occurrence in a soil amended with plastic debris. *Chemosphere, 262*, 127784. https://doi.org/10.1016/j.chemosphere.2020.127784

Wang, C., Zhao, J., & Xing, B. (2020). Environmental source, fate, and toxicity of microplastics. *Journal of Hazardous Materials, 407*, 124357. https://doi.org/10.1016/j.jhazmat.2020.124357

Wang, J., Luo, Y., Teng, Y., Ma, W., Christie, P., & Li, Z. (2013). Soil contamination by phthalate esters in Chinese intensive vegetable production systems with different modes of use of plastic film. *Environmental Pollution (Barking, Essex : 1987), 180C*, 265–273. https://doi.org/10.1016/j.envpol.2013.05.036

Wang, J., Lv, S., Zhang, M., Chen, G., Zhu, T.-B., Zhang, S., Teng, Y., Christie, P., & Luo, Y. (2016). Effects of plastic film residues on occurrence of phthalates and microbial activity in soils. *Chemosphere, 151*, 171–177. https://doi.org/10.1016/j.chemosphere.2016.02.076

Wu, P., Huang, J., Yang, Y., Zhang, Y., He, F., Chen, H., Quan, G., Yan, J., Li, T., & Gao, B. (2019). Environmental occurrences, fate, and impacts of microplastics. *Ecotoxicology and Environmental Safety, 184*. https://doi.org/10.1016/j.ecoenv.2019.109612

Wu, X., Lu, J., Du, M., Xu, X., Beiyuan, J., Sarkar, B., Bolan, N., Xu, W., Xu, S., Chen, X., Wu, F., & Wang, H. (2021). Particulate plastics-plant interaction in soil and its implications: A review. *Science of The Total Environment, 792*, 148337. https://doi.org/10.1016/j.scitotenv.2021.148337

Xiao, K., Liang, S., Wang, X., & Chen, C. (2018). Current state and challenges of full-scale membrane bioreactor applications: A critical review. *Bioresource Technology, 271*. https://doi.org/10.1016/j.biortech.2018.09.061

Xu, Z., Xiong, X., Zhao, Y., Xiang, W., & Wu, C. (2019). Pollutants delivered every day: Phthalates in plastic express packaging bags and their leaching potential. *Journal of Hazardous Materials, 384*, 121282. https://doi.org/10.1016/j.jhazmat.2019.121282

Yang, D., Cho, J. S., Choi, K. R., Kim, T. Y., & Lee, S. Y. (2017). Systems metabolic engineering as an enabling technology in accomplishing sustainable development goals. *Microbial Biotechnology, 10*. https://doi.org/10.1111/1751-7915.12766

Yang, Y., Liu, W., Zhang, Z., Grossart, H.-P., & Gadd, G. (2020). Microplastics provide new microbial niches in aquatic environments. *Applied Microbiology and Biotechnology, 104*, 1–11. https://doi.org/10.1007/s00253-020-10704-x

Yoon, M., & Jeon, H. (2012). Biodegradation of polyethylene by a soil bacterium and AlkB cloned recombinant cell. *Journal of Bioremediation & Biodegradation, 03*. https://doi.org/10.4172/2155-6199.1000145

Yoshida, S., Hiraga, K., Takehana, T., Taniguchi, I., Yamaji, H., Maeda, Y., Toyohara, K., Miyamoto, K., Kimura, Y., & Oda, K. (2016). A bacterium that degrades and assimilates poly(ethylene terephthalate). *Science, 351*, 1196–1199. https://doi.org/10.1126/science.aad6359

Yuan, J., Ma, J., Yiran, S., Zhou, T., Zhao, Y., & Yu, F. (2020). Microbial degradation and other environmental aspects of microplastics/plastics. *Science of The Total Environment, 715*, 136968. https://doi.org/10.1016/j.scitotenv.2020.136968

Ying, Z., Ernest, K. Y., & Amarjeet S. B. (2005). A Review of Plastic Waste Biodegradation, *Critical Reviews in Biotechnology, 25:4*, 243–250. https://doi.org/10.1080/07388550500346359

Yamada-Onodera, K., Mukumoto, H., Katsuyaya, Y., Saiganji, A., & Tani, Y. (2001). Degradation of Polyethylene by a fungus, Penicillium simplicissimum YK. *Polymer Degradation and Stability - POLYM DEGRAD STABIL. 72*, 323–327. https://doi.org/10.1016/S0141-3910(01)00027-1

Zhang, S., Wang, J., Yan, P., Hao, X., Xu, B., Wang, W., & Aurangzeib, M. (2020). Non-biodegradable microplastics in soils: A brief review and challenge. *Journal of Hazardous Materials., 409*, 124525. https://doi.org/10.1016/j.jhazmat.2020.124525

Zhou, Y., Kumar, M., Sarsaiya, S., Sirohi, R., Awasthi, S., Raveendran, S., Parameswaran, B., Pandey, A., Bolan, N., Zhang, Z., Singh, L., Kumar, S., & Awasthi, M. (2021). Challenges and opportunities in bioremediation of micro-nano plastics: A review. *Science of The Total Environment., 802*, 149823. https://doi.org/10.1016/j.scitotenv.2021.149823

Zinicovscaia, I. (2016). Conventional methods of wastewater treatment. In *Cyanobacteria for bioremediation of wastewaters*. https://doi.org/10.1007/978-3-319-26751-7_3

Zubris, K., & Richards, B. (2005). Synthetic fibers as an indicator of land application of sludge. *Environmental pollution (Barking, Essex: 1987), 138*, 201–211. https://doi.org/10.1016/j.envpol.2005.04.013

Chapter 17
Micro and Nanoplastics in Agricultural Soils: Challenges and Future Directions

María Antonieta Riera and Medardo Anibal Zambrano-Arcentales

Abstract Plastics are a family of materials with great applicability, including agriculture. Large plastics (macroplastics) are used as inputs for agricultural activity. When using biofertilizers, plastics at micro and nanometric scales are also introduced (micro and nanoplastics). Its presence in crops soils is imminent. So are the effects it causes in the place where they are found. This chapter represents a contribution to the existing situation in agricultural soils, in the presence of this almost invisible threat. Reference is made to the entry of these particles during agriculture and their incorporation into the food chain. The damage caused by micro and nanoplastics to both health and the environment is pointed out, highlighting some remediation and mitigation measures. It ends by highlighting the challenge represented by the use of plastic in agricultural practices and the responsibility of nations and producers, given the existing problems.

Keywords agriculture · microplastics · nanoplastics · agricultural soils.

17.1 Introduction

The appearance of plastic has revolutionized different forms of work, including agricultural practices. Although its use has improved crop performance and care, it has the disadvantage of introducing tiny particles, known as micro and nanoplastics. The damage caused by the presence of these emerging contaminants is still under investigation. However, there is scientific evidence that affirms its occurrence not only to human health but also to the fauna of the ecosystem. These aspects are discussed in this chapter. It begins with an overview of plastic in agriculture and then

M. A. Riera (✉)
Facultad de Ciencias Matemáticas, Físicas y Químicas, Departamento de Procesos Químicos, Alimentos y Biotecnología. Universidad Técnica de Manabí, Portoviejo, Ecuador

Doctorado en Ingeniería Industrial, Universidad Nacional de Cuyo, Mendoza, Argentina
e-mail: maria.riera@utm.edu.ec

M. A. Zambrano-Arcentales
Facultad de Ciencias Físicas y Matemáticas, Universidad de Chile, Santiago de Chile, Chile

N. R. Maddela et al. (eds.), *Micro and Nanoplastics in Soil*,
https://doi.org/10.1007/978-3-031-21195-9_17

explains how the decomposition of these plastics into micro and nanoplastics is incorporated into the human food chain. The damage caused by micro and nanoplastics is exposed, in addition to the recommended prevention and remediation methods. It ends with a look to the future, pointing out the need to establish concrete actions to face this almost imperceptible enemy without putting agricultural activity at risk.

17.2 Distribution of Plastics in Agriculture

Plastic is a material that, given its versatility, durability, and low cost, has application in a wide variety of sectors. Agriculture is one of them, being introduced in 1948 when using cellophane to cover greenhouses in the United States. Later, polyvinyl chloride (PVC) was used in Japan for the same purpose. In the 1990s, this practice was adopted in developing countries. Currently, its use has spread to different parts of the world (Scarascia-Mugnozza et al., 2011; Zenner de Polanía & Peña Baracaldo, 2013).

The applications are diverse (Table 17.1), although the main purpose is in the construction of greenhouses, padding, or mulching. The purposes are also varied.

Table 17.1 Plastics and their applications in agriculture

Description	Use	Plastic type	Reference
Mulching	Modify crop microclimate.	Low density polyethylene (LDPE).	Berger et al. (2013), Montemayor-Trejo et al. (2018), Qi et al. (2020b).
Pocket/ Coverage	Cover fruits or crops with plastic bags or film to protect them from factors that detract from their quality.	LDPE films, polypropylene (PP).	Zenner de Polanía and Peña Baracaldo (2013)
Greenhouses	Protect crops from heavy rain or wind, as well as from sunlight and intense heat.	Polyethylene films (PE).	Orzolek (2017)
Nets	Protective top or side crop covers.	High density polyethylene (HDPE) PP.	Castellano et al. (2008)
Irrigation pipes	Drip irrigation for greater efficiency in water consumption, incorporation of fertilizers to the crop.	LDPE en tuberías de diámetro pequeño y PVC para las de mayor tamaño.	Orzolek (2017)
Tunnels and micro-tunnels	Facilítate intensive crop production on a small area of land. They are used to increase production cycles, improve the quality and yield of crops, protect against pests and adverse weather.	LDPE films	Ángel-Hernández et al. (2017), Jett (2017)

They include the protection of crops against birds and insects, protection against climatic threats (heavy rains, hail, snow), or modification of environmental conditions by reducing solar radiation and evapotranspiration (Castellano et al., 2008; Zenner de Polanía & Peña Baracaldo, 2013).

In today's the world, the production and consumption of plastic have maintained a growing trend since its inception. This behavior is the same in the agricultural sector. In 2012, world consumption of 4.4 million tons (Mt) of plastic films for greenhouses, mulching, and silage was obtained, projecting an increase of 69% for 2019 of the concept (Sintim & Flury, 2017). It is estimated that the total annual consumption of HDPE and PP in agriculture is 30,000 and 46,500 tons, respectively (Castellano et al., 2008). Currently, and during the last decade (Fig. 17.1), the share of the agricultural sector in the plastics market is close to 3% (PlasticsEurope, 2021). Although it is a percentage value that has been maintained over time, it represents an increase in the tons consumed (Statista, 2021).

The massive use of plastics in agriculture brings multiple benefits for the development and yield of crops. At the same time, it causes environmental damage related to the removal and disposal of the waste generated. Plastics after use are contaminated with soil, stones, and biological waste, which makes it difficult to recycle. One of the most common solutions to this problem is burning at the generation site (Prosperi et al., 2018) or disposal in open-air dumps.

They are non-biodegradable synthetic materials with very long decomposition periods ranging from decades to centuries, making their permanence in the disposal site prolonged. During this time, the materials break down into smaller particles due to the action of different agents, releasing microplastics (MPs) and nanoplastics (NPs) that remain in the soil or accumulate in the sea (Sintim & Flury, 2017; Chen et al., 2021).

MPs and NPs are smaller plastic particles than macroplastics, with sizes (1 μm–5 mm) and (<100 nm) respectively, present in the environment due to anthropogenic activities. They are a type of emerging pollutant that emerged in recent decades and, due to their long-term problem, they are of great interest worldwide

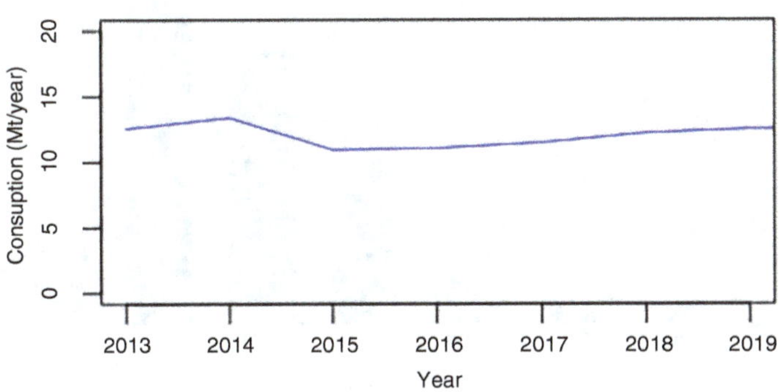

Fig. 17.1 Consumo estimado de plástico en el sector agrícola. Elaborado a partir de cálculos realizados con información reportada por Statista (2021) and Plastics Europe (2021)

(Ivleva, 2021; Huang et al., 2021). In recent years, a growing number of investigations on MPs/NPs have been reported worldwide (Fig. 17.2).

Although most of the research related to MPs and NPs is oriented to the presence of these particles as water contaminants, they also represent a threat to soils. The development of modern practices in agriculture, such as the introduction of agricultural membranes, the use of plastic materials to increase crop yields, the recycling of biological sludge (biosolids), and changes in irrigation methods (Fig. 17.3), generate this type of contaminants with a proven presence in the soil (Huang et al., 2021).

The main route of entry of these particles in agricultural land is the use of biosolids generated in the treatment of wastewater. Although these have a high organic content, they also have MPs and NPs, in addition to other synthetic materials, heavy metals, pharmaceutical products, and engineered nanoparticles, which contaminate the soil (Tian et al., 2022; Junhao et al., 2021; Mohajerani & Karabatak, 2020). Studies show that wastewater treatment plants eliminate up to 99% of the microplastics present in the water, which are then concentrated in the biosolids generated. Through material balances, it is calculated that biosolids contain about 300,000 plastic particles per kg, which are subsequently introduced into agricultural soils (Castan et al., 2021; Talvitie et al., 2017).

When making a comparison of this agricultural practice, it was also found that the soils where biological sludge was applied exceeded the concentration of MPs by a great difference, with an average of 2130 and 3060 MPs/kg of light and heavy density, respectively. Regarding those that were not treated with biosolids, whose contents were 930 and 1100 MPs/kg of light and heavy density, respectively (Van den Berg et al., 2020). According to this and based on data analysis, a minimum and maximum addition of tons of MPs/year to cropland of 26,156–151,137, 21,249–122,780, 13,660–78,930, 1518–8770 and 1241–7170, in the European

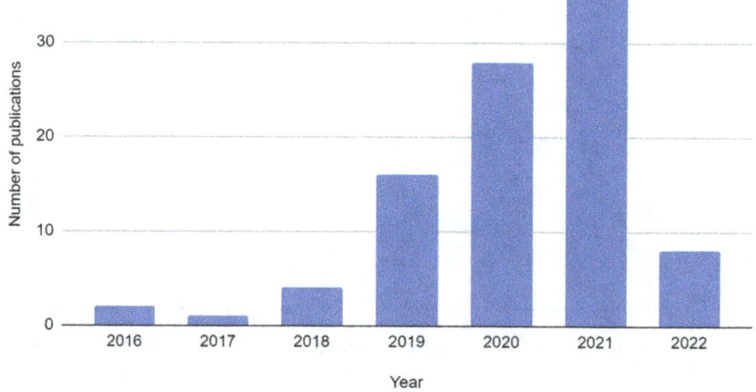

Fig. 17.2 Number of publications related to contamination by MPs/NPs in agriculture. (*Source*: WoS, January 2022. Code: Thesaurus = (agriculture AND (microplastics OR nanoplastics)))

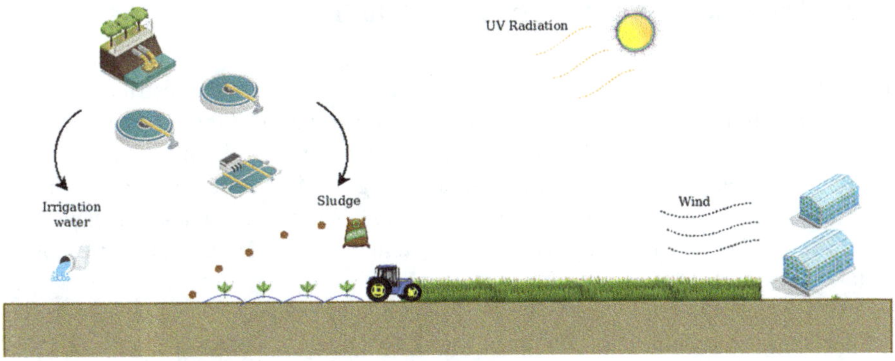

Fig. 17.3 Introduction of MPs and NPs in agricultural soils

Union, the United States, China, Canada, and Australia, for the application of biosolids (Mohajerani & Karabatak, 2020).

Another source is the wastewater used in the irrigation of crops, which may have primary MPs when added to cosmetic products (shampoo, soap, others) during its manufacture. The action of the wind is also a variable that contributes to the transport of these particles. Added to this are the processes of infiltration of the flow of rainwater or irrigation from top to bottom, which transfer MPs/NPs to empty places in the soil (Bullard et al., 2021; Huang et al., 2021). Finally, there are the MPs and NPs that originate from the fragmentation or degradation of the plastics used in agriculture, as a consequence of exposure to different agents, which are the effects of UV radiation, mechanical abrasion of the fauna of the soil, and agricultural practices (Tian et al., 2022).

The rate of generation of these contaminants depends on aspects such as chemical structure, morphology, rigidity, thickness, anisotropy, and density of the polymer, given its use and agent to which the origin of the plastic is exposed. Previous research estimates that when a plastic bag or bottle is torn open, 14.000 to 75.000 PMs are generated, with a mass of 0.8–1.4 ng and a dimension of 10 μm^3 per microplastic. Similarly, the release of 147 to 475 particles/cm^2 with sizes from 0.02 to 0.10 mm is calculated when exposing biodegradable, oxodegradable, and nondegradable mulch films to a cumulative UV radiation of 2,1 MJ/m^2 (Yang et al., 2021; Huang et al., 2021, Sobhani et al., 2020; Andrady, 2017).

The amount of micro and nanoplastics generated in a country's agriculture depends largely on the consumption of plastic in the area, the agricultural practices used, and external factors such as exposure to ultraviolet rays and other agents. Regardless of the volume produced, the main concern of this problem is the environmental threat it represents. These particles affect the functioning of the ecosystem due to the accumulation of heavy metals or other potentially harmful organic contaminants in agricultural soils, which subsequently migrate to food and threaten the health of human beings and animal life (Castan et al., 2021; Qi et al., 2020a).

17.3 Incorporation Into the Food Chain

The MPs/NPs generated in the different agricultural activities are distributed in the air by the action of the wind, in the water in the irrigation of the crops, and in the soil at various depth levels. This creates an environment rich in these particles that can enter the plants or adhere to their surfaces (Fig. 17.4). MPs are usually found on the surface of roots and seeds, while NPs, being smaller particles that have lost crystallinity and even molecular weight (Astner et al., 2019), can enter plant metabolism.

In soils contaminated with MPs/NPs, evidence of transport of these particles in plant tissues through the roots has been found. Similarly, it has been determined that the distribution of these particles during plant growth depends on the size of these contaminants (Ullah et al., 2021). Through the air, MPs are deposited on the surface of the leaves (Liu et al., 2020), finding between 7 and 19 particles/mm.[2] In addition, a large part of vegetable crops is destined for animal feed, opening the risk of bioaccumulation in the food chain (Wu et al., 2021).

The presence of MPs/NPs has been reported in foods for fresh consumption such as carrots, cucumbers, lettuce, oranges, pears, as well as in foods to be processed or intended for animal feed such as barley, wheat, corn, and rice (Campanale et al., 2022). MPs/NPs have also been found in processed foods such as beer, sugar, and honey, and milk (Liebezeit & Liebezeit, 2013, 2014; Shi et al., 2021). A higher concentration of microplastics is recorded in fruits such as apples and pears and a lower concentration in carrots, lettuce, potatoes, and broccoli. It is believed that the highest concentration of MPs/NPs in fruits is due to the greater vascularization of the pulp, the size and complexity of the roots, as well as the age of the trees (Oliveri Conti et al., 2020).

The daily intake of microplastics based on the consumption of fruits and vegetables has been reported with an average of up to 4.62 and 1.41 particles/kg/day in adults and children, respectively (Oliveri Conti et al., 2020). Other works report different particle sizes, types, and concentrations of MPs/NPs in food (Campanale et al., 2022; Toussaint et al., 2019). However, because they belong to the group of emerging contaminants, there is no clear definition of the characteristics that should be measured or identified concerning MPs/NPs in food. This factor makes it

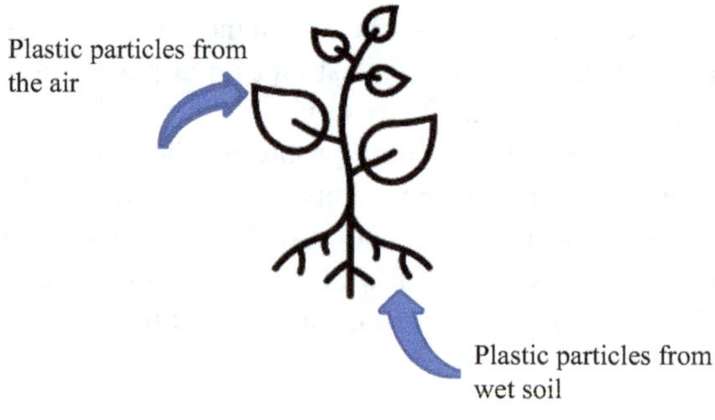

Fig. 17.4 Incorporation of plastic particles in terrestrial plants

difficult to reach a consensus on characteristics (size, shape, and composition), identification and measurement techniques, as well as more precise keywords to facilitate systematic reviews (Toussaint et al., 2019).

17.4 Alteration of Ecosystems and Possible Damage to Health

The presence of plastic particles in the soil can affect the natural transport of nutrients between organisms and microorganisms that live in the soil affecting plant growth. Not only because of their presence as particles that can be ingested by organisms or enter the metabolism of microorganisms but also because of their adsorption capacity with organic compounds present in agricultural soils (Hüffer & Hofmann, 2016). Non-covalent affinity has been found in MPs/NPs particles with pesticides, with a negative correlation with the size of the particles due to the increase in the contact area (Mo et al., 2021). This adsorption capacity is reversed when entering environments or metabolism where the compounds adsorbed on the surface of the MPs/NPs will be released by desorption. It has been shown that MPs/NPs can affect the microbial growth of bacteria and fungi in soils, altering growth rates, microbial diversity, hyphal colonization capacity, and even cell death (Iqbal et al., 2020).

In a study with Chinese cabbage leaves, a lower concentration of starch, chlorophyll, and a higher concentration of soluble sugars were found compared to controls, when the plants were exposed to high concentrations of MPs (Yang et al., 2021). In another investigation, it is mentioned that NPs in the soil can decrease plant growth, reducing the number of leaves and amount of biomass, in addition to affecting nutritional properties such as pigments, antioxidants, and proteins (Campanale et al., 2022).

MPs are easily ingested by earthworms, affecting their activity. Doing so increases oxidative stress that try to remove these particles. A decrease in their locomotion has also been observed, possibly due to internal damage to their body, so they would be in a state of lethargy to reduce damage (Boughattas et al., 2021). In other research, MPs/NPs were found to increase the buildup of phenanthrene, a toxic compound, in your body. This happens since the NPs inhibit the action of the bacteria responsible for the decomposition of said substance. The same study mentions that the toxic effect described could affect humans if these substances reach them through the food chain (Xu et al., 2021).

Regarding human health, using data from the U.S. population, it has been estimated that people incorporate between 39.000 and 52.000 plastic particles per year through their diet (Cox et al., 2019). If these figures are underestimated, concern about the accumulation of these particles in the body is increasing. There is little evidence on specific effects on health by MPs/NPs, however, some research points to the elimination of certain types of plastics used in agriculture because they contain large amounts of phthalate esters (PAEs). These compounds are known as "environmental hormones" and have a detrimental effect on human reproductive health, high toxicity, carcinogenicity, and other toxic responses (Qi et al., 2020a).

The interactions of MPs/NPs in human organs continue to be a pending task in science, but the potential health effects can be estimated through adsorption models and compared with experiments carried out on animals (Mofijur et al., 2021). The adsorption capacity of MPs can interact with gut proteins causing inflammation and changes in the microbiota (Elmassry et al., 2020). They can cause DNA damage and cellular toxicity in human liver cells. NPs, having a large contact area, can release persistent organic pollutants, previously adsorbed, in the human body (Mofijur et al., 2021; Vethaak & Legler, 2021). NPs can reach the circulatory system through the intestinal barrier and reach various body tissues where they would accumulate (Crawford & Quinn, 2017).

In summary, the risks to human health are associated with the problematic characteristics of PMs/NPs: a variety of particles with the capacity to adsorb toxic organic pollutants from agricultural soils; ease to be incorporated into food externally and internally, and desorption capacity of pollutants with subsequent decomposition into smaller particles and diffusion of these in the body (Fig. 17.5).

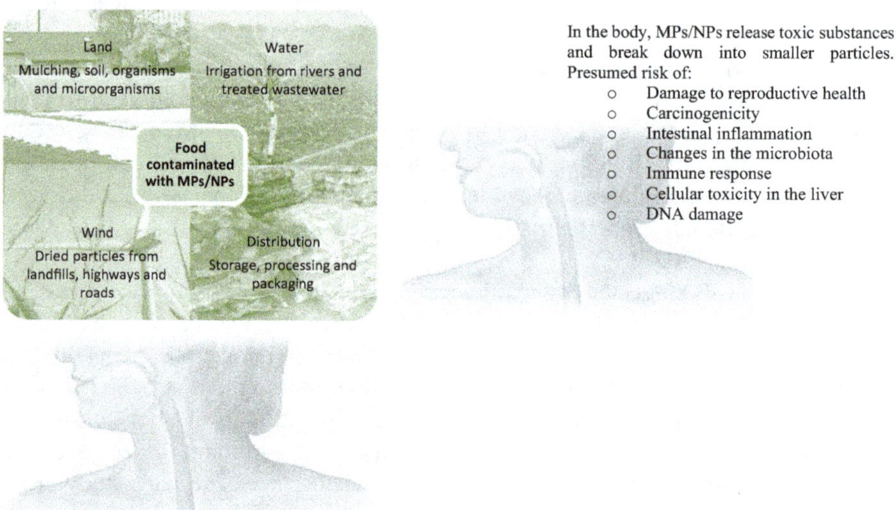

Fig. 17.5 Risks of contamination of MNP in agriculture and possible damage to health

17.5 Methods of Prevention and Remediation of Contaminated Soils

The invention of plastics came to revolutionize agriculture, improving the performance of this activity. However, it also brings a visible problem such as the accumulation of plastics in the soil. The emerging appearance of smaller contaminants from this material has motivated research for the development of alternatives that do not affect agricultural activity but at the same time minimize environmental damage (Fig. 17.6).

The use of bioplastics, especially biobased and biodegradable ones, could be an alternative for agricultural development that avoids the accumulation of MPs and NPs in soils (Qi et al., 2020a). Biodegradable plastic liners based on polybutylene adipate terephthalate (PBAT), polylactic acid (PLA), and starch can replace PE mulch as a sustainable and environmentally friendly solution (Liu et al., 2021).

The disadvantages of these biobased materials in comparison to polyethylene mulches are their low mechanical resistance and high production cost. Additionally, they must meet the demands of the crop, it must not be toxic or persist in the environment (Sintim & Flury, 2017). Although it is known that bioplastics do not affect seed germination, they could cause an inhibitory or stimulating effect on root and stem growth, delaying this process (Liwarska-Bizukojc, 2021).

An issue to be resolved in the use of bioplastics is the generation of bioplastic microparticles (MBPs) from biobased plastics. Similar behavior has been identified between MBPs and MPs. An example of this is polyhydroxyalkanoate (PHAs), a type of biodegradable plastic, which, when exposed in the form of a film in the aquatic environment, formed BMPs such as MPs (Shruti & Kutralam-Muniasamy, 2019). The total decomposition of a biodegradable bioplastic occurs under certain

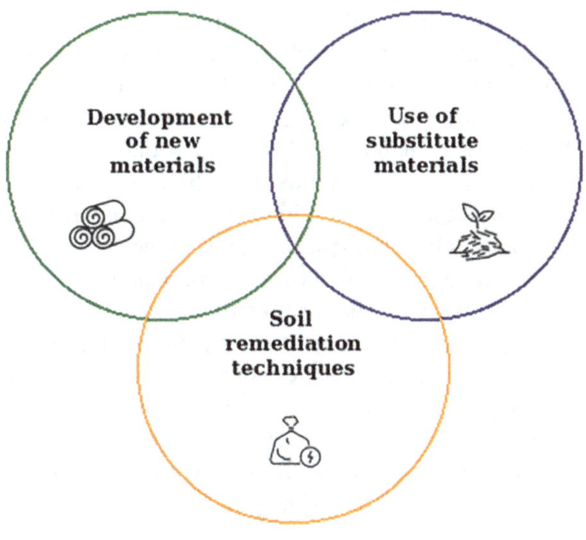

Fig. 17.6 Methods of prevention and remediation of contaminated agricultural soils

conditions, which is why research should be promoted to determine sorption properties, toxicity, concentration, and time in the soil of BMPs originating from PHAs, PLA, and PBAT (Fojt et al., 2020).

It is necessary to strengthen research for the development of biomaterials that meet the needs from the technical and environmental point of view, to improve the characteristics of bioplastics for their applicability in the agricultural sector, reduce production costs, guarantee the biodegradation of the materials in the environments used and ensure the maximum reduction of the risks of toxicity present today, by the introduction of conventional plastics in current agricultural practices.

Another alternative to deal with the problem analyzed is the implementation of sustainable production methods in agricultural practices. The plastic used in mulching can be replaced with dry materials such as leaves, grass, branches, crop residues, straw, among others. Studies have evaluated the incorporation of rice straw (an agricultural residue) as a vegetable cover in crops. The results obtained show an increase in the physicochemical properties of the soils, control of weeds, and water savings in crop irrigation (Monzó Pérez, 2020; Gordillo Manssur et al., 2018).

Lastly, there are the remediation actions of soils contaminated with MPs and NPs. Some technologies for the removal of these particles include metal oxide-assisted photodegradation of microplastics, as well as microbial and enzymatic approaches. The selection of an efficient removal methodology for MPs and NPs depends on the performance, sensitivity, economic viability, and volume handling capacity provided by each of these (Jaiswal et al., 2022; Patil et al., 2022).

17.6 Actions of Governments in the Face of the Problematic

The problem of contamination of MPs and NPs in agricultural soils requires the adoption of measures promoted by the governments of the world. Some countries have established laws and policies aimed at reducing the consumption of plastic bags, monitoring and use of the waste generated, as well as raising user awareness (Laskar & Kumar, 2019). Likewise, there are policies to promote the post-consumption collection of containers and their subsequent recycling through the generation of recovery notes or levying taxes on the use of landfills for this waste (Hopewell et al., 2009). These are restrictions on the use and disposal of packaging material, with strategies applicable to plastics used in the agricultural sector.

In this regard, governments, in alliance with companies, promote the development of biodegradable plastics for the manufacture of mulching, imposing taxes or establishing prohibitions on the use of synthetic plastics (Serrano-Ruiz et al., 2021). There are also programs created by governments to encourage the collection of waste by farmers. In Korea, for example, networks for the collection and treatment of plastic waste were established through alliances between the private sector and local governments, creating a subsidy for farmers who participate in the collection of plastic waste, mainly from mulches (Chang & Kim, 2018).

As for biosludge, some governments have limited its use. For example, Germany banned sewage sludge as fertilizer from the year 2029. Likewise, Australia, New Zealand, and the United States established limits for the treatment applied to obtain it. At present, it is probably not feasible to eliminate the use of biosludge in agriculture, but stricter regulation is needed that contemplates its quality in terms of concentrations of contaminants (Milojevic & Agnieszka, 2021).

If the use of synthetic plastics continues, both the government and producers are required to control the commercialization of plastic films with a high load of pollutants (Qi et al., 2020a). State intervention is vital for the development of public policies that encourage research for the development of biodegradable plastics capable of replacing those traditionally used in agriculture. Robust policies must be established to encourage the agricultural producer to collect the plastic waste generated and motivate the entrepreneur to recycle it. Reasonable limits must also be set for the use of biosludge, within a framework of sustainability.

In summary, governance and specific regulations, it is necessary to improve waste management based on a deep understanding of the risks associated with the presence of PMs and NPs in agricultural soils. It represents a challenge that must be assumed jointly and responsibly between governments, businessmen, and academia, to solve the emerging problem caused by these particles and respond to the 2030 SDGs proposed by the UN (Castan et al., 2021; Walker, 2021).

References

Andrady, A. (2017). The plastic in microplastics: A review. *Marine Pollution Bulletin, 119*, 12–22. https://doi.org/10.1016/j.marpolbul.2017.01.082

Ángel-Hernández, M. D., Zermeño-Gonzalez, A., Melendres-Alvarez, A. I., Campos-Magaña, S. G., Cadena-Zapata, M., & Del Bosque-Villarreal, G. A. (2017). Characteristics of a tunnel cover effect on radiation, chlorophyll and zucchini yield. *Revista mexicana de ciencias agrícolas, 8*(5), 1127–1142. https://doi.org/10.29312/remexca.v8i5.113

Astner, A. F., Hayes, D. G., O'Neill, H., Evans, B. R., Pingali, S. V., Urban, V. S., & Young, T. M. (2019). Mechanical formation of micro- and nano-plastic materials for environmental studies in agricultural ecosystems. *Science of the Total Environment, 685*, 1097–1106. https://doi.org/10.1016/j.scitotenv.2019.06.241

Berger, S., Kim, Y., Kettering, J., & Gebauer, G. (2013). Plastic mulching in agriculture—Friend or foe of N2O emissions? *Agriculture, Ecosystems & Environment, 167*, 43–51. https://doi.org/10.1016/j.agee.2013.01.010

Boughattas, I., Hattab, S., Zitouni, N., Mkhinini, M., Missawi, O., Bousserrhine, N., & Banni, M. (2021). Assessing the presence of microplastic particles in Tunisian agriculture soils and their potential toxicity effects using Eisenia Andrei as bioindicator. *Science of the Total Environment, 796*, 148959. https://doi.org/10.1016/j.scitotenv.2021.148959

Bullard, J., Ockelford, A., O'Brien, P., & McKenna, C. (2021). Preferential transport of microplastics by wind. *Atmospheric Environment, 245*, 118038. https://doi.org/10.1016/j.atmosenv.2020.118038

Campanale, C., Galafassi, S., Savino, I., Massarelli, C., Ancona, V., Volta, P., & Uricchio, V. F. (2022). Microplastics pollution in the terrestrial environments: Poorly known diffuse sources and implications for plants. *Science of the Total Environment, 805*, 150431. https://doi.org/10.1016/j.scitotenv.2021.150431

Castan, S., Henkel, C., Hüffer, T., & Hofmann, T. (2021). Microplastics and nanoplastics barely enhance contaminant mobility in agricultural soils. *Communications Earth & Environments, 2*, 193. https://doi.org/10.1038/s43247-021-00267-8

Castellano, S., Scarascia Mugnozza, G., Russo, G., Briassoulis, D., Mistriotis, A., Hemming, S., & Waaijenberg, D. (2008). Plastic nets in agriculture: A general review of types and applications. *Applied Engineering in Agriculture, 24*(6), 799–808. https://doi.org/10.13031/2013.25368

Chang, Y., & Kim, G. (2018). Agriculture and food sciences research. Forecasting waste agricultural plastics generation in the Republic of Korea and its policy implications. *Agriculture and Food Sciences Research, 5*(2), 68–72. https://doi.org/10.20448/journal.512.2018.52.68.72

Chen, Y., Kumar Awasthi, A., Wei, F., Tan, Q., & Li, J. (2021). Single-use plastics: Production, usage, disposal, and adverse impacts. *Science of the Total Environment, 752*, 141772. https://doi.org/10.1016/j.scitotenv.2020.141772

Cox, K., Covernton, G. A., Davies, H. L., Dower, J. F., Juanes, F., & Dudas, S. E. (2019). Human consumption of microplastics. *Environmental Science & Technology, 53*(12), 7068–7074. https://doi.org/10.1021/acs.est.9b01517

Crawford, C. B., & Quinn, B. (2017). The emergence of plastics. In C. B. Crawford & B. Quinn (Eds.), *Microplastic pollutants* (pp. 1–17). Elsevier. https://doi.org/10.1016/C2015-0-04315-5

Elmassry, M. M., Zayed, A., & Farag, M. A. (2020). Gut homeostasis and microbiota under attack: Impact of the different types of food contaminants on gut health. *Critical Reviews in Food Science and Nutrition*, 1–26. https://doi.org/10.1080/10408398.2020.1828263

Fojt, J., David, J., Přikryl, R., Řezáčová, V., & Kučerík, J. (2020). A critical review of the overlooked challenge of determining micro-bioplastics in soil. *Science of the Total Environment, 745*, 140975. https://doi.org/10.1016/j.scitotenv.2020.140975

Gordillo Manssur, F., Cos Farías, V., & Romero Blacio, F. (2018). Residues of rice cultivation as mulching for the cultivation of pumpkin *(cucurbita moschata l). Revista Universidad y Sociedad, 10*(1), 105–110. Available in: http://scielo.sld.cu/scielo.php?script=sci_arttext&pid=S2218-36202018000100105

Hopewell, J., Dvorak, R., & Kosior, E. (2009). Plastics recycling: Challenges and opportunities. *Philosophical Transactions of the Royal Society, B364*, 2115–2126. https://doi.org/10.1098/rstb.2008.0311

Huang, D., Tao, J., Cheng, M., Deng, R., Chen, S., Yin, L., & Li, R. (2021). Microplastics and Nanoplastics in the environment: Macroscopic transport and effects on creatures. *Journal of Hazardous Materials, 407*, 124399. https://doi.org/10.1016/j.jhazmat.2020.124399

Hüffer, T., & Hofmann, T. (2016). Sorption of non-polar organic compounds by micro-sized plastic particles in aqueous solution. *Environmental Pollution, 214*, 194–201. https://doi.org/10.1016/j.envpol.2016.04.018

Iqbal, S., Xu, J., Allen, S. D., Khan, S., Nadir, S., Arif, M. S., & Yasmeen, T. (2020). Unraveling consequences of soil micro- and nano-plastic pollution on soil-plant system: Implications for nitrogen (N) cycling and soil microbial activity. *Chemosphere, 260*, 127578. https://doi.org/10.1016/j.chemosphere.2020.127578

Ivleva, N. (2021). Chemical analysis of microplastics and nanoplastics: Challenges, advanced methods, and perspectives. *Chemical Reviews, 121*(19), 11886–11936. https://doi.org/10.1021/acs.chemrev.1c00178

Jaiswal, K. K., Dutta, S., Banerjee, I., Pohrmen, C. B., Singh, R. K., Das, H. T., Dubey, S., & Kumar, V. (2022). Impact of aquatic microplastics and nanoplastics pollution on ecological systems and sustainable remediation strategies of biodegradation and photodegradation. *Science of the Total Environment, 806*(3), 151358. https://doi.org/10.1016/j.scitotenv.2021.151358

Jett, L. W. (2017). High tunnels. In M. D. Orzolek (Ed.), *A guide to the manufacture, performance, and potential of plastics in agriculture* (pp. 1–20). Elsevier. https://doi.org/10.1016/B978-0-08-102170-5.00001-4

Junhao, C., Xining, Z., Xiaodong, G., Li, Z., Qi, H., & Siddique, K. H. M. (2021). Extraction and identification methods of microplastics and nanoplastics in agricultural soil: A review. *Journal of Environmental Management, 294*, 112997. https://doi.org/10.1016/j.jenvman.2021.112997

Laskar, N., & Kumar, U. (2019). Plastics and microplastics: A threat to environment. *Environmental Technology & Innovation, 14*, 100352. https://doi.org/10.1016/j.eti.2019.100352

Liebezeit, G., & Liebezeit, E. (2013). Non-pollen particulates in honey and sugar. *Food Additives & Contaminants: Part A, 30*(12), 2136–2140. https://doi.org/10.1080/19440049.2013.843025

Liebezeit, G., & Liebezeit, E. (2014). Synthetic particles as contaminants in German beers. *Food Additives & Contaminants. Part A, Chemistry, Analysis, Control, Exposure & Risk Assessment, 31*(9), 1574–1578. https://doi.org/10.1080/19440049.2014.945099

Liu, K., Wang, X., Song, Z., Wei, N., & Li, D. (2020). Terrestrial plants as a potential temporary sink of atmospheric microplastics during transport. *Science of the Total Environment, 742*, 140523. https://doi.org/10.1016/j.scitotenv.2020.140523

Liu, E., Zhang, L., Dong, W., & Yan, C. (2021). Biodegradable plastic mulch films in agriculture: Feasibility and challenges. *Environmental Research Letters, 16*, 011004. https://doi.org/10.1088/1748-9326/abd211

Liwarska-Bizukojc, E. (2021). Effect of (bio)plastics on soil environment: A review. *Science of the Total Environment, 795*, 148889. https://doi.org/10.1016/j.scitotenv.2021.148889

Milojevic, N., & Agnieszka, C. K. (2021). Agricultural use of sewage sludge as a threat of microplastic (MP) spread in the environment and the role of governance. *Energies, 14*(19), 6293. https://doi.org/10.3390/en14196293

Mo, Q., Yang, X., Wang, J., Xu, H., Li, W., Fan, Q., Gao, S., Yang, W., Gao, C., Liao, D., Li, Y., & Zhang, Y. (2021). Adsorption mechanism of two pesticides on polyethylene and polypropylene microplastics: DFT calculations and particle size effects. *Environmental Pollution, 291*, 118120. https://doi.org/10.1016/j.envpol.2021.118120

Mofijur, M., Ahmed, S. F., Rahman, S. M. A., Arafat Siddiki, S. Y., Islam, A. B. M. S., Shahabuddin, M., Ong, H. C., Mahlia, T. M. I., Djavanroodi, F., & Show, P. L. (2021). Source, distribution and emerging threat of micro- and nanoplastics to marine organism and human health: Socioeconomic impact and management strategies. *Environmental Research, 195*, 110857. https://doi.org/10.1016/j.envres.2021.110857

Mohajerani, A., & Karabatak, B. (2020). Microplastics and pollutants in biosolids have contaminated agricultural soils: An analytical study and a proposal to cease the use of biosolids in farmlands and utilise them in sustainable bricks. *Waste Management, 107*, 252–265. https://doi.org/10.1016/j.wasman.2020.04.021

Montemayor-Trejo, J. A., Suárez-González, E., Munguía-López, J. P., Segura-Castruita, M. A., Mendoza Villarreal, R., & Woo-Reza, J. L. (2018). Acolchados plásticos para la producción de maíz (Zea mays L.) forrajero en la Comarca Lagunera. *Revista mexicana de ciencias agrícolas, 9*(spe20), 4107–4115. https://doi.org/10.29312/remexca.v0i20.982

Monzó Pérez, J. M. (2020). *Efecto de una cubierta de paja de arroz sobre la respiración basal del suelo*. [Doctoral Thesis, Universidad Politécnica de Valencia]. https://riunet.upv.es/handle/10251/159249

Oliveri Conti, G., Ferrante, M., Banni, M., Favara, C., Nicolosi, I., Cristaldi, A., Fiore, M., & Zuccarello, P. (2020). Micro- and nano-plastics in edible fruit and vegetables. The first diet risks assessment for the general population. *Environmental Research, 187*, 109677. https://doi.org/10.1016/j.envres.2020.109677

Orzolek, M. D. (2017). Introduction. In M. D. Orzolek (Ed.), *A guide to the manufacture, performance, and potential of plastics in agriculture* (pp. 1–20). Elsevier. https://doi.org/10.1016/B978-0-08-102170-5.00001-4

Patil, S. M., Rane, N. R., Bankole, P. O., Krishnaiah, P., Ahn, Y., Park, Y. K., Kumar Yadav, K., Amin, M. A., & Jeon, B. H. (2022). An assessment of micro- and nanoplastics in the biosphere: A review of detection, monitoring, and remediation technology. *Chemical Engineering Journal, 430*(2), 132913. https://doi.org/10.1016/j.cej.2021.132913

PlasticsEurope. (2021). *Plastics-the facts 2020.* [Archivo PDF]. https://plasticseurope.org/es/wp-content/uploads/sites/4/2021/11/Plastics_the_facts-WEB-2020_versionJun21_final-1.pdf

Prosperi, M., Sisto, R., Lombardi, M., & Zhu, X. (2018). Production of bioplastics for agricultural purposes: A supply chain study. *Rivista di Studi sulla Sostenibilita, 1*, 119–136. https://doi.org/10.3280/RISS2018-001010

Qi, R., Jones, D. L., Li, Z., Liu, Q., & Yan, C. (2020a). Behavior of microplastics and plastic film residues in the soil environment: A critical review. *Science of the Total Environment, 703*, 134722. https://doi.org/10.1016/j.scitotenv.2019.134722

Qi, Y., Ossowicki, A., Yang, X., Huerta Lwanga, E., Dini-Andreote, F., Geissen, V., & Garbeva, P. (2020b). Effects of plastic mulch film residues on wheat rhizosphere and soil properties. *Journal of Hazardous Materials, 387*, 121711. https://doi.org/10.1016/j.jhazmat.2019.121711

Scarascia-Mugnozza, G., Sica, C., & Russo, G. (2011). Plastic materials in european agriculture: Actual use and perspectives. *Journal of Agricultural Engineering - Riv. di Ing. Agr., 3*, 15–28. Available in: https://www.agroengineering.org/index.php/jae/article/download/jae.2011.3.15/26/

Serrano-Ruiz, H., Martin-Closas, L., & Pelacho, A. M. (2021). Biodegradable plastic mulches: Impact on the agricultural biotic environment. *Science of the Total Environment, 750*, 141228. https://doi.org/10.1016/j.scitotenv.2020.141228

Shi, W., Healy, M. G., Ashekuzzaman, S. M., Daly, K., Leahy, J. J., & Fenton, O. (2021). Dairy processing sludge and co-products: A review of present and future re-use pathways in agriculture. *Journal of Cleaner Production, 314*, 128035. https://doi.org/10.1016/j.jclepro.2021.128035

Shruti, V. C., & Kutralam-Muniasamy, G. (2019). Bioplastics: Missing link in the era of microplastics. *Science of the Total Environment, 697*, 134139. https://doi.org/10.1016/j.scitotenv.2019.134139

Sintim, H. Y., & Flury, M. (2017). Is biodegradable plastic mulch the solution to agriculture's plastic problem? *Environmental Science & Technology, 51*(3), 1068–1069. https://doi.org/10.1021/acs.est.6b06042

Statista. (2021). *Producción de plástico a nivel mundial de 1950 a 2019.* https://es.statista.com/estadisticas/636183/produccion-mundial-de-plastico/

Sobhani, Z., Lei, Y., Tang, Y., Wu, L., Zhang, X., Naidu, R., Megharaj, M., & Fang, C. (2020). Microplastics generated when opening plastic packaging. *Scientific Reports, 10*(1), 4841. https://doi.org/10.1038/s41598-020-61146-4

Talvitie, J., Mikola, A., Koistinen, A., & Setälä, O. (2017). Solutions to microplastic pollution – Removal of microplastics from wastewater effluent with advanced wastewater treatment technologies. *Water Research, 123*, 401–407. https://doi.org/10.1016/j.watres.2017.07.005

Tian, L., Jinjin, C., Ji, R., Ma, Y., & Yu, X. (2022). Microplastics in agricultural soils: Sources, effects, and their fate. *Current Opinion in Environmental Science & Health, 25*, 100311. https://doi.org/10.1016/j.coesh.2021.100311

Toussaint, B., Raffael, B., Angers-Loustau, A., Gilliland, D., Kestens, V., Petrillo, M., Rio-Echevarria, I. M., & Van den Eede, G. (2019). Review of micro- and nanoplastic contamination in the food chain. *Food Additives & Contaminants: Part A, 36*(5), 639–673. https://doi.org/10.1080/19440049.2019.1583381

Ullah, R., Tsui, M. T., Chen, H., Chow, A., Williams, C., & Ligaba-Osena, A. (2021). Microplastics interaction with terrestrial plants and their impacts on agriculture. *Journal of Environmental Quality, 50*(5), 1024–1041. https://doi.org/10.1002/jeq2.20264

Van den Berg, P., Huerta-Lwanga, E., Corradini, F., & Geissen, V. (2020). Sewage sludge application as a vehicle for microplastics in eastern Spanish agricultural soils. *Environmental Pollution, 261*, 114198. https://doi.org/10.1016/j.envpol.2020.114198

Vethaak, A. D., & Legler, J. (2021). Microplastics and human health. *Science, 371*(6530), 672–674. https://doi.org/10.1126/science.abe5041

Walker, T. R. (2021). (Micro)plastics and the UN sustainable development goals. *Current Opinion in Green and Sustainable Chemistry, 30*, 100497. https://doi.org/10.1016/j.cogsc.2021.100497

Wu, R. T., Cai, Y. F., Chen, Y. X., Yang, Y. W., Xing, S. C., & Liao, X. D. (2021). Occurrence of microplastic in livestock and poultry manure in South China. *Environmental Pollution, 277*, 116790. https://doi.org/10.1016/j.envpol.2021.116790

Xu, G., Liu, Y., Song, X., Li, M., & Yu, Y. (2021). Size effects of microplastics on accumulation and elimination of Phenanthrene in earthworms. *Journal of Hazardous Materials, 403*, 123966. https://doi.org/10.1016/j.jhazmat.2020.123966

Yang, M., Huang, D. Y., Tian, Y. B., Zhu, Q. H., Zhang, Q., Zhu, H. H., & Xu, C. (2021). Influences of different source microplastics with different particle sizes and application rates on soil properties and growth of Chinese cabbage (Brassica Chinensis L.). *Ecotoxicology and Environmental Safety, 222*, 112480. https://doi.org/10.1016/j.ecoenv.2021.112480

Zenner de Polanía, I., & Peña Baracaldo, F. (2013). Plastic products in agriculture: Benefice and ambient cost: A review. *Revista U.D.C.A Actualidad & Divulgación Científica, 16*(1), 139–150. Available in: http://www.scielo.org.co/scielo.php?script=sci_arttext&pid=S0123-42262013000100017&lng=en&tlng=es

Index

N. R. Maddela et al. (eds.), *Micro and Nanoplastics in Soil*,
https://doi.org/10.1007/978-3-031-21195-9

GPSR Compliance
The European Union's (EU) General Product Safety Regulation (GPSR) is a set of rules that requires consumer products to be safe and our obligations to ensure this.

If you have any concerns about our products, you can contact us on

ProductSafety@springernature.com

In case Publisher is established outside the EU, the EU authorized representative is:

Springer Nature Customer Service Center GmbH
Europaplatz 3
69115 Heidelberg, Germany

www.ingramcontent.com/pod-product-compliance
Lightning Source LLC
Chambersburg PA
CBHW071205140526
45046CB00003B/477
* 9 7 8 3 0 3 1 2 1 1 9 4 2 *